# Test Bank
for

# *Discovering the Universe*

*Fifth Edition*

WILLIAM J. F. WILSON
T. ALAN CLARK

W. H. Freeman and Company
New York

ISBN: 0-7167-3578-4



Printed in the United States of America

First printing 1999

# Contents

# Preface

This Test Bank has been assembled to accompany the fifth edition of *Discovering the Universe,* by Neil F. Comins and William J. Kaufmann. As such, it is designed to meet the need for multiple-choice questions for a one-term course in basic astronomy. The Test Bank has evolved over time at the authors' home institution, the University of Calgary, where it supplied questions for weekly assignments, three in-class closed-book tests, and a final examination. Along with take-home observing exercises, these tests provide the assessment for final course grades.

The Test Bank is also available from the publisher on a CD-ROM that is both Macintosh and PC-compatible. It is created with Diploma test-generating software developed by Brownstone Research Group. Diploma is a comprehensive question-editing and -handling program that allows course instructors to choose questions chapter-by-chapter from the Test Bank, modify existing questions, or create new ones. The selected questions can be printed as an assignment or a test, and the order of the answers can be randomized after the questions are chosen. For more information, please contact your local W. H. Freeman and Company sales representative.

The Test Bank follows the textbook very closely. As far as the authors are aware, all questions have answers that are contained within the text, or can be derived or deduced from information in the text.

Each question has a reference (separate from the question itself) to the section in the textbook where the answer to the question can be found. This reference is for the information of the instructor and does not appear on the assignments and tests that are created by Diploma for distribution to the students. The reference is to either the Section number within the chapters of the book or to specific material within boxes or tables in the textbook. Thus, if a student does not understand, or disagrees with, the answer to a particular question on a returned assignment or test, the instructor can refer the student directly to the relevant section of the textbook.

The questions in the Test Bank have been designed to meet the varied needs of a first-year general course in astronomy and to provide goals and incentives for learning. There are questions suitable for either *take-home assignments* or *closed-book tests*. This distinction is not made in the Test Bank, however, and the instructor should carefully assess the suitability of a question for a specific use. For example, if the question has a reference to a section, figure, or table in the textbook, this question is unsuitable for a closed-book test. Other questions asking for specific information (e.g., the diameter of Neptune) are designed as an incentive for the student to look at the textbook and are not suitable for closed-book tests. The closed-book questions are those that contain none of these features; they can, of course, be used equally well in take-home assignments. One particular advantage of the computerized version of the Test Bank is the ability to modify and edit questions to suit the course and teaching style of the individual instructor. Questions can easily be added with Diploma.

Questions suitable for *closed-book* tests will contain no numerical calculations, no reference to material in the text or specific details or facts, unless these are deemed useful for the student's understanding of the question.

Question suitable for *open-book, take-home* assignments have specific aims, in addition to assessing a student's knowledge. Some of these are:

(1) to guide the student through simple calculations with the use of commonplace or relatively trivial examples.

(2) to initiate and foster careful reading of the textbook and its material. (This has proven to be a major feature of the Test Bank, at least with conscientious students. There will always be the "group" effort where collective answers are produced, but even this generates active and lively discussion, often ending in the instructor's office!)

(3) to provide a knowledge of background facts and reinforce the concepts discussed in class.

(4) to initiate and influence thought by providing occasional provocative questions on philosophical matters.

The resulting familiarity with the textbook and, hence, with the course material, has often been cited by students as an advantage to the use of regularly scheduled take-home assignments. This is particularly true when computer scoring ensures that results can be communicated rapidly to large classes and students can keep a record of their progress. In practice, take-home assignments will carry a rather low weighting in the overall assessment for a course.

Used alone or as supplementary material to other essay-type questions, the Test Bank provides incentives and stimulates out-of-class discussion while allowing the instructor to assess student knowledge and understanding. At the authors' home institution, one-hour tests typically include 50 questions, which are completed early by the majority of students, while a two-hour final examination includes 120–150 questions.

It is a pleasure to acknowledge the stimulation and encouragement provided by successive classes of students at the University of Calgary. They have always been quick to point out ambiguities and anomalies in questions and have been super-critical of misleading or "trick" questions. It is also a pleasure to thank colleagues who have suggested questions or who have corrected or amended existing ones.

W.J.F.W.
T.A.C.
July, 1999

# Build Your Foundation I: Discovering Astronomy

1. Astronomy
   A) Provides the scientific basis for astrology and the preparation of horoscopes but is not very useful otherwise.
   B) has not yet influenced our lives significantly, although future developments like mining on the Moon and on asteroids may become important.
   C) has produced many discoveries which have led to important results, including the development of Newtonian mechanics upon which much of modern technology is based.
   D) studies only far-away objects and has no application to our lives here on Earth.
   Answer: C Source: Foundations I-1

2. Which scientific discipline can be considered to be truly universal?
   A) Geology B) Biological science C) Social studies D) Astronomy
   Answer: D Source: Foundations I-1

3. Which unique quality distinguishes human beings from other living things?
   A) The will to fight to resolve conflicts.
   B) Curiosity, the need to explore and understand our surroundings.
   C) The continuing need to test the ability and strength of the leaders.
   D) Social behavior, the desire to protect the young, the old and the weak.
   Answer: B Source: Foundations I-1

4. Why were ancient peoples on Earth particularly interested in the positions of the Moon and the Sun in the sky?
   A) Because they believed that gods inhabited these celestial objects and they were anxious not to displease them.
   B) Because ocean tides were directly affected by these astronomical objects.
   C) Because wind patterns necessary for sailing vessels were governed by Moon and Sun.
   D) Because the positions of these objects, particularly the Moon, in the sky could be used for navigation.
   Answer: B Source: Foundations I-1

5. Which of the following statements best represents the overall rationale for scientific investigation?
   A) Reality is comprehensible, and a limited number of fundamental principles govern the nature and behavior of the universe.
   B) There are certain patterns in nature from which future events can be predicted, but there are no underlying basic principles or laws.
   C) The Universe is a hodgepodge of unrelated things behaving in unpredictable ways, but we must continue to observe it in case this behavior threatens to the Earth.
   D) The behavior of the whole Universe is governed by man's observation of it, in a way that hides the fundamental truth.

   Answer: A   Source: Foundations I-1

6. An underlying theme of astronomy is that
   A) the Universe is a hodgepodge of unrelated things, behaving in arbitrary and unexplainable ways.
   B) the fundamental physical laws which govern the Universe change in a predictable way with increasing distance from the Earth.
   C) the entire Universe is governed by a single set of fundamental physical laws.
   D) the fundamental physical laws differ randomly from galaxy to galaxy, but they can be learned for a given galaxy by detailed observation.

   Answer: C   Source: Foundations I-1

7. A scientist observes a new phenomenon that disagrees with his explanation or hypothesis. Following the scientific method, he should
   A) discard the observation as erroneous.
   B) modify his hypothesis.
   C) reject those observations which do not agree with the theory.
   D) wait until someone develops an adequate explanation before announcing his observation.

   Answer: B   Source: Foundations I-1

8. In science, if new and reliable observations disagree with a well-established theory, then
   A) the theory must be discarded immediately.
   B) the theory must be modified to account for the observations, and if this is not possible then the theory must be discarded.
   C) this shows the futility of attempting to understand the Universe. The observations should be classified for future reference and the theory retained as the best explanation of the phenomenon.
   D) the observations must be discarded.

   Answer: B   Source: Foundations I-1

9. The scientific method is a major force in science, and theories about physical phenomena have been developed to ensure that
   A) they agree with what we find in experiments and observations.
   B) they agree with the wisdom of the ancients.
   C) results from experiments can be adjusted to agree with carefully-constructed theoretical ideals.
   D) they are so good and our faith in them is so strong that we never need to test them against observations.
   Answer: A Source: Foundations I-1

10. In following the principles of the scientific method, a theory proposed to explain a given phenomenon must
    A) predict new and different experiments which will extend the scope of the theoretical understanding but need not explain all the previous observations since no theory is expected to explain everything completely.
    B) explain all previous and reliable observations in a consistent manner but need not suggest new tests for the theory since a theory should be complete in itself.
    C) explain all known and reliable observations and predict new experiments and observations.
    D) agree with, and build upon, previous theories but need not explain all observations since some of these may be erroneous.
    Answer: C Source: Foundations I-1

11. The "rules" that govern the behavior of matter and that have been exploited in engineering endeavors are known as
    A) Murphy's laws.
    B) the laws of physics.
    C) the laws of chance.
    D) the laws of social behavior.
    Answer: B Source: Foundations I-1

12. Which fundamental belief about the Universe, established by the Greeks and adopted by the early Christian church, was shattered by Galileo's observation of moons orbiting Jupiter?
    A) The belief that our Moon was the only moon orbiting around a planet in our solar system.
    B) The belief that everything in the Universe orbits the Earth.
    C) The belief that everything in the Universe orbits the Sun.
    D) The belief that everything in the visible Universe must orbit the center of the Milky Way galaxy.
    Answer: B Source: Foundations I-1

13. According to the *scientific method*, a hypothesis which is proposed to explain a particular physical phenomenon is considered to be wrong if
    A) leading scientists in the world believe that it is wrong.
    B) it is in conflict with the results of just one reliable and repeatable observation.
    C) it disagrees with the accepted theory at the time of the proposal.
    D) it appears to defy logic and logical reasoning.
    Answer: B Source: Foundations I-2

14. In applying the *scientific method* to the study of our natural surroundings, scientists are
    A) formulating hypotheses or models which describe the present observations of nature, and which predict possible further tests for these models.
    B) developing a theoretical view of the Universe which incorporates all previous ideas and myths as part of an overall philosophy.
    C) slowly amassing a vast bank of observations of nature which, some time in the future, will be assembled into the correct description of the Universe.
    D) discovering by observation the absolute truth about limited areas of science, and are therefore slowly building up the correct view of the Universe.
    Answer: A Source: Foundations I-2

15. Which of the following is a critical component of the scientific method?
    A) The automatic rejection of a theory when new results disagree with its predictions.
    B) The testing of predictions from a scientific theory or theoretical model.
    C) The rejection of new scientific results when they disagree with the presently accepted theory.
    D) The belief that a theory accepted by leading scientists is correct.
    Answer: B Source: Foundations I-2

16. A scientific theory is accepted as the best description of a certain phenomenon if
    A) its prediction can be repeatedly and independently checked by observation.
    B) has been developed by a scientist with a solid reputation such as Albert Einstein or Sir Isaac Newton.
    C) it has obtained the stamp of approval of an internationally renowned scientific body such as the Royal Society of London or the International Astronomical Union.
    D) it accounts for a full range of historical observations, even though recent observations cast doubt upon the theory.
    Answer: A Source: Foundations I-2

17. The most important aspect of exploring other planets is that
    A) it allows us to understand our own planet more thoroughly.
    B) it extends Man's thoughts on to a plane which is totally removed from our limited existence here on Earth.
    C) it has become vital for us to develop these planets as sources of raw materials.
    D) the study of the solar system is our greatest source of information on the formation of the Universe.
    Answer: A Source: Foundations I-2

18. The average distance from the Earth to the Sun, 149,600,000 km, can be written in shorthand notation as
    A) $1.496 \times 10^8$ km.
    B) $1.496 \times 10^7$ km.
    C) $1.496 \times 10^6$ km.
    D) $1.496 \times 10^9$ km.
    Answer: A Source: Toolbox I-1

19. The diameter of the hydrogen atom, 0.000,000,000,11 meters, can be written in shorthand notation as
    A) $1.1 \times 10^{-9}$ m. B) $1.1 \times 10^{-10}$ m. C) $1.1 \times 10^{-11}$ m. D) $1.1 \times 10^{-8}$ m.
    Answer: B Source: Toolbox I-1

20. The mean distance of Jupiter from the Sun, 778,300,000 km, can be written in shorthand notation as
    A) $7.783 \times 10^6$ km. B) $7.783 \times 10^7$ km. C) $7.783 \times 10^8$ km. D) $7.783 \times 10^9$ km.
    Answer: C Source: Toolbox I-1

21. There are 1000 mm in one meter. This means that a distance of 5 mm is
    A) $5 \times 10^{-2}$ meter. B) $2 \times 10^{-4}$ meter. C) $5 \times 10^3$ meter. D) $5 \times 10^{-3}$ meter.
    Answer: D Source: Toolbox I-1

22. 0.0064 meter is
    A) 64 mm. B) 0.64 mm. C) 640 mm. D) 6.4 mm.
    Answer: D Source: Toolbox I-1

23. $10 \times 10^5 =$
    A) 1 billion. B) 1 million. C) 10 thousand. D) 10 million.
    Answer: B Source: Toolbox I-1

24. $10^0$ (10 to the power 0)
A) 0. B) undetermined, not a real number. C) 1. D) 10.
Answer: C Source: Toolbox I-1

25. The number five hundred thousand is written in powers-of-ten notation as
A) $5 \times 10^4$. B) $5 \times 10^5$. C) $(5 \times 10^2) \times 10^{-3} = 5 \times 10^{-1}$. D) $5 \times 10^{-5}$.
Answer: B Source: Toolbox I-1

26. $10^{-2} \times 10^2 =$
A) 1. B) 1/100. C) 0. D) 10,000.
Answer: A Source: Toolbox I-1

27. $(3 \times 10^4)^4 =$
A) $9 \times 10^{16}$. B) $2.7 \times 10^{16}$. C) $8.1 \times 10^{17}$. D) $8.1 \times 10^{16}$.
Answer: C Source: Toolbox I-1

28. $(1/2)^3 =$
A) 0.167. B) 0.5. C) 8. D) $1.25 \times 10^{-1}$.
Answer: D Source: Toolbox I-1

29. $(0.5)^2 =$
A) 2.5. B) 0.25. C) 1.0. D) $2.5 \times 10^{-2}$.
Answer: B Source: Toolbox I-1

30. $10^5 \times 10^8 =$
A) $10^{20}$. B) $10^3$. C) $10^{40}$. D) $10^{13}$.
Answer: D Source: Toolbox I-1

31. $10^6/10^9 =$
A) $10^{-3}$. B) $10^{15}$. C) $10^3$. D) $10^{54}$.
Answer: A Source: Toolbox I-1

32. $(9.0 \times 10^5)/(1.5 \times 10^3) =$
A) $7.5 \times 10^{-2}$. B) 600. C) $3 \times 10^{-2}$. D) $13.5 \times 10^{15}$.
Answer: B Source: Toolbox I-1

33. $10^6/10^6 =$
A) 10. B) 0. C) $10^{12}$. D) 1.
Answer: D Source: Toolbox I-1

34. $2.5 \times 10^4 \times 2.5 \times 10^{-4} =$
A) 6.25. B) 5.0. C) $1.0 \times 10^6$. D) $5.0 \times 10^{-8}$.
Answer: A Source: Toolbox I-1

35. One billionth divided by one millionth is equal to
A) $10^{-3}$. B) $10^{15}$. C) $10^3$. D) $10^{-15}$.
Answer: A Source: Toolbox I-1

36. $(2 \times 10^2)^3 =$
A) $8 \times 10^5$. B) $6 \times 10^6$. C) $8 \times 10^6$. D) $6 \times 10^5$.
Answer: C Source: Toolbox I-1

37. By what approximate factor, in powers-of-ten notation, is a human being (height about 2 m) larger than the nucleus of a hydrogen atom, or proton (diameter about $10^{-15}$ m)?
A) $2 \times 10^{13}$. B) $2 \times 10^{30}$. C) $2 \times 10^{15}$. D) $2 \times 10^{-15}$.
Answer: C Source: Toolbox I-1

38. By how many powers of ten is the diameter of the Sun ($1.4 \times 10^5$ km) greater than the length of a beetle (14 mm)?
A) 5. B) 10. C) 8. D) 11.
Answer: B Source: Toolbox I-1

39. In this age of space exploration, Man has now traveled to the Moon. By how many orders of magnitude, (powers of ten) was this journey greater than that of Columbus when he traveled from the Old World to the Americas?
A) 3 orders of magnitude or $10^3$.
B) 6 orders of magnitude or $10^6$.
C) 2 orders of magnitude or $10^2$.
D) 1 order of magnitude, or $10^1$.
Answer: C Source: Toolbox I-1

40. The speed of light is
A) $3 \times 10^{10}$ m/sec. B) $3 \times 10^8$ m/sec. C) $3 \times 10^5$ m/sec. D) $3 \times 10^{12}$ m/sec.
Answer: B Source: Toolbox I-1

41. One light-year is a distance of approximately
A) $1.5 \times 10^8$ km. B) $3 \times 10^5$ km. C) $6.3 \times 10^4$ km. D) $9.5 \times 10^{12}$ km.
Answer: D Source: Toolbox I-2

42. 1 AU or 1 astronomical unit is defined as
A) the distance at which the Earth-Sun distance will subtend an angle of 1 arc second.
B) the distance traveled by light in 1 year.
C) the distance traveled by light in 1 second.
D) the mean distance between the Sun and the Earth.
Answer: D Source: Toolbox I-2

43. An Astronomical Unit (AU) is
A) a unit of length, the average distance between the Sun and the Earth.
B) a standard unit of length, defined as the wavelength of light from krypton gas.
C) a unit of time, equal to the time taken for the Earth to orbit the Sun once.
D) a unit of mass, equal to one solar mass.
Answer: A Source: Toolbox I-2

44. A light-year is a measure of
A) arc length along an orbit.
B) time.
C) distance.
D) the expansion rate of the universe.
Answer: C Source: Toolbox I-2

45. 1 light-year is
A) the time taken for the Earth to orbit the Sun once.
B) the distance between the Earth and Sun.
C) the time taken for light to travel from the Sun to Earth.
D) the distance that light travels in 1 year.
Answer: D Source: Toolbox I-2

46. If a radio message were sent toward the nearest star to the Sun and a reply were sent back immediately upon receipt of the message by intelligent beings from a planet near that star, how long after transmission would we have to wait for a reply?
A) 20,000 years. B) 1 light year. C) 1 year. D) 8.6 years.
Answer: D Source: Toolbox I-2 and Appendices

47. If a supernova was first seen in the year of Christ's birth (the Star of Bethlehem?) and its distance from the Earth was 2 kpc, approximately when did the supernova actually explode?
A) 2 B.C. B) 6520 A.D. C) 6520 B.C. D) 2000 B.C.
Answer: C Source: Toolbox I-2

48. Light from fires, lit at the time of the Battle of Hastings in England in 1066 AD, has traveled out into space at the speed of light. How far has this light now reached in space, compared to the distances to the 20 brightest stars (see Appendix, Table 5, Kaufmann & Comins, *Discovering the Universe,* 5th Ed.)?
A) Beyond Sirius and Betelguese, but not beyond Rigel and Deneb.
B) Beyond Sirius, Betelguese and Rigel, but not beyond Deneb.
C) Beyond Sirius, but not beyond Betelgeuse, Rigel and Deneb.
D) Beyond Sirius, Betelguese, Rigel, and Deneb.
Answer: B Source: Toolbox I-2

49. Suppose that, at the same time on the same night, we see two supernovas (exploding stars) explode in the night sky. If one is in the Andromeda galaxy, two million light-years away from us and the other is in the galaxy M82, six million light-years away from us, which of the following statements is correct concerning the actual explosion times of these supernovas?
A) We know both stars exploded at the same time, because we saw the explosions at the same time.
B) The supernova in the Andromeda galaxy actually occurred after the one in M82.
C) We cannot tell which star actually exploded first, because they are so far away.
D) The supernova in the Andromeda galaxy actually occurred before the one in M82.
Answer: B Source: Toolbox I-2

50. In 2000AD, an inhabitant of a planet orbiting a distant star observes the flash of the first nuclear explosion on Earth, which occurred in July 1945. Approximately how far away is his solar system from Earth?
A) 1.7 pc. B) 55 pc. C) 179.3 pc. D) 17 pc.
Answer: D Source: Toolbox I-2

51. An inhabitant of a planet orbiting the star Betelguese observes the flash from the first nuclear weapon on Earth, exploded in July 1945. If this extraterrestrial being were to send a signal to Earth immediately to confirm this sighting by exploding a bomb of equivalent brightness, and we were watching his planet, when would we expect to see this flash?
A) 2565 AD B) 2255 AD C) 2040 AD D) 1976 AD
Answer: A Source: Toolbox I-2

52. The time taken for light to travel from a galaxy which is 10 Mpc away is
A) $3.07 \times 10^6$ years. B) $10^7$ years. C) $3.26 \times 10^7$ years. D) 32.6 years.
Answer: C Source: Toolbox I-2

53. Light takes about 8.3 minutes to travel from the Sun to the Earth. If so, then approximately how much longer would light take to travel from Jupiter to Earth when Jupiter is at conjunction (appearing closest to the Sun in our sky) than when it is at opposition (on the opposite side of the Earth from the Sun)? (Assume that light can come over or under the Sun at conjunction). (A diagram might help!)
A) 16.6 minutes.
B) 43.2 minutes.
C) 8.3 minutes.
D) There is insufficient information to answer this question.
Answer: A Source: Toolbox I-2

54. Approximately how long does it take light to travel from the fingertips of your extended arm to your eye?
A) 2.5 billionths of a second (2.5 nsec).
B) 2.5 trillionths of a second ($2.5 \times 10^{-12}$ sec).
C) 2.5 millionths of a second (2.5 msec).
D) Zero time, since light is transmitted instantaneously.
Answer: A Source: Toolbox I-2

55. How long would it take for a light flash from a laser to travel from the Earth to the Moon and back, if it were reflected from the retro-reflectors left on the surface of the Moon by astronauts?
A) 1.28 sec. B) 2.56 microsec. C) 1.28 millisec. D) 2.56 sec.
Answer: D Source: Toolbox I-2

56. The Crab Nebula (Fig. 12-18, Kaufmann & Comins, *Discovering the Universe*, 5th Ed.) is the result of a supernova explosion of a star which occurred at a distance of about 1.84 kpc from Earth. If people first saw the explosion in the year 1054 A.D., when did the explosion actually occur?
A) 1054 AD, of course. B) 786 BC C) About 4946 BC D) About 1048 AD
Answer: C Source: Toolbox I-2

57. How far from the Sun would you be if the angle between the Earth and the Sun was one second of arc as you looked back from space? (Assume that the Earth-Sun line is at right angles to your line of sight.)
A) 1 pc. B) 1 Mpc. C) 1 AU. D) 1 ly.
Answer: A Source: Toolbox I-2

58. The distance from the Earth to the star Betelgeuse (in the constellation Orion) has been measured as 160 pc. Expressed in light-years, this is approximately
A) 49 light-years.
B) 5200 light-years.
C) 520 light-years.
D) $1.0 \times 10^7$ light-years.
Answer: C Source: Toolbox I-2

59. The distance to the star t Scorpii has been measured as 750 ly. Expressed in parsecs, this is approximately
A) 230 pc. B) $4.7 \times 10^7$ pc. C) 23 pc. D) 2445 pc.
Answer: A Source: Toolbox I-2

60. The star g Aquilii is 340 ly from Earth. Expressed in parsecs, this is approximately
A) 105 pc. B) 0.340 kpc. (kiloparsecs) C) 750 pc. D) 1100 pc.
Answer: A Source: Toolbox I-2

61. The star z Puppis is about 750 pc from the Earth. Expressed in light-years, this is approximately
A) $4.7 \times 10^7$ ly. B) 1500 ly. C) 230 ly. D) 2445 ly.
Answer: D Source: Toolbox I-2

62. An astronomer finds an object at a distance of 6.8 AU from the Earth. Which type of object is this likely to be?
A) A comet in our solar system.
B) A distant galaxy.
C) A star in our Galaxy.
D) An artificial satellite orbiting the Earth.
Answer: A Source: Toolbox I-2

63. An astronomer finds an object at a distance of 5.6 pc from the Earth. At this distance, what is this object most likely to be?
    A) A distant galaxy.
    B) A star in our Galaxy.
    C) An asteroid in our solar system.
    D) An artificial satellite orbiting the Earth.
    Answer: B Source: Toolbox I-2

64. An astronomer discovers an object at a distance of 28 Mpc from the Earth. Based on the distance, which of the following is this object most likely to be?
    A) A distant galaxy.
    B) A comet in our solar system.
    C) An artificial satellite orbiting the Earth.
    D) A star in our Galaxy.
    Answer: A Source: Toolbox I-2

65. What is the distance between the Earth and the nearest star? (THINK carefully about your answer!) (Check Chapter 9 and Table 4 of the Appendix, Kaufmann & Comins, *Discovering the Universe,* 5th Ed.)
    A) 1 A.U. B) 4.3 parsecs. C) 4.3 light-years. D) 5.2 A.U.
    Answer: A Source: Toolbox I-2

66. Most observational astronomers spend the majority of their research time
    A) re-analyzing old data in the light of new theories, using new mathematical techniques.
    B) devising new and different theories to explain their latest observations
    C) observing at the telescope or at the telescope control console, if the instrument is space-borne.
    D) analyzing and organizing data, planning new observational programs or designing new instruments.
    Answer: D Source: Foundations I-3

67. Which of the following space exploits has not yet been accomplished by humans?
    A) Discovery of active volcanoes on other worlds.
    B) Walking on the surface of the Moon.
    C) Landing of a spacecraft on Saturn's surface.
    D) Analysis of Martian soil.
    Answer: C Source: Foundations I-4

68. In the history of planetary exploration,
    A) only the near-neighbor planets to Earth, Venus and Mars, have been visited by spacecraft.
    B) all of the planets except Pluto have been closely viewed by spacecraft.
    C) only the closer planets, Venus, Mars, Jupiter and Saturn, have been closely examined by space vehicles.
    D) all of the planets have now been explored by the close approach of spacecraft.
    Answer: B Source: Foundations I-4

69. Which of the following does NOT bring us information about the distant universe?
    A) Sound waves.
    B) High-speed nuclei of atoms called cosmic rays.
    C) Massless (or almost massless) atomic particles called neutrinos.
    D) Meteorites.
    Answer: A Source: Foundations I-4

# Chapter 1: Discovering the Night Sky

1. In modern astronomy, the constellations are
   A) a small number of well-defined groups of stars in our sky.
   B) 12 specific regions through which the planets and Moon appear to move in our sky.
   C) 88 sky regions covering the whole sky.
   D) specific patterns of stars which point to certain directions, useful for navigation.
   Answer: C Source: Section 1-1

2. The constellations used to describe the night sky are
   A) alignments of stars which indicate specific directions in the sky.
   B) 12 regions of sky through which the Moon and planets move across our sky.
   C) small well-defined groups of stars.
   D) regions of the sky, 88 in number, covering the whole sky.
   Answer: D Source: Section 1-1

3. Describing a star as being in the constellation Cygnus (the Swan) tells a modern astronomer that the star is
   A) in a distant galaxy located in a particular direction from Earth.
   B) inside our solar system.
   C) somewhere in a particular region of sky having definite boundaries.
   D) one of a set of bright stars which make up a particular "picture" in the sky.
   Answer: C Source: Section 1-1

4. Which of the following statements correctly describes the relationship between stars and constellations?
   A) Only stars close to the ecliptic (the Earth's orbital plane) are located in constellations.
   B) Only the brighter stars are in constellations.
   C) Only those stars which were visible to the ancient Greeks are located in constellations.
   D) Every star is located in a constellation.
   Answer: D Source: Section 1-1

5. If a star is described as being in the constellation Leo, a modern astronomer knows that it is
   A) somewhere in a particular region of sky having definite boundaries
   B) inside a region of the sky bounded by two lines of Right Ascension in the sky.
   C) in a distant galaxy located in a particular direction from Earth.
   D) one of a few individual bright stars making up a picture (of a lion) in the sky
   Answer: A Source: Section 1-1

6. The constellation whose stars are used as pointers to the North Celestial Pole in the northern hemisphere is
   A) Ursa Minor, the Little Bear, containing the bright star, Polaris.
   B) Bootes, the shepherd, containing the bright star, Arcturus.
   C) Leo, the lion, containing the bright star, Regulus.
   D) Ursa Major, the Big Dipper.
   Answer: D Source: Section 1-1

7. If the stars Polaris and Arcturus (as shown in Fig. 1-2, Kaufmann & Comins, *Discovering the Universe,* 5th Ed.), are seen to be 71° apart, how far away from Polaris is the closest star in Ursa Major?
   A) 2.5° B) 25° C) 7.1° D) 250°
   Answer: B Source: Section 1-1

8. On what time-scale will the specific pattern of stars in a particular constellation appear to change from our view upon Earth because of celestial motions?
   A) A year, because of Earth's orbital motion.
   B) A few hours because of Earth's rotation.
   C) Millions of years, since stars move very slowly with respect to each other.
   D) Thousands of years, because of motions of individual stars.
   Answer: D Source: Section 1-2

9. If the unaided human eye is sensitive enough to see about 6,000 of the stars in the entire sky, about how many stars would be seen at one time on a given night from a single location where the horizon is completely visible around the observer?
   A) 3,000.
   B) 6,000, of course.
   C) only a small fraction of 6,000, say 1,000, because the rest are hidden by the Earth.
   D) It depends upon the observer's latitude; observers at the poles will see 6,000, while equatorial observers will see only ½ of this number, or 3,000.
   Answer: A Source: Chapter 1: Introduction

10. How much of the overall sky is above the celestial equator, that is, in the northern hemisphere?
   A) More than one half, because of the precession of the poles.
   B) Exactly one half.
   C) All of it, by definition.
   D) Less than one half, because of the tilt of the equator to the ecliptic plane.
   Answer: B Source: Section 1-2

11. Which of the following lines or points is always directly over your head, no matter where on the Earth you go?
    A) The zenith.
    B) The celestial equator.
    C) The ecliptic.
    D) 90° north declination.
    Answer: A Source: Section 1-2

12. The zenith defines a direction
    A) toward the Sun at noon
    B) vertically above a point on the equator
    C) vertically above an observer
    D) vertically above the North Pole
    Answer: C Source: Section 1-2

13. Which of the following directions does or do NOT always remain fixed in place relative to an observer's horizon?
    A) The summer solstice
    B) The north celestial pole
    C) The zenith
    D) The points where the celestial equator contacts the horizon
    Answer: A Source: Section 1-2

14. Which of the following points REMAINS FIXED in the sky relative to an observer's horizon?
    A) The direction to a distant star (e.g., Betelgeuse, in Orion)
    B) The vernal equinox
    C) The north celestial pole
    D) The winter solstice
    Answer: C Source: Section 1-2

15. Which of the following directions REMAINS FIXED in the sky relative to an observer's horizon?
    A) The autumnal equinox
    B) The zenith
    C) The direction to the Moon at noon, over one month
    D) The direction to the Sun at noon, over one year
    Answer: B Source: Section 1-2

16. If you point toward the zenith today and point there again 45 days later, you will have pointed twice in the same direction relative to
    A) the fixed stars. B) the Sun. C) the horizon. D) the Moon.
    Answer: C Source: Section 1-2

17. Declination of a star is a measure of its
    A) position above the observer's horizon, measured from the horizon.
    B) position, measured along the celestial equator.
    C) position north or south of the celestial equator, along a great circle passing through the N and S celestial poles.
    D) time of rising in the eastern sky.
    Answer: C Source: Section 1-2

18. Two celestial coordinates, which together describe a star's position precisely in our sky, are
    A) Right ascension and sidereal time.
    B) Longitude and latitude.
    C) Right ascension and declination.
    D) Sidereal time and latitude.
    Answer: C Source: Section 1-2

19. The difference in declination angles between the North and South celestial poles is
    A) variable, depending on the season. B) 180°. C) 23.5°. D) 90°.
    Answer: B Source: Section 1-2

20. In the right ascension coordinate direction, 1 hour corresponds to what equivalent angle along the celestial equator?
    A) 360° B) 1° C) 15° D) variable, depending on the declination of the star
    Answer: C Source: Section 1-2

21. The declination of a star in our sky is defined as
    A) the angle between the celestial equator and the star, measured along a great circle passing through both celestial poles.
    B) the angle between the Sun and the star, measured along the ecliptic plane.
    C) the angle between the position of the center of the galaxy and the star, measured along the ecliptic plane.
    D) the angle between the Vernal Equinox and the star, measured along the celestial equator.
    Answer: A Source: Section 1-2

22. For an observer at a fixed location on the Earth, the angle between the north celestial pole and an observer's horizon depends on
    A) the time of day.
    B) the observer's longitude (east or west of Greenwich).
    C) the time of year.
    D) the observer's latitude (north or south of the equator).
    Answer: D Source: Section 1-2

23. The angle between an observer's horizon and the celestial north pole is governed by
    A) longitude. B) latitude. C) sidereal time. D) local time.
    Answer: B Source: Section 1-2

24. The elevation angle between the northern horizon of a fixed observer and the N celestial pole is equal to
    A) the right ascension of the vernal equinox.
    B) the observer's longitude.
    C) a variable value, depending on the time of year.
    D) the observer's latitude.
    Answer: D Source: Section 1-2

25. A comet which is moving northwards from the equator toward the north celestial pole can be described as having its
    A) declination decrease with time.
    B) declination increase with time.
    C) right ascension increase with time.
    D) right ascension decrease with time.
    Answer: B Source: Section 1-2

26. The nightly motion of objects across our the sky is caused by
    A) revolution of Earth around the Sun.
    B) rotation of the Earth on its axis.
    C) the motion of the solar system around the galaxy.
    D) rotation of the whole celestial sphere of stars around the fixed Earth.
    Answer: B Source: Section 1-3

27. Which way are you moving with respect to the stars because of the rotation of the Earth?
    A) Southward. B) Westward. C) Northward. D) Eastward.
    Answer: D Source: Section 1-3

28. The most readily observed east-to-west motion of objects in the night sky is caused by the
    A) the motion of the Moon and planets across the sky.
    B) the relative motions of stars with respect to each other in the sky.
    C) the revolution of the Earth around the Sun.
    D) the rotation of the Earth on its axis.
    Answer: D Source: Section 1-3

29. The pattern of stars that is visible from one position on the Earth gradually shifts from east to west across the sky over one night. This is caused by
    A) the rotation of the Earth about its own north-south axis.
    B) the motion of the Moon and planets across the sky.
    C) the motion of the Earth around the Sun.
    D) atmospheric motions and winds.
    Answer: A Source: Section 1-3

30. Diurnal motion of objects in the sky is caused by
    A) motion of the Moon across the sky.
    B) rotation of the Earth on its axis.
    C) precession of the Earth's axis.
    D) revolution of Earth around the Sun.
    Answer: B Source: Section 1-3

31. The most easily observed motion in the night sky is produced by
    A) the rotation of the Earth on its axis.
    B) the motion of stars with respect to each other in the sky.
    C) the revolution of the Earth around the Sun.
    D) the motion of the planets along their orbits around the Sun.
    Answer: A Source: Section 1-3

32. The phrase "diurnal motion" refers to
    A) the gradual motion of the constellations from east to west across the sky each night, resulting in different constellations being visible at 4 am than at 10 pm on any given night.
    B) the apparent motion of the Sun along the ecliptic over the course of a year.
    C) the change in position of the Moon in the sky as it runs through its phases over the course of a month.
    D) the slow change in position of the constellations from east to west from night to night, resulting in different constellations being visible at 11 pm in May than at 11 pm in December.
    Answer: A Source: Section 1-3

33. Over the course of one night, an observer at any given location on the Earth sees the constellations gradually shift across the sky from east to west. This is caused primarily by
    A) the motion of the Earth around the Sun.
    B) the rotation of the Earth around its own axis.
    C) the inherent rotation of the Universe.
    D) the wind.
    Answer: B Source: Section 1-3

34. If you watch (or photograph) stars near the north celestial pole for a period of several hours, in what basic pattern do they appear to move?
    A) Almost straight lines, rising from the horizon toward the zenith.
    B) Ellipses, with the north pole at one focus.
    C) Circles, with the north celestial pole at the center.
    D) Spirals, as the stars move while the Earth rotates.
    Answer: C Source: Section 1-3

35. When we watch the nighttime sky, we find that
    A) stars and constellations slowly rise in the east, pass overhead, and set in the west.
    B) all stars and constellations reach their highest point in the sky at midnight.
    C) stars and constellations slowly rise in the west, pass overhead, and set in the east.
    D) the stars and constellations remain fixed in our sky, not rising or setting in a time as short as one night because they are so far away.
    Answer: A Source: Section 1-3

36. During a given night, some stars will be observed to pass through (from one side to the other of)
    A) the celestial equator. B) the zodiac. C) the vernal equinox. D) the zenith.
    Answer: D Source: Section 1-3

37. Which way are you moving with respect to the background stars because of the revolution of the Earth *in its orbit around the Sun*?
    A) Westward B) Northeastward C) Northwestward D) Eastward
    Answer: D Source: Section 1-3

38. As the Earth rotates, the point above the head of a person (her zenith) standing on the equator sweeps out
    A) the celestial equator.
    B) the ecliptic plane.
    C) a great circle path between the north and south poles
    D) a variable path across the sky within the region of the zodiac, crossing but not on the celestial equator.
    Answer: A Source: Section 1-3

39. The Sun rises due East in the sky when viewed
    A) from any site upon Earth on the first day of summer and the first day of winter.
    B) from any site along the Earth's equator at midsummer and midwinter.
    C) from any site upon Earth ONLY on the first day of spring and the first day of fall.
    D) from any site on the equator on every day of the year.
    Answer: C Source: Section 1-3

40. For a particular observer in the northern hemisphere, a given star in the sky will reach its highest point when it passes through
    A) the celestial equator.
    B) the region of the sky due south.
    C) the zodiac.
    D) the ecliptic plane.
    Answer: B Source: Section 1-3

41. A given star in the sky will reach its highest point for a particular observer in the southern hemisphere when it passes through
    A) the celestial equator.
    B) the zodiac.
    C) the region of the sky due north.
    D) the ecliptic plane.
    Answer: C Source: Section 1-3

42. Where would you have to be in order to see the south celestial pole on your horizon?
    A) At the south pole of Earth.
    B) At the north pole of Earth.
    C) On the equator.
    D) About 1° away from the south pole, to allow for the Earth's precession.
    Answer: C Source: Section 1-3

43. Over a period of one year, what fraction of the overall sky will an observer on the equator be able to see?
    A) A variable amount, depending upon which year.
    B) A variable amount, depending upon the person's longitude.
    C) 100%.
    D) 50%.
    Answer: C Source: Section 1-3

44. Over the period of one complete year, an observer at the South Pole would be able to see what fraction of the overall sky?
    A) 100%.
    B) A variable amount, depending upon his longitude.
    C) A variable amount, depending upon which year.
    D) 50%.
    Answer: D Source: Section 1-3

45. From the Earth's north pole
    A) the whole of the celestial sphere will be visible at some time during the year.
    B) only stars which are within 66.5° of the north celestial pole can be seen.
    C) only half the celestial sphere can be seen, on any clear night.
    D) only stars which are 23.5° above the celestial equator can be seen.
    Answer: C Source: Section 1-3

46. Where would you have to be in order to see the north celestial pole directly over your head (ie. in your zenith)?
    A) On the equator.
    B) At the north pole of Earth.
    C) At the south pole of Earth.
    D) At a position about 1° away from the south pole, to account for precession.
    Answer: B Source: Section 1-3

47. If you were standing on the equator, which of the following positions in the sky would pass through your zenith at some time in one 24-hour period?
    A) The ecliptic pole, or perpendicular to the ecliptic plane.
    B) The north celestial pole, which is the perpendicular to the celestial equator.
    C) Vernal equinox, or 0 hours right ascension, 0°declination.
    D) The position of the Sun at summer solstice.
    Answer: C Source: Section 1-3

48. The star grouping Leo (the lion) extends for about 30° along a region close to the celestial equator. At low to mid-latitudes, roughly how long will it take Leo to rise above the horizon?
    A) 30 minutes B) 30 seconds C) 2 hours D) 5 hours
    Answer: C Source: Section 1-3

49. The Sun appears to be about 1/2° in diameter. On the equator, approximately how long does it take for the Sun to set, from first contact with the horizon to the Sun completely set below that horizon?
    A) 2 seconds B) About 1 hour C) 4 minutes D) 2 minutes
    Answer: D Source: Section 1-3

50. When viewed from Earth, Venus subtends an angle of about 1 arc minute (1/60°) when it is at the closest point to Earth in its orbit. If you were watching Venus set in the west over a clear horizon (e.g., ocean), how long would it take from first reaching the horizon for Venus to complete disappear below the horizon?
    A) About 4 seconds
    B) About 4 minutes
    C) Almost instantaneous (much less than 1 second)
    D) 1/4 sec
    Answer: A Source: Section 1-3

51. In angular measurements used in astronomy, how many right angles are there in a full circle?
A) 4 B) 2 C) 6 D) 1
Answer: A Source: Toolbox 1-1

52. If the Moon subtends about 30 arc minutes in the sky and is at about 400,000 km from Earth at the time, what is its approximate diameter?
A) 60 km B) 35,000 km C) 350 km D) 3,500 km
Answer: D Source: Toolbox 1-1

53. Astronauts on the Moon look back at the Earth, a distance of about 400,000 km away. If the cities of Washington, D.C. and New York are separated by about 300 km, what will be the angle between them when viewed from the Moon?
A) 2.5 arc minutes B) 2.5 degrees C) 1300 degrees D) 3/4 degree
Answer: A Source: Toolbox 1-1

54. The Moon's angular diameter in our sky is measured to be half a degree. From this, we can find
A) the diameter of the Moon in kilometers even if we have no other information about the Moon.
B) the diameter of the Moon in kilometers if we know the Moon's distance.
C) the bulk density of the Moon (the average number of kilograms per cubic meter of Moon material).
D) the distance to the Moon even if we have no other information about the Moon.
Answer: B Source: Toolbox 1-1

55. The Moon's angular diameter in our sky is measured to be half a degree. From this, we can find
A) the diameter of the Moon in kilometers even if we have no other information about the Moon.
B) the diameter of the Moon in kilometers if we know the Moon's distance.
C) the distance to the Moon even if we have no other information about the Moon.
D) the bulk density of the Moon (the average number of kilograms per cubic meter of Moon material).
Answer: B Source: Section 1-5

56. Astronauts on the Moon look back at Earth, whose diameter is about 12,800 km. Since the Earth-Moon distance is about 400,000 km and the Moon's diameter is about 3,500 km, how much bigger or smaller will the Earth appear in their sky than the Moon does in our sky?
A) 3.7 times. B) 0.27 times. C) The same, obviously. D) 220 times.
Answer: A Source: Toolbox 1-1

57. The number of degrees in a full circle is
A) 360. B) 60. C) 3600. D) 57.3.
Answer: A Source: Toolbox 1-1

58. The angle between your zenith and your horizon is
A) 180°. B) 90°. C) 45°. D) 57.3°.
Answer: B Source: Toolbox 1-1

59. The angle of 60° between the line from the Sun to Jupiter and the line from the Sun to a Trojan group of asteroids (see Fig. 8-7, Kaufmann & Comins, *Discovering the Universe*, 5th Ed.) is what fraction of a full circle?
A) 1/6 B) 1/2 C) 1/3 D) 1/5
Answer: A Source: Toolbox 1-1

60. How many Moon diameters would fit between the so-called "pointer stars" in Ursa Major, the Big Dipper, shown in Fig. 1-2 of Kaufmann and Comins, *Discovering the Universe*, 5th Ed?
A) 10 B) 2 C) 15 D) 5
Answer: A Source: Toolbox 1-1

61. How long will it take, in solar time, for the Big Dipper, or Ursa Major, the well-known northern constellation, to return to the same position in an observer's sky?
A) 23 hours 56 minutes
B) 24 hours exactly
C) 365 1/4 days
D) 24 hours 4 minutes
Answer: A Source: Section 1-3

62. If observed carefully night by night, a particular star will be seen to
A) rise about 4 minutes earlier every night.
B) rise about 4 minutes later every night.
C) rise at a varying time every night, sometimes earlier, sometimes later than a specified time, because of Earth's differing orbital speed.
D) rise at the same time every night.
Answer: A Source: Section 1-3

63. Any star (except the Sun), when viewed from low and mid-latitudes, will rise in the east
    A) at the same time each evening.
    B) at about 4 minutes later each evening.
    C) about an hour later each evening.
    D) at about 4 minutes earlier each evening.
    Answer: D Source: Section 1-3

64. A particular star is seen to cross the horizon at 10 P.M. (22:00 hours) on a particular night. When would this star cross the horizon on the next night, from the same location?
    A) 10:00 P.M.
    B) 10:04 P.M.
    C) This star will not rise the next night, and will be seen again only after 1 year
    D) 9:56 P.M.
    Answer: D Source: Section 1-3

65. On Dec. 1 at 10 P.M., the bright star Procyon will just be rising on the eastern horizon. When would this star rise on Christmas Day (24 days later)?
    A) 8:24 P.M. B) 9:36 P.M. C) 11:36 P.M. D) 10 P.M.
    Answer: A Source: Section 1-3

66. The bright star Procyon is seen to rise on the Eastern horizon at 10 P.M. on Dec. 1. At approximately what time will this star rise 7 days later, on Dec. 8th?
    A) 10:28 P.M. B) 10 P.M. C) 9:32 P.M. D) 9:53 P.M.
    Answer: C Source: Section 1-3

67. If you are standing upon the Earth's equator, your zenith (the vertical direction above your head), over a period of one year, will take up which of the following alignments?
    A) It will always remain at a fixed angle of 23.5° to the spin axis of Earth.
    B) It will be at an angle to the Earth's spin axis that will vary between 0 and 23.5° over a period of 6 months.
    C) It will always be parallel to the Earth's spin axis.
    D) It will always be perpendicular to the Earth's spin axis.
    Answer: D Source: Section 1-3

68. An arc second is a measure of
    A) time taken for the Earth to move through 1° along its orbit.
    B) angle.
    C) time interval, the time between oscillations of a standard clock.
    D) length along the circumference of a circle.
    Answer: B Source: Toolbox 1-1

69. 1 arc second is equal to
A) 1/360 of a full circle.
B) 1/60 degree.
C) 1/3600 degree.
D) 1/60 of a full circle.
Answer: C Source: Toolbox 1-1

70. 1 arc minute is equal to
A) 1/60 degree. B) 1/60 arc second. C) 1/3600 degree. D) 1/60 of a full circle.
Answer: A Source: Toolbox 1-1

71. The number of arc seconds in 1 degree is
A) 60. B) 3600. C) $2.06 \times 10^5$. D) 360.
Answer: B Source: Toolbox 1-1

72. Which ONE of the following statements about angle is CORRECT?
A) 30 arc minutes is 1/2 degree.
B) 50 arc minutes is 1/2 degree.
C) 30 arc minutes is 1/2 arc second.
D) 180 arc minutes is 1/2 of a full circle.
Answer: A Source: Toolbox 1-1

73. The Crab Nebula (shown in Fig. 12-18, Kaufmann & Comins, *Discovering the Universe,* 5th Ed.), has a diameter of about 10 light years and is at a distance of 6,300 light years. What angle will this supernova remnant subtend in our sky?
A) 630 arc seconds
B) 5.5 arc minutes
C) $1.6 \times 10^{-3}$ arc seconds
D) 32.7 arc seconds
Answer: B Source: Toolbox 1-1

74. If the Moon subtends about 30 arc minutes in the sky and is at about 400,000 km from Earth at the time, what is its approximate diameter?
A) 60 km B) 3,500 km C) 350 km D) 35,000 km
Answer: B Source: Toolbox 1-1

75. When viewed from the Earth on a particular night, Jupiter subtends an angle of 42 arc sec. This angle is
A) about three-quarters of an arc minute.
B) less than half of an arc minute.
C) more than an arc minute, although less than a degree.
D) more than a degree.
Answer: A Source: Toolbox 1-1

76. If Venus has an angular diameter of 30 arc seconds when viewed from Earth at a particular time, how does this compare with the typical angular diameter of the Moon?
A) 60 times larger. B) 1/3600 as large. C) 1/60 as large. D) 1/30 as large.
Answer: C Source: Toolbox 1-1

77. The angle subtended at an observer by a city transit bus (length 9 m) at a distance of 1000 m is close to
A) 9/1000 arc second.
B) 9000 arc minutes.
C) 1/2 degree.
D) 1000/9 arc second.
Answer: C Source: Toolbox 1-1

78. What basic pattern do stars follow over a period of hours as they are observed (or photographed) near the north celestial pole?
A) Wobbly circles, because of the precession of the Earth's axis.
B) Almost straight lines, because of the motion of the Earth in its orbit.
C) Circles with the north pole at the center.
D) Ellipses, with the north pole at the focus.
Answer: C Source: Section 1-3

79. When we watch the nighttime sky, we find that
A) stars and constellations slowly rise in the east, pass overhead, and set in the west.
B) stars and constellations slowly rise in the west, pass overhead, and set in the east
C) stars and constellations remain fixed in our sky, not rising or setting in a time as short as one night because they are so far away.
D) a particular star will rise and set several times in the East over the course of one night because of Earth's rotation.
Answer: A Source: Section 1-3

80. A time zone on the Earth, defined for convenience as that region over which civil time is the same at all locations, extends over what range of longitude, on average?
A) 1° B) 15° C) 90° D) 30°
Answer: B Source: Section 1-4

81. In general, a sundial is NOT a good timekeeper because
A) the Earth's orbital speed around the ecliptic is variable.
B) the Earth's rotation rate changes throughout the year.
C) the sky is often cloudy.
D) the Sun's large angular diameter produces a fuzzy shadow.
Answer: A Source: Section 1-4

82. In winter, clocks in New York (maintaining civil time, or mean solar time) compared to those in California, will be
A) 3 hours behind. B) 3 hours ahead. C) 2 hours ahead. D) the same.
Answer: B Source: Section 1-4

83. Of the following years, which was not, or will not be, a leap year?
A) 1988 AD B) 1800 AD C) 1996 AD D) 2000 AD
Answer: B Source: Section 1-5

84. Leap years, containing an extra day, are necessary because
A) the Earth's speed of revolution around its orbit varies throughout the year.
B) the rotation rate of Earth around its axis varies, leading to days of different length during the year.
C) the length of a year is not an exact number of days.
D) 365 days is not exactly divisible by 4.
Answer: C Source: Section 1-5

85. Leap-year corrections in the calendar are necessary in order to account for
A) the slow drift in the direction of the Earth's spin axis.
B) the fact that 1 year is not exactly 365 days.
C) the fact that the actual rotation time of the Earth with respect to the stars is not exactly 24 hours.
D) the wobble of the Earth's axis.
Answer: B Source: Section 1-5

86. The person who introduced the leap year into our calendar was
A) Pope Gregory XIII. B) Sir Isaac Newton. C) Julius Caesar. D) Ptolemy.
Answer: C Source: Section 1-5

87. The most recent correction to the calendar to keep the yearly date in tune with the seasons (resulting in the present calendar) was instituted by
A) Sir Isaac Newton. B) Pope Gregory XIII. C) Julius Caesar. D) Galileo.
Answer: B Source: Section 1-5

88. The present-day calendar, which includes leap years and century year exclusion, was modified by
A) Pope Gregory XIII in 1582.
B) Karl Marx in 1917.
C) Julius Caesar in 50 BC.
D) Ptolemy in the 2nd century, BC.
Answer: A Source: Section 1-5

89. The true orbital period of the Earth around the Sun, defined as the time taken to complete one orbit with respect to the background stars, is
    A) one Gregorian year.
    B) one tropical year.
    C) one solar day.
    D) one sidereal year.
    Answer: D Source: Section 1-5

90. The time which elapses before the Sun returns to the same point in space in the solar system compared to the background stars in our sky is
    A) one solar day.
    B) one sidereal year.
    C) one Gregorian year.
    D) one tropical year.
    Answer: B Source: Section 1-5

91. The time that elapses before the Sun returns to the position of the vernal equinox in our sky is
    A) one solar day.
    B) one tropical year.
    C) one Gregorian year.
    D) one sidereal year.
    Answer: B Source: Section 1-5

92. The length of the Earth's year at the present time is
    A) approximately 365¼ days.
    B) exactly 365¼ days.
    C) exactly 365 days
    D) varies between 365 ands 366 days over a period of about 400 years.
    Answer: A Source: Section 1-5

93. The ecliptic can be defined as
    A) the path traced out by the Moon in our sky in one month against the background stars.
    B) the path traced out by the Sun in our sky over one year against the background stars.
    C) the extension of the Earth's equator onto the sky.
    D) the plane which is perpendicular to the Earth's spin axis.
    Answer: B Source: Section 1-6

94. In which direction does the Sun appear to move along the ecliptic over the course of a year, relative to the background stars?
    A) Toward the west
    B) Toward the southwest
    C) Toward the northwest
    D) Toward the east
    Answer: D Source: Section 1-6

95. The Sun's apparent path across our sky against the background stars (which would be seen if the sunlit sky were not light) is known as
   A) the great circle.
   B) the ecliptic.
   C) the celestial equator.
   D) the celestial meridian.
   Answer: B Source: Section 1-6

96. If the Sun passes directly over your head on at least one day per year, then you are standing
   A) within 23½° of the equator.
   B) exactly on the equator.
   C) anywhere on the Earth (no limitation).
   D) within 66½° of the equator.
   Answer: A Source: Section 1-6

97. The Sun in our sky, if we could see stars in the daytime, would
   A) remain stationary against the background stars.
   B) appear to move eastward against the background stars at a rate of 1° per day.
   C) appear to move westward against the background stars at a rate of 1° per day.
   D) appear to move eastward against the background stars at a rate of 15° per day.
   Answer: B Source: Section 1-6

98. The ecliptic crosses the celestial equator
   A) at two points, known as solstices.
   B) at two points, known as equinoxes.
   C) at one point only, known as the vernal equinox.
   D) at the meridian.
   Answer: B Source: Section 1-6

99. The tilt angle of the Earth's spin axis to the direction perpendicular to the ecliptic (known as the ecliptic pole)
   A) varies rapidly through the year from +23.5° to -23.5°.
   B) is 90° and fixed.
   C) is 0° and fixed, since this defines the ecliptic plane.
   D) is 23.5° and fixed.
   Answer: D Source: Section 1-6

100. Over an interval of 6 months, the tilt of the Earth's spin axis with respect to the background stars will change by
   A) 180°. B) 23.5°. C) 47°. D) 0°.
   Answer: D Source: Section 1-6

101. At what approximate value of declination was the Sun on March 21 this year?
A) 180° B) 0° C) No unique value D) 23.5°
Answer: B Source: Section 1-6

102. At what approximate value of declination was the Sun on June 21 this year?
A) -23.5° B) 0° C) 90° D) 23.5°
Answer: D Source: Section 1-6

103. At what approximate value of declination was the Sun on September 22 this year?
A) 180° B) 23.5° C) 0° D) 90°
Answer: C Source: Section 1-6

104. At what approximate value of declination was the Sun on December 21 this year?
A) 0° B) -23.5° C) 90° D) 23.5°
Answer: B Source: Section 1-6

105. At what approximate value of Right Ascension was the Sun this year on March 21?
A) No particular value
B) 1 hour 0 minutes
C) 0 hours 0 minutes
D) 12 hours 0 minutes
Answer: C Source: Section 1-6

106. At what approximate value of Right Ascension was the Sun this year on June 21?
A) 6 hours 0 minutes.
B) 0 hours 0 minutes.
C) 12 hours 0 minutes.
D) 18 hours 0 minutes.
Answer: A Source: Section 1-6

107. At what approximate value of right ascension was the Sun this year on December 21?
A) 18 hours 0 minutes
B) 12 hours 0 minutes
C) 0 hours 0 minutes
D) 6 hours 0 minutes
Answer: A Source: Section 1-6

108. The CHANGE in the right ascension of the Sun between June 21 and September 22 is approximately
A) 6 hours 0 minutes.
B) 0 hours 0 minutes.
C) 12 hours 0 minutes.
D) 18 hours 0 minutes.
Answer: A Source: Section 1-6

109. The CHANGE in the right ascension of the Sun between June 21 and December 21 is approximately
A) 6 hours 0 minutes.
B) 0 hours 0 minutes.
C) 12 hours 0 minutes.
D) 18 hours 0 minutes.
Answer: C Source: Section 1-6

110. The CHANGE in the declination of the Sun between December 21 and June 21 is approximately
A) 47°. B) 23.5°. C) 180°. D) 90°.
Answer: A Source: Section 1-6

111. The CHANGE in the declination of the Sun between March 21 and June 21 is approximately
A) 23.5°. B) 90°. C) 47°. D) 180°.
Answer: A Source: Section 1-6

112. If the Earth's spin axis were to be perpendicular to the plane of its orbit (the ecliptic), seasonal variations on the Earth
A) would be much more severe.
B) would have the same severity but each season would last twice as long.
C) would be nonexistent.
D) would remain the same as they are at present.
Answer: C Source: Section 1-6

113. The Earth would NOT have seasons if
A) its axis of rotation were perpendicular to its equatorial plane.
B) the observer's vertical axis were perpendicular to the Earth's orbital plane.
C) its axis of rotation were perpendicular to its orbital plane.
D) its equatorial plane were perpendicular to its orbital plane.
Answer: C Source: Section 1-6

114. Seasonal variations on a planet's surface are caused by
A) the variation of the planet's distance from the Sun during its passage round its orbit.
B) volcanoes which erupt periodically because of tidal interactions and obscure the atmosphere of planets.
C) clouds which periodically form and disappear as the planet orbits the Sun.
D) the tilt of the planet's spin axis with respect to the perpendicular to its orbital plane.
Answer: D Source: Section 1-6

115. One essential condition for "seasons" to occur on a planet is that
A) the planet have a thick atmosphere.
B) the planet have its axis perpendicular to its orbital plane.
C) the planet have its equator tilted with respect to its orbital plane.
D) the planet's distance from the Sun varies.
Answer: C Source: Section 1-6

116. Summertime in the northern hemisphere is when
A) the clearest skies occur, because of climate changes.
B) sunlight falls more directly upon this hemisphere, heating it more than average.
C) the Earth is closest to the Sun in its elliptical orbit.
D) sunlight heats the atmosphere the most by passing through it at an oblique angle.
Answer: B Source: Section 1-6

117. Summertime in the northern hemisphere is when
A) the Sun is closest to the Earth.
B) the northern hemisphere is tilted toward the Sun.
C) the Moon is closest to the Earth.
D) the Sun is closest to the ecliptic.
Answer: B Source: Section 1-6

118. In the northern hemisphere, summertime occurs when
A) the Earth is closest to the Sun in its elliptical orbit.
B) the Earth's equator is parallel to the plane of its orbit.
C) sunlight falls less directly on this hemisphere, spreading the heat out over a greater area.
D) sunlight falls more directly on this hemisphere, heating it more than at other times of the year.
Answer: D Source: Section 1-6

119. Winter in the northern hemisphere occurs when
A) the Earth is furthest from the Sun in the elliptical orbit.
B) the Earth's axis is at its largest angle with respect to the ecliptic plane because of precession.
C) the Earth is furthest from the ecliptic plane.
D) sunlight falls most obliquely upon that region of Earth.
Answer: D Source: Section 1-6

120. The lowest amount of solar energy per square meter is incident upon the surface of Earth in the northern hemisphere on or about
A) January 5th, mid-winter.
B) December 21st, the beginning of winter.
C) September 21st, the beginning of fall or autumn.
D) March 21st, the end of winter.
Answer: B Source: Section 1-6

121. At what time of the year will the shadow of a vertical pole (a sundial) at any site in the northern hemisphere be the shortest?
A) Dawn, June 21st, at the beginning of summer.
B) Noon, August 5th, mid-summer.
C) Noon, June 21st, at the beginning of summer.
D) Noon, December 21st, at the beginning of winter.
Answer: C Source: Section 1-6

122. At what time of the year in the northern hemisphere will your shadow in sunlight at midday be shortest?
A) The first day of spring, or about March 21
B) The first day of summer, or about June 21
C) Mid-summer, or about August 5
D) Mid-winter, or early January
Answer: B Source: Section 1-6

123. At what average speed does the Sun appear to move across our sky with respect to the stars in order to move through one full circle in one year?
A) About 13° per day.
B) About 24° per day, or very close to 1° per hour.
C) About 1° per day.
D) The Sun never appears to move with respect to the stars in the sky.
Answer: C Source: Section 1-6

124. If our daytime sky were not so bright, how fast would we see the Sun move across our sky with respect to the stars, as it moves through one full circle in one year?
A) About 1° per day.
B) About 15° per hour.
C) Exactly 24° per day or 1° per hour.
D) The Sun would never appear to move with respect to the stars in the sky.
Answer: A Source: Section 1-6

125. A remarkable composite photograph of the Sun's motion through a full year, with 1 photograph per week, is shown in Fig. 1-12 of Kaufmann and Comins, *Discovering the Universe*, 5th Ed. The diagonal streaks show the motion of the Sun as it rises on selected days, because of the Earth's rotation. If the Sun moved through a full range of declination of 47° from -23.5° to +23.5° over the whole year, what was the declination of the Sun when the exposure of the central streak was taken?
A) +14.9° B) -8.6° C) 0°, on the celestial equator D) +8.6°
Answer: D Source: Section 1-6

126. Because of the tilt of the spin axis of Earth to the plane of the Earth's orbit (the ecliptic plane), sunrise in the winter months in the mid-latitude northern hemisphere occurs in which direction in the observer's sky?
A) Southwest. B) Northeast. C) The Sun *always* rises due west. D) Southeast.
Answer: D Source: Section 1-6

127. If you were standing on the equator, which of the following positions in the sky would pass directly over your head (i.e., through your zenith) at some time in one 24-hour period? (See Fig.1-14, Kaufmann & Comins, *Discovering the Universe*, 5th Ed.)
A) The ecliptic pole, or the perpendicular to the direction of the ecliptic plane.
B) The north celestial pole, or the perpendicular to the direction of the celestial equator.
C) The position of the Sun at summer solstice.
D) The vernal equinox, or the zero point of the Right Ascension on the celestial equator.
Answer: D Source: Section 1-6

128. If you stand at latitude 10° N, how many times during the year will the Sun pass precisely through overhead?
A) Twice B) Never C) Once D) Every day for a half a year
Answer: A Source: Section 1-6

129. Where on Earth would you have to stand in order for the Sun to pass directly overhead (ie. pass through your zenith) at some time during the year?
A) Within the tropics, or within +/- 23.5° of the equator.
B) There is no restriction, since this happens at any latitude at some time during the year.
C) Within the Arctic Circle.
D) ONLY on the equator, and nowhere else.
Answer: A Source: Section 1-6

130. Where would you have to be, in either the northern or southern hemispheres, in order for the Sun to remain below the horizon for a 24-hour period for at least a part of a year?
A) Nowhere, since the Sun is always visible at some time of the day, anywhere on Earth.
B) Only at 90°, or at the poles.
C) Above about 23.5° latitude.
D) Above about 66.5° latitude.
Answer: D Source: Section 1-6

131. What will be the lowest latitude above which one would see the Sun for a full 24 hours on at least one day per year?
A) 52° B) 90° C) 66.5° D) 23.5°
Answer: C Source: Section 1-6

132. From the Earth's North Pole, how long will the Sun remain above the horizon once it first appears at the beginning of spring?
A) About 6 hours.
B) About 6 months.
C) Less than an hour.
D) Exactly 12 hours.
Answer: B Source: Section 1-6

133. If the horizon is considered to be split into northern and southern parts by the East-West line, can the Sun ever rise in the southern part of the sky when viewed from a mid-latitude site in the northern hemisphere?
A) Yes, for exactly half a year.
B) No, since the site is in the northern hemisphere.
C) Yes, but only for a few days around mid-summer.
D) Yes, for most of the year, since the observing site is in the northern hemisphere.
Answer: A Source: Section 1-6

134. If the horizon is considered to be split into northern and southern parts by the East-West line, can the Sun ever rise in the northern part of the sky when viewed from a mid-latitude site in the southern hemisphere?
A) Yes, for exactly half a year.
B) No, since the site is in the southern hemisphere.
C) Yes, but only for a few days around mid-summer.
D) Yes, for most of the year, since the observing site is in the southern hemisphere.
Answer: A Source: Section 1-6

135. The Arctic Circle is a line around the Earth at a latitude of
A) 23.5° N. B) 66.5° N. C) 66.5° S. D) variable, averaging 66.5°.
Answer: B Source: Section 1-6

136. The Arctic Circle is defined as a line on Earth where
 A) the Sun is always 23.5° or more above or below the horizon.
 B) the Sun never shines, at any time of the year.
 C) the Sun can be seen for 24 hours on at least one day of the year.
 D) the Sun always shines, winter or summer.
 Answer: C Source: Section 1-6

137. The "Land of the Midnight Sun" is so-named because
 A) the Sun is above the horizon for a full 24 hours at certain times of the year.
 B) the Sun passes overhead at least once during the year from this region.
 C) twilight is bright and lasts all night thorough the summer months since the Sun never gets far below the horizon from these locations.
 D) the full Moon is always up whenever the Sun sets, maintaining light skies throughout the summer months.
 Answer: A Source: Section 1-6

138. The zodiac is
 A) a band of sky extending 8° on each side of the celestial equator.
 B) a band of sky 16° wide centered on the ecliptic.
 C) a constellation representing a boat in the sky.
 D) a band of sky 8° wide centered on the ecliptic.
 Answer: B Source: Section 1-6

139. The vernal equinox is one time of the year when the Sun
 A) is at its lowest point in the sky at midday.
 B) crosses the celestial equator.
 C) crosses the Moon's orbital path in the sky.
 D) crosses the ecliptic plane.
 Answer: B Source: Section 1-6

140. If you were standing on the South Pole (with the south celestial pole in your zenith) at the time of the vernal equinox, where would you see the Sun all day?
 A) In your zenith.
 B) On your horizon.
 C) Well below your horizon.
 D) 23.5° above the horizon.
 Answer: B Source: Section 1-6

141. The vernal equinox is that time of the year when
   A) the Sun crosses the equatorial plane, or celestial equator, moving north.
   B) the Sun crosses the equatorial plane or celestial equator, moving south.
   C) the Earth is at the closest point to the Sun in its elliptical orbit.
   D) the Sun crosses the ecliptic plane.
   Answer: A Source: Section 1-6

142. When the Sun is at one of the equinoxes
   A) the day is longer than the night in one hemisphere of the Earth and shorter in the other hemisphere.
   B) day and night are of equal length only for people on the equator.
   C) people on the equator have perpetual daylight.
   D) day and night are of equal length everywhere on the Earth.
   Answer: D Source: Section 1-6

143. The equinoxes are located at the intersections of
   A) the ecliptic and the celestial equator.
   B) the horizon and the celestial equator.
   C) the ecliptic and the horizon.
   D) the ecliptic and the Moon's orbit.
   Answer: A Source: Section 1-6

144. Twice per year, when day and night are equal in length, the Sun is at one of two positions in the sky known as equinoxes. These points are the intersections of which two planes in the sky?
   A) Ecliptic and arctic circle
   B) Celestial meridian and celestial equator
   C) Ecliptic and celestial meridian
   D) Celestial equator and ecliptic
   Answer: D Source: Section 1-6

145. The Autumnal Equinox is that time of the year when
   A) the Earth is at the closest point to the Sun in its elliptical orbit.
   B) the Sun passes through the galactic plane.
   C) the Sun crosses the equatorial plane, moving south.
   D) the Sun crosses the ecliptic plane, moving north.
   Answer: C Source: Section 1-6

146. On the day of the vernal equinox (approximately March 21st each year), which of the following conditions holds?
   A) The Sun passes through an observer's zenith only on this day each year.
   B) The Sun rises at its most northerly point on the horizon on this day.
   C) The length of daylight is longest on this day.
   D) Both day and night are almost exactly 12 hours long at all locations on the Earth.
   Answer: D Source: Section 1-6

147. The approximate date around March 21 represents which season to people living in New Zealand?
A) Beginning of spring.
B) Beginning of autumn.
C) Beginning of summer.
D) Beginning of winter.
Answer: B Source: Section 1-6

148. At the summer solstice in the Northern Hemisphere, the Sun
A) is at its lowest angle above the southern horizon at midday, for the whole year.
B) is on the celestial equator.
C) is nearest to the Earth.
D) reaches its highest angle in the sky for the whole year.
Answer: D Source: Section 1-6

149. If you were at the South Pole of the Earth for a full year, what would be the highest angle reached by the Sun above your horizon (at midday, of course)?
A) 23.5°.
B) 90°.
C) 0°, it would only reach the horizon.
D) It would never reach above the horizon, the South Pole being always in darkness.
Answer: A Source: Section 1-6

150. If you were standing on the South Pole at the time of the autumnal equinox, where would you expect the Sun to be at midday?
A) Well below your horizon.
B) On your horizon.
C) 23.5° above the horizon.
D) In your zenith.
Answer: B Source: Section 1-6

151. Where would you expect to see the Sun in your sky if you were at the North Pole at the beginning of fall (about Sept. 24th)?
A) Below your horizon all day.
B) On the horizon.
C) At your zenith.
D) At about 23.5° above your horizon all day.
Answer: B Source: Section 1-6

152. What would be the position and motion of the Sun on December 21st from the South Pole on Antarctica?
   A) It would pass across the sky from the horizon at midnight to reach an angle of 23.5° above the horizon at midday and then return to the horizon.
   B) It would rise in the east at 6am and set in the west at 6pm, reaching 47° above the horizon at midday.
   C) It would move completely around the sky in 24 hours while maintaining an angle of 23.5° above the horizon..
   D) It would remain below the horizon for the whole 24 hours.
   Answer: C Source: Section 1-6

153. Because of precession, how long will it be before the spin axis of the Earth points toward the present Pole Star again?
   A) 26,000 years B) 13,000 years C) 9 years D) At least 1 million years
   Answer: A Source: Section 1-7

154. Precession makes the spin axis of the Earth move in a slow coning pattern, the end of the axis covering a complete circle in a period of
   A) 26,000 years. B) 1 year. C) 2600 years. D) 26 million years.
   Answer: A Source: Section 1-7

155. Precession is
   A) a very slow coning motion of the Earth's axis of rotation.
   B) the occasional reversal in geological time of the direction of the spin axis of Earth.
   C) the daily rotational motion of the Earth.
   D) the motion of the Earth along its orbital path.
   Answer: A Source: Section 1-7

156. Precession is
   A) the slow coning motion of the spin axis of the Earth, similar to that of a spinning top.
   B) another name for a parade.
   C) the motion of the Earth along its orbital path during a year.
   D) the daily spinning motion of the Earth, producing the apparent motion of the Sun and the stars.
   Answer: A Source: Section 1-7

157. Precession of the Earth's spin axis results in
    A) a gradual shift of the vernal equinox along the ecliptic.
    B) a daily shift in the position of the overhead position of an observer relative to the celestial equator (for an observer at a fixed location on Earth).
    C) a gradual change in the angle between the ecliptic and the celestial equator.
    D) changes of the positions of the constellations which are visible at night from Earth over the period of one year.
    Answer: A Source: Section 1-7

158. Precession of the Earth's axis of rotation is caused by
    A) changes in the rate of rotation (length of the day) of the Earth caused primarily by the gravitational pull of the Moon.
    B) changes in the shape of the Earth's orbit due to the gravitational pull of the Moon.
    C) changes in the shape of the Earth's orbit due to the gravitational pull of the Sun.
    D) the gravitational pull of the Moon and the Sun on the equatorial bulge of the Earth.
    Answer: D Source: Section 1-7

159. The reason for the slow drift of the position of the Vernal Equinox through our sky against the background stars over long periods of time is
    A) the movement of the Sun in the Milky Way Galaxy.
    B) the motion of the Earth in its orbit.
    C) the overall movement of local stars in our sky.
    D) the precession of the spin axis of the Earth.
    Answer: D Source: Section 1-7

160. The phenomenon of precession of the Earth's spin axis is caused by
    A) the tidal ebb and flow of ocean waters upon Earth.
    B) the variation of the spin rate of Earth.
    C) the gravitational pull of Moon and Sun upon the Earth's equatorial bulge.
    D) the varying intensity, and hence pressure, of sunlight upon Earth throughout the year.
    Answer: C Source: Section 1-7

161. Polaris, the "pole star," is at present
    A) exactly perpendicular to the ecliptic plane (ecliptic pole).
    B) above the Earth's magnetic pole.
    C) precisely at the north celestial pole.
    D) within 1° of the north celestial pole.
    Answer: D Source: Section 1-7

162. As the Earth rotates, the apparent motion of the pole star, Polaris, in a period of a day, is
   A) a small circle with a radius of less than 1° in about 24 hours.
   B) a slow but noticeable drift across the sky.
   C) a circle with a radius of 23.5° in a period of 24 hours.
   D) zero, there is no motion of the pole star, by definition.
   Answer: A Source: Section 1-7

163. If the polar axis of Earth moves through a full circle in 26,000 years as a result of precession, (see Fig. 1-17, Kaufmann & Comins, *Discovering the Universe*, 5th Ed.) how long will it take for the line between the center of the circle and the spin axis to move through 180° (i.e., to the other side of the circle)?
   A) about 100 years B) 26,000 years C) 1 year D) 13,000 years
   Answer: D Source: Section 1-7

164. The tropical year is different from the sidereal year because
   A) the Sun moves through the galaxy.
   B) the Earth precesses on its axis.
   C) the Earth moves in its orbit.
   D) the Earth's orbit is elliptical.
   Answer: B Source: Section 1-7

165. To which constellation will the North Celestial Pole be closest in the year 14,000 AD? (see Fig 1-17, Kaufmann & Comins, *Discovering the Universe*, 5th Ed.)
   A) Lyra.
   B) Ursa Minor, since the North Celestial Pole never moves, by definition.
   C) Draco.
   D) Cepheus.
   Answer: A Source: Section 1-7

166. The precessional motion of the north celestial pole of Earth is a circle of 47° diameter across the northern sky over a period of 26,000 years. The equivalent motion of the south celestial pole is
   A) a much smaller circle, covered in 26,000 years.
   B) the south celestial pole does not move at all during precession.
   C) a circle of 47° diameter, covered in 26,000 years.
   D) a circle of 47° diameter, covered in a much shorter time, about 1000 years.
   Answer: C Source: Section 1-7

167. The present position of the Vernal Equinox in the sky is in the constellation
   A) Aquarius. B) Aries. C) Pisces. D) Ursa Minor.
   Answer: C Source: Section 1-7

168. Why do the astronomical coordinates of Right Ascension and Declination of a star change systematically night by night?
   A) They do NOT change, since they are positions on a fixed star chart.
   B) Because of the motion of the Earth in its orbit around the Sun.
   C) Because of precession of the Earth's spin axis.
   D) Because of the bending of light by the Earth's atmosphere.
   Answer: C Source: Section 1-7

169. Why do we see different phases of the Moon?
   A) Because the half of the Moon which is illuminated by the Sun becomes more or less visible from Earth as the Moon orbits it.
   B) Because Moon rotation brings more or less of the illuminated hemisphere into view from Earth.
   C) Because the Earth's shadow gradually moves over the Moon's surface as the Moon orbits the Earth.
   D) Because the Moon's distance from Earth changes as it moves in its elliptical orbit, thereby changing its apparent brightness.
   Answer: A Source: Section 1-8

170. Which of the following is the correct sequence of appearances of Moon phases in the sky?
   A) New moon, full moon, waxing crescent, waning crescent.
   B) Waxing crescent, first quarter, waxing gibbous, full moon.
   C) New moon, waning crescent, first quarter, full moon.
   D) Full moon, waxing gibbous, third quarter, waning crescent.
   Answer: B Source: Section 1-8

171. The one major difference between the Sun and the Moon in our sky is that
   A) the spectrum of their light is very different.
   B) their diameters subtend very different angles.
   C) the Sun emits light while the Moon merely scatters and reflects it.
   D) their motions across the sky in the course of a day are very different.
   Answer: C Source: Section 1-8

172. Which way will the "horns," or sharp ends of the crescent, of the Moon point in the sky when the Moon is on the western horizon at sunset, at a phase 3 days beyond new moon?
   A) Toward the Sun, westward.
   B) The Moon is NOT crescent-shaped at this phase, 3 days beyond new.
   C) Away from the Sun, eastward.
   D) At right angles to the Sun direction, northward.
   Answer: C Source: Section 1-8

173. From the Northern Hemisphere, where in the sky would you expect to see true astronomical New Moon?
A) In a direction opposite to that of the Sun.
B) In a direction at right angles to that of the Sun.
C) The Moon is not visible at New Moon.
D) Always in the south.
Answer: C Source: Section 1-8

174. The phase of the Moon when the Sun and Moon have the same right ascension is
A) gibbous. B) full moon. C) first quarter. D) new moon.
Answer: D Source: Section 1-8

175. If the Moon is between the Sun and the Earth and almost in line with the Sun, we call its phase
A) gibbous.
B) You cannot fool me, the Moon never goes between the Sun and the Earth.
C) full moon.
D) new moon.
Answer: D Source: Section 1-8

176. At approximately what time will the new moon rise?
A) Midnight B) Sunset C) Midday D) Close to sunrise
Answer: D Source: Section 1-8

177. On a given evening, you notice that the sunlit portion of the Moon has a crescent shape. This simple observation tells you
A) that the Moon is closer to the Sun than is the Earth at that time.
B) that the line from the Earth to the Moon is exactly at right angles to the Sun-Earth line.
C) that the Moon is further from the Sun than is the Earth at that time.
D) nothing at all about the position of the Moon in space compared to that of Earth and Sun.
Answer: A Source: Section 1-8

178. In order for the Moon to appear as a crescent shape to an observer
A) the observer must be on the opposite side of the Sun to the Moon.
B) the Moon must be closer to the Sun than is the observer, and somewhere between Sun and the observer.
C) the observer and the Moon must be at equal distances from the Sun.
D) the observer must be closer to the Sun than to the Moon.
Answer: B Source: Section 1-8

179. The crescent moon just after a new moon will be seen from Earth only
A) near the zenith.
B) in the western sky.
C) in the northern sky.
D) in the eastern sky.
Answer: B Source: Section 1-8

180. At approximately what time will the crescent moon just after New Moon be seen?
A) Sunrise B) Sunset C) Midday D) Midnight
Answer: B Source: Section 1-8

181. Which of its Moon's phases is most easily seen during the daytime?
A) Quarter. B) New. C) The Moon is NEVER visible in daylight. D) Full.
Answer: A Source: Section 1-8

182. When will the first quarter Moon rise, approximately?
A) Midnight B) 6 P.M. C) Noon D) 6 A.M.
Answer: C Source: Section 1-8

183. The phase of the Moon when Sun and Moon are separated by 6 hours of right ascension is always
A) either first or third quarter. B) crescent. C) new moon. D) full moon.
Answer: A Source: Sections 1-2 and 1-8

184. How much of the total surface of the Moon is illuminated by the Sun when it is at quarter phase?
A) One quarter. B) All of it. C) One half. D) Very little.
Answer: C Source: Section 1-8

185. When the Moon is in its gibbous phase, the positions of Moon, Earth and Sun are such that
A) the Moon is further from the Sun than is the Earth.
B) the relative distances of Earth and Moon from the Sun are irrelevant, since this phase can occur at any time.
C) the Sun and Moon are 90° apart as seen from the Earth.
D) the Moon is closer to the Sun than is the Earth.
Answer: A Source: Section 1-8

186. At what time does a full moon rise, approximately?
A) At sunrise. B) At noon. C) At sunset. D) At midnight.
Answer: C Source: Section 1-8

187. Full moon always occurs
    A) when the Moon is further from the Sun than is the Earth.
    B) when the Moon is at right angles to the direction of the Sun.
    C) when the Moon is closer to Sun than is the Earth.
    D) on the first of every month.
    Answer: A Source: Section 1-8

188. When the Sun and Moon are separated by 12 hours of right ascension, the phase of the Moon is always
    A) third quarter. B) either first or third quarter. C) full moon. D) first quarter.
    Answer: C Source: Section 1-8

189. When does the third quarter moon rise?
    A) About 6 P.M. B) About 6 A.M. C) Close to noon. D) Close to midnight.
    Answer: D Source: Section 1-8

190. If the Moon is located at the vernal equinox on the first day of spring, what then is the phase of the Moon?
    A) Third quarter B) Full C) New D) First quarter
    Answer: C Source: Section 1-8

191. On a particular day, the Sun is at the summer solstice and the Moon is at the vernal equinox. The lunar phase on that day is
    A) full. B) quarter. C) not predictable from this information alone. D) new.
    Answer: B Source: Sections 1-6 and 1-8

192. On a particular day, the Sun is at the vernal equinox and the Moon is at the autumnal equinox. The lunar phase on this particular day is
    A) new. B) quarter. C) not predictable from this information alone. D) full.
    Answer: D Source: Sections 1-6 and 1-8

193. Which of the following planets will be seen from the Earth as crescent-shaped at certain times in their orbits?
    A) Jupiter B) Venus C) Uranus D) Mars
    Answer: B Source: Section 1-8

194. Which of the following "planets" will never be seen from Earth as a crescent?
    A) Mercury B) Mars C) The Moon D) Venus
    Answer: B Source: Section 1-8

195. If you were on Mars, which of the following "planets" would never be seen as a crescent?
A) Earth's Moon B) Venus C) Jupiter D) Earth
Answer: C Source: Section 1-8

196. If you were on the Moon, which of the following "planets" could occasionally appear as crescent-shaped?
A) Earth B) Mars C) Jupiter D) the asteroid Ceres
Answer: A Source: Section 1-8

197. When the Moon is in its gibbous phase, the right ascensions of the Sun and the Moon differ by
A) 6 hours. B) less than 6 hours. C) more than 6 hours. D) 0 hours.
Answer: C Source: Sections 1-2 and 1-8

198. The gibbous phase of the Moon occurs between the two positions
A) of first quarter and full moon.
B) when Moon and Sun are at right angles to each other, in the section of the orbit which includes the New Moon position.
C) of new moon and first quarter.
D) of third quarter and new moon.
Answer: A Source: Section 1-8

199. A full moon will always be at its highest in our sky at about
A) midday. B) sunrise. C) sunset. D) midnight.
Answer: D Source: Section 1-8

200. Full moon can be on the horizon
A) at any time, day or night, with no restriction.
B) only at sunrise or sunset.
C) only at midnight.
D) only at midday.
Answer: B Source: Section 1-8

201. The Moon is visible in the sky in the daytime from most places upon Earth
A) almost never; only during solar eclipses when the sky is dark.
B) at some time on every day, but it is difficult to see because of the blue sky.
C) about half the time, or for two weeks in every month.
D) only at full Moon phases, when it is very bright.
Answer: C Source: Section 1-8

202. How does the Moon rotate in order to keep one face pointed toward the Earth at all times, as seen in Fig. 1-18, Kaufmann and Comins, *Discovering the Universe*, 5th Ed.?
A) It does not rotate at all.
B) It rotates once per day.
C) It rotates once per year.
D) It rotates once per month.
Answer: D Source: Section 1-8

203. The Moon is seen to keep one face toward the Earth at all times. If viewed from a point directly above the plane of the planetary system, how does it have to rotate in order to maintain this alignment?
A) It must rotate once per day, to maintain its direction toward the Earth.
B) It must not rotate at all, since we always see the same face from Earth.
C) It must rotate once per month, or once per orbit around the Earth.
D) It must rotate once per year as the Earth and Moon orbit the Sun together.
Answer: C Source: Section 1-8

204. One synodic month is longer than one sidereal month by about
A) 2.2 days. B) 1 week. C) 1 hour. D) 4 minutes.
Answer: A Source: Section 1-8

205. Why is the period between two successive full moons NOT equal to the Moon's orbital period, or sidereal month?
A) Because the Earth-Moon system is also orbiting the Sun.
B) These two time intervals are not related, since full moon time depends upon the Moon's rotation period about its own axis.
C) Because the Moon's orbit is elliptical, and the Moon therefore moves irregularly round the Earth.
D) Because the Moon's orbit is inclined at about 5° to the Earth's orbital plane.
Answer: A Source: Section 1-8

206. The fact that the Earth-Moon system orbits the Sun (covering 30° per month) while the Moon orbits the Earth means that, compared to one lunar orbital (sidereal) period, the time between successive full moons, the synodic month, is
A) about 2 days shorter.
B) about two days longer.
C) longer.
D) about twice as long.
Answer: C Source: Section 1-8

207. The length of time for the Moon to move from new moon to new moon is known as one synodic month. Compared to one full orbital period with respect to the star background, or one sidereal month, this synodic month is
A) about twice as long.
B) about two days longer.
C) about 2 days shorter.
D) exactly the same length.
Answer: B Source: Section 1-8

208. If you were on the Moon at the dividing line between dark and light, (the terminator) at a particular time, say sunrise, how long would it be before this dividing line returned to your position?
A) 23 hours 56 minutes. B) 27½ days. C) 29½ days. D) 365¼ days.
Answer: C Source: Section 1-8

209. The Moon will appear to an observer at mid-latitudes on Earth to rise in the east, in solar time, at about (you might attempt to verify this by observation)
A) 4 minutes earlier each evening.
B) the same time each evening.
C) 4 minutes later each evening.
D) one hour later each evening.
Answer: D Source: Section 1-8

210. The motion of the Moon across our sky, against the background of stars, is approximately
A) 1° per day. B) 13° per day. C) its own diameter per day. D) 15° per hour.
Answer: B Source: Section 1-8

211. The motion of the Moon across our sky in one hour, as seen against the background of stars, is approximately
A) 1/10 degree. B) 13°. C) its own diameter (½°). D) 4°.
Answer: C Source: Section 1-8

212. The direction of motion of the Moon in our sky, against the background of stars (see Fig. 1-18, Kaufmann & Comins, *Discovering the Universe*, 5th Ed.), is
A) always eastward.
B) sometimes eastward but mostly westward.
C) always westward, in concert with the stars.
D) mostly eastward, but occasionally, (at full moon) westward.
Answer: A Source: Section 1-8

213. In which way does the Moon move day by day in the sky, against the background of stars, when viewed from Earth?
A) toward the west.
B) in no particular direction and with no particular pattern
C) toward the east.
D) toward the North in summer and the South in winter.
Answer: C Source: Section 1-8

214. Which of the following statements is correct for eclipses in the Sun-Earth-Moon system?
A) A total eclipse of the Sun occurs only at new Moon.
B) A total eclipse of the Sun occurs only at full Moon.
C) A total eclipse of the Moon occurs only at new Moon.
D) A total eclipse of the Sun occurs only at first quarter Moon.
Answer: A Source: Section 1-9

215. Which of the following statements is NOT correct for eclipses in the Sun-Earth-Moon system?
A) Eclipses of Moon and Sun do not occur at quarter Moon phases.
B) A total eclipse of the Moon occurs only at full Moon.
C) A total eclipse of the Sun occurs only at new Moon.
D) A total eclipse of the Sun occurs only at full Moon.
Answer: D Source: Section 1-9

216. A NECESSARY condition for lunar or solar eclipses is that
A) the Moon be close to or crossing the ecliptic plane.
B) the Sun be on or close to the ecliptic plane.
C) the Earth be on the ecliptic plane.
D) the Sun be on the celestial equator.
Answer: A Source: Section 1-9

217. There is about a 5° angle between the orbit of the Moon and the
A) plane of the Earth's equator.
B) plane of the Sun's equator.
C) plane of the ecliptic, or the Earth's orbit.
D) spin axis of the Earth.
Answer: C Source: Section 1-9

218. What is the approximate inclination of the Moon's orbit to the ecliptic plane?
A) 0° B) 17° C) 5° D) 23.5°
Answer: C Source: Section 1-9

219. The Moon's path across our sky
 A) is confined to a band of sky around the ecliptic, the zodiac.
 B) is always along the ecliptic plane, by definition.
 C) can be anywhere in our sky.
 D) is confined to regions north of the celestial equator.
 Answer: A Source: Section 1-9

220. The line of nodes of the Moon's orbit is the line of intersection of the orbit with
 A) the celestial meridian through Greenwich, England.
 B) the ecliptic plane.
 C) the celestial equator.
 D) the observer's celestial meridian.
 Answer: B Source: Section 1-9

221. The line of nodes of the Moon's orbit is
 A) the line between the Earth and Moon when the Moon is furthest from the ecliptic plane.
 B) the major axis (longest diameter) of the Moon's elliptical orbit.
 C) the line of intersection between the Moon's orbit and the Earth's orbit (the ecliptic plane).
 D) the line joining the points of the Moon's nearest (perigee) and furthest (apogee) distances from the Earth.
 Answer: C Source: Section 1-9

222. A solar eclipse occurs on Earth when
 A) the Sun passes in front of the Moon.
 B) the Moon passes behind the Sun.
 C) the Moon casts a shadow upon the Earth.
 D) the Earth casts a shadow on the Moon .
 Answer: C Source: Section 1-11

223. A solar eclipse can occur ONLY when
 A) the Moon comes between the Earth and the Sun.
 B) the Sun comes between the Moon and the Earth.
 C) the Earth comes between the Moon and the Sun.
 D) the Sun, Moon, and Earth form a precise right-angled triangle.
 Answer: A Source: Section 1-11

224. Which of the following conditions holds for relative distances during a solar eclipse?
   A) The Moon is closer to the Sun than is the Earth.
   B) The Earth is closer to the Sun than is the Moon.
   C) Moon and Earth are at the same distance from the Sun.
   D) Since the condition for a solar eclipse is independent of relative distances of Earth and Moon from the Sun, either the Moon or the Earth can be closest to the Sun.
   Answer: A Source: Section 1-11

225. What is the phase of the Moon during a total solar eclipse?
   A) Crescent B) First Quarter C) New D) Full
   Answer: C Source: Section 1-11

226. The phase of the Moon at the time of solar eclipse
   A) will be third quarter.
   B) will be new.
   C) will be full.
   D) can be any phase, new, quarter or full.
   Answer: B Source: Section 1-11

227. We can occasionally see a total eclipse of the Sun on Earth because
   A) the angular sizes of Sun and Moon, when viewed from Earth, are almost the same.
   B) the physical sizes of Sun and Moon are almost the same.
   C) the Moon is cooler than the Sun.
   D) both the Moon and Sun move precisely along the ecliptic plane.
   Answer: A Source: Section 1-11

228. You travel to an exotic place to observe a total solar eclipse in December and someone on this trip tells you that the next eclipse to occur on Earth will be a lunar eclipse in March. Is this likely to be true?
   A) Yes, but only once about every 1000 years.
   B) Yes.
   C) No.
   D) It could be true, since such an eclipse can occur when the Sun's position with respect to the celestial equator is changing rapidly.
   Answer: C Source: Section 1-11

229. A lunar eclipse does not occur at every full moon because
   A) the path of the Sun is inclined at an angle of 5° to the ecliptic plane.
   B) the orbit of the Moon is not a perfect circle.
   C) the plane of the Moon's orbit is at an angle to the plane of the Earth's orbit.
   D) a lunar eclipse cannot occur after sunset.
   Answer: C Source: Section 1-10

230. Eclipses of the Moon can occur
   A) only once per year.
   B) only during two specific periods in any year.
   C) twice per month.
   D) once every month.
   Answer: B Source: Section 1-10

231. If the plane of the Moon's orbit were to be the same as the ecliptic plane, there would be a lunar eclipse
   A) once every month. B) every day. C) twice per month. D) only twice per year.
   Answer: A Source: Section 1-10

232. If the Moon in its orbit around the Earth moves alternately between Earth and Sun and behind the Earth from the Sun, why then do we not see solar and lunar eclipses every month?
   A) Because the Moon's motion in its orbit is so slow that it only reaches eclipse position every six months.
   B) Because the Moon's orbital plane is at right angles to the ecliptic.
   C) Because the Moon's orbital plane is slightly inclined to the ecliptic.
   D) Because the Moon's orbital plane is inclined slightly to the celestial equator, which is the path of the Sun across our sky.
   Answer: C Source: Section 1-10

233. What is the phase of the Moon during a total lunar eclipse?
   A) Gibbous B) New C) Full D) First Quarter
   Answer: C Source: Section 1-10

234. The maximum number of eclipses (both solar and lunar) that can occur in one calendar year is
   A) 2 B) 5 C) 7 D) 1
   Answer: C Source: Section 1-9

235. A lunar eclipse is caused by
   A) the Moon passing into the shadow of the Earth.
   B) the Earth moving into the Moon's shadow.
   C) the Moon passing behind the Sun.
   D) the Sun passing behind the Moon.
   Answer: A Source: Section 1-10

236. A lunar eclipse can occur ONLY when
   A) the Earth comes between the Moon and the Sun.
   B) the Sun, Moon and Earth form a right-angle triangle.
   C) the Moon comes between the Earth and the Sun.
   D) the Sun comes between the Moon and the Earth.
   Answer: A Source: Section 1-10

237. Eclipses of the Moon can only occur
   A) in the spring and fall seasons, when the Sun is on the ecliptic plane.
   B) at Full Moon.
   C) in June and December, when the Sun is near the solstices.
   D) at New Moon.
   Answer: B Source: Section 1-10

238. At the time of lunar eclipse, the phase of the Moon
   A) is full. B) is first quarter. C) is new. D) can be any phase.
   Answer: A Source: Section 1-10

239. In a penumbral lunar eclipse,
   A) some points on the Moon are totally shaded from the Sun while others are only partly shaded.
   B) no points on the Moon are shaded from the Sun, either totally or partially.
   C) all of the Moon is shaded from the Sun.
   D) all parts of the Moon are partly (not totally) shaded from the Sun.
   Answer: D Source: Section 1-10

240. A total lunar eclipse is visible in principle (assuming clear skies everywhere)
   A) only to people in a long, narrow and very specific path, much smaller than a hemisphere.
   B) only to people in a circular area on the Earth having a diameter equal to that of the Moon.
   C) to everyone upon the Earth.
   D) to everyone in one hemisphere of the Earth.
   Answer: D Source: Section 1-10

241. To someone upon Earth who is watching a total lunar eclipse
   A) the Moon is hidden behind the Sun.
   B) the Sun is relatively high in the sky, since the Earth - Moon line is at right angles to the Earth - Sun line.
   C) the Sun is hidden below the horizon.
   D) the Sun is hidden behind the Moon.
   Answer: C Source: Section 1-10

242. The Earth's shadow at a distance of the Moon's orbit from the Earth is
   A) considerably wider than the Moon.
   B) slightly less wide than the size of the Moon.
   C) almost exactly as wide as the Moon.
   D) extremely small, leaving only a narrow shadow band on the Moon during eclipse.
   Answer: A Source: Section 1-10

243. What is the maximum length of totality for a lunar eclipse?
   A) several hours.
   B) seven minutes.
   C) about two minutes.
   D) one hour and forty minutes.
   Answer: D Source: Section 1-10

244. When in total lunar eclipse, the Moon shows a reddish color because
   A) the red light is the residual thermal glow from a still-warm Moon, after the abrupt removal of the heat of the Sun.
   B) only the red part of the solar spectrum is deflected onto it by the Earth's atmosphere.
   C) the Moon is illuminated only by the residual glow from the dark side of the Earth, which is predominantly red.
   D) light from the northern and southern lights(the aurora) on Earth, which are predominantly red, illuminates the Moon.
   Answer: B Source: Section 1-10

245. The Moon not look completely dark when it is in the Earth's shadow during a total solar eclipse because
   A) atmospheric refraction bends red solar light onto the Moon.
   B) there is a remnant glow from the hot lunar surface.
   C) there are faint emissions from the tenuous lunar atmosphere, excited by solar wind bombardment.
   D) of light reflected from the clouds on the Earth, the earthshine.
   Answer: A Source: Section 1-10

246. Which of the following factors makes it far more likely that a person will have seen a total lunar eclipse than a total solar eclipse?
   A) Total solar eclipses occur much less frequently than total lunar eclipses.
   B) A total lunar eclipse occurs at full Moon when the Moon is bright and high in the sky while a total solar eclipse occurs at new Moon when the Moon is dark and low in the sky.
   C) A total lunar eclipse can be seen by people on most of the nighttime side of Earth while a specific total solar eclipse can only be seen by people within a narrow strip of the Earth's surface.
   D) The Moon appears brighter during a total lunar eclipse than does the Sun during a total solar eclipse.
   Answer: C Source: Section 1-10

247. The total phase of a particular solar eclipse will be seen
   A) only over a region of Earth within +/-23.5° of the Earth's equator, or in the tropics.
   B) from anywhere on the sunlit hemisphere of the Earth.
   C) only within a specific narrow strip across the Earth's surface.
   D) anywhere upon the surface of the Earth.
   Answer: C Source: Section 1-11

248. Where on Earth would you have to be in order to observe a particular total solar eclipse?
   A) On the dark side of the Earth.
   B) Within 250 km of the Earth's equator.
   C) Always within 23.5° of the equator (i.e., within the tropics).
   D) Within a narrow and specific strip of the Earth's surface, less than 250 km wide.
   Answer: D Source: Section 1-11

249. A total solar eclipse is visible (assuming clear skies everywhere)
   A) to everyone on the Earth.
   B) only to people in a long narrow path, much smaller than a hemisphere.
   C) only to people in a circular area on the Earth having a diameter equal to that of the Moon.
   D) to people anywhere in the sunlit hemisphere of the Earth.
   Answer: B Source: Section 1-11

250. Which of the following parameters will dictate whether a particular solar eclipse appears as a total or an annular eclipse to an observer on the center-line of the Moon's shadow?
   A) The time of day or night.
   B) The distance of the Moon from the Earth at the time of eclipse.
   C) The phase of the Moon, whether it is new, quarter or full.
   D) The distance of the Earth from the Sun at the time of eclipse.
   Answer: B Source: Section 1-11

251. During a particular solar eclipse (when the Moon and Sun are precisely in line), the eclipse can be either total (Sun completely covered) or annular (Sun not quite covered) when viewed from the eclipse center-line, because
   A) of the time of day at the viewing site, annular eclipses always occurring in early mornings and early evenings.
   B) the Moon's distance from Earth varies from eclipse to eclipse.
   C) the Moon has deep valleys on its surface.
   D) the Moon's orbit is inclined at several degrees to that of the Earth.
   Answer: B Source: Section 1-11

252. A person standing in the Moon's penumbra will see
   A) a total lunar eclipse.
   B) a partial lunar eclipse.
   C) a partial solar eclipse.
   D) a total solar eclipse.
   Answer: C Source: Section 1-11

253. In view of the elliptical orbits of Earth and Moon, which of the following conditions will result in the longest period of totality during a total solar eclipse.
   A) The Earth is closest to the Sun when the Moon is closest to the Earth.
   B) The Earth is furthest from the Sun when the Moon is furthest from the Earth.
   C) The Earth is closest to the Sun when the Moon is furthest from the Earth.
   D) The Earth is furthest from the Sun when the Moon is closest to the Earth.
   Answer: D Source: Section 1-11

254. What is the maximum time of totality for any total solar eclipse observed from the Earth's surface?
   A) About 7.5 minutes.
   B) A full 12 hour period.
   C) About 2 hours.
   D) Only a few seconds.
   Answer: A Source: Section 1-11

# Chapter 2: Gravitation and the Waltz of the Planets

1. So far as we know, the first person who claimed that natural phenomena could be described by mathematics was
   A) Pythagoras. B) Copernicus. C) Aristotle. D) Ptolemy.
   Answer: A Source: Introduction to Section 2-1

2. The ancient Greek thinker Pythagoras held the view that
   A) the Sun is at the center of the planetary system.
   B) natural phenomena are wonderful to watch but cannot be described by mathematics.
   C) triangles do not exist.
   D) natural phenomena can be described mathematically.
   Answer: D Source: Introduction to Section 2-1

3. What is the name for a theory that describes the overall structure of the universe in which the Earth is located?
   A) Field theory. B) Cosmology. C) Astronomy. D) Astrology.
   Answer: B Source: Introduction to Section 2-1

4. In the Greek era, it was almost universally believed that
   A) the pole star represented the center of the universe, about which the Earth and all other objects revolved
   B) the Sun was at the center of the universe
   C) the Milky Way represented the observable universe, with its center being the center of the universe
   D) the Earth was at the center of the universe
   Answer: D Source: Introduction to Section 2-1

5. The center, or fixed point, of the Greek model of the universe was
   A) the Sun's center
   B) a point mid-way between Earth and Sun
   C) the center of the galaxy
   D) close to the Earth's center
   Answer: D Source: Introduction to Section 2-1

6. The word "planet" is derived from a Greek term meaning
   A) bright night-time object
   B) astrological sign
   C) wanderer
   D) non-twinkling star
   Answer: C Source: Introduction to Section 2-1

7. The planets that were known before the telescope was invented were
   A) Venus, Jupiter, Saturn, Mars, and Pluto.
   B) Mercury, Venus, Mars, Jupiter, and Neptune.
   C) Jupiter, Mercury, Mars, Uranus, and Venus.
   D) Saturn, Venus, Mars, Mercury, and Jupiter.
   Answer: D Source: Introduction to Section 2-1

8. Planets move past the background stars as seen by someone on the Earth. What is the normal direction of this motion?
   A) From west to east because of the motion of the planet along its orbit.
   B) From east to west because of the motion of the planet along its orbit.
   C) From east to west because of the rotation of the Earth.
   D) From west to east because of the motion of the Earth along its orbit.
   Answer: A Source: Introduction to Section 2-1

9. The motions of the planets against the background stars in our sky can best be described as
   A) general eastward motion but with occasional stationary periods, with no motion at all.
   B) regular and uniform eastward motion.
   C) regular patterns with general eastward motion interrupted by periods of westward motion.
   D) regular patterns with general westward motion interrupted by periods of eastward motion.
   Answer: C Source: Introduction to Section 2-1

10. When observing planetary motions from the Earth, the phrase "direct motion" refers to
    A) a slow eastward motion of the planet from night to night compared to the background stars.
    B) the motion of the planet directly toward or away from the Earth in certain parts of the planet's orbit.
    C) a slow westward motion of the planet from night to night compared to the background stars.
    D) the apparent westward motion of the planet (and the sun, Moon, and stars) across the sky due to the rotation of the Earth.
    Answer: A Source: Introduction to Section 2-1

11. An apparent EASTWARD motion of a planet from night to night compared to the background stars (as viewed from Earth) is referred to as
    A) retrograde motion.
    B) direct motion.
    C) rising (if in the east) or setting (if in the west).
    D) precession.
    Answer: B Source: Introduction to Section 2-1

12. An apparent WESTWARD motion of a planet from night to night compared to the background stars (as viewed from Earth) is referred to as
    A) precession.
    B) retrograde motion.
    C) direct motion.
    D) rising (if in the east) or setting (if in the west).
    Answer: B Source: Introduction to Section 2-1

13. When observing planetary motions from the Earth, the phrase "retrograde motion" refers to
    A) motion of the planet away from the Earth during part of its orbit.
    B) a slow eastward motion of the planet from night to night compared to the background stars.
    C) a slow westward motion of the planet from night to night compared to the background stars.
    D) the apparent westward motion of the planet (and the Sun, Moon, and stars) across the sky due to the rotation of the Earth.
    Answer: C Source: Introduction to Section 2-1

14. The term retrograde motion for a planet refers to
    A) a reversal in the apparent direction of motion of a planet past the background stars as seen from the Earth.
    B) the motion of a planet which orbits around the Sun in the opposite direction to the motion of the other planets.
    C) the apparent east to west motion of a planet as seen by an observer on the Earth, due to the Earth's rotation.
    D) the motion of a planet around its deferrent in the geocentric model of the solar system.
    Answer: A Source: Introduction to Section 2-1

15. The term "retrograde motion" for a planet refers to
    A) the apparent motion of a planet's moon in the opposite direction to the motion of the planet itself during half of each orbit of the moon around the planet.
    B) a temporary reversal of the planet's normal east-to-west motion past the background stars as seen from the Earth.
    C) a temporary reversal of a planet's direction of spin about its axis of rotation.
    D) a temporary reversal of the planet's normal west-to-east motion past the background stars as seen from the Earth.
    Answer: D Source: Introduction to Section 2-1

16. The direction of retrograde motion for a planet as seen by an observer on the Earth is
    A) from west to east relative to the background stars.
    B) from east to west relative to the background stars.
    C) from east to west relative to objects on the person's horizon.
    D) from west to east relative to objects on the person's horizon.
    Answer: B Source: Introduction to Section 2-1

17. Retrograde motion of a planet is
    A) westward motion against the star background.
    B) westward motion with respect to the foreground on Earth.
    C) eastward motion with respect to the Moon.
    D) eastward motion against the star background.
    Answer: A Source: Introduction to Section 2-1

18. Retrograde motion of a planet against the background stars is always
    A) movement from east to west.
    B) movement from west to east.
    C) movement northward away from the ecliptic plane.
    D) the apparent motion of the planet away from the Earth.
    Answer: A Source: Introduction to Section 2-1

19. Retrograde motion of a planet refers to which motion, when viewed from Earth?
    A) The southward motion of the planet as it moves away from the northern sky.
    B) The setting of the planet in the west to any observer, caused by Earth rotation.
    C) The westward apparent motion with respect to the stars.
    D) The eastward apparent motion with respect to the stars.
    Answer: C Source: Introduction to Section 2-1

20. When the planet Mars is moving in a retrograde direction, its motion against the background stars is seen to be
    A) westward.
    B) eastward.
    C) stationary, with no motion against the stars.
    D) exactly perpendicular to the equator.
    Answer: A Source: Introduction to Section 2-1

21. In the path of Mars against the background stars shown in Figure 2-1, Kaufmann and Comins, *Discovering the Universe*, 5th Ed., the planet appears from Earth to move in a loop, moving westward for a period of time. What is the angle between the Earth-Sun line and the Earth-Mars line when the planet is half-way through the retrograde motion, on about Sept. 1?
    A) 90°. B) 0°. C) 180°. D) It can be any angle.
    Answer: C Source: Introduction to Section 2-1

22. Retrograde motion of a planet when viewed from the Earth is caused by the fact that
    A) The planet's orbit is elliptical.
    B) the Sun is moving.
    C) the planet's orbit is inclined at an angle to the Earth's orbit.
    D) the Earth is moving.
    Answer: D Source: Introduction to Section 2-1

23. Ptolemy's nationality was
    A) Polish. B) Italian. C) Greek. D) Egyptian.
    Answer: C Source: Section 2-1

24. The Greek mathematician, Ptolemy, devised
    A) a method to measure the Earth's radius.
    B) the first known sundial.
    C) a heliocentric model for the solar system.
    D) a geocentric model for the solar system.
    Answer: D Source: Introduction to Section 2-1

25. Ptolemy's model for the solar system was
    A) Earth-centered, with elliptical planetary orbits.
    B) Sun-centered, with planets moving in circles around it.
    C) Earth-centered, with Sun, Moon, and planets moving in ellipses in the sky.
    D) Sun-centered, with elliptical planetary orbits.
    Answer: A Source: Introduction to Section 2-1

26. A major contribution of Ptolemy to the development of astronomy was to
    A) derive a mathematical model for the solar system, in which planets move in epicycles and the epicycles orbited the Earth.
    B) originate the idea of a geocentric (Earth-centered) cosmology, which was later developed mathematically by Aristarchus.
    C) derive a mathematical model for the solar system, in which planets move around the Sun in circular orbits.
    D) derive a mathematical model for the solar system, in which planets move around the Earth in elliptical orbits, moving fastest when closest to the Earth.
    Answer: A Source: Toolbox 2-1

27. In the geocentric model of the solar system developed by Ptolemy,
    A) planets move in circular epicycles around the Sun while the Sun moves in a circular orbit around the Earth.
    B) planets move in circular epicycles while the centres of the epicycles move in circular orbits around the Earth.
    C) planets move at constant speeds in circular orbits around the Earth.
    D) planets move with varying speeds in elliptical orbits around the Earth.
    Answer: B Source: Toolbox 2-1

28. The epicycle, in the Greek planetary model, is
    A) the focus of the ellipse which is the orbit of the planet around the Earth.
    B) the circle centered on Earth, about which the center of the smaller circular motion moves.
    C) the off-center point in the planetary system, occupied by Earth.
    D) the small circle through which the planet moves, as the center of this circle orbits the Earth.
    Answer: D Source: Toolbox 2-1

29. In the geocentric model for the solar system developed by Ptolemy, to what does the word "epicycle" refer?
    A) The length of time from when the planet is furthest from Earth to the next time it is furthest from Earth.
    B) The large circle (orbit) which carries the planet around the Earth, while the planet itself is moving in a smaller circle.
    C) A small circle about which a planet moves while the centre of this circle moves around the Earth.
    D) One complete cycle of planetary motions after which the motions repeat themselves (almost) exactly.
    Answer: C Source: Toolbox 2-1

30. In Ptolemy's geocentric theory of the solar system, what name is given to the small circle around which the planet moves while the center of this circle orbits the Earth?
A) Deferent. B) Celestial equator. C) Epicycle. D) Ecliptic.
Answer: C Source: Toolbox 2-1

31. In Ptolemy's description of the solar system, the deferent is
A) a circular path along which a planet moves, while the center of this circular path itself moves in a circle around the Earth.
B) a circular path (around the Sun) along which the center of a planet's epicycle moves.
C) an elliptical path along which a planet moves around the Sun.
D) a circular path (around the Earth) along which the center of a planet's epicycle moves.
Answer: D Source: Toolbox 2-1

32. The deferent, in the Greek planetary model, is
A) the small circle about which the planet moves, as the center of the circle orbits the Earth.
B) the off-center point in the planetary system, occupied by Earth.
C) the circle about which each planet's epicycle center moves.
D) the part of the planet's orbit when it appears to move "backward" (i.e., westward) in the sky.
Answer: C Source: Toolbox 2-1

33. In the geocentric model for the solar system developed by Ptolemy, to what does the word "deferent" refer?
A) The distance of the center of the epicycle from the center of the Earth.
B) The large circle (orbit) which carries the planet around the Earth, while the planet itself is moving in a smaller circle.
C) The distance of offset between the center of the Earth and the center of a planet's orbit.
D) A small circle about which a planet moves while the centre of this circle moves around the Earth.
Answer: B Source: Toolbox 2-1

34. In Ptolemy's geocentric theory of the solar system, what name is given to the large circle (orbit) which carries the planet around the Earth?
A) Deferent. B) Epicycle. C) Celestial equator. D) Ecliptic.
Answer: A Source: Toolbox 2-1

35. In the geocentric universe, when is the planet closest to the Earth?
   A) During direct motion, eastward.
   B) There is no specific time in the orbit when the planet is closest to the Earth.
   C) During retrograde motion, westward.
   D) When the planet is crossing the deferent.
   Answer: C Source: Toolbox 2-1

36. According to the Ptolemaic model of the Solar system (Toolbox 2-1, Kaufmann & Comins, *Discovering the Universe*, 5th Ed.), during retrograde motion a planet would be
   A) closer to Earth than average.
   B) at varying distances from the Earth, sometimes closer and sometimes further away than the average distance.
   C) further away from Earth than average.
   D) always at the same distance from Earth, since the planet orbits the Earth in a circle in this.model
   Answer: A Source: Toolbox 2-1

37. The *Almagest* is
   A) a collection of ancient data and predictions of positions of the Sun, Moon, and planets, compiled by Ptolemy.
   B) a Renaissance book describing in detail the development of a heliocentric Universe, with elliptical orbits, the law of equal areas, and the harmonic law.
   C) a detailed multi-volume account of a heliocentric cosmology produced by ancient Greek astronomers.
   D) an ancient book describing the construction of Stonehenge.
   Answer: A Source: Toolbox 2-1

38. The *Almagest*, a collection of earlier data and description of calculations, was written by
   A) Erathosthenes. B) Kepler. C) Ptolemy. D) Copernicus.
   Answer: C Source: Toolbox 2-1

39. The purpose of describing planetary orbits in terms of epicycles and deferents was to account for
   A) the pattern of alternating direct and retrograde motion.
   B) the general motion of all objects toward the west in the sky each day.
   C) the fact that a planet's speed in its orbit is fastest when it is closest to the Sun.
   D) the pattern of alternating conjunctions and oppositions.
   Answer: A Source: Toolbox 2-1

40. The initial reason why the geocentric model for the solar system began to be discarded after the 15th Century AD was that
    A) the heliocentric model is conceptually simpler.
    B) observations by spacecraft proved that all planets orbit the Sun.
    C) the invention of the telescope provided observations which were in better agreement with a heliocentric model.
    D) Isaac Newton was able to derive all planetary motion from one universal law of gravity.
    Answer: A Source: Section 2-1

41. Nicolaus Copernicus was the first person to
    A) develop a mathematical model for a Sun-centered solar system.
    B) use a telescope to observe the sky at night.
    C) use ellipses to describe the orbits of the planets.
    D) describe planetary orbits using the force of gravity.
    Answer: A Source: Section 2-1

42. The person who developed the first mathematical model for a heliocentric cosmology was
    A) Kepler. B) Aristarchus. C) Ptolemy. D) Copernicus.
    Answer: D Source: Section 2-1

43. Copernicus lived
    A) after Ptolemy but before Kepler.
    B) after Kepler but before Ptolemy.
    C) after Newton but before Kepler.
    D) after Kepler but before Newton.
    Answer: A Source: Section 2-1

44. Copernicus' nationality was
    A) Italian. B) Greek. C) Egyptian. D) Polish.
    Answer: D Source: Section 2-1

45. The Copernican system for planetary motions is
    A) Earth-centered, with the planets, Sun, and stars mounted on crystal spheres, pivoted to allow the correct motions around the Earth.
    B) Earth-centered, with the planets moving in epicycles around the Earth.
    C) Sun-centered, with the planets moving in elliptical orbits, the Sun being at one focus of the ellipse.
    D) Sun-centered, with the planets moving in perfect circles around the Sun.
    Answer: D Source: Section 2-1

46. The contribution of Copernicus to the development of astronomy was a mathematical model for
    A) a geocentric cosmology in which the planets move in circular epicycles.
    B) a heliocentric cosmology in which the planets move in elliptical orbits.
    C) a heliocentric cosmology in which the planets move in circular orbits.
    D) the solar system in which the planets move under the gravitational influence of the Sun.
    Answer: C Source: Section 2-1

47. Copernicus used the fact that Mars can sometimes be seen high in our sky at midnight to conclude that
    A) Mars can come between the Earth and the Sun.
    B) the Earth can come between Mars and the Sun.
    C) the Sun can come between the Earth and Mars.
    D) Mars and the Sun can never be on the same side of the Earth at the same time.
    Answer: B Source: Section 2-1

48. Which of the following statements CORRECTLY describes why Copernicus decided that the orbits of Mercury and Venus are smaller than the orbit of the Earth?
    A) Both planets can sometimes be seen high in our sky at midnight.
    B) Both planets show a complete cycle of phases, like the Moon.
    C) Both planets occasionally pass through conjunction with the Sun, as seen from the Earth.
    D) Both planets stay fairly close to the Sun in our sky.
    Answer: D Source: Section 2-1

49. When Venus is at inferior conjunction,
    A) its speed in its orbit has its greatest value.
    B) it is at its greatest distance from the Earth.
    C) it is at its greatest angle from the Sun, as seen from the Earth.
    D) it is at its smallest distance from the Earth.
    Answer: D Source: Section 2-1

50. A planet at inferior conjunction is always
    A) on the opposite side of the sky from the Sun, as seen from the Earth.
    B) below the horizon and therefore invisible.
    C) closer to us than the Sun is.
    D) further away from us than is the Sun.
    Answer: C Source: Section 2-1

51. An inferior planet will be closest to the Earth when it is at
    A) greatest elongation.
    B) opposition.
    C) inferior conjunction.
    D) superior conjunction.
    Answer: C Source: Section 2-1

52. Which of the following planetary configurations or positions is impossible for a superior planet?
    A) Opposition. B) Inferior conjunction. C) Perihelion. D) Conjunction.
    Answer: B Source: Sections 2-1 and 2-3

53. Venus can occasionally pass in front of the Sun as seen from the Earth. It can do so only when it is at
    A) greatest elongation.
    B) superior conjunction.
    C) opposition.
    D) inferior conjunction.
    Answer: D Source: Section 2-1

54. According to the heliocentric theory, which of the following objects can never transit (pass in front of) the Sun as seen from the Earth?
    A) Venus. B) Mercury. C) the Moon. D) Mars.
    Answer: D Source: Section 2-1

55. Which of the following objects could transit (pass in front of) the Sun as seen from Saturn?
    A) Jupiter. B) Uranus. C) Pluto. D) Neptune.
    Answer: A Source: Section 2-1

56. As seen by an observer on Saturn (or one of its moons), which of the following planets can never pass through inferior conjunction?
    A) Venus. B) Neptune. C) Jupiter. D) Earth.
    Answer: B Source: Section 2-1

57. When Mercury is at its farthest distance from the Earth, it is at
    A) superior conjunction.
    B) opposition.
    C) inferior conjunction.
    D) greatest elongation.
    Answer: A Source: Section 2-1

58. An inferior planet will be furthest from Earth when it is at
    A) greatest elongation.
    B) inferior conjunction.
    C) opposition.
    D) superior conjunction.
    Answer: D Source: Section 2-1

59. When Venus is at superior conjunction,
   A) it is at its greatest angle from the Sun, as seen from the Earth.
   B) its speed in its orbit has its greatest value.
   C) it is at its greatest distance from the Earth.
   D) it is at its smallest distance from the Earth.
   Answer: C Source: Section 2-1

60. The best time(s) to see inferior planets from the Earth are when these planets are at positions of
   A) superior conjunction
   B) inferior conjunction
   C) opposition
   D) greatest elongation
   Answer: D Source: Section 2-1

61. Greatest elongation in a planetary orbit occurs when
   A) the angle from the Earth to the planet and then to the Sun has its greatest possible value.
   B) the angle from the Sun to the Earth and then to the planet is 90°.
   C) the angle from the Sun to the planet and then to the Earth is 90°.
   D) the angle from the Earth to the Sun and then to the planet is 90°.
   Answer: C Source: Section 2-1

62. In what direction is Venus moving when it is at greatest elongation?
   A) Directly toward or away from the Earth.
   B) Directly toward or away from the Sun.
   C) It is not possible to say, since the direction is different from one greatest elongation to the next.
   D) Perpendicular to the line from Venus to the Earth.
   Answer: A Source: Section 2-1

63. At what position in its orbit will an inferior planet appear to be moving (for a day or two) more or less directly toward the Earth? (See Fig. 2-3, Kaufmann & Comins, *Discovering the Universe*, 5th Ed., and draw a diagram if it will help.)
   A) Superior conjunction.
   B) Opposition.
   C) Greatest eastern elongation.
   D) Inferior conjunction.
   Answer: C Source: Section 2-1

64. Where and when would Venus be seen from Earth when it is at greatest elongation?
   A) just after sunset, in the west.
   B) at midnight, in the south.
   C) just before sunrise, in the east.
   D) just after sunset, in the east.
   Answer: A Source: Section 2-1

65. In which part of the sky will Venus appear at sunset when it is at greatest elongation?
    A) Western.
    B) Due south.
    C) It will not be visible, since it will be on the other side of the Sun.
    D) Eastern.
    Answer: A Source: Section 2-1

66. An inferior planet moves more or less directly toward the Earth at greatest eastern elongation. What does this mean when you are watching the night-by-night motion of this planet against the background stars?
    A) The planet appears to remain stationary for a few days.
    B) The planet appears to be moving at its fastest motion against the background.
    C) The planet is invisible in the sky, since it is moving toward the Earth.
    D) Its motion against the background will appear to be normal, that is, direct motion, since this apparent motion is always constant in a circular orbit.
    Answer: A Source: Section 2-1

67. What is the angle between the line from the Earth to Mercury and the line from Mercury to the Sun when Mercury is at greatest elongation?
    A) Anywhere between 0° and 180°, depending on the particular planetary alignment.
    B) 90°.
    C) 180°.
    D) 0°.
    Answer: B Source: Section 2-1

68. When Saturn is at its farthest distance from the Earth, it is at
    A) conjunction.
    B) greatest elongation (about 47° from the Sun).
    C) inferior conjunction.
    D) opposition.
    Answer: A Source: Section 2-1

69. When Mars is at opposition, it is
    A) rising at about midnight.
    B) high in the sky at sunset.
    C) high in the sky at midnight.
    D) high in the sky at noon.
    Answer: C Source: Section 2-1

70. Where and when would Jupiter be seen from Earth when it is at opposition?
   A) Just after sunset, on the western horizon.
   B) High in the south at midnight.
   C) Just before sunrise, on the eastern horizon.
   D) In the daytime sky.
   Answer: B Source: Section 2-1

71. When a planet is seen at opposition, it is ALWAYS
   A) at its closest point to the Earth.
   B) at its most distant point from the Sun.
   C) at its most distant point from the Earth.
   D) at its closest point to the Sun.
   Answer: A Source: Section 2-1

72. Which of the following planetary configurations is impossible for an inferior planet?
   A) opposition.
   B) inferior conjunction.
   C) greatest elongation.
   D) superior conjunction.
   Answer: A Source: Section 2-1

73. The planet Venus can never reach which planetary configuration, when viewed from Earth?
   A) superior conjunction.
   B) greatest elongation.
   C) opposition.
   D) inferior conjunction.
   Answer: C Source: Section 2-1

74. Which of the following planetary configurations is not possible for the planet Mercury?
   A) Inferior conjunction.
   B) Greatest elongation.
   C) Superior conjunction.
   D) Opposition.
   Answer: D Source: Section 2-1

75. What is the angle between the line from the Earth to Jupiter and the line from Earth to the Sun when Jupiter is at opposition?
   A) 0°.
   B) 180°.
   C) Anywhere between 0° and 180°, depending on the particular planetary alignment.
   D) 90°.
   Answer: B Source: Section 2-1

76. When Jupiter is at opposition, it will rise
   A) at midnight. B) at sunrise. C) at noon. D) at sunset.
   Answer: D Source: Section 2-1

77. Jupiter will be at which configuration when it is at the middle of its retrograde motion?
    A) conjunction
    B) Jupiter never undergoes retrograde motion since it is a superior planet
    C) opposition
    D) maximum eastern elongation
    Answer: C Source: Section 2-1

78. When Saturn is at its closest distance from the Earth, it is at
    A) inferior conjunction.
    B) greatest elongation (about 47° from the Sun).
    C) opposition.
    D) conjunction.
    Answer: C Source: Section 2-1

79. The sidereal period of a planet is defined as
    A) the time between two successive identical configurations (e.g., opposition to opposition).
    B) the time between two successive passages of the planet in front of a particular point in the sky (e.g., a star) as seen from the Sun.
    C) the time between two successive greatest elongations (e.g., greatest western elongation to greatest eastern elongation).
    D) the time between two successive passages of the planet in front of a particular point in the sky (e.g., a star) as seen from the Earth.
    Answer: B Source: Section 2-1

80. The synodic period of a planet is defined as
    A) the time between two successive identical configurations (e.g., opposition to opposition).
    B) the time between two successive passages of the planet in front of a particular point in the sky (e.g., a star) as seen from the Sun.
    C) the time between two successive greatest elongations (e.g., greatest western elongation to greatest eastern elongation).
    D) the time between two successive passages of the planet in front of a particular point in the sky (e.g., a star) as seen from the Earth.
    Answer: A Source: Section 2-1

81. The time period between two successive passages of a planet through the position of opposition is
    A) its precessional period.
    B) its sidereal period.
    C) its synodic period.
    D) one year.
    Answer: C Source: Section 2-1

82. The synodic period of a superior planet as it moves around the Sun, as viewed from Earth, is defined as
    A) the time between conjunction and opposition on any orbit.
    B) the time between two successive appearances of the planet at its highest point in the observer's sky.
    C) the time between two successive passages through identical configurations, for example, two successive conjunctions.
    D) the time between two successive passages of Earth through the vernal equinox.
    Answer: C Source: Section 2-1

83. The time interval between two successive repeated positions of a planet with respect to the Sun and the Earth in its orbit, such as conjunction to conjunction, is known as
    A) the planet's synodic period.
    B) the planet's sidereal period.
    C) 1 day.
    D) 1 year.
    Answer: A Source: Section 2-1

84. The time period between two successive passages of a planet past a particular star as seen from the Sun is
    A) its sidereal period.
    B) its precessional period.
    C) its synodic period.
    D) its rotational period.
    Answer: A Source: Section 2-1

85. What is the difference between the synodic and sidereal periods of a planet?
    A) There is no difference, they are one and the same time period; the synodic period is the name used in the geocentric theory, while the sidereal period is the name used in the heliocentric theory.
    B) The synodic period refers to the planet's period with respect to the Earth's motion, while the sidereal period is the true period with respect to the background stars.
    C) The synodic period refers to the planet's rotation around its axis, while the sidereal period is the time for one orbit.
    D) The synodic period refers to the planet's motion with respect to the background stars, while the sidereal period is the true period with respect to the Earth's motion.
    Answer: B Source: Section 2-1

86. A planet's sidereal year is different from its synodic year because
    A) the Earth moves.
    B) the planet's speed changes along its elliptical orbit.
    C) the planet rotates about its own axis in addition to its orbital motion.
    D) the planet moves.
    Answer: A Source: Section 2-1

87. The greatest inaccuracy in Copernicus' theory of the solar system was that
   A) he did not allow for retrograde motion.
   B) he assumed that the planets move in elliptical orbits with constant speeds rather than variable speeds.
   C) he placed the planets in circular orbits.
   D) he placed the planets on epicycles, the centers of which followed orbits around the Sun.

   Answer: C Source: Section 2-1

88. The reason why Copernicus' heliocentric theory soon came to be regarded as preferable to the geocentric theory of Ptolemy is that
   A) the heliocentric theory accounted for the same observed motions of the planets as the geocentric theory, but did so in a much simpler way.
   B) the heliocentric theory used complex constructions called epicycles and deferrents to account for the observed motions of the planets, and so was considered more reliable than the geocentric theory.
   C) the heliocentric theory accounted for retrograde motion, which the geocentric theory was unable to explain.
   D) the heliocentric theory accounted for the same observed motions of the planets as the geocentric theory, but did so much more accurately.

   Answer: A Source: Section 2-1

89. The phenomenon of parallax is
   A) the change in apparent position of a nearby object as the observer moves, compared to background objects.
   B) the apparent change in angular size of an object as it moves toward or away from an observer.
   C) the change in direction of motion of a planet from retrograde to direct motion.
   D) the change in the apparent position of an object compared to background objects, as a result of the motion of the object.

   Answer: A Source: Section 2-2

90. Tycho Brahe demonstrated that the supernova of 1572 was not a nearby event (close to the Earth) by
   A) showing that it did not pass in front of the Sun at conjunction.
   B) proving that it did not show parallax over the course of one night.
   C) proving that it did not move past the background stars like a planet in our solar system.
   D) showing that it did not get brighter and fainter as the Earth moved toward and away from it over the course of a year.

   Answer: B Source: Section 2-2

91. The major contribution of Tycho Brahe to the development of modern astronomy was to
    A) prove that planetary orbits are ellipses.
    B) observe the phases of Venus.
    C) measure planetary positions very accurately.
    D) use parallax to prove that the Earth moves around the Sun.
    Answer: C Source: Section 2-2

92. Tycho Brahe
    A) developed the first detailed heliocentric model for the solar system, which replaced the geocentric model of Ptolemy.
    B) improved the refracting telescope, which allowed him to extend Galileo's observations of the sky.
    C) made accurate measurements of planetary positions, which Kepler later used to find the shapes of planetary orbits.
    D) developed a reflecting telescope, which used a curved mirror to focus the light.
    Answer: C Source: Section 2-2

93. The person who compiled the large set of accurate observations of planetary positions which formed the basis for proving that planets move in elliptical orbits around the Sun was
    A) Ptolemy. B) Nicolaus Copernicus. C) Johannes Kepler. D) Tycho Brahe.
    Answer: D Source: Section 2-2

94. Before deriving the shapes of planetary orbits, Johannes Kepler worked as an assistant to
    A) Ptolemy. B) Tycho Brahe. C) Nicolaus Copernicus. D) Galileo Galilei.
    Answer: B Source: Section 2-2

95. Kepler as a young man became the assistant to
    A) Tycho Brahe B) Nicolaus Copernicus C) Ptolemy D) Sir Isaac Newton
    Answer: A Source: Section 2-2

96. The Danish astronomer Tycho Brahe had a young assistant who became famous himself some time later. His name was
    A) Ptolemy B) Johannes Kepler C) Nicolaus Copernicus D) Galileo Galilei
    Answer: B Source: Section 2-2

97. The person who was an assistant to the Danish astronomer, Tycho Brahe, before becoming famous himself was
    A) Johannes Kepler.
    B) Nicolaus Copernicus.
    C) Galileo Galilei.
    D) Sir Isaac Newton.
    Answer: A Source: Section 2-2

98. The major contribution of Johannes Kepler to the development of modern astronomy was to
    A) prove that planetary orbits are ellipses.
    B) develop the first mathematical heliocentric model of the solar system.
    C) use parallax to prove that the Earth moves around the Sun.
    D) observe the satellites (moons) of Jupiter.
    Answer: A Source: Section 2-3

99. The person who first showed that planetary orbits are ellipses was
    A) Copernicus. B) Kepler. C) Newton. D) Galileo.
    Answer: B Source: Section 2-3

100. The model of the solar system that Johannes Kepler proposed was
    A) Sun-centered, with elliptical planetary orbits.
    B) Sun-centered, with planets moving in circles around it.
    C) Earth-centered, with planets moving epicycles.
    D) Earth-centered, with the Sun, Moon, and planets moving in ellipses.
    Answer: A Source: Section 2-3

101. Kepler's first law states that a planet moves around the Sun
    A) in a circle, with the Sun at the center.
    B) in an elliptical orbit, with the Sun at the center of the ellipse.
    C) in an elliptical orbit, with the Sun on the minor axis of the ellipse.
    D) in an elliptical orbit, with the Sun at one focus.
    Answer: D Source: Section 2-3

102. Kepler's first law states:
    A) The orbit of a planet about the Sun is an ellipse with the Sun at the center.
    B) The orbit of a planet about the Sun is an oval with the Sun at the center.
    C) The orbit of a planet about the Sun is an ellipse with the Sun at one focus.
    D) The orbit of a planet about the Sun is a circle with the Sun at the center.
    Answer: C Source: Section 2-3

103. Mars moves in an elliptical orbit around the Sun. The location of the Sun relative to this ellipse is
    A) at the focus which is closer to the point where Mars moves the fastest.
    B) at one end of the major axis of the ellipse.
    C) at the focus which is closer to the point where Mars is moving the slowest.
    D) at the exact center of the ellipse.
    Answer: A Source: Section 2-3

104. The eccentricity of a planet's orbit describes
A) its motion at any specific point in its orbit as seen from Earth, i.e. whether direct, retrograde or stationary.
B) its tilt with respect to the plane of the Earth's orbit (the ecliptic plane).
C) the tilt of the planet's spin axis with respect to its orbital plane.
D) its shape compared to that of a circle.
Answer: D Source: Section 2-3

105. If an object has an orbit around the Sun that has an eccentricity of 0.1, then the orbit is
A) a straight line.
B) exactly circular.
C) almost circular, but not quite.
D) a long, thin ellipse.
Answer: C Source: Section 2-3

106. If an object has an orbit around the Sun that has an eccentricity of 0.8, then the orbit is
A) exactly circular.
B) a straight line.
C) a long, thin ellipse.
D) almost circular, but not quite.
Answer: C Source: Section 2-3

107. The distance from the perihelion point to the aphelion point of a planetary orbit is
A) the semimajor axis.
B) the minor axis.
C) the major axis.
D) equal to the distance between the foci.
Answer: C Source: Section 2-3

108. In an ellipse, the major axis is a distance measured
A) along the circumference, between the closest point to and the furthest point from one focus.
B) along the longer diameter, passing through the foci of the ellipse.
C) along the shorter diameter.
D) from focus to focus.
Answer: B Source: Section 2-3

109. The semimajor axis of an ellipse is
A) the distance from the center to one side of the ellipse, along the shortest diameter of the ellipse.
B) half the distance between the foci of the ellipse.
C) the distance from the center of the ellipse to one end, along the largest diameter of the ellipse.
D) the distance from one focus to any point on the circumference of the ellipse.
Answer: C Source: Section 2-3

110. To which point in a planetary orbit does the word "perihelion" refer?
   A) The precise center of the orbit.
   B) The point farthest from the Sun.
   C) The point closest to the Sun.
   D) The "other" focus (the one not occupied by the Sun).
   Answer: C  Source: Section 2-3

111. To which point in a planetary orbit does the word "aphelion" refer?
   A) The point farthest from the Sun.
   B) The "other" focus (the one not occupied by the Sun).
   C) The point closest to the Sun.
   D) The precise center of the orbit.
   Answer: A  Source: Section 2-3

112. Which point in a comet's orbit is closest to the Sun?
   A) Greatest elongation.  B) Perihelion.  C) Aphelion.  D) Inferior conjunction.
   Answer: B  Source: Section 2-3

113. At which point in a planet's elliptical orbit is it furthest from the Sun?
   A) Quadrature.  B) Superior conjunction.  C) Aphelion.  D) Perihelion.
   Answer: C  Source: Section 2-3

114. Where is a planet when it is moving most rapidly in its orbit?
   A) At aphelion.
   B) At the focus of its orbit.
   C) At perihelion.
   D) Approaching the closest distance to the Sun.
   Answer: C  Source: Section 2-3

115. At what point in a planetary orbit is the planet's speed the slowest?
   A) Approaching the closest distance to the Sun.
   B) At aphelion.
   C) At the focus of its orbit.
   D) At perihelion.
   Answer: B  Source: Section 2-3

116. In any one day the line joining a planet to the Sun will sweep through some particular angle, as seen from the Sun. Where is the planet when this angle has its smallest value?
   A) At greatest elongation.
   B) At aphelion.
   C) At inferior conjunction.
   D) At perihelion.
   Answer: B  Source: Section 2-3

117. Kepler's second law states that a planet moves fastest when it
   A) passes through the minor axis.
   B) is closest to the Sun.
   C) is at conjunction.
   D) is furthest from the Sun.
   Answer: B Source: Section 2-3

118. Kepler's second law states:
   A) A line joining a planet to the Sun sweeps out equal areas in equal times.
   B) A line joining a planet to the Sun moves equal distances along the planet's orbit in equal times.
   C) A line joining a planet to the Sun points in the same direction at all times.
   D) A line joining a planet to the Sun sweeps through equal angles in equal times.
   Answer: A Source: Section 2-3

119. If the line joining a planet to the Sun sweeps out a particular area in one day, then in two days it will sweep out
   A) more than twice the area if the planet is approaching perihelion and less than twice the area of it leaving perihelion.
   B) less than twice the area if the planet is approaching perihelion and more than twice the area if it is leaving perihelion.
   C) half the area.
   D) exactly twice the area.
   Answer: D Source: Section 2-3

120. Kepler's third law, the harmonic law, provides a relationship between a planet's
   A) Orbital period and mass.
   B) Orbital period and orbital eccentricity.
   C) Orbital eccentricity and length of semimajor axis.
   D) Orbital period and length of semimajor axis.
   Answer: D Source: Section 2-3

121. Kepler's third law tells us that
   A) the cube of a planet's period in yr equals the square of its semimajor axis in AU.
   B) the period of a planet in yr equals its semimajor axis in AU.
   C) the square of a planet's period in yr equals the cube of its semimajor axis in AU.
   D) the square of a planet's period in yr equals the fourth power of its semimajor axis in AU.
   Answer: C Source: Section 2-3

122. Kepler's third law can be described in which of the following ways?
   A) The larger the orbit, the longer it takes for the planet to complete one revolution.
   B) The smaller the orbit, the longer it takes for the planet to complete one revolution.
   C) The time to complete one revolution of its orbit is dependent upon the size or radius of the planet.
   D) The smaller the radius of a planet, the more rapidly it rotates on its axis.
   Answer: A Source: Section 2-3

123. Kepler's third law applies
   A) accurately only close to the Sun, and becomes less accurate with increasing distance from the Sun.
   B) to all situations where two objects orbit each other solely under the influence of their mutual gravitational attraction.
   C) only to situations similar to planets orbiting the Sun, where the mass of the orbiting body is small compared to the mass of the object being orbited.
   D) only to planets orbiting the Sun.
   Answer: B Source: Section 2-3

124. Kepler's third law of planetary motion relates the period $P$ (in sidereal years) to the length of the semimajor axis $a$ (in astronomical units) in which way?
   A) $P = a^2$. B) $P^3 = a^2$. C) $P = 1/a^2$. D) $P^2 = a^3$.
   Answer: D Source: Section 2-3

125. In the simplified version of Kepler's third law, $P^2 = a^3$, the units of the orbital period $P$ and the semimajor axis of the ellipse, $a$ must be, respectively
   A) years and light-years.
   B) years and astronomical units.
   C) seconds and meters.
   D) years and meters.
   Answer: B Source: Section 2-3

126. A space probe is put into a circular orbit around the Sun at a distance of exactly 2 AU from the Sun. According to Kepler's third law how long does it take this probe to orbit around the Sun once?
   A) 1 year.
   B) 1.6 years (cube root of 4).
   C) 8 years.
   D) 2.8 years (square root of 8).
   Answer: D Source: Section 2-3

127. A distant asteroid is discovered which takes 50 years to orbit once around the Sun. According to Kepler's third law, what is the average distance of this asteroid from the Sun?
A) 353 AU (square root of 125,000).
B) 13.6 AU (cube root of 2500).
C) 2500 AU.
D) 50 AU.
Answer: B Source: Section 2-3

128. If a tenth planet (tentatively predicted to exist, on the basis of perturbations in the orbits of Uranus and Neptune) were to be discovered with a sidereal period of 125 years, what would be the radius of its orbit (assumed to be circular)?
A) 8.55 AU. B) 125 AU. C) 1 AU. D) 25 AU.
Answer: D Source: Section 2-3

129. If a tenth planet (tentatively predicted to exist, on the basis of perturbations in the orbits of Uranus and Neptune) were to be discovered with a sidereal period of 200 years, what would be the radius of its orbit (assumed to be circular)?
A) 34.2 AU. B) 2828 AU. C) 342 AU. D) 200 AU.
Answer: A Source: Section 2-3

130. According to Kepler's law, the approximate sidereal period of an asteroid moving around the Sun in the asteroid belt is
A) 46.8 years. B) 4.68 years. C) 2.8 years. D) 1.99 years.
Answer: B Source: Section 2-3

131. If a planet were to exist in our solar system in a circular orbit with a radius of 3 AU, about how long would it take to orbit the Sun once?
A) 5.2 years. B) 27 years. C) 3 years. D) 2.1 years.
Answer: A Source: Section 2-3

132. Suppose an asteroid is discovered with an elliptical orbit, a period of exactly 1 year, and perihelion 0.5 AU from the Sun. Using Kepler's third law, how far from the Sun is this asteroid when at aphelion? (Drawing a diagram of the orbit, including the Sun, will help.)
A) 2.5 AU. B) 1.0 AU. C) 1.5 AU. D) 2.0 AU.
Answer: C Source: Section 2-3

133. Halley's Comet returns to the Sun's vicinity every 76 years in an elliptical orbit. According to Kepler's third law, what will be the semimajor axis of this orbit?
A) 50.000 AU. B) 17.5 AU. C) 1 AU. D) 0.59 AU.
Answer: B Source: Section 2-3

134. If Halley's Comet has an elliptical orbit with a semimajor axis of 17.5 AU, approximately how far out into the planetary system does it reach at its furthest point (aphelion)? (Assume that perihelion distance from the Sun is negligible; be careful, the question needs some thought.)
A) Between the orbits of Neptune and Pluto.
B) Between the orbits of Saturn and Uranus.
C) Between the orbits of Uranus and Neptune.
D) Beyond the orbit of Pluto.
Answer: A Source: Section 2-3

135. A comet is observed to return to the vicinity of the Sun on a long elliptical orbit with a period of 31.7 years. What is the semimajor axis of is orbit?
A) 10 AU. B) 31.7 AU. C) 178.5 AU. D) 1000 AU.
Answer: A Source: Section 2-3

136. An asteroid-like object (or a cometary nucleus) is seen to be orbiting the Sun in a circular path with a period of 120 years. What is its distance of closest approach to Earth (the separation distance when the object is at opposition)? (Careful with your answer; a diagram might be helpful.)
A) 23.33 AU. B) 25.33 AU. C) 1314.5 AU. D) 24.33 AU.
Answer: A Source: Section 2-3

137. In which country did Galileo make his astronomical discoveries?
A) Poland. B) Greece. C) Italy. D) England.
Answer: C Source: Section 2-4

138. Who was the first astronomer to use a telescope for viewing the sky?
A) Ptolemy. B) Galileo. C) Newton. D) Brahe.
Answer: B Source: Section 2-4

139. Which of the following astronomers was a contemporary of (lived at the same time as) Galileo?
A) Newton. B) Ptolemy. C) Copernicus. D) Kepler.
Answer: D Source: Section 2-4

140. Galileo's early observations of the sky with his newly made telescope included
A) the discovery of the phases of Venus.
B) the discovery of retrograde motion in planets.
C) the discovery of Pluto.
D) the discovery of of Jupiter's magnetosphere.
Answer: A Source: Section 2-4

141. What did Galileo see when he observed Venus through his telescope?
   A) Four satellites (moons) orbiting Venus.
   B) Nothing interesting, since Venus is perpetually cloud covered.
   C) A set of rings.
   D) Phases like the Moon.
   Answer: D Source: Section 2-4

142. What did Galileo see when he observed Venus through his telescope?
   A) Venus has phases like the Moon, and also like the Moon is almost constant in angular size.
   B) Venus has phases like the Moon, and has its largest angular diameter at gibbous phase.
   C) Venus has phases like the Moon, and has its largest angular diameter at crescent phase.
   D) Venus has an angular size, which increases and decreases markedly, but does NOT show phases (e.g., crescent).
   Answer: C Source: Section 2-4

143. When viewed from Earth, which of the following planets shows the greatest change in angular size over one planetary orbit?
   A) Jupiter B) Venus C) Mercury D) Uranus
   Answer: B Source: Section 2-4

144. As Venus orbits the Sun, its angular size as viewed from Earth appears to change by what factor, from smallest to largest? (See Fig. 2-8, Kaufmann & Comins, *Discovering the Universe*, 5th Ed.)
   A) 58X.
   B) 1.72X.
   C) 5.8X.
   D) Its angular size does not change, only its brightness.
   Answer: C Source: Section 2-4

145. A planet will be observed (through a telescope) to have a crescent shape whenever it is close to which point as seen from the Earth? (It may help to draw a diagram and show the sunlit side of the planet on it.)
   A) superior conjunction B) perihelion C) opposition D) inferior conjunction
   Answer: D Source: Section 2-4

146. At which point, as seen from the Earth, will a planet appear exactly half-lit (looking like the first or last quarter Moon)? (It may help to draw a diagram and show the sunlit side of the planet on it.)
A) inferior conjunction
B) superior conjunction
C) greatest elongation
D) opposition
Answer: C Source: Section 2-4

147. Which of the following statements CORRECTLY states the significance of Galileo's observation that Venus shows phases?
A) The phases showed that, like the Moon, Venus is always much closer to the Earth than the Sun is.
B) Since the phases were NOT correlated with angular size, they actually supported the geocentric theory more than the heliocentric theory.
C) The phases were interesting, but did not have any particular significance other than that.
D) The phases were correlated with angular size in a way which supported the heliocentric theory.
Answer: D Source: Section 2-4

148. Venus shows changes in angular size and also shows phases similar to those of the Moon. According to Galileo, who first saw this, these observations showed that
A) Venus moves in epicycles.
B) the Moon really orbits Venus, and not the Earth at all.
C) Venus orbits the Sun.
D) Venus orbits the Earth in an elliptical orbit.
Answer: C Source: Section 2-4

149. How many moons of Jupiter did Galileo see?
A) 12
B) None, he was unable to see them with the naked eye.
C) 4
D) Only one.
Answer: C Source: Section 2-4

150. Which of the following statements CORRECTLY states the significance of Galileo's observation that Jupiter has satellites (moons)?
 A) It showed that Jupiter must be four times the size of the Earth (since Jupiter has four moons and the Earth has one).
 B) It showed that Jupiter must orbit around the Sun, not around the Earth.
 C) It was interesting, but had no particular significance other than that.
 D) It showed that bodies can orbit an object other than the Earth.
 Answer: D Source: Section 2-4

151. One of the major contributions of Galileo to the development of modern astronomy was to
 A) prove that planetary orbits are ellipses.
 B) use parallax to prove that the Earth moves around the Sun.
 C) discover the satellites (moons) of Jupiter.
 D) develop the first mathematical heliocentric model of the solar system.
 Answer: C Source: Section 2-4

152. What did Galileo see when he observed Jupiter through his telescope?
 A) Four satellites (moons) orbiting Jupiter.
 B) A set of rings.
 C) Phases like the Moon.
 D) Nothing interesting, since Jupiter is perpetually cloud covered.
 Answer: A Source: Section 2-4

153. What was the MOST IMPORTANT difference between the development of Isaac Newton's theory of planetary motion and that of Johannes Kepler?
 A) Newton lived in a freer political climate, whereas Kepler risked house arrest if his theory opposed the Bible or Aristotle.
 B) Newton based his theory on accurate telescopic observations, whereas Kepler used observations made by eye.
 C) Newton developed his theory from basic physical assumptions, whereas Kepler simply adjusted his theory to fit the data.
 D) Newton lived in England, which is famous for clear skies, whereas Kepler lived on the Continent, which is notorious for bad weather.
 Answer: C Source: Section 2-5

154. Which one of the following relationships represents Newton's MOST IMPORTANT contribution to physics?
A) The relationship between force and motion.
B) The relationship between gravity and distance.
C) The relationship between mass and weight.
D) The relationship between velocity and acceleration.
Answer: A Source: Section 2-5

155. What was the MOST IMPORTANT contribution of Newton to the development of astronomy?
A) He invented the reflecting telescope.
B) He showed that astronomical phenomena can be explained using only basic physics and mathematics.
C) He showed that planetary orbits are ellipses, with the planet moving fastest when closest to the Sun.
D) He was the first person to observe the sky through a telescope.
Answer: B Source: Section 2-5

156. Why were Newton's three laws so important to astronomy?
A) They showed that planets can move around the Sun by themselves forever, without coming to rest.
B) They provided a physical basis which did not conflict with the Bible, Aristotle, or Plato.
C) They showed that acceleration always results from a change in velocity.
D) They showed why objects released from rest always fall to the ground.
Answer: A Source: Section 2-5

157. According to Newton's first law,
A) the rate of change of speed of an object is larger, the larger the force acting upon the object.
B) an applied force always causes a change in the direction of travel of an object.
C) if no force is acting upon an object then the object's speed and direction of travel will both be constant.
D) an applied force always causes a change in the speed of an object.
Answer: C Source: Section 2-5

158. In which direction would the Earth move if the Sun's gravitational force were suddenly removed from it?
   A) In a straight line along a tangent to its circular orbit.
   B) Continue in a circular orbit, because of its spin.
   C) In a straight line toward the Sun.
   D) In a straight line directly away from the Sun.
   Answer: A Source: Section 2-5

159. Which path would a planet (like Earth!) take if the force of gravity from the Sun were to be suddenly removed?
   A) The planet would begin to move in a long ellipse with the Sun at one focus.
   B) The planet would move in a straight line outwards, directly away from the Sun's position.
   C) The planet would move in a straight line tangential to its present orbit.
   D) The planet would stop moving altogether since there would now be no gravity acting upon it.
   Answer: C Source: Section 2-5

160. How many forces need to be applied to a body in space in order to keep it moving with a constant velocity?
   A) None.
   B) Two unequal forces.
   C) One force, in a direction opposite to the direction of motion.
   D) One force, in the direction of motion.
   Answer: A Source: Section 2-5

161. To specify an object's velocity completely, we need to specify
   A) its direction of travel.
   B) its speed.
   C) the rate of change of its acceleration.
   D) its speed and direction of travel.
   Answer: D Source: Section 2-5

162. To specify an object's velocity, we need to specify
   A) how fast it is moving, and also in which direction it is moving.
   B) only how fast it is moving.
   C) how fast it is moving, and also its mass.
   D) only in which direction it is moving.
   Answer: A Source: Section 2-5

163. The careful description of the velocity of a moving object at a particular time requires that one define
   A) the speed and acceleration of the object.
   B) the mass, speed, and position of the object.
   C) the speed and the direction of the moving object.
   D) only the speed of the object.
   Answer: C Source: Section 2-5

164. A body whose velocity is constant
   A) has a positive acceleration.
   B) has a negative acceleration.
   C) can have any non-zero acceleration.
   D) has zero acceleration.
   Answer: D Source: Section 2-5

165. The acceleration of an object is defined as
   A) the rate of change of its direction of travel.
   B) the rate of change of its velocity.
   C) the rate of change of its position.
   D) the rate of changes of its speed.
   Answer: B Source: Section 2-5

166. The acceleration of a moving body is defined as
   A) the rate of change of kinetic energy with time
   B) the rate of change of position with time
   C) the rate of change of mass with time
   D) the rate of change of velocity with time
   Answer: D Source: Section 2-5

167. The acceleration of a body is defined its rate of change of
   A) velocity B) mass C) weight D) position
   Answer: A Source: Section 2-5

168. A certain object in space is accelerating. From this, we know for certain that
   A) its speed OR its direction or travel is changing, or possibly both.
   B) its speed OR its direction of travel is changing, but NOT both.
   C) BOTH its speed and its direction of travel are changing.
   D) its speed is changing.
   Answer: A Source: Section 2-5

169. Which of the following four objects or persons is NOT accelerating?
 A) An Olympic swimmer exerting considerable force to maintain a constant speed in a straight line through the water.
 B) An apple falling to the ground from an apple tree.
 C) A bicyclist gradually slowing down on a straight road while coasting toward a stop sign.
 D) A motorcyclist travelling around a circular racetrack at a constant speed.
 Answer: A Source: Section 2-5

170. Newton's second law states that
 A) force equals mass divided by acceleration.
 B) force equals mass times acceleration.
 C) acceleration equals force times mass.
 D) weight equals force times acceleration.
 Answer: B Source: Section 2-5

171. Newton's second law of motion states that
 A) a body acted upon by a force will accelerate constantly while the force is applied.
 B) a force is always required in order to keep an object in motion.
 C) objects orbiting the Sun always move in elliptical orbits.
 D) a body acted upon by a force will move at a constant speed while the force is applied.
 Answer: A Source: Section 2-5

172. According to Newton's second law of motion,
 A) an object acted on by a constant force does not move.
 B) an object acted on by a constant force moves with a constant velocity.
 C) an object acted on by a constant force moves with a constant speed, although the direction may vary.
 D) an object acted on by a constant force moves with a constant acceleration.
 Answer: D Source: Section 2-5

173. According to Newton's laws, a force must be acting whenever
 A) time passes.
 B) the direction of an object's motion changes.
 C) an object's position changes.
 D) an object moves with some speed.
 Answer: B Source: Section 2-5

174. Two spaceships that have different masses but rocket engines of identical force are at rest in space. If they fire their rockets at the same time, which ship will speed up faster?
A) they will not speed up at all, but move at a constant speed since they are in space and the rocket has nothing against which to push
B) the one with the higher mass
C) they will increase speed at the same rate, since they have identical rocket engines
D) the one with the lower mass
Answer: D Source: Section 2-5

175. How many forces need to be applied to a body in space in order to keep it moving with a constant velocity?
A) none
B) two unequal forces
C) one force, in the direction of motion
D) one force, in a direction opposite to the direction of motion
Answer: A Source: Section 2-5

176. Newton stated that a constant force, continuously applied to a body in space, will give it
A) a headache
B) a constant velocity
C) a constant acceleration
D) a change of position from one state of rest to another state of rest
Answer: C Source: Section 2-5

177. Newton stated that if a force were applied to an object in space, the resultant acceleration would depend upon
A) the size of the object.
B) the mass of the object.
C) the initial position of the object.
D) the initial speed of the object.
Answer: B Source: Section 2-5

178. An unbalanced force acting on an object will ALWAYS cause it to
A) change its speed or its direction of travel or both.
B) change its speed.
C) change its direction of travel.
D) change its acceleration.
Answer: A Source: Section 2-5

179. An object orbiting the Sun in a circle can be said to be
   A) always accelerating.
   B) moving under the action of equal and opposite forces.
   C) moving at a constant velocity.
   D) weightless.
   Answer: A   Source: Section 2-5

180. Which of the following statements about an asteroid moving in a circular orbit around the Sun is UNTRUE?
   A) It is moving at a constant speed.
   B) It is accelerating.
   C) It is moving in a flat plane.
   D) It is moving at a constant velocity.
   Answer: D   Source: Section 2-5

181. An accelerating body must at all times
   A) have a changing direction of motion.
   B) be moving.
   C) have a changing velocity.
   D) have an increasing velocity.
   Answer: C   Source: Section 2-5

182. Which of the following statements is a CORRECT version of Newton's third law?
   A) Whenever some object A exerts a force on some other object B, B must exert a force of equal magnitude on A in the opposite direction.
   B) Whenever some object A exerts a force on some other object B, B must exert a force of equal magnitude on A in the same direction.
   C) Whenever two forces act, they must be equal in magnitude and opposite in direction.
   D) Whenever any object feels some force it must also feel another force of equal magnitude in the opposite direction from some other source.
   Answer: A   Source: Section 2-5

183. According to Newton's third law, if a force is acting on an object then
   A) the object must move in a circular path.
   B) there must be some other force also acting on the object, with the same magnitude but in the opposite direction.
   C) the object must accelerate.
   D) there must be some other force acting on a different object, with the same magnitude but in the opposite direction.
   Answer: D   Source: Section 2-5

184. Which of the following pairs of forces is an example of an action-reaction pair by Newton's third law?
  A) For a dog pulling on its leash, the force of the dog on the leash and the force of the leash on the dog.
  B) For a baseball player sliding into first base, the force of the baseball player on his shoes and the friction force of the ground on his shoes.
  C) For a tugboat pulling a barge, the force of the water on the barge and the force of the barge on the tugboat.
  D) For the solar system, the force of the Sun on the Earth and the force of the Sun on Mars.
  Answer: A Source: Section 2-5

185. A horse is dragging a loaded sled across a field. Which of the following pairs of forces is an action-reaction pair by Newton's third law?
  A) The force of the horse on the sled and the force of the ground on the horse.
  B) The force of the horse on the sled and the force of the ground on the sled.
  C) The force of the horse on the sled and the force of the sled on the horse
  D) The force of the sled on the ground and the force of the horse's hooves on the ground.
  Answer: C Source: Section 2-5

186. A diver weighing 138 pounds has just dived up and out from the high board and is doing a back flip before starting to descend toward the water. How much force does the diver exert on the Earth while doing the back flip?
  A) 138 pounds.
  B) Much less than 138 pounds (but more than zero), because the diver has so much less mass than the Earth.
  C) Zero.
  D) Much more than 138 pounds, since the Earth is so much more massive than the diver.
  Answer: A Source: Section 2-5

187. The old story about the person who sneezed so hard that he fell off his barstool is an exaggerated illustration of which physical law?
  A) The law of inertia, Newton's first law.
  B) The law forbidding the partaking of alcohol during physical experiments.
  C) Newton's law of universal gravitation.
  D) Newton's third law, of equal and opposite forces of action and reaction.
  Answer: D Source: Section 2-5

188. The Earth exerts a force on you as you stand on its surface. What is the size of the force exerted on the Earth by you, when compared to the above force?
A) zero, you do not exert a force on the Earth.
B) the same.
C) twice as large, because of the Earth's rotation.
D) very small, because your mass is small compared to that of Earth.
Answer: B Source: Section 2-5

189. A person standing on a bathroom scale sees a reading on the scale of 148 pounds. This person is acted on by
A) two forces of equal size acting in opposite directions.
B) only one force, of 296 pounds (the sum of 148 pounds from the Earth and 148 pounds by the scale).
C) no forces at all.
D) only one force (148 pounds, as shown by the reading on the scale).
Answer: A Source: Section 2-5

190. A rocket that is accelerated by the force from the ejection of large quantities of hot gases represents an example of which physical law, originally stated by Newton?
A) His third law of motion, concerning action and reaction forces.
B) His law of elliptical motion of planets.
C) His first law of motion, concerned with state of rest or uniform motion.
D) His law of universal gravitation.
Answer: A Source: Section 2-5

191. The action and reaction forces referred to in Newton's third law of motion
A) act upon the same body.
B) act upon different bodies.
C) need not be equal in magnitude, but must act in the same straight line.
D) must be equal in magnitude, but need not act in the same straight line.
Answer: B Source: Section 2-5

192. The strength of gravity on Mars is about 40% of that on the Earth. If you were to take a standard red brick to Mars, which property of the brick would be significantly different on Mars than on the Earth?
A) Its weight. B) Its volume. C) Its color. D) Its mass.
Answer: A Source: Section 2-5

193. The strength of gravity on Mars is about 40% of that on the Earth. If your mass on the Earth is 60 kg, what would your mass be on Mars?
A) Zero. B) 60 kg. C) 24 kg. D) 150 kg.
Answer: B Source: Section 2-5

194. The strength of gravity on Mars is about 40% of that on the Earth. If you were to visit Mars, what would happen to your mass and weight compared to when you were on Earth?
A) Your weight would be the same but your mass would be less.
B) Your weight and mass would both be unchanged from when you were on Earth.
C) Your weight and mass would both be less than when you were on Earth.
D) Your mass would be the same but your weight would be less.
Answer: D Source: Section 2-5

195. Compared with your mass on Earth, your mass out in space among the stars would be
A) zero. B) huge. C) negligibly small. D) the same.
Answer: D Source: Section 2-5

196. On the Moon, where gravity is 1/6 of that upon Earth, which of the following activities would an astronaut NOT find easier to carry out?
A) Slowing down and stopping.
B) Long jumping.
C) High jumping.
D) Running.
Answer: A Source: Section 2-5

197. I have a massive purple object in my laboratory. If I were to take it to the Moon, which of its characteristics will be guaranteed to change?
A) Mass. B) Color. C) Weight. D) Density (mass per unit volume).
Answer: C Source: Section 2-5

198. Two solid spheres of equal size and mass are joined by a spring-loaded, extendible rod. The whole unit is spinning like a baton (end-over-end) five times a second. Suddenly the rod stretches, pushing the spheres farther apart. What change will take place as a result of this?
A) The rate of spin of the spheres and rod will decrease.
B) The angular momentum of the spheres and rod will decrease.
C) The rate of spin of the spheres and rod will increase.
D) The angular momentum of the spheres and rod will increase.
Answer: A Source: Section 2-5

199. The angular momentum of an object depends ONLY on
   A) the mass of the object, how the mass is distributed, how fast the object is spinning, and how strong a gravitational field the object is in.
   B) the mass of the object, how the mass is distributed, and how fast the object is spinning.
   C) the mass of the object and how fast it is spinning.
   D) how fast the object is spinning.
   Answer: C  Source: Section 2-5

200. The REQUIRED condition for the angular momentum of an object to be conserved (i.e., to not change) is that
   A) the size of the object does not change.
   B) the object is not acted on by an outside force.
   C) the speed of the object does not change.
   D) the object is not acted on by an internal force (e.g., the muscles in a skater's arms).
   Answer: B  Source: Section 2-5

201. Suppose that a planet of the same mass as the Earth were orbiting the Sun at a distance of 10 AU. The gravitational force on this planet due to the Sun would be
   A) 100 times that which the Sun exerts on the Earth.
   B) 1/100 of that which the Sun exerts on the Earth.
   C) 10 times that which the Sun exerts on the Earth.
   D) 1/10 of that which the Sun exerts on the Earth.
   Answer: B  Source: Section 2-6

202. Suppose that a planet of 10 times the mass of the Earth were orbiting the Sun at the same distance as the Earth (1 AU). The gravitational force on this planet due to the Sun would be
   A) 1/10 of that on the Earth due to the Sun.
   B) 1/100 of that on the Earth due to the Sun.
   C) 10 times that on the Earth due to the Sun.
   D) 100 times that on the Earth due to the Sun.
   Answer: C  Source: Section 2-6

203. If the mass of the Sun were doubled, the gravitational force on Jupiter due to the Sun would
   A) be 4 times its present value.
   B) be 16 times its present value.
   C) be twice its present value.
   D) stay the same.
   Answer: C  Source: Section 2-6

204. The force of gravity between two objects is proportional to
A) the difference of their masses.
B) the ratio of their masses.
C) the product of their masses.
D) the sum of their masses.
Answer: C Source: Section 2-6

205. If two massive bodies, initially held at rest in space, are released, then they will begin to
A) move away from each other with constant acceleration.
B) move toward one another.
C) orbit one another in circles.
D) move in elliptical orbits around one another.
Answer: B Source: Section 2-6

206. If an astronaut landed on a planet of the same radius as the Earth but four times the mass, then the astronaut's weight on the planet would be
A) sixteen times her weight on the Earth.
B) twice her weight on the Earth.
C) four times her weight on the Earth.
D) the same as on Earth, because weight is independent of location.
Answer: C Source: Section 2-6

207. Suppose two asteroids are located at the same distance from the Sun. One asteroid has twice the mass of the other. According to Newton's law of gravitation (and ignoring all forces except that from the Sun),
A) the more massive asteroid feels half the force that the other does.
B) the more massive asteroid feels twice the force that the other does.
C) both asteroids feel the same force, because gravity acts equally on all objects.
D) neither feel any force because they are weightless in space.
Answer: B Source: Section 2-6

208. How much gravitational force acts on an astronaut in the Space Shuttle in a circular orbit 300 km above the Earth's surface?
A) Almost (but not quite) as much as when the astronaut is standing on the surface of the Earth.
B) Almost zero, but not quite.
C) Zero - the astronaut is weightless.
D) Exactly the same as when the astronaut is standing on the surface of the Earth.
Answer: A Source: Section 2-6

209. A person orbiting the Earth in the Space Shuttle feels weightless because
   A) two forces are acting on her in opposite directions, so they cancel and produce the same effect as if no force at all were acting.
   B) only one force (gravity) acts on her, but gravity also accelerates the Shuttle so the Shuttle does not push up on her to create the feeling of weight.
   C) her mass is zero in space, and weight requires mass.
   D) no forces act on her.
   Answer: B Source: Section 2-6

210. The first person derive the elliptical shape of planetary orbits from basic physics and mathematics (not from observations of planetary positions) was
   A) Newton. B) Copernicus. C) Kepler. D) Galileo.
   Answer: A Source: Section 2-5

211. Who was the discoverer of the planet Uranus?
   A) Clyde Tombaugh, in 1930.
   B) William Herschel, in 1781.
   C) Galileo Galilei, in 1610.
   D) Johann Galle, in 1846.
   Answer: B Source: Section 2-6

212. How was the planet Uranus discovered?
   A) Accidentally during a telescopic survey of the sky.
   B) No one knows-it has been known since ancient times.
   C) By mathematical prediction using Newton's laws.
   D) It happened to pass close to Jupiter in the sky, and was discovered by an astronomer studying Jupiter.
   Answer: A Source: Section 2-6

213. How was the planet Neptune discovered?
   A) Accidentally during a telescopic survey of the sky.
   B) No one knows-it has been known since ancient times.
   C) By mathematical prediction using Newton's laws.
   D) It happened to pass close to Jupiter in the sky, and was discovered by an astronomer studying Jupiter.
   Answer: C Source: Section 2-6

214. Which planet was discovered mathematically by using Newton's laws before it was discovered observationally by using the telescope?
   A) Pluto. B) Mercury. C) Neptune. D) Uranus.
   Answer: C Source: Section 2-6

215. Which planet was discovered after its position had been predicted by the application of Newtonian mechanics to the deviation in the motion of another planet?
A) Uranus. B) Pluto. C) the minor planet, Ceres. D) Neptune.
Answer: D Source: Section 2-6

216. In the years after Newton published his laws of motion, it was found that the observed positions of the planet Uranus did not match the predictions of Newton's theory. The reason for this turned out to be
A) the perturbing effect of Neptune, which at that time had not yet been discovered.
B) the perturbing effect of Pluto, which at that time had not yet been discovered.
C) the weakening of gravity with increasing distance from the Sun.
D) the perturbing effects of Jupiter and Saturn, whose masses at that time were not accurately known.
Answer: A Source: Section 2-6

217. In the years after Newton published his laws of motion, it was found that the observed positions of the planet Uranus did not match the predictions of Newton' theory. The reason for this turned out to be
A) the gravitational influence of a previously unknown planet.
B) the perturbing effects of previously-unknown satellites (moons) of Uranus
C) inaccuracies in the observations
D) inaccuracies in Newton's laws (later corrected by general relativity)
Answer: A Source: Section 2-6

218. Who predicted the existence of the planet Neptune before it was discovered observationally?
A) Newton and Halley.
B) Adams and Leverrier.
C) Kepler and Brahe.
D) Kepler and Galileo.
Answer: B Source: Section 2-6

219. The first major astronomical prediction of Newton's theory of gravitation to be confirmed by observation was
A) the discovery of Neptune.
B) the return of Halley's comet.
C) the discovery of Pluto.
D) the supernova of 1572.
Answer: B Source: Section 2-6

220. Suppose that an object is discovered moving around the Sun once every 120 years. Which of the following paths is a possible orbit for this object?
A) Hyperbola. B) Ellipse. C) Straight line. D) Parabola.
Answer: B Source: Section 2-6

221. Which objects are often found to follow parabolic orbits?
A) Comets.
B) No objects-all orbits have to be ellipses.
C) Asteroids.
D) Small satellites of the outer planets.
Answer: A Source: Section 2-6

222. What is a hyperbola?
A) A long, skinny ellipse.
B) An ellipse with equal major and minor axes.
C) A straight line extending to infinity in one direction.
D) An infinitely long curve which never closes on itself.
Answer: D Source: Section 2-6

223. In which order (from earliest to latest) did the following people make their major contributions to astronomy?
A) Kepler, Copernicus, Brahe, Newton.
B) Copernicus, Brahe, Kepler, Newton.
C) Copernicus, Kepler, Brahe, Newton.
D) Brahe, Copernicus, Newton, Kepler.
Answer: B Source: Chapter 2

224. The correct order of "appearance" of the following "actors" on the "stage of scientific discovery" is
A) Copernicus, Newton, Kepler, Einstein, Ptolemy
B) Ptolemy, Copernicus, Kepler, Newton, Einstein
C) Ptolemy, Copernicus, Newton, Kepler, Einstein
D) Ptolemy, Kepler, Copernicus, Newton, Einstein
Answer: B Source: Chapter 2

225. Which of the following most closely expresses the principle of "Occam's Razor" as it applies to theoretical explanations of physical phenomena?
A) The theory requiring the least number of assumptions is the most likely explanation.
B) The theory with the longest history is the most likely explanation.
C) The theory requiring the largest number of assumptions is the most likely explanation.
D) The newest theory is most likely the correct one.
Answer: A Source: Insight 2-1

226. The concept called "Occam's Razor" tells us that
  A) when two theories describe the same phenomena equally accurately, always choose the simpler theory.
  B) the theory that describes phenomena more accurately is more likely to be correct.
  C) when two theories describe the same phenomena equally accurately, always choose the theory with the greater complexity.
  D) the theory that is applicable to the greatest range of phenomena is more likely to be correct.

  Answer: A   Source: Insight 2-1

# Chapter 3: Light and Telescopes

1. When visible light passes through a prism of glass, as shown in Fig 3-1, which wavelengths of light are deflected most by the glass?
   A) Intermediate wavelengths, the green color.
   B) The longer wavelengths.
   C) The shorter wavelengths.
   D) All wavelengths are deviated by the same amount.
   Answer: C Source: Section 3-1

2. When light passes through a prism of glass
   A) the different colors or wavelengths of light are separated in angle by the prism.
   B) the different colors are caused by multiple reflections in the prism and interference between the resulting beams.
   C) the prism adds colors to different parts of the broadly scattered beam coming out of it.
   D) the prism absorbs colors from different parts of the broad beam coming out of the prism, leaving the complementary colors which we see.
   Answer: A Source: Section 3-1

3. In 1670, Isaac Newton performed a crucial experiment on the nature of light when he
   A) demonstrated that the colors which make up white light are intrinsic, and not produced by the glass through which the light passes.
   B) proved mathematically that light can be described by oscillating electric and magnetic fields.
   C) showed the wave nature of light by passing light through two slits and obtaining a pattern of bright interference bands on a screen.
   D) showed that light which passes through a prism has a spectrum of colors added to it by the prism.
   Answer: A Source: Section 3-1

4. Who first showed that light does not travel at infinite speed?
   A) Joseph von Fraunhofer in 1814
   B) Isaac Newton in 1704
   C) James Clerk Maxwell in 1864
   D) Ole Roemer in 1675
   Answer: D Source: Section 3-1

5. What phenomenon did Roemer use to show that light does not travel at infinite speed?
   A) When a pulse of light is reflected back to the Earth from the Moon, there is a 2½ second delay between the outgoing pulse and the reflected pulse.
   B) Eclipses of Jupiter's satellites by the planet appear to occur later when the Earth is farther away from Jupiter, due to the travel time for the light over the extra distance.
   C) When a light beam is sent through an opaque, rotating disk with holes in it and then reflected back from a mirror, the returning beam does not return through the same hole it went out through, due to the travel time for light.
   D) When light is sent to a mirror on a distant hill, there is a time delay between sending out the light and receiving back the reflected beam.

   Answer: B Source: Section 3-1

6. Roemer in 1675 measured the speed of light by
   A) measuring how long it took the light from stars located at different distances to reach the Earth.
   B) reflecting light from a mirror rotating at a known speed and measuring the angle of deflection of the light beam.
   C) timing eclipses of Jupiter's satellites by the planet, which appeared to occur later when Earth was further from Jupiter.
   D) opening a shutter on a lantern on a hilltop and measuring the time taken for light from an assistance's shuttered lantern to return.

   Answer: C Source: Section 3-1

7. What prevented Ole Roemer from calculating a value for the speed of light from his measurements of the delays in eclipse times of Jupiter's moons?
   A) The dimensions of the solar system, particularly the length of 1 AU, was unknown.
   B) The distance between the Earth's orbit and Jupiter's orbit was unknown.
   C) The clocks available at that time were not sufficiently accurate.
   D) Telescopes at that time were not sharp enough to show Jupiter's moons clearly, preventing accurate timings.

   Answer: A Source: Section 3-1

8. How much time elapsed from when Roemer discovered that light does not travel at infinite speed to when the speed of light was first measured accurately?
   A) Almost a thousand years.
   B) No time at all - it was Roemer who measured it!
   C) 28 years.
   D) More than a century, but less than two centuries.

   Answer: D Source: Section 3-1

9. The speed of light in space is
   A) infinite, traveling through space instantaneously.
   B) variable, depending on the speed of its source, but very large (on average, $3 \times 10^8$ meters per second).
   C) $3 \times 10^8$ meters per second, independent of the speed of the source.
   D) $3 \times 10^{10}$ meters per second, independent of the speed of the source.
   Answer: C Source: Section 3-1

10. A scientist reports the detection of an atomic particle which came toward his experiment from outer space at a speed of $4 \times 10^5$ km $s^{-1}$. What conclusion can we draw from this report?
    A) He has made an error in his experiment, since such a speed is considered to be impossible by all previous experiments.
    B) This "particle" must have been a photon or quantum of electromagnetic radiation of very high energy in order to have traveled this fast.
    C) This result is acceptable since atomic particles can travel this fast, whereas larger bodies are limited to $3 \times 10^5$ m $s^{-1}$.
    D) This is an acceptable result for a particle originating from outer space, since particle speed from such regions is unlimited.
    Answer: A Source: Section 3-1

11. When the Galileo spacecraft reached Jupiter on December 7, 1995, Jupiter was almost at conjunction with the Sun. Given that it takes 8 1/3 minutes for light to travel a distance of 1 AU, how long did it take Galileo's signals to reach the Earth from Jupiter? (It might be useful to draw a diagram.)
    A) 60 minutes B) 52 minutes C) 35 minutes D) 43 minutes
    Answer: B Source: Section 3-1

12. The average distance of Pluto from the Sun is 40 AU. How long does it take for light to travel across the solar system from one side of Pluto's orbit to the other?
    A) 8 min B) 22 hr C) 5½ hr D) 11 hr
    Answer: D Source: Section 3-1

13. 1 nanometer (nm) is
    A) $10^{-6}$ m B) $10^{-12}$ m C) $10^{-9}$ m D) $10^6$ m
    Answer: C Source: Section 3-2

14. Visible wavelengths of electromagnetic radiation have a range of wavelengths of
    A) 1 to 100 nm. B) 400 to 700 nm. C) 90 to 130 nm. D) 800 to 1900 nm.
    Answer: B Source: Section 3-2

15. Which wavelength region of the electromagnetic spectrum is taken up by visible light?
    A) 400 nm to 700 nm
    B) 100 nm to 400 nm
    C) 20 to 120 nm
    D) 1200 nm to 1500 nm
    Answer: A Source: Section 3-2

16. The wavelength of green light is about 500 nanometers. What is this in meters?
    A) 500 thousands of a meter
    B) 500 trillionths of a meter (1 trillion = 1,000,000 millions)
    C) 500 billionths of a meter (1 billion = 1000 millions)
    D) 500 millionths of a meter
    Answer: C Source: Section 3-2

17. Who first proved that light is a wave?
    A) Albert Einstein B) Isaac Newton C) Thomas Young D) James Clerk Maxwell
    Answer: C Source: Section 3-2

18. How did Thomas Young prove that light behaves as a wave?
    A) He showed that light striking a metal surface releases electrons from the metal, with violet light producing more energetic electrons than red light.
    B) He showed that light passing through two narrow, closely spaced slits produces a pattern of light and dark bands on a screen by wave interference.
    C) He showed that the speed of light decreases when it enters a denser, transparent medium.
    D) He showed that the speed of light in a vacuum is the same for all observers.
    Answer: B Source: Section 3-2

19. Who was the first person to suggest that light is an electromagnetic wave?
    A) Thomas Young B) Isaac Newton C) James Clerk Maxwell D) Albert Einstein
    Answer: C Source: Section 3-2

20. In the 1860s, Maxwell derived a set of mathematical equations that described electromagnetic waves that could have different wavelengths. These waves, which include visible light, have since been shown to
    A) exist only over a wavelength range from infrared to ultraviolet radiation.
    B) have no short wavelength limit, but waves cannot exist with wavelengths longer than about the diameter of the Earth.
    C) have no upper wavelength limit, but waves cannot exist with wavelengths smaller than an atom.
    D) have no wavelength limit, either short or long.
    Answer: D Source: Section 3-2

21. Electromagnetic radiation moving through space with the speed of light consists of
    A) oscillating magnetic fields which over time and distance change to oscillating electric fields and back again.
    B) oscillating electric fields, with magnetic fields occasionally accompanying them, moving in the same direction.
    C) oscillating electric and magnetic fields moving in opposite directions along the same line in space.
    D) oscillating electric and magnetic fields, always inseparable, always having the same frequency and wavelength, and traveling in the same direction.
    Answer: D Source: Section 3-2

22. In what way does a photon of blue light NOT differ from a photon of yellow light, in a vacuum?
    A) Speed B) Energy C) Color D) Wavelength
    Answer: A Source: Section 3-2

23. In the full wavelength range of electromagnetic radiation, visible light occupies what proportion of this possible range?
    A) almost the full range between radio and X-rays.
    B) two narrow but separate ranges between ultra-violet and infra-red radiations, the red and the blue, which mix to give all the other colors.
    C) about half of the possible range.
    D) a very narrow range.
    Answer: D Source: Section 3-2

24. The wavelength of infrared radiation is longer than visible light and is usually measured in units of micrometers. 1 micrometer (mm) is
    A) $10^{-3}$ m B) $10^{-6}$ m C) $10^{6}$ m D) $10^{-9}$ m
    Answer: B Source: Section 3-2

25. Violet light differs from red light in that
    A) violet light has a shorter wavelength than red light.
    B) violet light travels more slowly (through a vacuum) than red light.
    C) violet light has a longer wavelength than red light
    D) violet light travels more quickly (through a vacuum) than red light.
    Answer: A Source: Section 3-2

26. One difference between violet light and red light is that
    A) violet light has a shorter wavelength than red light.
    B) photons of violet light have less energy than photons of red light.
    C) violet light travels faster than red light.
    D) violet light is hotter than red light.
    Answer: A Source: Section 3-2

27. What is the relationship between color and wavelength for light?
    A) Wavelength decreases from violet to red.
    B) Wavelength depends only on brightness and is independent of color.
    C) Wavelength increases from violet to red.
    D) Wavelength increases from violet to yellow-green, then decreases again to red.
    Answer: C Source: Section 3-2

28. How does the wavelength of visible light compare to the wavelengths of other forms of electromagnetic radiation?
    A) Longer than x-rays but shorter than gamma rays.
    B) Longer than ultraviolet but shorter than x-rays.
    C) Longer than infrared but shorter than radio waves.
    D) Longer than x-rays but shorter than radio waves.
    Answer: D Source: Section 3-2

29. Visible light occupies what position in the whole electromagnetic spectrum?
    A) Between infrared and ultraviolet.
    B) Between ultraviolet and x-rays.
    C) Between radio and infrared radiation.
    D) Between infrared and microwave.
    Answer: A Source: Section 3-2

30. Which ONE of the following statements is TRUE?
    A) Visible light takes up only a very small part of the total range of wavelengths in the electromagnetic spectrum.
    B) Visible light takes up all of the electromagnetic spectrum.
    C) Visible light takes up most (but not all) of the total range of wavelengths in the electromagnetic spectrum.
    D) Visible light is NOT part of the electromagnetic spectrum.
    Answer: A Source: Section 3-2

31. Which of the following lists of different types of electromagnetic radiation is correctly ordered, from shortest wavelength to longest?
A) Radio, infrared, ultraviolet, gamma rays
B) Gamma rays, ultraviolet, radio, infrared
C) Gamma rays, ultraviolet, infrared, radio
D) Radio, ultraviolet, infrared, gamma rays
Answer: C Source: Section 3-2

32. What is the wavelength of electromagnetic radiation whose frequency in $10^6$ cycles per second ($10^6$ Hz or 1000 kHz, the frequency of ordinary A.M. radio)?
A) 3 mm B) 300 meters C) 3 cm D) 3 meters
Answer: B Source: Section 3-2

33. What is the wavelength of radiation emitted by an FM radio station transmitting at a frequency of 100 MHz?
A) 1 meter B) 0.03 meters C) 3 meters D) 300 meters
Answer: C Source: Section 3-2

34. In a radio wave transmitter (such as that used by a radio or TV station), when the frequency of the signals is increased
A) the wavelength is decreased.
B) the speed of transmission of the waves is increased.
C) the wavelength remains constant.
D) the wavelength and speed of transmission both increase.
Answer: A Source: Section 3-2

35. Which is the correct sequence of electromagnetic radiation, in order of increasing energy of the photons (or quanta)?
A) Visible light, microwave, radio waves, infrared rays.
B) Gamma rays, radio waves, x-rays, infrared rays.
C) Visible light, UV radiation, x-rays, gamma rays.
D) Radio waves, microwaves, gamma rays, UV radiation.
Answer: C Source: Section 3-2

36. Which of the following types of electromagnetic radiation has the longest wavelength?
A) Infrared radiation B) Radio waves C) Ultraviolet light D) Microwaves
Answer: B Source: Section 3-2

37. Which of the following types of electromagnetic radiation has the shortest wavelength?
A) Violet light B) X-rays C) Ultraviolet light D) Gamma rays
Answer: D Source: Section 3-2

38. In terms of wavelengths, gamma rays are
A) intermediate between radio and infrared waves.
B) the longest wavelength electromagnetic waves.
C) intermediate between x-rays and ultraviolet waves.
D) the shortest wavelength electromagnetic waves.
Answer: D Source: Section 3-2

39. Which of the following is an electromagnetic wave?
A) Sound wave B) Microwave C) Cosmic ray proton D) Gravitational wave
Answer: B Source: Section 3-2

40. Which of the following wave effects is NOT electromagnetic in nature?
A) Microwaves B) Radio waves C) Seismic waves D) Gamma rays
Answer: C Source: Section 3-2

41. Electromagnetic radiation emitted by a planet has a wavelength of 10 micrometers (1 mm = $10^{-6}$ m). What name is given to this type of electromagnetic radiation?
A) Gamma rays B) Infrared radiation C) Visible light D) Radio radiation
Answer: B Source: Section 3-2

42. Suppose an astronomical satellite observes the Orion Nebula at a wavelength of 1250 nm. In what wavelength range is this satellite observing?
A) X-rays B) Ultraviolet C) Infrared D) Visible light
Answer: C Source: Section 3-2

43. Suppose an astronomical satellite observes the Crab Nebula at a wavelength of 0.85 nm. In what wavelength range is this satellite observing?
A) Gamma rays B) Infrared C) X-rays D) Ultraviolet
Answer: C Source: Section 3-2

44. The Lyman-alpha spectral line of hydrogen has a wavelength of 121.6 nm. In which wavelength band does this line occur?
A) Infrared B) Visible light C) Ultraviolet D) X-rays
Answer: C Source: Section 3-2

45. What is the one fundamental difference between x-rays and radio waves?
A) Their wavelengths are very different.
B) They always come from different sources.
C) Their speeds in outer space are different.
D) Radio waves are always wavelike, while x-rays always behave like particles.
Answer: A Source: Section 3-2

46. X-rays and light
    A) are different because x-rays are made up of waves, whereas light is made up of particles.
    B) are the same thing except that x-rays have a longer wavelength than light.
    C) are different because x-rays are made up of particles, whereas light is made up of waves.
    D) are the same thing except that x-rays have a shorter wavelength than light.
    Answer: D Source: Section 3-2

47. The diameter of the Earth is about 13,000 km. How many times the diameter of the Earth does light travel in one second?
    A) 46 B) 0.043 C) 23 D) 23,077
    Answer: C Source: Section 3-2

48. A particular photon has a wavelength of 450 nm and a second one has a wavelength of 580 nm. Which of the following statements is true for the energies of these two photons?
    A) All photons have the same energy, regardless of wavelength.
    B) The photon from the higher-intensity light source has the higher energy (regardless of wavelength).
    C) The 450-nm photon has the higher energy.
    D) The 580-nm photon has the higher energy.
    Answer: C Source: Section 3-3

49. A particular photon of ultraviolet (UV) light has a wavelength of 200 nm and a photon of infrared (IR) light has a wavelength of 2000 nm. What is the energy of the UV photon compared to the IR photon?
    A) It has 1/100 of the energy of the IR photon.
    B) It has 100 times more energy than the IR photon.
    C) It has 1/10 of the energy of the IR photon.
    D) It has 10 times more energy than the IR photon.
    Answer: D Source: Section 3-3

50. If two photons in a vacuum have different energies, what can we say about the wavelengths of these photons?
    A) We can't say anything; wavelength depends only on color and not on energy.
    B) They have the same wavelength; all photons have the same wavelength, regardless of energy.
    C) The higher-energy photon has the shorter wavelength.
    D) The higher-energy photon has the longer wavelength.
    Answer: C Source: Section 3-3

51. Radio waves travel through space at what speed?
    A) Much faster than the speed of light
    B) Slightly faster than the speed of light, since their wavelength is longer.
    C) At the speed of light, $3 \times 10^8$ m/s.
    D) Much slower than the speed of light.
    Answer: C Source: Section 3-4

52. Which of the following can travel at the speed of light in a vacuum?
    A) Light, atoms, x-rays, sub-atomic particles (e.g., electrons).
    B) Only light; all other electromagnetic waves travel slower than light speed.
    C) Light, radio waves, x-rays and gamma rays.
    D) Light, infrared, ultraviolet and sub-atomic particles (e.g., electrons).
    Answer: C Source: Section 3-4

53. An electrical spark, such as lightning, generates electromagnetic radiation over a wide range of wavelengths. How much longer will a pulse of radio energy take to travel between two detector stations 100 m apart than a pulse of ultra-violet radiation from the same spark?
    A) Much shorter since long wavelength radiations travel faster.
    B) Just a little longer, since the high frequency UV radiation travels faster than the low frequency radio waves.
    C) The time will be identical since both pulses travel at the speed of light.
    D) Much longer since radio waves have much longer wavelengths and therefore travel slower.
    Answer: C Source: Section 3-4

54. The two ranges of electromagnetic radiation for which the Earth's atmosphere is reasonably transparent are
    A) visible and radio radiation.
    B) UV and radio waves.
    C) x-rays and visible radiation.
    D) visible and far infrared radiation.
    Answer: A Source: Section 3-4

55. The Earth's atmosphere is transparent to which of the following types of electromagnetic radiation?
    A) Long infrared wavelengths
    B) X-rays
    C) Radio waves
    D) Short ultraviolet wavelengths
    Answer: C Source: Section 3-4

56. Which one of the following wavelength regions MUST be observed from space, since no energy in this region reaches the ground?
A) Infrared B) Radio C) Visible light D) X-rays
Answer: D Source: Section 3-4

57. Who was the first astronomer to build and use a telescope to observe the night sky?
A) Copernicus B) Galileo C) Tycho Brahe D) Newton
Answer: B Source: Section 3-5

58. The major reason why astronomers seek after funds to build larger telescopes is
A) to provide magnified images of stars.
B) to measure a wider spectrum of light from stars.
C) to bring stars closer to the Earth.
D) to collect more light from distant objects.
Answer: D Source: Section 3-5

59. The main reason for building large optical telescopes on the Earth's surface is
A) for national prestige, with no scientific reason.
B) to magnify images of objects and produce higher resolution photographs.
C) to collect more light from faint objects.
D) to bring astronomical objects closer for more detailed examination by scientists.
Answer: C Source: Section 3-5

60. What is the refraction of light?
A) The change in direction of a light ray as it crosses from a less dense, transparent material to a more dense one.
B) The absorption of light as it traverses a dense, transparent material.
C) The breaking of white light into its composite colors.
D) The change in direction of a light ray as it reflects from a more dense material than the one in which it is traveling.
Answer: A Source: Section 3-5

61. Refraction is
A) the bending of light as it enters a transparent material at an angle more perpendicular to the surface of the material.
B) the change in direction of light when it bounces off a smooth surface.
C) the bending of light around the sharp edge of an obstacle.
D) the change in the color of light when it enters a transparent but colored material such as glass.
Answer: A Source: Section 3-5

62. Which way does a light ray bend when it enters the plane, smooth surface of a dense transparent material (from air or a vacuum) at an angle to the perpendicular?
    A) It does not bend at all, since the material is transparent.
    B) It bends away from the perpendicular.
    C) It always bends so as to travel perpendicular to the surface.
    D) It bends toward the perpendicular.
    Answer: D Source: Section 3-5

63. When light in air or a vacuum enters the plane surface of a dense, transparent medium at an angle to the perpendicular to this surface, which way does the light ray bend?
    A) It does not bend at all, since the material is transparent.
    B) It bends toward the perpendicular.
    C) It bends so that whatever the incoming angle, the light always travels along the perpendicular to the surface.
    D) It bends away from the perpendicular.
    Answer: B Source: Section 3-5

64. When a light ray in air or a vacuum hits the surface of a piece of perfectly smooth, flat glass at an angle and enters the glass, it will
    A) reverse its direction and return back along its original path.
    B) not change its direction at all.
    C) bend toward the perpendicular to the surface.
    D) bend away from the perpendicular to the surface.
    Answer: C Source: Section 3-5

65. When a ray of light enters the flat smooth surface of a dense transparent material at an angle to the perpendicular, which way does it bend?
    A) Toward the perpendicular.
    B) Away from the perpendicular.
    C) It does not bend at all, since the material is transparent.
    D) It always bends so as to travel perpendicular to the surface.
    Answer: A Source: Section 3-5

66. As a light ray leaves a glass surface, traveling from the glass back into air at an angle to the surface, the light ray
    A) travels in a straight line without bending.
    B) bends away from the perpendicular to the surface.
    C) bends toward the perpendicular to the surface.
    D) bends so as to travel exactly along the perpendicular to the surface.
    Answer: B Source: Section 3-5

67. Which of the following statements correctly describes the refraction of light?
    A) A light ray reverses its direction of travel after striking a mirror surface.
    B) The path of a ray of light bends as the light enters or leaves a dense transparent medium such as glass.
    C) A ray of light spreads out after passing through an opening, because of the wave nature of light.
    D) A ray of light is partially absorbed as it enters the denser material.
    Answer: B Source: Section 3-5

68. How does the speed of light in a substance such as glass compare to the speed of light in a vacuum?
    A) The speed of light in glass is slower than the speed of light in a vacuum.
    B) The speed of light in glass can be faster or slower than the speed of light in a vacuum, depending on the type of glass.
    C) The speed of light is the same in both, equal to 299,792.458 km/s.
    D) The speed of light in glass is faster than the speed of light in a vacuum.
    Answer: A Source: Section 3-5

69. Compared to the speed of visible light in a vacuum (or in space), its speed in water or in glass is
    A) greater.
    B) much greater.
    C) less.
    D) exactly the same, since the speed of light cannot vary.
    Answer: C Source: Section 3-5

70. The speed of light
    A) always has the same value in a vacuum, but is faster in transparent materials such as water and glass.
    B) always has the same value in a vacuum, but is slower in transparent materials such as water and glass.
    C) has the same constant value in all situations.
    D) always has the same value in a vacuum but its speed inside a transparent material depends upon the temperature of the material and can be faster or slower than its speed in a vacuum.
    Answer: B Source: Section 3-5

71. Light, when it passes from air into a dense but transparent material such as glass,
    A) maintains its speed, as required by the laws of physics.
    B) changes its speed to that of sound in the medium.
    C) slows down.
    D) speeds up.
    Answer: C Source: Section 3-5

72. Light enters the smooth flat surface of a block of glass from the vacuum of space (e.g., a spacecraft window). What is the speed of light inside the glass compared to that in vacuum?
    A) Faster, since the glass is denser than the vacuum.
    B) The same, since the speed of light is always constant by definition.
    C) Slower, since the glass is denser than the vacuum.
    D) The speed will depend upon the angle that the light ray makes with the perpendicular to the surface, becoming slower as the angle increases.
    Answer: C Source: Section 3-5

73. Light enters a plane-parallel block of glass from the vacuum of space and leaves again into space through the opposite side (e.g., in a spaceborne instrument). What is the speed of light after it leaves the glass?
    A) Slower speed, since it has passed through the glass and been slowed by it.
    B) The exit speed will depend upon the thickness of the glass, becoming slower, the thicker the glass.
    C) Faster than when it entered since it has passed through the dense glass.
    D) The same speed as it had when it entered the glass.
    Answer: D Source: Section 3-5

74. Which characteristic property of a glass lens is the most important in bending light rays to form a focused image?
    A) The diameter or size of the lens.
    B) The thickness of the center of the lens.
    C) The curvature and shape of its surfaces.
    D) The color of the glass.
    Answer: C Source: Section 3-5

75. What is the main optical element of a refracting telescope?
    A) A curved mirror B) A plane (flat) mirror C) A prism D) A lens
    Answer: D Source: Section 3-5

76. A refracting telescope is the type that uses as its main optical element
    A) a mirror.
    B) a prism of glass.
    C) a lens.
    D) a combination of many small plane mirrors.
    Answer: C Source: Section 3-5

77. Which type of telescope uses a lens as the main optical element?
    A) A radio telescope.
    B) A Cassegrain telescope.
    C) A Newtonian telescope.
    D) A refracting telescope.
    Answer: D Source: Section 3-5

78. At what distance from the objective lens in a refracting telescope is the image formed (i.e. where would the photographic film be placed)?
    A) Its diameter.
    B) Its focal length.
    C) Twice its focal length.
    D) Immediately behind the lens to collect the most light.
    Answer: B Source: Section 3-5

79. A typical refracting telescope is made up of
    A) a long-focal-length lens at the front and a short-focal-length lens at the rear (next to your eye as you look through the telescope).
    B) two mirrors, one concave and the other convex.
    C) a mirror that gathers and focuses the light, and a lens next to your eye to examine the image.
    D) a short-focal-length lens at the front and a long-focal-length lens at the rear (next to your eye as you look through the telescope).
    Answer: A Source: Section 3-5

80. How far apart should the objective lens and the eyepiece lens be in a refracting telescope? (See Figure 3-10.)
    A) A distance equal to the focal length of the objective lens.
    B) A distance equal to the focal length of the eyepiece lens.
    C) A distance equal to the sum of the focal lengths of the two lenses.
    D) A distance equal to the focal length of the objective lens minus the focal length of the eyepiece lens.
    Answer: C Source: Section 3-5

81. The light-gathering power of a telescope is related directly to
    A) the focal length of its primary mirror or lens.
    B) the image quality of its optics (resolution).
    C) the ratio of the focal lengths of its primary element (mirror or lens) and its eyepiece.
    D) the area of its primary mirror or lens.
    Answer: D Source: Section 3-6

82. Many amateur astronomers have telescopes with mirrors 20 cm (1/5 m) in diameter. In comparison, the largest astronomical telescope in the world is the Keck telescope, with a diameter of 10 m. How many times larger is the light-gathering power of the Keck telescope than a 20-cm telescope?
    A) 7 times larger
    B) 50 times larger
    C) 2,500 times larger
    D) 125,000 times larger
    Answer: C Source: Section 3-6

83. For many years, the Palomar telescope (5 m diameter) in California was the largest telescope in the world; now that honor falls to one of the two Keck telescopes (each of diameter 10 m) in Hawaii. How many times larger is the light-gathering power of the Keck telescope than the Palomar telescope?
    A) 1.4 times larger B) 8 times larger C) 4 times larger D) 2 times larger
    Answer: C Source: Section 3-6

84. Compared with the 2.5-m diameter Mount Wilson telescope, the 10-m diameter Keck telescope has a greater light-gathering power, by a factor of
    A) 4.. B) 16. C) 2. D) 1; they have the same light gathering capability.
    Answer: B Source: Section 3-6

85. How many times more light will be collected by the optics of one side of a pair of binoculars ($7 \times 50$, with magnification of 7 and an objective lens aperture diameter of 50 mm) compared to that collected by an average human eye, with a typical aperture diameter of 5 mm?
    A) 10 B) 100 C) 2500 D) 1.4 or 7/5
    Answer: B Source: Section 3-6

86. How many times more light can the 5-meter telescope at Mount Palomar collect than the unaided human eye (with a diameter of 5 mm)?
    A) $10^6$ B) 1000 C) 5,000 D) 10,000
    Answer: A Source: Section 3-6

87. How many times more light can the 10-m diameter Keck telescope on Mauna Kea in Hawaii collect than an average unaided human eye, with a typical aperture diameter of 5 mm?
A) $2.5 \times 10^5$ B) $2 \times 10^4$ C) $4 \times 10^6$ D) 2,000
Answer: C Source: Section 3.6

88. A particular reflecting telescope has an objective mirror with a focal length of 1.2 m and an eyepiece lens of focal length 6 mm. What is the magnifying power of this telescope?
A) 200x B) 20x C) 2000x D) 5x
Answer: A Source: Section 3-6

89. A department store sells an "astronomical telescope" with an objective lens of 30 cm focal length and an eyepiece lens of focal length 5 mm. What is the magnifying power of this telescope?
A) 150x B) 6x C) 15x D) 60x
Answer: D Source: Section 3-6

90. A particular reflecting telescope has a primary mirror 0.4 m in diameter and 2.0 m focal length, and an eyepiece lens 1.0 cm in diameter and 0.5 cm focal length. What is the magnifying power of this telescope?
A) 80x B) 400x C) 40x D) 4x
Answer: B Source: Section 3-6

91. A refracting telescope has an objective lens of focal length 80 cm and a diameter of 10 cm and an eyepiece of focal length 5 cm and diameter 1 cm. What is the magnifying power of this telescope?
A) 10x B) 16x C) 8x D) 80x
Answer: B Source: Section 3-6

92. Which of the following characteristics of an astronomical telescope is the most important for determining the angular resolution?
A) The diameter of the objective lens or mirror.
B) The magnifying power of the telescope.
C) The focal length of the objective lens or mirror.
D) The focal length of the eyepiece.
Answer: A Source: Section 3-6

93. In telescopes, the angular resolution is worse for
    A) smaller diameter lenses or mirrors and shorter wavelength light (or other electromagnetic radiation)
    B) smaller diameter lenses or mirrors and longer wavelength light (or other electromagnetic radiation)
    C) larger diameter lenses or mirrors and shorter wavelength light (or other electromagnetic radiation)
    D) larger diameter lenses or mirrors and longer wavelength light (or other electromagnetic radiation)
    Answer: B Source: Section 3-6

94. In single-telescope astronomical systems, either optical or radio,
    A) the longer the focal length of the primary mirror, the sharper the image.
    B) the larger the main mirror aperture in general, the sharper the image.
    C) the smaller the main-mirror aperture in general, the sharper the image.
    D) the longer the wavelength, the sharper the image.
    Answer: B Source: Section 3-6

95. In a telescope, to what does the term "aberration" refer?
    A) The absorption of light by the glass in the lenses.
    B) The magnifying power of the telescope.
    C) A fundamental limitation of any telescope to resolve very small details in the image.
    D) A defect in design that blurs or distorts the image.
    Answer: D Source: Section 3-7

96. A lens with spherical surfaces suffers from spherical aberration because
    A) the starlight is distorted by turbulence in the Earth's atmosphere.
    B) different parts of the lens focus the light at different distances from the lens.
    C) different colors are focused at different distances from the lens.
    D) the lens sags under its own weight, distorting the image.
    Answer: B Source: Section 3-7

97. How is spherical aberration USUALLY overcome in the objective lens of a refracting telescope?
    A) By making the lens very thin, with a long focal length.
    B) By coating the lenses with an anti-reflective coating.
    C) By grinding the lens surfaces to a parabolic shape.
    D) By using a filter that allows only a small range of wavelengths (colors) through the telescope.
    Answer: A Source: Section 3-7

98. Which of the following types of telescopes is most seriously affected by chromatic aberration?
    A) A Newtonian telescope
    B) A Cassegrain telescope
    C) A radio telescope
    D) A refracting telescope
    Answer: D Source: Section 3-7

99. Chromatic aberration is the failure of a telescope objective to bring all colors of light to the same focus, and appears
    A) only in a reflecting telescope.
    B) in both reflecting and refracting telescopes.
    C) in all telescopes, since it is caused by a basic property of light.
    D) only in a refracting telescope.
    Answer: D Source: Section 3-7

100. When visible light passes through a prism or a lens made of glass, which wavelengths of light are deflected most by the glass? (See Figs. 3-1 and 3-13, Kaufmann & Comins, *Discovering the Universe*, 5th Ed.)
    A) Intermediate wavelengths, the green color.
    B) The longer wavelengths.
    C) The directions of all wavelengths are changed by the same amount.
    D) The shorter wavelengths.
    Answer: D Source: Sections 3-1 and 3-7

101. What causes chromatic aberration in the objective lens of a telescope?
    A) Different colors are refracted through different angles at each surface of the lens.
    B) The shape of the lens surface distorts the image.
    C) Reflection of light back and forth inside the lens blurs the image.
    D) Different colors suffer different amounts of absorption by the glass in the lens.
    Answer: A Source: Section 3-5

102. What is chromatic aberration in a telescope?
    A) Light reflecting from the sides of the telescope comes to a focus at different points inside the telescope.
    B) Light striking the lens at different distances from the center of the lens comes to a focus at different points inside the telescope.
    C) The objective lens varies in thickness from center to edge, and light passing through these different thicknesses suffers different amounts of absorption.
    D) Light of different colors comes to a focus at different points inside the telescope.
    Answer: D Source: Section 3-7

103. Chromatic aberration occurs in a refracting telescope when
   A) the lenses bend under their own weight, distorting the final image.
   B) some wavelengths are scattered out of the telescope by imperfections in the glass in the lenses.
   C) the different colors of light do not all come to the same focal point.
   D) light from some wavelengths is absorbed by the lenses, distorting the colors of objects.
   Answer: C Source: Section 3-7

104. How is chromatic aberration corrected in a refracting telescope?
   A) By using a prism to recombine the colors into white light.
   B) By grinding the lens surfaces into a parabolic shape.
   C) By using a compound lens made of two different types of glass.
   D) By painting the inside of the telescope jet black to minimize reflections.
   Answer: C Source: Section 3-7

105. Chromatic aberration, the inability of a simple lens to focus all colors of light to the same focal point, is reduced in astronomical telescopes by
   A) adding a carefully shaped mirror to the lens to produce a lens-mirror combination.
   B) combining two lenses of different shape, made from the same kind of glass.
   C) combining two lenses of different shape, made from different kinds of glass.
   D) bending the lens in its mounting, to correct the aberration.
   Answer: C Source: Section 3-7

106. The largest refracting telescope in the world is the 102 cm (40 in.) telescope at Yerkes Observatory, built in 1897. Why have no larger refracting telescopes ever been built?
   A) Larger lenses give too much magnification.
   B) Larger lenses sag too much under their own weight.
   C) Larger lenses are too thick, and suffer too much spherical aberration.
   D) Larger lenses have too much chromatic aberration.
   Answer: B Source: Section 3-7

107. When a ray of light strikes a smooth mirror surface at an angle from the perpendicular, the ray is reflected
   A) at various angles, depending on wavelength.
   B) on the "other" side of the perpendicular, but at the same angle as the incoming ray.
   C) so that it travels along the perpendicular to the surface of the mirror.
   D) back along its original (incoming) path.
   Answer: B Source: Section 3-8

108. Who developed the first reflecting astronomical telescope?
A) Isaac Newton B) Copernicus C) Galileo D) Ptolemy
Answer: A Source: Section 3-8

109. A Newtonian telescope uses
A) only one mirror.
B) two curved mirrors.
C) one curved mirror and one flat mirror at a 45° angle to the curved mirror.
D) several mirrors, some curved, some flat, to guide light to a fixed focus with respect to the Earth.
Answer: C Source: Section 3-8

110. A Newtonian telescope consists of the following types of mirrors:
A) one concave focusing mirror.
B) one concave and one flat mirror.
C) two lenses, to produce an image the correct way around.
D) two curved mirrors, one concave, the second convex.
Answer: B Source: Section 3-8

111. A Newtonian telescope consists of the following types of mirrors:
A) two lenses, to produce an image the correct way around.
B) one concave focusing mirror.
C) one concave and one flat mirror.
D) two curved mirrors, one concave, the second convex.
Answer: C Source: Section 3-8

112. How is a Newtonian reflecting telescope constructed?
A) A concave primary mirror and flat, diagonal secondary mirror.
B) A concave primary mirror and a concave secondary mirror that reflects light back through a hole in the primary mirror.
C) A concave primary mirror and a convex secondary mirror that reflects light back through a hole in the primary mirror.
D) With a series of mirrors that channel the light to a remote location.
Answer: A Source: Section 3-8

113. What is the magnification of a Newtonian telescope which has a primary mirror of diameter 0.25m and focal length of 4m, when used with an eyepiece of focal length 25 mm and an optical diameter of 2.5 mm?
A) 1600 B) 100 C) 10 D) 160
Answer: D Source: Section 3-8

114. Why does a reflecting telescope used at prime focus NOT suffer from chromatic aberration?
    A) Because all wavelengths of light are reflected by the same amount irrespective of color.
    B) Because the light has passed through only one lens.
    C) Because the lens is perfectly formed such that all colors of light travel through the lens along the same path.
    D) Because the aluminum coating on the mirror absorbs light from all wavelengths except the one of interest to the astronomer.
    Answer: A Source: Section 3-8

115. The best shape for the cross-section of a large astronomical mirror in order to produce the sharpest images of very distant objects is
    A) a perfectly flat, smooth surface.
    B) a spherical shape.
    C) an elliptical shape.
    D) a parabolic shape.
    Answer: D Source: Section 3-8

116. The prime focus of a reflecting telescope is reached by light after
    A) reflection at two mirrors.
    B) refraction at one lens.
    C) one reflection.
    D) refraction through one lens, and reflection at one mirror.
    Answer: C Source: Section 3-8

117. How is a Cassegrain reflecting telescope constructed?
    A) A concave primary mirror and convex secondary mirror that reflects light back through a hole in the primary mirror.
    B) A concave primary mirror and flat, diagonal secondary mirror.
    C) With a series of mirrors that channel the light to a remote location.
    D) A concave primary mirror and concave secondary mirror that reflects light back through a hole in the primary mirror.
    Answer: A Source: Section 3-8

118. When light from the concave primary mirror of a telescope is reflected by a small secondary mirror through a hole in the primary to its focus, this is called the
    A) Newtonian focus. B) prime focus. C) Coude focus. D) Cassegrain focus.
    Answer: D Source: Section 3-8

119. A particular telescope is set up to use the Coude focus. What arrangement of mirrors is used to bring the light to this focus?
   A) A series of mirrors that channel the light to a remote and fixed location.
   B) A concave primary mirror and convex secondary mirror that reflects light back through a hole in the primary mirror.
   C) A concave primary mirror and concave secondary mirror that reflects light back through a hole in the primary mirror.
   D) A concave primary mirror and flat, diagonal secondary mirror.
   Answer: A Source: Section 3-8

120. If you want to build a telescope having the LEAST possible amount of chromatic aberration, you should use
   A) as thin a front lens as you can.
   B) mirrors instead of lenses.
   C) a front lens that has been coated with a special material to reduce refraction.
   D) a front lens that is composed of two closely–spaced lenses made of different kinds of glass.
   Answer: B Source: Section 3-9

121. The prime focus cage of a telescope whose primary mirror is 3 meters in diameter has a diameter of 0.5 meters. What fraction of the incoming light is obstructed by this cage?
   A) about 1/6, or 17% B) about 3% C) about 36% D) about 44%
   Answer: B Source: Section 3-9

122. A spherical mirror suffers from spherical aberration because
   A) the starlight is distorted by turbulence in the Earth's atmosphere.
   B) different parts of the mirror focus the light at different distances from the mirror.
   C) different colors are focused at different distances from the mirror.
   D) the mirror sags under its own weight, distorting the image.
   Answer: B Source: Section 3-9

123. Spherical aberration can be corrected in a reflecting telescope by
   A) grinding the mirror to a more accurate spherical curve.
   B) grinding the mirror to a parabolic shape.
   C) grinding the mirror to an elliptical shape.
   D) using light of only one color.
   Answer: B Source: Section 3-9

124. The reason that the primary mirror of an astronomical telescope is often polished to a parabolic shape is
   A) that it is lighter and easier to mount in a telescope, since it requires less glass than an equivalent spherical mirror.
   B) that it is easier to produce, even though the resulting mirror will produce more spherical aberration than an equivalent spherical mirror.
   C) to avoid spherical aberration, by bringing parallel rays to a single focus.
   D) to avoid the chromatic aberration which would be produced by an equivalent spherical mirror.
   Answer: C Source: Section 3-9

125. Which of the following statements is NOT correct in describing a disadvantage of large refracting telescopes when compared to large reflecting telescopes?
   A) Refracting telescopes suffer from chromatic aberration and reflecting telescopes do not.
   B) Sagging of the primary optical element under its own weight is a problem with refracting telescopes but not with reflecting telescopes.
   C) Air bubbles in the glass are more of a problem in refracting telescopes than in reflecting telescopes.
   D) Refracting telescopes suffer from spherical aberration and reflecting telescopes do not.
   Answer: D Source: Section 3-9

126. How is spherical aberration eliminated in a reflecting telescope?
   A) By making the primary mirror very thin.
   B) By using the Cassegrain focus of the telescope.
   C) By grinding the primary mirror surface to a parabolic shape.
   D) By bending the photographic film into a concave shape.
   Answer: C Source: Section 3-9

127. What are the basic parts of a Schmidt telescope?
   A) A spherical, concave primary mirror and a glass corrector plate.
   B) A parabolic, concave primary mirror and a convex secondary mirror.
   C) A parabolic, concave primary mirror and a concave secondary mirror
   D) A parabolic, concave primary mirror and a flat, diagonal secondary mirror.
   Answer: A Source: Section 3-9

128. Which combination of optics is used in the Schmidt Telescope for producing high-quality wide-angle images of the sky?
   A) A single, short-focal-length spherical mirror.
   B) Two lenses of different glasses, one convex, the other concave, to reduce chromatic aberration.
   C) A parabolic mirror of short focal length, used at prime focus.
   D) A spherical primary mirror with a thin correcting lens ahead of it.
   Answer: D  Source: Section 3-9

129. Which of the following telescopes consists of a concave, spherical primary mirror and a glass corrector plate to correct for the spherical aberration in the primary?
   A) A refracting telescope.
   B) A Newtonian telescope.
   C) A Schmidt telescope.
   D) A Cassegrain telescope.
   Answer: C  Source: Section 3-9

130. What is the purpose of the glass corrector plate in the front of a Schmidt telescope?
   A) It corrects for sagging of the primary mirror under its own weight when the telescope is turned at an angle to the vertical.
   B) It corrects for spherical aberration in the primary mirror.
   C) It reflects unwanted wavelengths away from the primary mirror.
   D) It corrects for chromatic aberration in the primary mirror.
   Answer: B  Source: Section 3-9

131. Schmidt telescopes are used mostly for
   A) photographing wide areas of the night sky.
   B) photography of planets, using high magnification.
   C) high-magnification views of faint, distant galaxies and quasars.
   D) tracking artificial satellites in Earth orbit.
   Answer: A  Source: Section 3-9

132. The main use of the Schmidt-type telescope, which uses a spherical primary mirror and a refracting "corrector plate," is for
   A) wide angle photography, for sky surveys.
   B) single-star spectroscopy.
   C) narrow field-of-view photography and photometry.
   D) radio astronomy.
   Answer: A  Source: Section 3-9

133. The major cause of blurring and unsharp images of objects observed through very large telescopes, at the extreme limit of magnification, is
   A) the poor optical polish achievable on large mirrors.
   B) the clumsiness of the telescope operator.
   C) the poor tracking capabilities of modern telescopes.
   D) air turbulence in the Earth's atmosphere.
   Answer: D Source: Section 3-10

134. What does the word "seeing" mean to an astronomer using a telescope?
   A) The combined effects of aberrations in the telescope, particularly spherical and chromatic aberration.
   B) The twinkling and blurring of the image due to air currents in the Earth's atmosphere.
   C) The amount of haze or thin cloud in the atmosphere, that affects the brightness of the telescope image.
   D) The effect of wind, that can cause the telescope to vibrate, blurring the image.
   Answer: B Source: Section 3-10

135. If one observatory site is described as having better "seeing" than another observatory site, what is it that is better at the better site?
   A) The sky is more transparent (less haze) at that site, giving brighter images.
   B) There is less air turbulence at that site, causing less twinkling and blurring of the images.
   C) There are more clear nights at that site.
   D) The winds are lighter at the better site, reducing vibrations in the telescope.
   Answer: B Source: Section 3-10

136. It is difficult to improve the angular resolution of optical telescopes located on the surface of the Earth beyond a certain limit because
   A) we would need to build larger telescopes and this is very expensive.
   B) the star images are distorted more by air turbulence than by the telescope optics.
   C) spherical mirrors suffer from too much aberration.
   D) large telescopes are always reflecting telescopes and these suffer from too much chromatic aberration.
   Answer: B Source: Section 3-10

137. The sharpest images of stars that can be formed by a single, large, ground-based, visible-light telescope are of the order of what angular size?
   A) About $10^{-3}$ arc second.
   B) About 0.2 arc second.
   C) About 1 degree.
   D) About 1 arc minute.
   Answer: B Source: Section 3-10

138. The best Earth-based sites for modern large astronomical telescopes, in order to reduce the effects of bad seeing upon astronomical images, are
   A) on the tops of high mountains, above a large fraction of the disturbing atmosphere.
   B) near to large cities, where the warm air from human activity serves to stabilize the overlying atmosphere.
   C) at sea level, where the air is less turbulent.
   D) on the down-wind side of mountain ranges, where smooth airflow produces clear air and stable images.
   Answer: A Source: Section 3-10

139. What factor has seriously reduced the effectiveness of the 5-m Hale telescope on Mt. Palomar, formerly the largest telescope in the world, over the last few years?
   A) Bending of the mirror surface from repeated exposure to the cold night air.
   B) Cracking of the mirror by the Los Angeles earthquake.
   C) Tarnishing of the mirror surface by air pollution.
   D) Light pollution from nearby cities.
   Answer: D Source: Section 3-10

140. The Hubble Space Telescope, placed in orbit from the Space Shuttle in 1990, has a primary mirror of diameter
   A) 6 m. B) 1 m. C) 57 m. D) 2.4 m.
   Answer: D Source: Section 3-11

141. What is the angular resolution of the Hubble Space Telescope?
   A) 1 arc second B) 0.001 arc second C) 0.1 arc second D) 0.01 arc second
   Answer: C Source: Section 3-11

142. Over what range of wavelengths does the Hubble Space Telescope operate?
   A) Visible only
   B) Only visible and infrared
   C) Only visible and ultraviolet
   D) Infrared, visible and ultraviolet
   Answer: D Source: Section 3-11

143. After the Hubble Space Telescope was launched, it was found to suffer seriously from
   A) spherical aberration.
   B) too much angular resolution.
   C) jittery images caused by strong stratospheric winds.
   D) chromatic aberration.
   Answer: A Source: Section 3-11

144. What characteristic of the 10-m Keck telescope makes it fundamentally different from older telescopes such as the one on Mt.Palomar?
    A) It is actually four closely spaced telescopes that can be operated together as if they were a single telescope.
    B) Its primary mirror is a large, very thin piece of glass that can be deformed rapidly to compensate for seeing.
    C) It is located in Earth orbit, away from all the distortions of the Earth's atmosphere.
    D) Its primary mirror is made up of 36 separate mirror segments.
    Answer: D Source: Section 3-12

145. The primary mirror in the 10-m Keck telescope is made from
    A) six individual mirrors mounted so that they all bring the light to the same focus.
    B) one large hollow honeycomb of glass, lightweight but relatively thick to prevent sagging.
    C) 36 individual mirrors mounted so that they all bring the light to the same focus.
    D) one large, thin slab of glass whose surface is continuously adjusted by computer-controlled motors.
    Answer: C Source: Section 3-12

146. In a telescope that uses adaptive optics to correct for atmospheric distortion of images, or "seeing,"
    A) the light rays are focused electronically, without the use of lenses or mirrors.
    B) computer-controlled motors adjust the shape of either the primary mirror or a smaller mirror within the optics many times per second.
    C) computer-controlled motors rapidly adjust the orientation of the individual mirrors in a multiple-mirror telescope (MMT).
    D) a corrector lens compensates for image distortion by having its shape electronically controlled.
    Answer: B Source: Section 3-12

147. Why was adaptive optics developed?
    A) To compensate for spherical aberration.
    B) To prevent distortion of mirrors by the vacuum of space.
    C) To compensate for image distortion caused by the Earth's atmosphere.
    D) To prevent distortion by sagging in very thin, light-weight mirrors.
    Answer: C Source: Section 2-12

148. The technique thhat astronomers are using to increase the amount of detail that can be seen or photographed through telescopes is
   A) increased size, where mirrors are made far larger than was possible before (e.g., 8, 10, or even more meters in diameter).
   B) multiple mirror telescopes, where several mirrors are mounted together in a single telescope to simulate the performance of a single, very large mirror.
   C) anti-reflective coatings, where the mirror is coated with a substance such as fluorite to reduce the amount of reflected light.
   D) adaptive optics, where the tilt and shape of mirrors in the telescope are changed many times per second to compensate for atmospheric turbulence.
   Answer: D Source: Section 3-12

149. What fraction of the light falling on a piece of photographic film is typically wasted (does NOT contribute to the formation of the image)?
   A) 18% B) 45% C) 98% D) 2%
   Answer: C Source: Section 3-13

150. What is a CCD (charge-coupled device)?
   A) A detector in which a small electric current is controlled by a bimetallic strip that expands and contracts in response to infrared radiation.
   B) An array of small light-sensitive cells that can be used in place of photographic film to obtain a picture.
   C) An array of electrodes used on spacecraft to detect charged particles such as electrons and protons.
   D) An electronic filter to single out one wavelength or set of wavelengths for studying astronomical objects.
   Answer: B Source: Section 3-13

151. The charged coupled device (CCD), now used extensively for astronomical imaging, works on what principle?
   A) Light from the image modifies the plastic material on a disk, that can then be read on a standard video compact disk (CD) player.
   B) Light from the image is detected by new, high sensitivity, fine grain, automatically processed film.
   C) Light generates electrical charge on a computer-readable multi-element array of detectors.
   D) A single optical detector generates an electrical signal as it is scanned quickly across the astronomical image.
   Answer: C Source: Section 3-13

152. The detector that in many instances has replaced the photographic plate for astronomical photography is
   A) the PMT (photomultiplier tube).
   B) the diffraction grating
   C) the CCD (charge-coupled device).
   D) the interferometer.
   Answer: C Source: Section 3-13

153. The fraction of incoming photons recorded by a charge-coupled device (CCD) is greater than that recorded by a typical photographic plate by what factor?
   A) About ten times
   B) More than 100 times
   C) About twice
   D) About forty times
   Answer: D Source: Section 3-13

154. Who first discovered that radio radiation can be received from astronomical objects?
   A) Karl Jansky B) Thomas Young C) Edwin Hubble D) Isaac Newton
   Answer: A Source: Section 3-14

155. The first person to detect and identify radio emissions coming from deep space was
   A) Karl Jansky B) Arno Penzias C) Thomas Edison D) Galileo
   Answer: A Source: Section 3-14

156. What was the first extraterrestrial radio source to be detected?
   A) Jupiter
   B) The Sun
   C) The center of our Galaxy
   D) Cassiopeia A (a supernova remnant)
   Answer: C Source: Section 3-14

157. The first astronomical radio source, identified by Karl Jansky during measurements of sources of radio noise, was
   A) the Orion Nebula.
   B) Jupiter.
   C) the Crab Nebula.
   D) the center of our Galaxy.
   Answer: D Source: Section 3-14

158. The first radio energy to be detected by man from outer space came from the direction of
   A) the center of our Milky Way Galaxy, near Sagittarius.
   B) the Sun.
   C) the quasar 3C 273.
   D) the Moon.
   Answer: A Source: Section 3-14

159. One major difference between radio waves and light is that
 A) radio waves are electromagnetic whereas light waves are not.
 B) radio waves have shorter wavelength than light.
 C) radio waves have lower frequencies than light waves.
 D) light waves are electromagnetic whereas radio waves are not.
 Answer: C Source: Section 3-14

160. Which of the following non-visible radiations from outer space was the first to be detected?
 A) gamma rays B) radio waves C) ultraviolet (UV) light D) x-rays
 Answer: B Source: Section 3-15

161. The first astronomical radio source was detected and identified in the year
 A) 1967. B) 1945. C) 1897. D) 1932.
 Answer: D Source: Section 3-15

162. How does angular resolution for a given diameter of telescope depend upon wavelength?
 A) Angular resolution worsens as wavelength increases.
 B) Angular resolution improves as wavelength increases.
 C) Angular resolution may improve or worsen as wavelength increases, depending on other factors such as intensity and spectral range (e.g., optical, infrared, radio).
 D) Angular resolution depends only on the diameter of the telescope and is independent of wavelength.
 Answer: A Source: Section 3-14

163. In telescopes, the angular resolution is worse for
 A) smaller diameter lenses or mirrors and shorter wavelength light (of other electromagnetic radiation).
 B) smaller diameter lenses or mirrors and longer wavelength light (or other electromagnetic radiation).
 C) larger diameter lenses or mirrors and shorter wavelength light (or other electromagnetic radiation).
 D) larger diameter lenses or mirrors and longer wavelength light (of other electromagnetic radiation).
 Answer: B Source: Section 3-14

164. In single-telescope astronomical systems, either optical or radio,
 A) the larger the main mirror aperture in general, the sharper the image.
 B) the smaller the main-mirror aperture in general, the sharper the image.
 C) the longer the wavelength, the sharper the image.
 D) the longer the focal length of the primary mirror, the sharper the image.
 Answer: A Source: Section 3-14

165. Why is the angular resolution of a 20-m diameter radio telescope worse than that of a 0.5-m diameter optical telescope?
 A) Optical mirrors suffer from chromatic aberration.
 B) Angular resolution gets worse as wavelength increases.
 C) Angular resolution gets worse as wavelength decreases.
 D) Angular resolution gets worse as mirror size increases.
 Answer: B Source: Section 3-14

166. How does the angular resolution of the 305-m Arecibo radio telescope in Puerto Rico compare with the angular resolution of the 10-m Keck optical telescope on Mauna Kea, Hawaii?
 A) The 305-m telescope has much better angular resolution than the 10-m Keck telescope.
 B) They have about the same angular resolution, since angular resolution is limited by turbulence in the Earth's atmosphere, not by mirror diameter.
 C) It is not possible to compare their angular resolution, since they work in different wavelength ranges.
 D) The 305-m telescope has much worse angular resolution than the 10-m Keck telescope.
 Answer: D Source: Section 3-14

167. How is interferometry used in radio astronomy?
 A) Signals from two or more different radio telescopes are combined to produce a single image of greater angular resolution than from either telescope alone.
 B) Signals from different places around the dish of a radio telescope are combined to produce interference and increase the angular resolution of the telescope.
 C) Signals from a radio source are combined with a time delay that causes interference and increases the angular resolution of the telescope.
 D) Signals from different parts of a radio source are combined to increase the angular resolution of the telescope.
 Answer: A Source: Section 3-14

168. How does the angular resolution (sharpness) of the images obtained from radio telescopes using very long baseline interferometry (VLBI) compare to the typical angular resolution of optical telescopes?
A) VLBI images have about the same resolution as the best optical images, about 0.2 arc seconds.
B) VLBI images are very much less sharp because of the longer wavelength.
C) VLBI images are about 10 times less sharp than the best optical images.
D) VLBI images are much sharper.
Answer: D Source: Section 3-14

169. The PRIMARY reason for spreading many radio telescopes across a large area and combining the signals at a central station is
A) to produce much sharper images of radio sources.
B) to avoid interference between signals from the separate telescopes.
C) to collect more radiation than would be possible with the same telescopes clustered together.
D) to ensure that cloudy weather only affects a few of the telescopes, leaving the others to continue observing.
Answer: A Source: Section 3-14

170. What is the main reason for using several radio telescopes together as an interferometer?
A) To ensure that observations are uninterrupted by the failure of one or two telescopes.
B) To obtain much better angular resolution (sharpness) of the images.
C) To ensure that at least one of the telescopes is in a radio interference-free zone.
D) To collect more radiation from very faint sources.
Answer: B Source: Section 3-14

171. What is the maximum angular resolution (sharpness of the image) obtainable with radio telescopes on the Earth?
A) 0.1 arc second using the 27-dish Very Large Array of radio telescopes at Socorro, New Mexico.
B) 100 arc seconds using the 305-m Arecibo telescope in Puerto Rico.
C) 0.0001 arc second using the Very Long Baseline Array.
D) 0.00001 arc second using VLBI.
Answer: D Source: Section 3-14

172. Which substance in the Earth's atmosphere is the main absorber of infrared radiation from space?
A) Ozone B) Methane C) Water vapor D) Carbon dioxide
Answer: C Source: Section 3-15

173. The main absorber in the atmosphere for infrared radiation, which impedes observations of astronomical infrared objects, is
   A) electrons and ionized atoms in the Earth's ionosphere.
   B) dust.
   C) $N_2$ and $O_2$, the major constituents of the atmosphere.
   D) water vapor, $H_2O$.
   Answer: D Source: Section 3-15

174. The main reason for carrying a telescope and scientific equipment in an aircraft in order to carry out infrared astronomy is
   A) to obtain longer observing times on specific sources by moving in the direction of the Earth's rotation.
   B) to avoid stray IR radiation from the warm Earth and its occupants.
   C) to obtain photographs of higher resolution than can be obtained on the ground.
   D) to avoid the absorption of the IR radiation by water vapor.
   Answer: D Source: Section 3-15

175. A high mountaintop such as Mauna Kea in Hawaii is a good site for an infrared observatory because it is
   A) closer to the stratospheric ozone layer.
   B) above most of the moisture in the Earth's atmosphere.
   C) above most of the turbulence in the Earth's atmosphere.
   D) further from city lights.
   Answer: B Source: Section 3-15

176. Telescopes are placed in space to view distant galaxies primarily
   A) to avoid having to steer the telescope against the Earth's motion.
   B) to avoid the absorption of the light or other radiations in the atmosphere of Earth.
   C) to avoid the light pollution from the Earth's populated areas.
   D) to get closer to the observed objects.
   Answer: B Source: Section 3-15

177. Astronomy from space vehicles is particularly useful because
   A) the telescope is in a gravity-free state, the mirror is not distorted by gravitational stress, and can produce sharper images.
   B) the telescope is above the Earth's absorbing and distorting atmosphere and can measure radiation over a very wide wavelength range.
   C) the telescope moves smoothly in a constant orbit, and can produce sharp photographs.
   D) the telescope is in a clean, dust-free environment and scattered light is much reduced.
   Answer: B Source: Section 3-15

178. What is the main reason for placing astronomical telescopes and detectors on satellites?
A) To get closer to the objects being viewed.
B) To avoid light pollution from cities and other built-up areas.
C) To get above the absorption in the Earth's atmosphere.
D) To avoid dust and haze in the Earth's atmosphere.
Answer: C Source: Section 3-15

179. Which was the first satellite to be dedicated to infrared astronomy?
A) IRAS B) IUE C) EUVE D) The Hubble Space Telescope
Answer: A Source: Section 3-15

180. The IRAS satellite, placed in orbit in 1983 to survey the whole sky, detected which kind of electromagnetic radiation?
A) Infrared B) X-rays C) UV radiation D) Visible light
Answer: A Source: Section 3-15

181. Which was the first satellite to be dedicated to ultraviolet astronomy?
A) IRAS B) The Hubble Space Telescope C) IUE D) EUVE
Answer: C Source: Section 3-15

182. In which wavelength range does the astronomical satellite known as the Compton Observatory observe?
A) Uultraviolet (UV) B) Gamma rays C) X-rays D) Infrared (IR)
Answer: B Source: Section 3-15

# Chapter 4: The Origin and Nature of Light

1. When a rod of metal is heated intensely, its predominant color will
   A) change from blue through white, then orange and finally red, when it becomes red-hot at its hottest.
   B) be white, all colors mixed together, as the intensity of light increases.
   C) remain predominantly red, as the intensity of light increases.
   D) change from red, through orange to white, and then to blue.
   Answer: D Source: Section 4-1

2. Pieces of metal are heated by varying amounts in a flame. The hottest of these will be the one that shows which color most prominently?
   A) Blue B) Yellow C) Red D) Black
   Answer: A Source: Section 4-1

3. What changes would you expect to see in the resulting spectrum of emitted light from a piece of metal when it is heated slowly in an intense flame from 500 K to 1500 K?
   A) The intensity of radiation would increase only slightly, while the color would change from red through white to blue.
   B) The intensity of radiation would increase greatly, while its color would remain a dull red.
   C) The intensity of radiation would increase greatly and its color would change from red through white to blue.
   D) The intensity of radiation would increase greatly, while the color would change from blue through white to red.
   Answer: C Source: Section 4-1

4. The temperature scale most often used by scientists is the
   A) Kelvin scale. B) Fahrenheit scale. C) Celsius scale. D) Richter scale.
   Answer: A Source: Back of Book

5. The Kelvin scale measures
   A) mass per unit volume, or density, with water having a value of 1.0.
   B) temperature in Celsius-sized degrees above absolute zero.
   C) temperature referenced to zero at the freezing point of water.
   D) temperature in Fahrenheit-sized degrees above absolute zero.
   Answer: B Source: Back of Book

6. What is the main reason that astronomers (and other scientists) almost always use the Kelvin (absolute) temperature scale, rather than the Celsius or Fahrenheit scales?
   A) The size of each degree (or unit) of temperature is more convenient.
   B) The scale has a physically meaningful absolute zero of temperature.
   C) The temperature of freezing and boiling water are easier to remember.
   D) Calculations are easier on the Kelvin scale.
   Answer: B Source: Back of Book

7. A typical but very cool star might have a temperature of 3100 degrees Celsius. On the Kelvin scale, this is about
   A) the same number 3100 K, since Kelvin and Celsius scales are the same
   B) 3373 K.
   C) 3068 K, since 0 K, the freezing point of water, is 32°C.
   D) 2827 K.
   Answer: B Source: Back of Book

8. A scientist reports that his measurement of the temperature of the surface of a newly discovered planet is -20 K. What conclusion can you draw from this report?
   A) The planet is a very long way from the Sun.
   B) The planet has no atmosphere.
   C) The result is erroneous since one cannot have negative absolute temperature.
   D) The scientist measured only the dark side of the planet, away from the Sun.
   Answer: C Source: Back of Book

9. The normal temperature of the melting point of water ice is
   A) 0 K B) 293 K C) 100 K D) 273 K
   Answer: D Source: Back of Book

10. On the absolute Kelvin temperature scale, the temperature of freezing water is about
    A) +373 K. B) -273 K. C) +273 K. D) 0 K.
    Answer: C Source: Back of Book

11. The temperature of boiling water at ordinary pressures on the absolute scale, Kelvin, is
    A) 100 K. B) 273 K. C) 373 K. D) 212 K.
    Answer: C Source: Back of Book

12. The range of temperatures in the Kelvin (absolute) scale between the freezing point and boiling point of water is
    A) 100 degrees. B) 10 degrees. C) 273 degrees. D) 212 degrees.
    Answer: A Source: Back of Book

13. A scientist measures the temperature change between freezing water and boiling water with a thermometer calibrated in the Kelvin or absolute scale. How many degrees Kelvin (K) will he measure?
A) 100 B) 180 C) 373 D) 273
Answer: A Source: Back of Book

14. The temperature at the top of the clouds on Jupiter is about 165 K. In degrees Celsius, this is
A) -165°C. B) -108°C. C) 0°C. D) 438°C.
Answer: B Source: Back of Book

15. The minimum temperature reached on the surface of Mars, -140°C, is represented on the absolute (Kelvin) temperature scale as
A) -133 K. B) 133 K. C) 140 K. D) 153 K.
Answer: B Source: Back of Book

16. To a physicist, a blackbody is defined as an object which
A) absorbs all radiation which falls upon it.
B) always appears to be black, whatever its temperature.
C) always emits the same spectrum of light, whatever its temperature.
D) reflects all radiation which falls upon it, never heating up and always appearing black.
Answer: A Source: Section 4-1

17. An ideal "blackbody" in physics and astronomy is an object which
A) emits only infrared light, and hence appears black to the eye.
B) does not emit or absorb any electromagnetic radiation.
C) absorbs all electromagnetic radiation, but emits none.
D) absorbs and emits electromagnetic radiation at all wavelengths.
Answer: D Source: Section 4-1

18. A blackbody is an idealized object in physics and astronomy which
A) reflects no light, and emits light in a manner determined by its temperature.
B) does not emit or reflect any radiation.
C) reflects and emits radiation in a manner which is completely determined by its temperature.
D) reflects and emits light with the same intensity at all wavelengths.
Answer: A Source: Section 4-1

19. A perfect blackbody is so named because
    A) it reflects all radiation falling upon it but emits none of its own.
    B) it always emits the same amount and color of radiation irrespective of its temperature.
    C) it absorbs all radiation falling upon it and reflects none.
    D) it never emits radiation.
    Answer: C Source: Section 4-1

20. Which of the following situations will most closely exhibit the characteristics of a perfect blackbody, as defined by physicists?
    A) A small hole in a hollow steel sphere.
    B) The polished surface of a block of gold.
    C) A sheet of steel, painted black.
    D) The surface of a piece of transparent glass.
    Answer: A Source: Section 4-1

21. Two physicists, one on the East Coast and one on the West Coast of North America, discover that they have constructed ideal blackbodies in their laboratories. The two blackbodies are made from very different materials. Without conducting tests, they know that the radiation emitted by these two objects will be
    A) different from each other because they were made from different materials.
    B) different, because the amount of light falling on them is likely to be different in the two laboratories.
    C) identical if the blackbodies are at the same temperature, but not otherwise.
    D) identical to each other if the blackbodies have the same size, even if their temperatures are different.
    Answer: C Source: Section 4-1

22. The hot, dense gas existing on the Sun
    A) emits energy only at certain wavelengths, line emission from hydrogen gas, and not at other wavelengths.
    B) emits energy at all wavelengths, with a peak at one particular wavelength (color).
    C) emits energy with the same intensity at all wavelengths, the Earth's atmosphere absorbing radiation at both short and long visible wavelengths to produce the observed spectrum.
    D) emits energy mostly at the longest and shortest wavelengths, with a minimum in between.
    Answer: B Source: Section 4-2

23. The star Vega has a higher surface temperature than the Sun. If so, then, with IR referring to infrared and UV referring to ultraviolet,
    A) Vega emits less IR and less UV flux than the Sun.
    B) Vega emits less IR and more UV flux than the Sun.
    C) Vega emits more IR and more UV flux than the Sun.
    D) Vega emits more IR and less UV flux than the Sun.
    Answer: C Source: Section 4-2 and Figure 4-2

24. Which of the following statements is true, if the star Betelgeuse has a lower surface temperature than the Sun? (Assume that IR means infrared radiation and UV means ultraviolet radiation.)
    A) Betelgeuse emits more IR and more UV flux than the Sun.
    B) Betelgeuse emits less IR and more UV flux than the Sun.
    C) Betelgeuse emits less IR and less UV flux than the Sun.
    D) Betelgeuse emits more IR and less UV flux than the Sun.
    Answer: C Source: Section 4-2 and Figure 4-2

25. If all stars are considered to be perfect blackbodies, then the following relationship should hold regarding the energy flux, or energy emitted per unit area, from stars.
    A) All stars of the same temperature should emit the same energy flux.
    B) All stars of the same composition (made of exactly the same material) should emit the same energy flux.
    C) All stars of the same size should emit the same energy flux.
    D) All stars of the same mass should emit the same energy flux.
    Answer: A Source: Section 4-2

26. The Stefan-Boltzmann law, which relates the energy per unit area $F$ emitted by an object to its temperature $T$, $F = sT^4$, is obeyed by what kind of object?
    A) A blackbody, a perfect absorber and emitter of energy at all wavelengths.
    B) A red-colored object, which absorbs blue light but reflects red light.
    C) Only hot gases, whose atoms emit and absorb only specific colors (e.g., neon tubes).
    D) All objects, whatever their color or reflective properties.
    Answer: A Source: Section 4-2 and Toolbox 4-1

27. The energy flux, $F$, from a star is
    A) the amount of energy emitted by each square meter of the star's surface each second.
    B) the amount of incoming light reflected by the star each second.
    C) the total energy emitted by the star over its lifetime.
    D) the amount of energy emitted by the entire star each second.
    Answer: A Source: Toolbox 4-1

28. Which of the following equations represents the dependence of the total energy flux, $F$, of radiation emitted per unit area by a "blackbody" (e.g., star) upon its temperature $T$, with s a constant?
A) $F = s/T$ B) $F = sT^4$ C) $FT^4 = s$ D) $F^4 = sT$
Answer: B Source: Toolbox 4-1

29. When a solid body is heated to a temperature $T$, the total energy flux from this body per second per unit area $F$ at all wavelengths, is given by (where s is a constant)
A) $F = sT$ B) $F = s/T^2$ C) $FT = s$ D) $F = sT^4$
Answer: D Source: Toolbox 4-1

30. By what factor does the total energy emitted per unit time at all wavelengths from an object increase if its temperature increases by a factor of 3 (e.g., from room temperature to 900 K)?
A) 27 B) 81 C) 9 D) 3
Answer: B Source: Section 4-2 and Toolbox 4-1

31. A piece of iron is heated from 400 to 800 K (127 to 527°C). The total energy emitted per second by this iron will increase by a factor of
A) 16 B) 296.5 C) 4 D) 2
Answer: A Source: Section 4-2 and Toolbox 4-1

32. The temperature of the surface of the Sun is 5800 K. What would be the surface temperature of a star which emits twice the energy flux (watts per square meter) that the Sun emits?
A) 8200 K. B) 11,600 K. C) 4880 K. D) 6900 K.
Answer: D Source: Section 4-2 and Toolbox 4-1

33. An example of an object which emits no radiation at all is
A) any object at the temperature of outer space.
B) any blackbody at any temperature.
C) any object made of ice, at 0°C.
D) any object at a temperature of 0 K.
Answer: D Source: Section 4-2 and Toolbox 4-1

34. Wien's law, relating the peak wavelength $l_{max}$ of light emitted by a dense object to its temperature $T$, can be represented by
A) $l_{max} = \text{constant}/T^2$
B) $l_{max} = \text{constant} \times T^4$
C) $l_{max}/T = \text{constant}$
D) $l_{max}T = \text{constant}$
Answer: D Source: Toolbox 4-1

35. The laws governing the energy flux $F$ and wavelength of maximum intensity $l_{max}$ of emitted radiation from a hot, dense body whose temperature is $T$ are given by (where s and $a$ are constants)
A) $F = sT^4$, $l_{max}T = a$
B) $F = sT^2$, $l_{max}T = a$
C) $F = sT^4$, $l_{max} = aT$
D) $F = sT$, $l_{max} = a/T^4$
Answer: A Source: Toolbox 4-1

36. Cepheid variable stars pulsate regularly in size. During the contraction part of the cycle, when the star's temperature is increasing, the peak wavelength of the emitted radiation will
A) shift towards longer or shorter wavelengths at random as the temperature changes.
B) remain unchanged.
C) shift from the visible to the UV part of the spectrum.
D) shift from the UV to the visible part of the spectrum.
Answer: C Source: Section 4-2

37. As a newly formed star continues to contract, its temperature increases while the chemical nature of the gas does not change. What will happen to the peak wavelength of its emitted radiation?
A) It will not change, since it is not dependent upon temperature.
B) It will move toward shorter wavelengths (e.g., IR to visible).
C) It will move toward longer wavelengths (e.g., visible to IR).
D) It will remain constant, since the chemical state of the gas does not change.
Answer: B Source: Section 4-2

38. As a new star evolves from cool dust and gas to a hot star, the peak wavelength of its spectrum of emitted electromagnetic radiation will
A) increase from the visible to infrared wavelengths.
B) change from the infrared to the visible wavelengths.
C) change from the ultraviolet to the visible range.
D) remain the same.
Answer: B Source: Section 4-2

39. When a solid body (or a dense gas such as a star) is cooled from a temperature of several thousand degrees, the "color" or wavelength of maximum emission of radiation will
    A) move initially towards the red end of the spectrum, then move back towards the blue end of the spectrum as the intensity of radiation fades and eventually becomes invisible to the eye.
    B) move steadily towards the red end of the spectrum.
    C) remain constant, being dependent only upon the original color of the body.
    D) move steadily towards the blue end of the spectrum.
    Answer: B Source: Section 4-2

40. The average temperature of Mars is less than that of Earth. What would be the relative wavelengths of the peak emissions of the infra-red spectra from these planets?
    A) The wavelength of peak emission from Mars will be at a longer wavelength than that from the Earth.
    B) The emission from the two planets will peak at the same wavelength but the radiation from Mars will be less intense than that from the Earth.
    C) It is not possible to predict the outcome of this experiment from the information given.
    D) The wavelength of peak emission from the Earth will be at the longer wavelength than that from Mars.
    Answer: A Source: Section 4-2

41. What will be the peak wavelength of electromagnetic radiation emitted by a piece of iron which is just melting at a temperature of 1808 K? (Use Wien's law, which relates the peak wavelength $l_{max}$ emitted by a body to its temperature *T*, and see Astronomer's Toolbox 4-1, and Fig. 4-2, Kaufmann & Comins, *Discovering the Universe*, 5th Ed.) (1 mm = $10^{-6}$m).
    A) 1.04 mm, very near infrared.
    B) 1.89 mm, near infrared.
    C) 1.6 mm, near infrared.
    D) 16 mm, intermediate infrared.
    Answer: C Source: Toolbox 4-1

42. A star whose surface temperature is 100,000 K will emit a spectrum whose peak wavelength is (see Astronomer's Toolbox 4-1 and Table 4-1, Kaufmann & Comins, *Discovering the Universe*, 5th Ed. and be careful with wavelength units)
    A) at infrared wavelengths.
    B) at visible wavelengths.
    C) at x-ray wavelengths.
    D) at ultraviolet wavelengths.
    Answer: D Source: Toolbox 4-1

43. If the human eye has evolved over time so that its peak wavelength sensitivity is about 0.5 mm (1 mm = $10^{-6}$ cm), what will be the temperature of a blackbody to which the eye will be most sensitive? (Use Wien's Law, Astronomer's Toolbox 4-1, Kaufmann & Comins, *Discovering the Universe*, 5th Ed.)
A) 5,800 K. B) 580 K. C) 14,240 K. D) 0.58 K.
Answer: A Source: Toolbox 4-1

44. You are asked to design an infra-red system to detect human beings (at a normal temperature of 310K) in darkness. What would the wavelength of peak sensitivity of your equipment need to be? (1 mm = $10^{-6}$ m). (Hint: Use Wien's law).
A) 90 mm. B) 0.00094 mm or 0.94 nm. C) 9.35 mm. D) 0.935 mm.
Answer: C Source: Toolbox 4-1

45. What is the approximate peak wavelength of radiation emitted by (live) human beings, who are (normally) at a temperature of about 310 K? (Hint: Use Wien's law, Kaufmann & Comins, *Discovering the Universe*, 5th Ed., Chapter 4, 1 mm = $10^{-6}$ m)
A) 3.1 mm. B) 9.4 mm. C) 0.94 mm. D) 94 mm.
Answer: B Source: Toolbox 4-1

46. Huge fluxes of x-rays are detected from the direction of Cygnus X-1 (see Figs. 13-5 and 13-6, Kaufmann & Comins, *Discovering the Universe*, 5th Ed.) with a spectrum which looks similar to that of a blackbody with a peak wavelength of 1.45 nm (1 nm = $10^{-9}$ m). These x-rays are probably emitted by matter being heated as it falls into a black hole. What is the temperature of this gas?
A) 4,205 K. B) $2 \times 10^4$ K. C) $2 \times 10^{-2}$ K. D) $2 \times 10^6$ K.
Answer: D Source: Toolbox 4-1

47. A small particle of interplanetary material is heated by friction from a temperature of 400K to 4000K as it falls into the atmosphere of the Earth and produces a meteor or a shooting star in our sky. If this object behaves like a perfect black body over this short time, how will its emitted radiation change as it is heated?
A) Its intensity will rise by a factor of 100 while the peak wavelength of emitted light will become shorter by a factor of 100, moving from the infra-red to the ultra-violet.
B) Its total emitted intensity will rise by a factor of 10,000 while its peak wavelength will become shorter by a factor of 10, from infra-red to red visible light.
C) Its total intensity will rise by a factor of 10, while its peak wavelength will become shorter by a factor of 10, moving from the infra-red to red visible light.
D) Its total intensity will rise by a factor of 10,000 while its peak wavelength will become longer by a factor of 10, moving from the visible to the infra-red or heat radiation.
Answer: B Source: Section 4-2 and Toolbox 4-1

48. In the revolution which overtook physics around 1900, the assumption which Planck made in order to solve the problem concerning the spectrum of radiation emitted by a hot blackbody was
    A) that radiation was emitted as continuous waves whose wavelength was inversely proportional to the temperature of the object.
    B) that radiation was made up of small, discrete packets or quanta of energy whose individual energies were all the same, independent of wavelength.
    C) that radiation from hot objects was emitted in small, discrete packets or quanta of energy, each quantum having an energy directly proportional to the wavelength of the light.
    D) that all radiation was emitted in small, discrete packets or quanta of energy whose individual energies were inversely proportional to the wavelength of the light.
    Answer: D Source: Section 4-2

49. The important breakthrough in theoretical physics which was first suggested by Planck to explain the shape of the spectrum of a hot body was
    A) the discovery of the formula $F = sT^4$, which can be used to calculate the total energy flux emitted by the hot body over all wavelengths.
    B) the idea that light is a form of electromagnetic energy transmitted at a constant speed and that there is a continuous spectrum of such waves from gamma rays to radio waves.
    C) the idea that light traveled at a constant speed, whatever the speed of the source.
    D) the concept that electromagnetic energy was emitted in small packets or quanta.
    Answer: D Source: Section 4-2

50. The light from a small amount of a particular chemical element, when heated in a flame, is found to consist of
    A) a pattern of narrow, bright emissions at specific wavelengths which are the same for all elements, only the relative intensities of each line differing for different elements.
    B) a continuous spectrum of light whose peak wavelength is specific to the particular element.
    C) a pattern of narrow, bright emissions at wavelengths which are specific to the element and different for each element.
    D) a continuous spectrum of light from which certain colors are missing or absorbed, the absorbed colors being different for different elements.
    Answer: C Source: Introduction to Section 4-3

51. Which very significant fact was discovered in the 1800's concerning the spectra produced by hot gases, such as elements heated on the solar surface (Fraunhofer, with the solar spectrum) or in a flame (Bunsen and Kirchhoff, with laboratory spectra)?
    A) Each chemical element produces its own characteristic pattern of spectral lines which remains fixed as the temperature increases.
    B) Chemical elements emit spectral lines which move continuously toward the blue end of the spectrum as the gas temperature increases.
    C) All chemical elements produce the same set of spectral emission or absorption lines but their relative emission intensities differ, hence elements are distinguishable from each other by their spectra.
    D) The higher the temperature, the greater the red shift of the emitted spectral lines.
    Answer: A Source: Introduction to Section 4-3

52. The chemical makeup of a star's surface is usually inferred
    A) by spectroscopy of the light emitted by the star.
    B) by taking a sample of that surface with a space probe.
    C) by theoretical methods, considering evolution of the star.
    D) by measuring the chemical elements present in the solar wind.
    Answer: A Source: Section 4-3

53. Where was the element helium first discovered, and by what technique? (See Chapter 4, Kaufmann & Comins, *Discovering the Universe*, 5th Ed.)
    A) on the Sun, from spectroscopy during a solar eclipse.
    B) in the laboratory, by examining the spectrum of heated chemicals in a flame.
    C) in radioactive rocks, by radioactive measurements of uranium deposits.
    D) in the upper atmosphere of Earth by studying the spectrum of the aurora, or northern lights.
    Answer: A Source: Section 4-3

54. The element Helium was first discovered and identified as a separate element
    A) in rocks containing radioactively-decaying elements such as Uranium.
    B) in natural gas originating underground, from the spectrum emitted from a flame of burning natural gas.
    C) inside meteorites which have come from outer space.
    D) on the Sun, from the emitted spectrum from its upper atmosphere.
    Answer: D Source: Section 4-3

55. Which of the following elements in the periodic table, shown in Figure 4-6 of Kaufmann and Comins, *Discovering the Universe*, 5th Ed., will have chemical properties which are MOST SIMILAR to those of nitrogen (N, atomic number 7)?
A) oxygen (O, atomic number 8)
B) carbon (C, atomic number 6)
C) phosphorus (P, atomic number 15)
D) chlorine (Cl, atomic number 17)
Answer: C Source: Section 4-3

56. A spectrograph is a scientific instrument which
A) spreads out light from a source into its component colors or spectrum.
B) measures the transparency of different types of glass at different colors by using glass prisms.
C) measures the effectiveness of the specular reflection of mirrors.
D) focusses all the colors of light from a star at one point in order that the total intensity of the star can be measured, for photometry.
Answer: A Source: Section 4-3

57. When light passes through a prism of glass
A) the different colors or wavelengths of light are separated in angle by the prism.
B) the prism adds colors to different parts of the broadly scattered beam coming out of it.
C) the prism absorbs colors from different parts of the broad beam coming out of the prism, leaving the complementary colors which we see.
D) the different colors are caused by multiple reflections in the prism and interference between the resulting beams.
Answer: A Source: Section 4-3

58. Which of the following items is considered to be the best one to use in a spectrograph, to disperse light into its spectrum of colors?
A) A CCD array.
B) A diffraction grating.
C) A two-element achromatic lens.
D) A glass prism.
Answer: B Source: Section 4-3

59. Which of the following points is NOT a characteristic of a prism compared to a diffraction grating, when used in a spectrograph?
A) A prism absorbs light unevenly over the spectrum (over the different colors).
B) A prism does not pass ultraviolet light.
C) A prism mixes the order of colors from the standard red-orange-yellow-green-blue-violet.
D) A prism spreads the light out unevenly in wavelength.
Answer: C Source: Section 4-3

60. What evidence do we have that the Sun contains the element iron?
    A) The peak wavelength of the continuum spectrum of sunlight is characteristic of the emission spectrum of iron, as seen when a piece of iron is heated in the laboratory.
    B) Solar spectra show absorption in spectral lines which are characteristic of iron and are unique to it.
    C) Magnetic fields exist in sunspots and on the Sun and these must be produced by iron in the same way that the Earth's magnetic field is generated.
    D) Scientists have collected meteorites which are almost pure iron which have originated in the Sun.

    Answer: B Source: Section 4-3

61. When astronomers look for evidence of hydrogen gas in the spectra of the Sun, the planets or the nearby stars, the positions of the spectral features or "lines" due to hydrogen
    A) will be in a pattern which depends upon the location of the planet or star and which can only be reproduced with difficulty in the laboratory.
    B) will be in a pattern in which the positions of the lines depends upon the temperature of the source.
    C) will always be in the same characteristic pattern as seen in the laboratory, a pattern unique to hydrogen.
    D) will be in the same pattern for the Sun and for planetary sources but very different for stars at larger distances because of absorption of light by the interstellar matter.

    Answer: C Source: Section 4-3

62. The spectrum of sunlight, when spread out by a spectrograph, has what characteristic appearance?
    A) A series of separate emission lines, characteristic of many elements, which overlap in certain regions of the spectrum to produce short sections of continuous color.
    B) A continuous band of color, crossed by innumerable emission lines.
    C) A continuous band of color, crossed by innumerable dark absorption lines.
    D) A continuous and uniform band of color from violet to deep red.

    Answer: C Source: Section 4-3

63. The dark absorption lines in the solar spectrum are caused
    A) by a hotter layer of gas which overlies the cooler solar surface, and which produces the absorption lines.
    B) by a cooler layer of gas overlying the hot solar surface, which contains many elements including H, He, Mg, Ca, Fe, etc.
    C) by a cooler layer overlying the hot solar surface, consisting solely of hydrogen gas which produces all the absorption lines.
    D) solely by absorption by atoms and molecules in the Earth's cool atmosphere.

    Answer: B Source: Section 4-3

64. Atoms in a thin, hot gas (such as a neon advertising sign), according to Kirchhoff's laws, emit light
   A) at specific wavelengths or colors, the pattern depending upon the element.
   B) at all wavelengths, the shape of the continuum spectrum depending upon the temperature of the gas.
   C) only at one specific, single wavelength or color.
   D) only at visible wavelengths.
   Answer: A Source: Section 4-4

65. An astronomer finds a source of light in space which emits light only in specific, narrow emission lines. Kirchhoff's laws lead him to which conclusion?
   A) The source is made up of a hot, dense gas.
   B) The source is made up of a hot, low-density gas.
   C) The source cannot consist of gases but must be a solid object.
   D) The source is made up of a hot, dense gas surrounded by a rarefied gas.
   Answer: B Source: Section 4-4

66. The gas in the interstellar space between stars is very tenuous ("thin") but can be heated to a very high temperature in the vicinity of a hot star. This hot, tenuous gas will emit
   A) light at certain wavelengths only ("spectral lines"), the wavelength of the lines being independent of gas temperature.
   B) light at certain wavelengths only ("spectral lines"), the wavelength of a given spectral line depending upon the gas temperature.
   C) no light at any wavelength, since hot thin gases do not emit light.
   D) light at all wavelengths, the continuous spectrum peaking at a certain wavelength or color, dependent upon temperature.
   Answer: A Source: Section 4-4

67. Atoms in a low-density, hot gas (e.g., in a fluorescent lamp or a neon tube) emit a spectrum which is
   A) continuous over all visible wavelengths, with maximum intensity in the red.
   B) a series of emissions at certain wavelengths which are the same for all atoms but which vary with temperature.
   C) a series of specific colors, unique to the type of atom in the tube, but fixed in position when gas temperature changes.
   D) a series of emissions at specific colors, whose wavelengths change in position as the gas temperature is changed.
   Answer: C Source: Section 4-4

68. If light from a hot, dense star passes through a cool cloud of gas (see Fig. 4-11, p. 89, Kaufmann & Comins, *Discovering the Universe*, 5th Ed.),
    A) only specific wavelengths of light will be removed from the spectrum.
    B) the atoms of the gas cloud will add energy to the overall spectrum, enhancing it at specific wavelengths to produce emission lines.
    C) the cool gas will not affect the spectrum of the star, since cool atoms cannot absorb light.
    D) the whole spectrum will be reduced in intensity.
    Answer: A Source: Section 4-4

69. According to Kirchhoff's Laws, the continuous spectrum of light from a hot star, after passing through a cool gas cloud,
    A) will contain additional emission lines from energy emitted by the atoms of the cool gas.
    B) will be enhanced at infrared wavelengths by a continuous spectrum emitted by the cool gas.
    C) will show absorption features where light has been absorbed by the atoms of the cool gas.
    D) will be unaffected, because atoms in the gas cloud are too cool to absorb or emit energy.
    Answer: C Source: Section 4-4

70. The star P Cygni (in the constellation Cygnus, the Swan) is surrounded by an extensive low-density atmosphere. Its spectrum of consists of a bright, continuous spectrum with many narrow, dark absorption lines and a few bright emission lines. The bright, continuous part of the spectrum is produced by
    A) the low-density atmosphere of the star emitting light in all directions.
    B) all parts of the star, the stellar surface and the atmosphere, equally.
    C) only the part of the low-density atmosphere which is between us and the surface of the star.
    D) the hot, dense, opaque gas of the star's surface.
    Answer: D Source: Section 4-4

71. The star P Cygni (in the constellation Cygnus, the Swan) is surrounded by an extensive low-density atmosphere. Its spectrum consists of a bright, continuous spectrum with many narrow, dark absorption lines and a few bright emission lines. The dark absorption lines are produced by
    A) all parts of the star, the stellar surface and the atmosphere, equally.
    B) the hot, low-density atmosphere of the star emitting light in all directions.
    C) only the part of the low-density atmosphere which is between us and the surface of the star.
    D) the hot, dense, opaque gas of the star's surface.
    Answer: C Source: Section 4-4

72. The star P Cygni (in the constellation Cygnus, the Swan) is surrounded by an extensive low-density atmosphere. Its spectrum consists of a bright, continuous spectrum with many narrow, dark absorption lines and a few bright emission lines. The bright emission lines are produced by
    A) only the part of the low-density atmosphere which is between us and the surface of the star.
    B) all parts of the star, the stellar surface and the atmosphere, equally.
    C) the hot, dense, opaque gas at or near the star's surface.
    D) the low-density atmosphere of the star emitting light in all directions.
    Answer: D Source: Section 4-4

73. The New Zealand physicist, Lord Rutherford, and his colleagues in England demonstrated the existence of the very small but massive nucleus inside every atom in which crucial experiment?
    A) Measurement of the spectrum of light emitted from a thin, heated metal sheet.
    B) The detection of the motion of electrons around the nucleus by the use of a very powerful microscope.
    C) The emission of electrons from a metal surface illuminated with UV light.
    D) The deflection, and occasional reflection backward, of energetic nuclear particles from a beam aimed at a thin metal sheet.
    Answer: D Source: Section 4-5

74. By bombarding atoms in a thin metal sheet with energetic particles, Rutherford found that most of the mass of an atom was concentrated inside a very small volume. What was the diameter of this "nucleus", as a fraction of that of the whole atom?
    A) 1/2 B) $1/10^6$ C) 1/10,000 D) 1/100
    Answer: C Source: Section 4-5

75. The majority of the mass of ordinary matter resides in
    A) the electromagnetic energy stored within the atom, from $E = mc^2$.
    B) the nuclei of atoms.
    C) the electron clouds around the nuclei of atoms.
    D) the electrons and the nuclei, shared about equally.
    Answer: B   Source: Section 4-5

76. The basic make-up of an atom is
    A) negative and positive charges mixed uniformly through the volume of the atom.
    B) small negatively charged particles orbiting around a central positive charge.
    C) small positively charged particles orbiting around a central negative charge.
    D) miniature planets, possibly with miniature people, gravitationally bound in orbits around a miniature star.
    Answer: B   Source: Section 4-5

77. The an atom consists of
    A) negatively charged electrons and positively charged protons mixed uniformly throughout the volume of the atom.
    B) negatively charged electrons moving around a very small but massive, positively charged nucleus.
    C) neutrons orbiting an electrically neutral nucleus composed of protons and electrons.
    D) positive protons, neutral neutrons and negative electrons orbiting a small but massive black hole.
    Answer: B   Source: Section 4-5

78. An atom is now known to consist of
    A) a small, massive, electrically charged core with electrons surrounding it.
    B) a positively charged crystal structure with electrons moving within it.
    C) a small, positively charged, black hole with electrons held in orbits around it by intense gravitational forces.
    D) a uniform distribution of positively charged matter with electrons embedded within it.
    Answer: A   Source: Section 4-5

79. The physical force which holds the components of an atom together is
    A) the electromagnetic attraction between the positive nucleus and the negative electrons.
    B) the nuclear force between protons, neutrons and electrons.
    C) the gravitational force between the massive nucleus and the much less massive electrons.
    D) the centrifugal force produced by their orbital motion on electrons moving around the nucleus.
    Answer: A   Source: Section 4-5

80. The diameter of the nucleus of an atom, compared to the overall diameter of the atom, is
A) about 99% B) 1/2000 C) 1/10,000 D) 1/10
Answer: C Source: Section 4-5

81. The diameter of the nucleus of a typical atom (as measured by Rutherford in the early 1900s) is
A) 1/2000 of the diameter of the atom.
B) $10^{-4}$ of the diameter of the atom.
C) about a half of that of the atom.
D) $10^{-3}$ of the diameter of the atom.
Answer: B Source: Section 4-5

82. Compared to the mass of an electron, the mass of a proton is
A) about the same.
B) almost 2000 times greater.
C) about twice as large.
D) about 1/2000 as large.
Answer: B Source: Section 4-5

83. The proton, the nucleus of the hydrogen atom, has a mass which exceeds that of the electron by approximately what factor?
A) 100 times. B) 2000 times. C) 2 times. D) $10^4$ times.
Answer: B Source: Section 4-5

84. By how much is a hydrogen atom heavier than a proton?
A) a hydrogen atom is lighter than a proton.
B) 2000 times heavier.
C) twice as heavy.
D) 1 part in 2000.
Answer: D Source: Section 4-5

85. The mass of the proton, the nucleus of the hydrogen atom, exceeds that of the electron by approximately what factor?
A) 2000 B) 2 C) 10,000 D) 100
Answer: A Source: Section 4-5

86. The neutron is the electrically-neutral particle which, along with the proton, makes up the atomic nucleus. The mass of the neutron, compared to that of the proton, is
A) 2000 times greater.
B) about 1/10 as large.
C) 1/2000 as large.
D) about the same.
Answer: D Source: Section 4-5

87. The electrically neutral particle, the neutron, which is one component of the atomic nucleus along with the proton, has a mass compared to that of the proton which is
A) about twice. B) 200 times greater. C) 2000 times less. D) about the same.
Answer: D Source: Section 4-5

88. The parameter of an atom which defines its unique position in the periodic table is
A) its size.
B) the number of protons in the nucleus.
C) its temperature.
D) the total number of protons and neutrons in the nucleus.
Answer: B Source: Section 4-5

89. The position of an element in the periodic table is directly related to
A) the number of protons in the atomic nucleus and hence to its positive charge.
B) the total number of protons and neutrons in the atomic nucleus.
C) the number of electrons in the atomic nucleus and hence to its negative charge.
D) the mass of the nucleus of the atom.
Answer: A Source: Section 4-5

90. The position of an element in the periodic table, its atomic number, is equal to
A) the number of the column in which the element is placed in the periodic table.
B) the number of neutrons in the nucleus of the atom.
C) the sum of the number of neutrons and protons in the nucleus of the atom.
D) the number of protons in the nucleus of the atom.
Answer: D Source: Section 4-5

91. The property of an neutral atom which defines its chemical behaviour and fixes its position in the periodic table is
A) the total number of protons and neutrons in its nucleus.
B) the number of protons in the nucleus.
C) its physical size.
D) the number of neutrons in the nucleus.
Answer: B Source: Section 4-5

92. Isotopes of a particular element in the periodic table have which nuclear property in common?
    A) the same number of neutrons but different numbers of protons in the nucleus.
    B) the same number of protons but different numbers of neutrons in the nucleus.
    C) the same total number of protons and neutrons in the nucleus.
    D) the same number of neutrons, but different numbers of protons and electrons in the nucleus.
    Answer: B Source: Section 4-5

93. One isotope of iron has an atomic number of 26. How many protons will there be in its nucleus?
    A) 26.
    B) 1 less than 26, or 25.
    C) 56.
    D) The atomic number is not related to the number of protons in the nucleus, which could be any number between 1 and 26.
    Answer: A Source: Section 4-5

94. How many neutrons are there in the nucleus of the isotope $^{17}O$ of oxygen?
    A) 9 B) 7 C) 17 D) 8
    Answer: A Source: Section 4-5

95. Iron occupies the 26th position in the periodic table and one isotope of iron, $^{57}Fe$, has an atomic mass of about 57. How many neutrons are in the nucleus of this atom?
    A) 26 B) 57 C) 31 D) 83
    Answer: C Source: Section 4-5

96. How many electrons will surround the nucleus of a neutral atom of the isotope $^{18}O$ of oxygen?
    A) 7 B) 10 C) 8 D) 18
    Answer: C Source: Section 4-5

97. How many electrons surround the nucleus of a neutral atom of Iron, the 26th element in the periodic table?
    A) 27 B) 52 C) 25 D) 26
    Answer: D Source: Section 4-5

98. The isotope $^{15}N$ has an atomic number of 7. The nucleus of this isotope contains
    A) 7 neutrons and 8 protons.
    B) 7 protons and 8 neutrons.
    C) 7 protons and 15 neutrons.
    D) 7 neutrons and 15 protons.
    Answer: B Source: Section 4-5

99. The isotope $^{20}Ne$ is at Position No. 10 in the periodic table. The nucleus of this isotope contains
A) 20 protons and 20 neutrons.
B) 9 protons and 11 neutrons.
C) 10 protons and 10 neutrons.
D) 11 protons and 9 neutrons.
Answer: C Source: Section 4-5

100. The atomic number of the isotope $^{238}U$ of uranium is 92. The nucleus of this isotope contains
A) 92 protons, 54 neutrons and 92 electrons.
B) 184 protons, 92 electrons and 54 neutrons.
C) 92 protons and 146 neutrons.
D) 146 protons and 92 neutrons.
Answer: C Source: Section 4-5

101. What is the total number of protons and neutrons in the nucleus of an atom of the fissionable isotope of uranium used in nuclear weapons, $^{235}U$, which has an atomic number of 92?
A) 327 B) 235 C) 143 D) 92
Answer: B Source: Section 4-5

102. Which of the following properties of the isotopes $^{15}N$ (nitrogen) and $^{15}O$ (oxygen) are the same or almost the same?
A) Only the nuclear mass.
B) Only the number of neutrons in the nucleus.
C) The nuclear mass AND the number of neutrons in the nucleus.
D) Only the number of protons in the nucleus.
Answer: A Source: Section 4-5

103. Considering the oxygen isotopes $O^{15}$ and $O^{16}$, which of the following properties is/are the same or almost the same for both isotopes?
A) Only the mass of the nucleus.
B) Only the number of neutrons in the nucleus.
C) The nuclear mass AND the number of neutrons in the nucleus.
D) Only the number of protons in the nucleus.
Answer: D Source: Section 4-5

104. If one neutron is added to a nucleus of an isotope of carbon, $C^{12}$, in a particular nuclear reaction, the result will be
A) $O^{12}$. B) $C^{11}$. C) $O^{13}$. D) $C^{13}$.
Answer: D Source: Section 4-5

105. When an atom of the radioactive carbon isotope $^{14}C$ decays into $^{14}N$ (nitrogen), what happens in the nucleus of the atom? (Think about the meaning of the number in the symbol for the isotope.)
A) A neutron becomes a proton.
B) A neutron is ejected from the nucleus.
C) A proton becomes a neutron.
D) The nucleus of the carbon atom captures a nucleus of a hydrogen atom (a proton).
Answer: A Source: Section 4-5

106. Which of the following combinations of particles in the nuclei of atoms does NOT make up the nucleus of an isotope of the same element?
A) 8 protons, 10 neutrons.
B) 8 protons, 9 neutrons.
C) 8 protons, 8 neutrons.
D) 7 protons, 9 neutrons.
Answer: D Source: Section 4-5

107. What is the half-life of a sample of matter containing a radioactive element?
A) Half the time needed for the radioactive decay rate to double.
B) Half of the time needed for all of the radioactive atoms in the sample to decay.
C) The time needed for half of the radioactive atoms in the sample to decay.
D) The time needed for the number of radioactive atoms in the sample to decay to 1/2.71828 of the number originally present.
Answer: C Source: Toolbox 4-2

108. Suppose that a particular sample of wood contains $10^{20}$ atoms of the radioactive isotope $^{14}C$, for which the half life is about 6000 years. Approximately how long will it take for all of the $^{14}C$ atoms to decay to $^{14}N$?
A) An infinite amount of time. B) 6000 years. C) 12,000 years. D) 3000 years.
Answer: A Source: Toolbox 4-2

109. Suppose that at some time a particular sample of radioactive material contains N radioactive atoms (e.g., N might be one billion radioactive atoms). How many radioactive atoms will be left after three half-lives have passed?
A) 1/6 N.
B) 1/8 N.
C) There will be no radioactive atoms left in the sample.
D) 1/2 N.
Answer: B Source: Toolbox 4-2

110. An ionized atom is one in which
A) a single proton has been removed from the atom.
B) an electron has been removed.
C) an extra electron has been captured into orbit around the nucleus.
D) a proton in its nucleus has been replaced by an electron.
Answer: B Source: Section 4-5

111. An atom in which one or more electrons has been removed is known as
A) an ion. B) an isotope. C) a molecule. D) an excited atom.
Answer: A Source: Section 4-5

112. Ionization of an atom occurs when
A) the nucleus is split, or fission occurs.
B) an electron is lifted from the ground state to an excited level.
C) an electron drops from a higher energy level to the ground state.
D) an electron is removed from the atom.
Answer: D Source: Section 4-5

113. An ionized hydrogen atom is simply
A) a proton. B) a neutron. C) a helium nucleus. D) an electron.
Answer: A Source: Section 4-5

114. The overall charge of a singly-ionized neon atom Ne, whose position in the periodic table, or atomic number is 10, in units of electron charge, is
A) -1. B) +1. C) +9. D) +10.
Answer: B Source: Section 4-5

115. O III is
A) doubly-ionized oxygen (an oxygen atom which has lost two electrons).
B) four-times ionized oxygen (an oxygen atom which has lost four electrons).
C) triply-ionized oxygen (an oxygen atom which has lost three electrons).
D) an oxygen isotope with three neutrons.
Answer: A Source: Section 4-5

116. S IV is
A) four-times ionized sulfur (a sulfur atom which has lost four electrons).
B) five-times ionized sulfur (a sulfur atom which has lost five electrons).
C) three-times ionized sulfur (a sulfur atom which has lost three electrons).
D) a sulfur isotope with four neutrons.
Answer: C Source: Section 4-5

117. How many electrons will be orbiting the nucleus of a singly-ionized oxygen atom?
A) 6 B) 7 C) 8 D) 9
Answer: B Source: Section 4-5

118. How many electrons will surround the nucleus of a triply-ionized magnesium atom, Mg IV?
A) 15 B) 9 C) 12 D) 11
Answer: B Source: Section 4-5

119. In Bohr's model of the hydrogen atom, light is emitted whenever
A) an electron jumps from a lower to an upper energy level or orbit.
B) an electron reverses its direction of motion in its orbit.
C) an electron spirals into the nucleus.
D) an electron jumps from an upper to a lower energy level or orbit.
Answer: D Source: Section 4-6

120. The specific colors of light emitted by an atom in a hot, thin gas (e.g., in a neon tube, a fluorescent bulb or a gas cloud in space) are caused by
A) protons jumping from level to level.
B) an electron dropping into the nucleus, producing small nuclear changes.
C) electrons jumping to lower energy levels, losing energy as they do so.
D) the vibrations of the electrons within the atom.
Answer: C Source: Section 4-6

121. What type of spectrum is given off by low-density, high-temperature hydrogen gas?
A) A uniform spectrum crossed by numerous dark absorption lines.
B) A series of emission lines with a constant wavelength spacing between them.
C) A series of emission lines, spaced in a mathematical sequence.
D) A uniform continuous spectrum containing all colors.
Answer: C Source: Sections 4-4 and 4-6

122. The specific sequence of spectral line series emitted by excited hydrogen atoms, in order of increasing wavelength range, is
A) Lyman, Paschen, Balmer.
B) Balmer, Lyman, Paschen.
C) Lyman, Balmer, Paschen.
D) Paschen, Balmer, Lyman.
Answer: C Source: Section 4-6

123. Atoms of hot hydrogen gas will emit the Balmer series of spectral lines at visible wavelengths when the electrons fall from all higher atomic energy levels to
A) the ground state, $n = 1$.
B) the next level down for each level (e.g., $n = 4$ to $n = 3$).
C) the ionization level, or $n$ = infinity.
D) the first excited level, $n = 2$.
Answer: D Source: Section 4-6

124. The Balmer series of visible spectral emissions from hydrogen gas arises from transitions in which electrons jump between energy levels
A) between adjacent levels (e.g., $n = 2$ to $n = 1$, $n = 3$ to $n = 2$, $n = 4$ to $n = 3$, etc.).
B) from all levels to the ground state, $n = 1$.
C) from higher levels to the second excited level, $n = 3$.
D) from higher levels to the first excited level, $n = 2$.
Answer: D Source: Section 4-6

125. When electrons jump from higher levels in hydrogen atoms to the level $n = 2$, the resulting spectrum will consist of
A) a sereies of IR spectral lines, the Paschen series.
B) a continuum of light with a maximum in the visible spectral range.
C) a series of UV lines, the Lyman series.
D) a series of visible spectral lines, the Balmer series.
Answer: D Source: Section 4-6

126. Light which originates in hydrogen atoms in which electrons have jumped from high levels to the level $n = 2$ will be part of which series of spectral lines?
A) Balmer.
B) Paschen.
C) There would be a continuum of light, not a series of lines.
D) Lyman.
Answer: A Source: Section 4-6

127. Hydrogen gas is heated to the point where there are electrons at atomic energy levels up the $n = 3$ level. When electrons return to the ground state, what possible emission lines from which spectral sequences will result? (Check Fig. 4-16, Kaufmann & Comins, *Discovering the Universe*, 5th ed.)
A) Balmer (visible) and Lyman (UV) series.
B) Lyman (UV) series only.
C) Paschen (IR), Balmer (visible), and Lyman (UV) series.
D) Balmer (visible) series only.
Answer: A Source: Section 4-6

128. The Balmer series of hydrogen has a series limit (at 364.6 nm) (see Fig. 4.13, Kaufmann and Comins, *Discovering the Universe*, 5th. Ed.) because hydrogen in the $n = 2$ state
   A) is ionized by photons of shorter wavelength than this limit.
   B) is excited to a higher discrete level by photons shorter than this wavelength.
   C) cannot absorb photons of longer wavelength than this limit.
   D) cannot absorb photons of shorter wavelength than this limit.
   Answer: A Source: Section 4-6

129. The sequence of ultraviolet emission lines emitted by hot hydrogen gas is known as the
   A) Planck series. B) Lyman series. C) Paschen series. D) Balmer series.
   Answer: B Source: Section 4-6

130. The Lyman series of ultraviolet spectral emission lines from hydrogen gas is produced by electrons jumping
   A) to the ground state from all other energy levels.
   B) to the first excited level from all higher levels.
   C) between adjacent energy levels, $n = 2$ to 1, $n = 3$ to 2, $n = 4$ to 3, etc.
   D) from the continuum level, to all other levels.
   Answer: A Source: Section 4-6

131. The strong ultraviolet spectral line emitted by hot hydrogen gas is known as the
   A) Paschen a line. B) 21-cm line. C) Balmer a line. D) Lyman a line.
   Answer: D Source: Section 4-6

132. If a continuous spectrum of ultra-violet radiation passes through a tube of cool hydrogen gas, what happens to its spectrum?
   A) All of the radiation passes through the tube unhindered since the hydrogen gas is cool and cannot absorb energy.
   B) Some of the radiation at all wavelengths is absorbed, reducing the intensity at all wavelengths uniformly.
   C) All of the radiation passes unhindered except the Lyman La wavelength, which is absorbed by the atoms.
   D) All of the radiation passes through the tube unhindered except at the specific wavelengths of the Lyman series, La, Lb etc., which are absorbed by the atoms.
   Answer: D Source: Section 4-6

133. An atom of hydrogen undergoes a collision with another atom in a hot gas, in which the energy of collision is about 11 eV. What is the likely outcome of this collision, in terms of the atom? (See Fig. 4-16, Kaufmann & Comins, *Discovering the Universe*, 5th Ed.)
A) The electron in the atom will be excited to the first excited level, $n = 2$. Its return to the ground state will produce a $L_a$ UV photon.
B) The electron in the atom will be excited to the second excited level, $n = 3$, and de-excitation will generate either a $L_b$ UV Lyman photon or an $H_a$ visible and a $L_a$ UV photon.
C) The electron of the hydrogen atom will be excited beyond the ionization level ($n$ = ¥); the atom will be ionized and the electron will leave the atom completely.
D) The electron in the atom will be excited to the first excited level, and de-excitation to the ground state will produce a visible photon of Balmer $H_a$ light.
Answer: A Source: Section 4-6

134. Where does the Paschen series of spectral lines from hydrogen gas appear in the electromagnetic spectrum?
A) In the infrared, with wavelengths longer than 700 nm.
B) In the radio range, with wavelengths longer than 0.01 m or 10 mm.
C) In the ultraviolet region, with wavelengths between 90 and 130 nm.
D) In the visible region, with wavelengths between 350 to 660 nm.
Answer: A Source: Section 4-6

135. The series of spectral absorption lines in the infrared part of the spectrum, known as the Paschen series, results from atomic transitions in hydrogen atoms in which electrons are lifted from which energy level to all higher atomic energy levels?
A) The $n = 2$ level.
B) The ionization level.
C) The $n = 3$ level.
D) The $n = 1$ level, the ground state, since all series start here.
Answer: C Source: Section 4-6

136. The series of spectral absorption lines in the infrared part of the spectrum which result from atomic transitions in hydrogen atoms in which electrons are lifted from the $n = 3$ level to all other atomic energy levels is known as
A) the Lyman series.
B) the Balmer series.
C) ionization transitions.
D) the Paschen series.
Answer: D Source: Section 4-6

137. The person who discovered the atomic nucleus was
 A) the New Zealand physicist, Sir Ernest Rutherford.
 B) the Danish physicist, Niels Bohr.
 C) the Swiss mathematician, Johann Balmer.
 D) the German optician, Joseph von Fraunhofer.
 Answer: A Source: Section 4-6

138. The person who discovered the relationship between the wavelengths of the spectral lines of hydrogen was
 A) the Swiss mathematician, Johann Balmer.
 B) the Danish physicist, Niels Bohr.
 C) the German optician, Joseph von Fraunhofer.
 D) the New Zealand physicist, Sir Ernest Rutherford.
 Answer: A Source: Section 4-6

139. The person who developed the orbital theory for electrons in the hydrogen atom was
 A) the New Zealand physicist, Sir Ernest Rutherford.
 B) the Danish physicist, Niels Bohr.
 C) the German optician, Joseph von Fraunhofer.
 D) the Swiss mathematician, Johann Balmer.
 Answer: B Source: Section 4-6

140. Three important milestones in our understanding of how atoms interact with light were (listed here in alphabetical order), Balmer's discovery of the mathematical relationship between the hydrogen spectral lines; Bohr's development of atomic orbital theory; and Rutherford's discovery of the atomic nucleus. In what order, from earliest to latest, did these scientists accomplish these milestones?
 A) Balmer, Bohr, Rutherford.
 B) Balmer, Rutherford, Bohr.
 C) Rutherford, Balmer, Bohr.
 D) Bohr, Rutherford, Balmer.
 Answer: B Source: Section 4-6

141. Because of the Doppler effect, the sound of the engines of an aircraft which flies over us
 A) changes from a low pitch or frequency to a higher pitch.
 B) starts at a high pitch or frequency, drops to a low frequency when the plane is overhead, then rises again as the plane moves away from us.
 C) changes from a high pitch or frequency to a lower pitch.
 D) remains at the same pitch or frequency but the intensity rises as the plane approaches and then falls again.
 Answer: C Source: Section 4-7

142. The Doppler effect is the change in the wavelength of light caused by the source
   A) being in an intense magnetic field.
   B) being within a high gravitational field.
   C) being embedded in a cloud of dust and gas.
   D) moving with respect to the observer.
   Answer: D Source: Section 4-7

143. The observed change in wavelength of light due to the Doppler effect occurs
   A) ONLY when the light source has a radial velocity (i.e., motion towards or away from the observer).
   B) whenever the light source is moving with respect to the observer (regardless of direction).
   C) ONLY when the temperature of an object changes.
   D) ONLY when the light source has proper motion (i.e., motion across the line of sight).
   Answer: A Source: Section 4-7

144. The Doppler effect is the
   A) change in the wavelength of peak emission of light when the source temperature changes.
   B) increase in the observed wavelength of light if the light source is moving towards you.
   C) increase in the observed wavelength of light if the source of light is moving away from you.
   D) splitting of spectral lines into two or more wavelengths because the source of the light is in a strong magnetic field.
   Answer: C Source: Section 4-7

145. According to the Doppler effect
   A) the wavelength of light is shifted to a shorter wavelength if the source of light is moving toward you.
   B) the wavelength of peak emission of light from a source changes as the temperature of the source changes.
   C) the wavelength of light is shifted to a longer wavelength if the source of the light is moving toward you.
   D) spectral lines are split into two or more wavelengths when the source of the light is in a strong magnetic field.
   Answer: A Source: Section 4-7

146. The spectrum of a star shows an equivalent set of dark absorption lines to those of the Sun, but with one exception. Every line appears at a slightly longer wavelength, shifted toward the red end of the spectrum. What conclusion can be drawn from this observation?
A) The star is moving rapidly away from Earth.
B) The temperature of the star's surface is higher than that of the Sun.
C) A cloud of cold gas and dust surrounds the star and is absorbing light from it.
D) The star is moving rapidly toward the Earth.
Answer: A Source: Section 4-7

147. When electromagnetic radiation (e.g., light) is Doppler-shifted by motion of the source away from the detector,
A) the measured wavelength is longer than the emitted wavelength.
B) the measured frequency of the radiation remains the same, but its wavelength is shortened, compared to the emitted radiation.
C) the speed of the radiation is less than the emitted speed.
D) the measured frequency is higher than the emitted frequency.
Answer: A Source: Section 4-7

148. The wavelength $l_0$ of light emitted by a moving object is detected as a different wavelength l by a stationary observer (with $v$ the object velocity, and $c$ the velocity of light), this Doppler shift being described by the equation
A) $l - l_0 = v - c$. B) $l.l_0 = vc$. C) $l/l_0 = v/c$. D) $(l-l_0)/l_0 = v/c$.
Answer: D Source: Section 4-7

149. A police radar bounces radio waves of wavelength 3 mm from the front of a speeding car and measures the Doppler shift in wavelength of the reflected waves. *Note that the shift is DOUBLED because of the reflection.* What will be the wavelength shift if the speeding car is moving at 60 mph (80 km per hour or 22.2 $ms^{-1}$) (in a 30 mph zone!)?
A) 6.75 m. B) 14.8 nm. C) 148 nm. D) 148 mm.
Answer: C Source: Section 4-7 and Toolbox 4-3

150. Hydrogen gas emits a strong spectral line of red light with a wavelength of 656.3 nm (the Balmer a line). This emission line is seen in the spectrum of a distant quasar, but at a wavelength of 721.9 nm. Applying Doppler's relation, how fast is this object moving with respect to Earth, in terms of the velocity of light, $c$?
A) $1/10\ c$. B) $1.1\ c$. C) $1/100\ c$. D) $10\ c$.
Answer: A Source: Toolbox 4-3

151. An astronomer measures the spectrum of an star and finds a spectral line at 499 nm wavelength. In the laboratory, this spectral line occurs at 500 nm. According to the Doppler effect, this object
A) is moving away from the Earth at 499/500 the speed of light.
B) is moving away from the Earth at 1/500 the speed of light.
C) is moving toward the Earth at 1/500 the speed of light.
D) is moving toward the Earth at 499/500 the speed of light.
Answer: C Source: Toolbox 4-3

152. An astronomer observing the spectrum of the Sun on the solar equator finds that the Balmer $H_b$ spectral line (l = 486 nm) is blueshifted by 0.0033 nm when measured at one edge of the Sun's disk compared to the same line at the center of the Sun's disk, and is redshifted by the same amount on the equator at the other side of the Sun's disk. If this Doppler shift is due to the Sun rotating (try drawing a diagram), then the rotational speed of the Sun at its equator is
A) 4 km/s. B) 2 km/s. C) 1 km/s. D) 0.5 km/s.
Answer: B Source: Toolbox 4-3

153. Proper motion is
A) the motion of a star or other object in any direction through space.
B) the motion of a star or other object across (at right angles to) the line of sight.
C) the motion of a star or other object along the line of sight, towards or away from us.
D) any motion allowed by the laws of physics.
Answer: B Source: Section 4-7

154. The proper motion of a star or other object is measured by
A) measuring the change in the angular diameter of the object.
B) measuring the intensity of the light in different regions of the blackbody curve.
C) observing the Doppler shift of spectral lines in the light from the object.
D) watching the object move compared to the background stars.
Answer: D Source: Section 4-7

155. The radial velocity of a star or other object is measured by
A) measuring the change in the angular diameter of the object.
B) watching the object move compared to the background stars.
C) measuring the intensity of the light in different regions of the blackbody curve.
D) observing the Doppler shift of spectral lines in the light from the object.
Answer: D Source: Section 4-7

# Build Your Foundation II: Solar System

1. The birthplace of the Sun and planets (and of other stars and maybe their planets) is thought to have been
   A) in the centers of galaxies.
   B) at the centers of supernova explosions.
   C) in black holes dotted about the universe.
   D) in cool gas and dust clouds.
   Answer: D Source: Section II-1

2. The most likely mechanism for the solar system is that
   A) planets were spun out of the Sun as smaller gas clouds and subsequently condensed.
   B) the Sun captured the planets as they drifted through space.
   C) the Sun and planets slowly condensed to their present form from a gas and dust cloud.
   D) the solar system was once a galaxy, from which the Sun and planets are the remnants, after evolution.
   Answer: C Source: Section II-1

3. What was the material from which the solar system formed?
   A) Debris from the explosion of a massive star.
   B) A nebula made mostly of heavy elements, but enriched in hydrogen and helium from the supernova explosions.
   C) A nebula made mostly of hydrogen and helium gas, but enriched in heavier elements from supernova explosions.
   D) A nebula made entirely of hydrogen and helium gas.
   Answer: C Source: Section II-1

4. The most abundant material in the universe is
   A) hydrogen. B) helium. C) carbon dioxide. D) water.
   Answer: A Source: Section II-1

5. The most common elements in the universe are
   A) equal amounts of hydrogen and helium, with small amounts of heavier elements.
   B) about equal amounts of all elements up to iron, but very little of any heavier elements.
   C) heavy elements, with smaller quantities of hydrogen and helium.
   D) mostly hydrogen, smaller quantities of helium and very small quantities of heavier elements.
   Answer: D Source: Section II-1

6. Hydrogen and helium together account for what percentage of the total mass of all the matter in the universe?
   A) 75% B) About 50% C) 90% D) 98%
   Answer: D Source: Section II-1

7. The fractional abundance of hydrogen by mass in the universe is about
   A) 75% B) 99.9% C) 98%. D) only about 20%
   Answer: A Source: Section II-1

8. What fraction of the mass of the universe is in the form of atoms other than hydrogen and helium?
   A) Much less than 1% B) 2% C) 10% D) 50%
   Answer: B Source: Section II-1

9. What fraction of the mass of the Earth is made up of the elements hydrogen and helium?
   A) 98% B) 2% C) About 70% D) Much less than 1%
   Answer: D Source: Section II-1

10. The composition of matter in the universe can be summarized by which statement?
    A) About half of the mass of the universe is in the form of rocks, molecules, and planetary material.
    B) All but 2% of the mass of the universe is hydrogen.
    C) All but 2% of the mass in the universe is hydrogen and helium.
    D) 2% of the mass of the universe is hydrogen and helium, the rest is of heavier elements.
    Answer: C Source: Section II-1

11. Where was all the hydrogen in the universe formed?
    A) In the dark clouds of dust and gas.
    B) In nuclear reactions in the cores of stars.
    C) In the Big Bang, at the very beginning of the universe.
    D) In supernovae (exploding stars).
    Answer: C Source: Section II-1

12. Where in the universe are heavy elements, with masses greater than that of helium, being produced at this time?
    A) In the dark clouds of dust and gas.
    B) In the central cores of stars.
    C) At the event horizons of massive black holes.
    D) In the surface layers of stars.
    Answer: B Source: Section II-1

13. Fusion is the process by which
    A) elements are transformed into heavier elements by nuclear reactions.
    B) massive protoplanetary cores pull gas onto themselves to create giant planets.
    C) clouds of interstellar gas and dust contract to form protostars.
    D) dust grains and ice crystals coalesce to form planetesimals.
    Answer: A Source: Section II-1

14. Most of the elements beyond H and He in the periodic table in our Sun and solar system most probably originated
    A) from fusion reactions in the centers of earlier stars.
    B) from the center of our own Sun, through fusion and later ejection as solar wind.
    C) in the original Big Bang of the universe.
    D) from chemical reactions in planetary atmospheres.
    Answer: A Source: Section II-1

15. What is thought to be the physical mechanism that was responsible for the present mix of chemical elements in the universe?
    A) All the known elements have been formed by the break-up (radioactivity) of the heavy elements formed in the initial Big Bang.
    B) H and He were formed in the Big Bang, while the heavier elements have been slowly forming by collisions in cold interstellar gas clouds.
    C) All of the known elements were formed in the Big Bang.
    D) H and He were formed in the Big Bang, while the heavier elements were made in the centers of stars.
    Answer: D Source: Section II-1

16. The birthplace of the Sun and planets (and of other stars and maybe their planets) is thought to have been
    A) in black holes dotted about the universe.
    B) in the centers of galaxies.
    C) in cool gas and dust clouds.
    D) at the centers of supernova explosions.
    Answer: C Source: Section II-2

17. What name is given to the concentration of mass that formed at the center of the solar nebula, and eventually became the Sun?
    A) The pseudosun. B) The antisun. C) The protosun. D) The solar hub.
    Answer: C Source: Section II-2

18. Why did the temperature start to rise at the center of the solar nebula?
    A) Supernova explosions were stirring up the material there and causing turbulence.
    B) Massive stars nearby were heating the nebula with their ultraviolet radiation.
    C) The nebula was contracting, which increased the speed of motion of the atoms in it.
    D) Fusion reactions were beginning in the core, releasing tremendous amounts of heat.
    Answer: C Source: Section II-2

19. Strong evidence for the existence of planetary systems in the process of formation around other stars comes from
    A) direct photography of actual planets near to other stars.
    B) detection of very regular pulses of radio energy from some stars.
    C) spectroscopic evidence of large quantities of molecules such as ammonia and methane, which can only exist in planetary atmospheres.
    D) photographs and infrared observations of disks of dust.
    Answer: D Source: Section II-2

20. Accretion is the process by which
    A) clouds of interstellar gas and dust contract to form protostars.
    B) massive protoplanetary cores pull gas onto themselves to create giant planets.
    C) elements are transformed into heavier elements by nuclear reactions.
    D) dust grains and ice crystals coalesce to form planetesimals.
    Answer: D Source: Section II-1

21. What are the three "common" substances that are believed to be important in planet formation?
    A) Solid, liquid, and gaseous hydrogen.
    B) Hydrogen, helium and neon gases.
    C) Electromagnetic radiation, electrical discharges (e.g., lightning), water.
    D) Rock, ices and gas.
    Answer: D Source: Section II-3

22. Which physical parameter, more than any other, most probably controlled the early evolution of the planetary system and dictated the characteristics of the planets that eventually formed?
    A) The temperature distribution within the nebula.
    B) The mix of chemical constituents.
    C) The density of hydrogen gas in the nebula.
    D) The overall rotation of the nebula.
    Answer: A Source: Section II-3

23. The most probable process for the formation or acquisition of the planets of the Sun is
    A) relatively slow growth of smaller objects by collisions and mutual gravitational attraction.
    B) the freezing of immense gas clouds by the cold temperature of space.
    C) the break-up of one single large companion body to the Sun, by tidal distortion.
    D) capture of planets from outer space by gravity.
    Answer: A Source: Section II-3

24. The most probable time sequence for the formation of the solar system was that
    A) the planets formed first out of a cold nebula of gas and dust, followed by the Sun, which formed when the gas had become much hotter.
    B) the Sun formed initially, and the planets and major moons were captured much later as they drifted by the Sun.
    C) the Sun formed first, the planets were spun off from the Sun, and the moons in turn were spun off from the planets.
    D) the Sun contracted first as a gas ball, and the planets and moons formed shortly afterwards by accretion and condensation.
    Answer: D Source: Section II-3

25. The most probable theory for the formation of the solar system is
    A) condensation of a nebula of cold gas and dust into the Sun and planets.
    B) condensation of a nebula of hot gas into the Sun and planets.
    C) an encounter, in which a passing star ripped off material from the Sun to form the planets.
    D) a capture theory in which the Sun, after formation, captured objects moving through space to form the planets.
    Answer: A Source: Section II-3

26. Suppose that you were to go back in time and explore the early solar nebula (during the formation of the solar system). If you were to travel outward from the protosun, the FIRST solid material you would encounter would be
    A) dust-sized grains of rocky material.
    B) snowflakes of frozen hydrogen and helium.
    C) snowflakes made of frozen water, methane, ammonia and carbon dioxide.
    D) dust-sized grains of frozen hydrogen, water ice and rocky minerals.
    Answer: A Source: Section II-3

27. The early phases of planetary formation into protoplanets were characterized by
    A) violent collapse of matter under gravity.
    B) condensation of hot gas clouds.
    C) the breaking apart of very large objects into planets.
    D) slow accretion of small particles by gravitational attraction and collision.
    Answer: D Source: Section II-3

28. The process of accretion in planetary formation is
    A) the relatively rapid gravitational collapse (in less than $10^6$ years) of gas clouds to form planets.
    B) the slow condensation by gravity of gas atoms into large dense gas clouds that are the pre-planetary masses.
    C) the slow acquisition from deep space by the giant planets of their complement of moons by gravitational capture.
    D) the slow accumulation of solid particles by gravity and collision into larger, solid objects.
    Answer: D Source: Section II-3

29. The formation of terrestrial-type of planets around a star is most likely to have occurred by what process?
    A) Break-up of a large disk of matter which formed around the star.
    B) Condensation of gas from the original star nebula.
    C) Accretion, or slow accumulation of smaller particles by mutual gravitational attraction.
    D) Capture of objects traversing the depths of space by the star.
    Answer: C Source: Section II-3

30. The reason for the vast amount of hydrogen in the interior of Jupiter is probably that
    A) the mass of the initial condensation of rocks at Jupiter's orbit was sufficient to attract vast amounts of gas to it.
    B) Jupiter became so hot in its interior that all kinds of atoms and molecules were melted down to the fundamental atom, hydrogen.
    C) nuclear fission of atoms in Jupiter's interior split all nuclei down to hydrogen nuclei early in its history.
    D) Jupiter formed from the initial gravitational contraction of hydrogen gas.
    Answer: A Source: Section II-3

31. According to modern theories, the most significant difference between the formation of the terrestrial and the large, outer planets is that
    A) both formed by accretion of rocky and icy planetesimals, but the terrestrial planets were close enough to the Sun that almost all of the ices escaped back to space after the planets formed.
    B) the terrestrial planets formed close to the Sun where there was an abundance of rock but no ice, whereas the outer planets formed far from the Sun where there was an abundance of hydrogen and ice but no rocky material.
    C) both formed by accretion of planetesimals but the outer planets became massive enough to also pull gas onto them directly from the solar nebula.
    D) the terrestrial planets formed by accretion of planetesimals, whereas the outer planets formed by direct condensation of gas from the solar nebula.
    Answer: C Source: Section II-3

32. The manner in which the terrestrial planets formed was
    A) gravitational condensation of gas followed by capture of solid planetesimals.
    B) accretion of planetesimals to form a core, followed by gravitational capture of gas from the solar nebula.
    C) gravitational condensation of hydrogen, helium, and dust in eddies or vortices in the solar nebula.
    D) accretion of solid planetesimals containing mostly rocky material.
    Answer: D Source: Section II-3

33. The steps in which the large, outer planets formed were
    A) gravitational condensation of hydrogen and helium gas, followed by capture of planetesimals.
    B) accretion of cold planetesimals containing large quantities of hydrogen and helium.
    C) gravitational condensation of hydrogen, helium, and dust in eddies or vortices in the outer solar nebula.
    D) accretion of planetesimals to form a core, followed by gravitational capture of hydrogen and helium gas.
    Answer: D Source: Section II-3

34. The time-scale over which material in the solar nebula accreted to form planets was about
    A) 4.6 billion years.
    B) 100 million years.
    C) 4.6 million years.
    D) 100 thousand years.
    Answer: B Source: Section II-3

35. At what point in time do we say that the protosun became the Sun?
A) When the temperature began to rise at its center.
B) When planetary formation was complete.
C) When it became hot enough to emit light and heat.
D) When thermonuclear fusion reactions began at its center.
Answer: D Source: Section II-3

36. What process had the greatest influence on the features of the Moon during the first billion years of its existence?
A) Impacts from space.
B) Erosion by an early, short-lived atmosphere.
C) Volcanoes.
D) Mountain building from geological activity.
Answer: A Source: Section II-3

37. The correct sequence of planets in our solar system, from the Sun outward, is
A) Mercury, Venus, Earth, Mars, Jupiter, Saturn, Uranus, Neptune.
B) Mercury, Venus, Mars, Earth, Jupiter, Saturn, Uranus, Neptune.
C) Mercury, Venus, Earth, Mars, Saturn, Uranus, Jupiter, Neptune.
D) Mercury, Earth, Venus, Mars, Jupiter, Saturn, Uranus, Neptune.
Answer: A Source: Section II-3

38. What are the main characteristics of our solar system?
A) Two large planets close to the Sun, three small planets next out, and four large planets farthest from the Sun.
B) Four small planets close to the Sun, four large planets far from the Sun, and one small planet farthest from the Sun.
C) Three small planets close to the Sun, four large planets far from the Sun, and two small planets farthest from the Sun.
D) Three small planets close to the Sun, five large planets far from the Sun, and one small planet farthest from the Sun.
Answer: B Source: Section II-4

39. The overall shape of the orbits of most of the planets in the solar system is
A) parabolic.
B) slightly elliptical, but nearly circular.
C) elliptical, very elongated.
D) perfectly circular.
Answer: B Source: Section II-4

40. The planets whose orbits are most noticeably elliptical are
   A) Uranus and Pluto.
   B) Mercury and Pluto.
   C) Uranus and Mars.
   D) Mercury and the Earth.
   Answer: B Source: Section II-4

41. One planet whose orbit carries it both inside and outside the orbital distance of another planet is
   A) Uranus. B) Pluto. C) Mercury. D) Mars.
   Answer: B Source: Section II-4

42. Most of the planets orbit the Sun on or close to
   A) the equatorial plane.
   B) the ecliptic plane.
   C) the plane containing both north and south celestial poles and the zenith at Greenwich, England.
   D) the plane of the Milky Way Galaxy.
   Answer: B Source: Section II-4

43. The planet whose orbit has the greatest inclination to the plane of the ecliptic is
   A) Pluto. B) Uranus. C) Saturn. D) Mercury.
   Answer: A Source: Section II-4

44. Compared to the orbital distance of Earth from the Sun, the equivalent orbital distances for the outer planets are
   A) greater than 5 times.
   B) between 2 and 5 times.
   C) between 2 and 20 times.
   D) greater than 10 times.
   Answer: A Source: Section II-4

45. In our solar system, which of the following planets is NOT a member of the terrestrial group?
   A) Mars B) Mercury C) Venus D) Jupiter
   Answer: D Source: Section II-4

46. In our solar system, which of the following planets is a member of the terrestrial group?
   A) Mars B) Jupiter C) Neptune D) Saturn
   Answer: A Source: Section II-4

47. Suppose that the Space Telescope, in 1998, discovers a series of planets with the following characteristics moving around a star that resembles our Sun: spherical, solid surfaces, mean densities about 4 times that of $H_2O$, radii about 4000 km, low density atmospheres. What would these planets be classified as, in comparison to our solar system?
A) Asteroids B) Outer planets C) Terrestrial planets D) Cometary nuclei
Answer: C Source: Section II-4

48. Suppose that observers using the Hubble Space Telescope detect around several solartype stars the presence of planets with the following characteristics: low density, large size, fluid surfaces, rapid rotation. How would these planets be classified, in terms of our solar system?
A) Comet nuclei B) Asteroids C) Outer planets D) Terrestrial planets
Answer: C Source: Section II-4

49. Which planet or planetary group occupies the next orbital position beyond Saturn?
A) Neptune B) Jupiter C) The asteroid belt D) Uranus
Answer: D Source: Section II-4

50. The mean density of a planet is
A) another way of describing its total mass.
B) its total mass divided by its total volume.
C) the amount of mass in unit volume of the material on its surface.
D) the mass of a unit volume (1 cubic meter) of the material at its core.
Answer: B Source: Section II-4

51. The average density of the large, outer planets is
A) close to the density of water.
B) close to the density of basaltic rocks on Earth.
C) much higher than the density of Earth rocks, due to the great gravitational compression of their interiors.
D) very much less than the density of water, because of the amount of hydrogen that they contain.
Answer: A Source: Section II-4

52. The average density of the outer "giant" planets compared to that of liquid water is
A) much less than that of water.
B) slightly higher, about 1.2 times.
C) very high, greater than 10 times.
D) higher, about 5 times.
Answer: B Source: Section II-4

53. The average density of which of the following planetary groups is close to that of water (1000 kg/m$^3$)?
    A) Mercury and Venus, because they are close to the Sun.
    B) The terrestrial planets, because they are of relatively low mass, and have been compressed very little by gravitational forces.
    C) The large, outer planets, because of their composition, H and He.
    D) The asteroids because they are very small objects.
    Answer: C Source: Section II-4

54. A curious fact about the structure of the planet Jupiter, compared to that of Earth, is that it has
    A) about the same mass but much higher density.
    B) much greater mass but much lower average density.
    C) much greater mass and greater average density.
    D) much greater mass but about the same density.
    Answer: B Source: Section II-4

55. The low average densities of the large, outer planets, which have high masses and hence high gravitational fields, is an indication of what fact about their interiors?
    A) They have hot, gaseous interiors, similar to cool stars.
    B) They are composed mainly of very light elements, such as H and He.
    C) Their interiors have not been condensed to liquid or solid form.
    D) Their interiors are composed of $H_2O$, $CH_4$ (methane), and $NH_3$ (ammonia).
    Answer: B Source: Section II-4

56. In order of increasing density, the large, outer planets would be listed as
    A) Neptune, Uranus, Saturn, Jupiter.
    B) Jupiter, Saturn, Uranus, Neptune.
    C) Saturn, Uranus, Jupiter, Neptune.
    D) Neptune, Saturn, Uranus, Jupiter.
    Answer: C Source: Section II-4

57. The large, outer planets have high masses and hence generate powerful gravitational fields, and yet have low average densities. What does this indicate about their interiors?
    A) They are composed mainly of very light elements, such as H and He.
    B) They have not condensed to liquid or solid form.
    C) The interiors are hot and gaseous, like those of cool stars.
    D) They are composed mainly of water.
    Answer: A Source: Section II-4

58. Which planet in our solar system has the lowest average density?
    A) Jupiter B) Uranus C) Earth D) Saturn
    Answer: D Source: Section II-4

59. The planet whose average density is less than that of water is
A) Jupiter. B) Neptune. C) Saturn. D) Earth.
Answer: C Source: Section II-4

60. The planet with the greatest mean density is
A) Earth. B) Jupiter. C) Neptune. D) Mercury.
Answer: A Source: Section II-4

61. The Earth has an average density of 5500 kg/$m^3$ while the density of rock on its surface is about 3000 kg/$m^3$. What conclusion can be reached about the Earth's core from this observation?
A) It is made of material far denser than surface rock.
B) It consists of lower density material than surface rock.
C) It must be very hot.
D) It must be composed of material with density about twice that of the surface material.
Answer: A Source: Section II-4

62. The smallest of the planets is
A) Neptune. B) Mercury. C) Pluto. D) Mars.
Answer: C Source: Section II-4

63. The smallest planet in our solar system is
A) Neptune. B) Mercury. C) Mars. D) Pluto.
Answer: D Source: Section II-4

64. Which is the largest planet in our solar system?
A) Jupiter B) Earth C) Uranus D) Saturn
Answer: A Source: Section II-4

65. The smallest terrestrial planet is
A) Mars. B) Neptune. C) Mercury. D) Ganymede.
Answer: C Source: Section II-4

66. The largest of the terrestrial planets is
A) Mars. B) Jupiter. C) Venus. D) Earth.
Answer: D Source: Section II-4

67. Which planet of our solar system has the highest mass?
A) Saturn B) Jupiter C) Uranus D) Earth
Answer: B Source: Section II-4

68. Which of the following statements is TRUE?
    A) The Earth is the most massive of the terrestrial planets.
    B) The Earth is the biggest of the planets.
    C) The average mass of terrestrial planets is close to the average mass of the large, outer planets.
    D) Jupiter has the highest average density of the planets.
    Answer: A Source: Section II-4

69. Which of the following characteristics is NOT typical of our planetary system?
    A) The orbits of most of the planets are almost circular.
    B) Most planets orbit the Sun in the same direction.
    C) Most of the planets have about the same physical size.
    D) Most of the planets have their spin axes aligned to within 30° to the perpendicular of the orbital plane.
    Answer: C Source: Section II-4

70. The word "albedo" refers to
    A) the amount of light reflected by a planet or other object.
    B) the fraction of the surface or atmosphere of a planet that is covered by cloud.
    C) the amount of light absorbed by a planet or other object.
    D) the ratio of infrared radiation to visible radiation emitted by a planet or other object.
    Answer: A Source: Section II-4

71. The amount of light reflected by a planet or other object is
    A) its color index. B) its apparent magnitude. C) its albedo. D) its emissivity.
    Answer: C Source: Section II-4

72. The albedo of Mercury is about 0.1. This means that
    A) Mercury is visible only 1/10 of the time.
    B) 1/10 of Mercury's surface is cloud-covered.
    C) Mercury reflects 1/10 of the sunlight falling on it.
    D) Mercury reflects 9/10 of the sunlight falling on it.
    Answer: C Source: Section II-4

73. Which planets have one or more satellites (or moons) orbiting them?
    A) All except the inner two planets.
    B) All except the inner two planets and Pluto.
    C) Only the Earth and the four large, outer planets.
    D) Only the Earth, Jupiter, and Saturn.
    Answer: A Source: Section II-4

74. Which planets do NOT have one or more natural satellites (moons) orbiting them?
    A) Mercury, Venus, and Mars.
    B) Mercury, Venus, Mars, and Pluto.
    C) Mercury and Mars.
    D) Mercury and Venus.
    Answer: D Source: Section II-4

75. The asteroid belt exists between the orbits of which planets?
    A) Mars and Jupiter
    B) Venus and Earth
    C) Earth and Mars
    D) Jupiter and Saturn
    Answer: A Source: Section II-5

76. The asteroid belt is made up of
    A) rocky bodies typically a few tens of kilometers in diameter.
    B) irregularly shaped bodies composed primarily of ices.
    C) large, rocky bodies typically about the size of our Moon.
    D) several planet-sized objects with dense methane atmospheres.
    Answer: A Source: Section II-5

77. The composition of a typical asteroid is
    A) rock and ice.
    B) rock and metal.
    C) pure ice, or perhaps ice with dust-sized grains of rock mixed in.
    D) ice with a liquid water core.
    Answer: B Source: Section II-5

78. A typical asteroid is (see Figure II-10)
    A) potato-shaped with large and small craters.
    B) spherical and deformed by thermal activity.
    C) spherical and densely covered with craters.
    D) potato shaped with a smooth, metallic surface.
    Answer: A Source: Section II-5

79. Most meteoroids are
    A) fragments of asteroids.
    B) rock-sized pieces of ice chipped off comets.
    C) small pieces of debris from the original solar nebula.
    D) centimeter-sized pieces of interstellar matter.
    Answer: A Source: Section II-5

80. Comets are typically
    A) chunks of ice that begin to vaporize if they pass close to the Sun.
    B) slushy mixtures of liquid and ice.
    C) gaseous bodies from which some of the gas is pushed out by the Sun to form a long tail.
    D) chunks of rock typically a few tens of kilometers in diameter.
    Answer: A Source: Section II-5

81. What is the basic difference between comets and asteroids?
    A) Comets are mostly composed of ices while asteroids are mainly composed of rocks.
    B) Comets are spherical while asteroids are mostly irregular in shape.
    C) Comets always emit their own light while asteroids only reflect sunlight.
    D) Comets always move in open orbits around the Sun and hence visit the Sun only once in their lifetime, while asteroids move in closed orbits.
    Answer: A Source: Section II-5

82. What was the FIRST direct evidence that some other solar-type stars might have planets?
    A) Warped disks of dust and gas around some young stars.
    B) The path of the star through space is slightly wavy, as if the star is being tugged by an orbiting planet.
    C) Faint pinpoints of light slowly circling the stars, as seen through the new 8- and 10-meter telescopes.
    D) Cyclic Doppler shift variations in the spectra of several stars.
    Answer: A Source: Section II-6

83. In the search for planets around other stars, which of the following possible lines of evidence has NOT yet been seen?
    A) A wavy path of a star through space, as if the star is being tugged by an orbiting planet.
    B) A warped disk of dust and gas around a young star.
    C) Faint pinpoints of light slowly circling a star.
    D) Cyclic Doppler shift variations in a star's spectrum.
    Answer: C Source: Section II-6

84. What is astrometry?
    A) Very precise measurement of lines in a star's spectrum (e.g., to measure the Doppler shift).
    B) Very precise measurement of a star's position on the sky (e.g., to measure its motion).
    C) Very precise measurement of a star's blackbody curve (e.g., to measure the star's temperature).
    D) Very precise measurement of a star's brightness (e.g., to measure light variations).
    Answer: B Source: Section II-6

85. How do we measure the mass of an extrasolar planet?
    A) Using Newton's law of gravity, using its measured distance from the star and its gravitational pull on the star.
    B) We can't make any firm estimate of the mass of an extrasolar planet with present technology.
    C) Using spectra to measure the planet's temperature and photometry to measure its brightness.
    D) By measuring the planet's angular diameter and hence its size, and using spectra to find its composition and hence density.
    Answer: A Source: Section II-6

86. What is the typical mass of the majority of extrasolar planets so far discovered?
    A) Roughly the mass of Jupiter, up to a few times Jupiter's mass.
    B) Close to the mass of the Earth.
    C) Quite large: between ten and a hundred times Jupiter's mass.
    D) Very small: between a tenth and a few tenths of the mass of the Earth.
    Answer: A Source: Section II-6

87. What is surprising about the extrasolar planets which have so far been discovered?
    A) Many of them are terrestrial planets like the Earth, but orbiting at distances characteristic of giant planets like Jupiter, where terrestrial planets cannot form.
    B) Many of them are giant planets like Jupiter, orbiting at distances characteristic of terrestrial planets like the Earth where giant planets cannot form.
    C) The majority of them rotate much faster than the planets in our solar system.
    D) More than half of them have strong lines of molecular oxygen in their spectra, a possible indication of life on these planets.
    Answer: B Source: Section II-6

# Chapter 5: Earth and Moon

1. The basic colors of Earth as seen from outer space (see Fig. 5-1, Kaufmann & Comins, *Discovering the Universe*, 5th ed.) are
   A) brown. B) blue, white and brown. C) grey and white. D) green and brown.
   Answer: B Source: Section 5-1

2. Which is the most abundant gas in the Earth's atmosphere?
   A) Carbon dioxide B) Nitrogen C) Hydrogen D) Oxygen
   Answer: B Source: Section 5-1

3. What is the ratio of nitrogen to oxygen in the Earth's atmosphere?
   A) 1 part nitrogen to 4 parts oxygen.
   B) 1 part nitrogen to 2 parts oxygen.
   C) 4 parts nitrogen to 1 part oxygen.
   D) Equal parts nitrogen and oxygen.
   Answer: C Source: Section 5-1

4. The major constituents of the Earth's atmosphere are
   A) 77% oxygen, 21% nitrogen.
   B) 95% carbon dioxide and some water vapor.
   C) 77% nitrogen, 21% oxygen.
   D) methane, ammonia, water vapor and carbon dioxide in about equal amounts.
   Answer: C Source: Section 5-1

5. Which of the planets fits the following description: "Cool, solid surface with an atmosphere of $N_2$ and $O_2$, and $H_2O$ clouds"?
   A) Earth B) Mars C) Venus D) Mercury
   Answer: A Source: Section 5-1

6. The planet whose atmosphere is composed primarily of nitrogen is
   A) Earth. B) Jupiter C) Mars. D) Venus.
   Answer: A Source: Section 5-1

7. The atmosphere that we are now breathing is
   A) the seventh atmosphere that the Earth has had.
   B) the third atmosphere that the Earth has had.
   C) the second atmosphere that the Earth has had.
   D) he first atmosphere that the Earth has had.
   Answer: B Source: Section 5-1

8. What were the dominant gases in the Earth's earliest atmosphere, after it first formed?
   A) Methane and ammonia.
   B) Nitrogen and oxygen.
   C) Carbon dioxide and nitrogen.
   D) Hydrogen and helium.
   Answer: D Source: Section 5-1

9. Why did the Earth's earliest atmosphere, composed primarily of hydrogen, not last long?
   A) The hydrogen soon became dissolved in the Earth's oceans.
   B) Biological activity very quickly combined the hydrogen with oxygen to form water.
   C) Hydrogen is highly reactive and soon became bound into chemical compounds in the Earth's rocks.
   D) Hydrogen is a very light gas and soon escaped into space.
   Answer: D Source: Section 5-1

10. What were the dominant gases in the Earth's second atmosphere (the one that replaced the Earth's earliest atmosphere)?
    A) Carbon dioxide and nitrogen.
    B) Methane and ammonia.
    C) Hydrogen and helium.
    D) Nitrogen and oxygen.
    Answer: A Source: Section 5-1

11. Roughly how heavy was the Earth's second atmosphere (i.e., how much gas was there in it) compared to today's atmosphere?
    A) Twice as heavy.
    B) Same as the present atmosphere.
    C) 100 times heavier.
    D) 1/10 as heavy.
    Answer: C Source: Section 5-1

12. Billions of years ago, the Earth's atmosphere was composed primarily of carbon dioxide. Where might you go on Earth to find a large fraction of this carbon dioxide today?
    A) Anywhere, since most of the carbon dioxide is still in the atmosphere, but nitrogen and oxygen have since been added to it.
    B) Nowhere, since most of the carbon dioxide escaped into space.
    C) To extinct volcanoes which are composed of rock from the Earth's interior.
    D) To the Rocky Mountains of North America, which are composed largely of limestone.
    Answer: D Source: Section 5-1

13. Billions of years ago, the Earth's atmosphere was composed primarily of carbon dioxide. What happened to much of this carbon dioxide?
    A) It is still in the atmosphere.
    B) It was dissolved into the Earth's oceans.
    C) It was lost to space.
    D) It was broken into carbon and oxygen by solar ultraviolet light.
    Answer: B Source: Section 5-1

14. In which of the following ways are Venus and Mars alike and yet are both markedly different from Earth?
   A) They are both perpetually shrouded in clouds.
   B) They both have either active or extinct volcanoes on their surface.
   C) Their surface temperatures are both much higher than that of Earth.
   D) Their atmospheres are both made up primarily of carbon dioxide.
   Answer: D Source: Section 5-1

15. The Earth's atmosphere differs from those of near-neighbor planets, Venus and Mars, in one important respect, in that
   A) it has a significant fraction of oxygen in it.
   B) it has a higher pressure than these other planetary atmospheres.
   C) the atmospheric temperature at the surface is much higher than those of the other two planets.
   D) it has a much larger fraction of $CO_2$ than either of the other two atmospheres.
   Answer: A Source: Section 5-1

16. One of the major differences between the Earth and its neighboring planets Venus and Mars is the lack of large quantities of $CO_2$ in its atmosphere near the surface of the planet. If all of these planets were originally formed with significant quantities of this gas in their atmospheres, where is this $CO_2$ on Earth at the present time?
   A) Concentrated high in the atmosphere, where it contributes to the greenhouse effect.
   B) Locked up in rocks such as limestone, formed by life-forms in the sea and on the Earth's surface.
   C) Dissociated by solar UV light into carbon and oxygen, which now exist in abundance as separate chemicals.
   D) Dissolved in seawater, a situation that cannot arise on Venus or Mars.
   Answer: B Source: Section 5-1

17. Which major constituent of the atmosphere of Venus and Mars is present in only very small amounts in the Earth's atmosphere?
   A) Nitrogen, $N_2$ B) Carbon dioxide, $CO_2$ C) Oxygen, $O_2$ D) Methane, $CH_4$
   Answer: B Source: Section 5-1

18. In what crucial way is the atmosphere of Earth very different from the atmospheres of Venus and Mars?
    A) Hydrogen content. This gas makes up a large fraction of the atmospheres of both Venus and Mars but it is absent from the Earth's atmosphere.
    B) Methane ($CH_4$) content. This gas is present in large quantities on Venus and Mars but is only a trace gas on Earth.
    C) Temperature near the surface. Earth's temperature is relatively cool whereas both Venus and Mars have very high atmospheric temperatures.
    D) $CO_2$ content. This gas is the dominant component on Venus and Mars but is a minor constituent on Earth.
    Answer: D Source: Section 5-1

19. Which of the following was NOT an important process in helping to remove carbon dioxide from the Earth's early atmosphere?
    A) Sedimentation of carbon compounds onto the ocean floors.
    B) Biological activity.
    C) Dissolving of carbon dioxide into the oceans.
    D) Escape of carbon dioxide into space.
    Answer: D Source: Section 5-1

20. The presence of oxygen in the Earth's atmosphere is thought to result directly from what type of process?
    A) Out-gassing of the oceans.
    B) Volcanic eruptions.
    C) Biological activity of plants and animals.
    D) The original condensation of interplanetary gas clouds.
    Answer: C Source: Section 5-1

21. The large amount of free oxygen in the Earth's present atmosphere is primarily a result of
    A) $CO_2$ becoming dissolved in the oceans, releasing $O_2$.
    B) biological processes such as photosynthesis.
    C) splitting of $CO_2$ into carbon and oxygen by solar ultraviolet light.
    D) out-gassing by volcanoes and other geological processes.
    Answer: B Source: Section 5-1

22. The molecular oxygen in the present Earth's atmosphere was most probably produced
    A) by biological activity such as photosynthesis from living things.
    B) at the formation of the Earth and has always been present.
    C) from volcanic eruptions as the primitive Earth cooled down.
    D) from seawater by out-gassing.
    Answer: A  Source: Section 5-1

23. Which one of the following statements is true for the Earth?
    A) Out-gassing by volcanic eruptions converted the carbon-dioxide-rich atmosphere into an oxygen-rich atmosphere, thus creating conditions in which life could develop.
    B) Life developed in a carbondioxiderich atmosphere, and then converted this into an oxygenrich atmosphere.
    C) The Earth has always had an oxygen-rich atmosphere, which was one reason life could develop on the Earth.
    D) The oxygen in our atmosphere was created by evaporation of sea water, allowing life to move from the seas onto dry land.
    Answer: B  Source: Section 5-1

24. Which of the following is not a region of the Earth's atmosphere or near-Earth environment?
    A) The stratosphere.
    B) The troposphere.
    C) The chromosphere.
    D) The magnetosphere.
    Answer: C  Source: Section 5-1

25. In which layer of the Earth's atmosphere is the ozone layer located?
    A) The stratosphere.
    B) The mesosphere.
    C) The troposphere.
    D) The thermosphere.
    Answer: A  Source: Section 5-1

26. What is ozone?
    A) A combination of oxygen, nitrogen, and electrons.
    B) Ionized oxygen atoms.
    C) One of a number of chlorofluorocarbons (CFCs).
    D) A molecule made of three oxygen atoms.
    Answer: D  Source: Section 5-1

27. The Earth's stratosphere is warmer than the layers above and below it because
    A) warm air heated by contact with the ground rises into the stratosphere and heats it.
    B) $CO_2$ in the stratosphere absorbs infrared light radiated outward by the ground.
    C) the methane released when we burn fossil fuel collects in this layer and absorbs infrared light.
    D) ozone in the stratosphere absorbs ultraviolet radiation from the Sun.
    Answer: D Source: Section 5-1

28. Why does the temperature in the stratosphere increase with increasing altitude?
    A) It is heated by solar infrared radiation absorbed by carbon dioxide and water vapor.
    B) Higher altitudes are closer to the Sun.
    C) It is heated by solar ultraviolet radiation absorbed by the ozone layer.
    D) Charged particles from the magnetosphere collide with atoms in the stratosphere, depositing energy.
    Answer: C Source: Section 5-1

29. The gas temperature in the stratosphere of the Earth's atmosphere reaches a maximum at about 50 km. The cause of this is
    A) heating by auroral activity higher in the atmosphere.
    B) absorption of solar ultraviolet radiation by ozone, $O_3$.
    C) ionization of $O_2$ and $N_2$ by solar ultraviolet radiation.
    D) turbulence, caused by wind and weather.
    Answer: B Source: Section 5-1

30. What is the most important reason why we need the ozone layer?
    A) It protects us from the solar wind.
    B) It provides a convenient dumping site for chlorofluorocarbon chemicals which are harmful to life.
    C) It allows longdistance radio communication by reflecting radio waves back to the Earth's surface.
    D) It shields us from harmful solar ultraviolet radiation.
    Answer: D Source: Section 5-1

31. The chemical constituent which absorbs UV radiation in the intermediate layers of the atmosphere of Earth (the stratosphere and mesosphere) thereby heating these layers to relatively high temperatures is
    A) $N_2$. B) ozone, $O_3$. C) $H_2O$ water vapor. D) $CO_2$.
    Answer: B Source: Section 5-1

32. Which of the following would NOT be a consequence of the destruction of the ozone layer on the Earth?
    A) Elimination of the rise in temperature in the stratosphere.
    B) Large-scale freezing of the oceans.
    C) Large-scale (perhaps total) destruction of life on the Earth.
    D) Drastic increase in ultraviolet radiation at the Earth's surface.
    Answer: B Source: Section 5-1

33. The major layers of the Earth's atmosphere, from the surface upward, in correct order, are
    A) troposphere, stratosphere, mesosphere, thermosphere.
    B) mesosphere, troposphere, thermosphere, stratosphere.
    C) stratosphere, mesosphere, thermosphere, troposphere.
    D) thermosphere, mesosphere, troposphere, stratosphere.
    Answer: A Source: Section 5-1

34. What is the basic structure of the Earth's atmosphere?
    A) Two layers: temperature decreasing with increasing altitude in the lower layer, then increasing with increasing altitude in the upper layer.
    B) Four layers of alternating temperature profiles: temperature decreasing, then increasing, then decreasing, then increasing with altitude.
    C) Smoothly increasing temperature with increasing altitude.
    D) Smoothly decreasing temperature with increasing altitude.
    Answer: B Source: Section 5-1

35. How does the temperature of the Earth's atmosphere vary with height over the range 0 to 80 km?
    A) It decreases steadily until it reaches a minimum at 80 km.
    B) It decreases, then increases, then decreases again.
    C) It increases, then decreases, then increases again.
    D) It rises steadily until it reaches a high and constant value above 80 km.
    Answer: B Source: Section 5-1

36. The lowest temperature in the Earth's atmosphere is about
    A) 273 K. B) 200 K. C) 218 K. D) 3 K.
    Answer: B Source: Section 5-1

37. The lowest temperature in the Earth's atmosphere occurs at an altitude of about
    A) 80 km. B) 50 km. C) 0 km (the surface of the Earth). D) 10 km.
    Answer: A Source: Section 5-1

38. The coldest layer of the Earth's atmosphere exists between which two main regions?
    A) Troposphere and Earth's surface.
    B) Stratosphere and mesosphere.
    C) Mesosphere and thermosphere.
    D) Troposphere and stratosphere.
    Answer: C Source: Section 5-1

39. The Earth's thermosphere is
    A) the layer of molten iron and nickel below the mantle.
    B) the region of the magnetosphere in which trapped high-energy charged particles spiral along magnetic field lines.
    C) the intermediate atmospheric layer in which ultraviolet light from the Sun is absorbed by ozone ($O_3$) molecules.
    D) the outermost atmospheric layer in which ultraviolet light from the Sun is absorbed by atoms, stripping them of one or more electrons.
    Answer: D Source: Section 5-1

40. Which is the lowest layer in the Earth's atmosphere?
    A) The stratosphere.
    B) The troposphere.
    C) The magnetosphere.
    D) The thermosphere.
    Answer: B Source: Section 5-1

41. In which layer of the Earth's atmosphere does all of the weather occur?
    A) The stratosphere.
    B) The mesosphere.
    C) The troposphere.
    D) The thermosphere.
    Answer: C Source: Section 5-1

42. The troposphere of a terrestrial planet is
    A) the atmospheric layer that contains the highest concentration of ozone.
    B) the uppermost layer of solid rock below the planet's crust.
    C) the atmospheric layer closest to the ground.
    D) the atmospheric layer above the mesosphere.
    Answer: C Source: Section 5-1

43. What is pressure?
    A) Force divided by the area over which the force acts.
    B) Force times the area over which the force acts.
    C) Same as force, but expressed in different units.
    D) Force times the distance over which the force acts.
    Answer: A Source: Section 5-1

44. What fraction of the total mass of the Earth's atmosphere is contained in the troposphere, the lowest layer of the Earth's atmosphere?
A) 75% B) 10% C) 50% D) 25%
Answer: A Source: Section 5-1

45. The pressure in the atmosphere of the Earth (or of any other planet) varies with height in what characteristic way?
A) It decreases by a fixed amount per unit height interval (20% of surface pressure every 5.5 km on Earth).
B) It decreases by a fixed fraction each interval of height (down by 1/2 every 5.5 km on Earth).
C) It decreases and increases several times with increasing height, following the temperature variation.
D) It remains constant with increasing height.
Answer: B Source: Section 5-1

46. Air pressure falls by roughly half for every 5.5 km increase in altitude. What, then, is the air pressure at the top of the troposphere, 11 km above the Earth's surface?
A) 25% of that at the Earth's surface.
B) 50% of that at the Earth's surface.
C) 33% of that at the Earth's surface.
D) Zero.
Answer: A Source: Section 5-1

47. Who first postulated that continents drift around over the Earth's surface?
A) Alfred Wegener.
B) J. Tuzo Wilson.
C) James Van Allen.
D) Charles Darwin.
Answer: A Source: Section 5-2

48. Which fact first gave Alfred Wegener the idea that continents can drift over the Earth's surface?
A) The ocean floors were much younger than the continents, indicating that they were still being formed.
B) The east coasts of N. and S. America fitted nicely against the west coasts of Europe and Africa.
C) A system of mountains and faults running up the middle of the Atlantic sea floor showed evidence that the crust was spreading apart there.
D) Volcanoes and earthquakes were localized into a welldefined "ring of fire" around the Pacific Ocean and other areas.
Answer: B Source: Section 5-2

49. How has the Mid-Atlantic Ridge been formed?
    A) The weight of sediments has caused the ocean floor to sink, and the ridge has resulted from slumping toward the center of this basin.
    B) Two crustal plates have collided, causing one plate to buckle, forming the ridge, while the other plate was thrust down beneath it.
    C) Two crustal plates have been sliding past each other in a transverse fault.
    D) Molten rock has pushed up from the Earth's interior and forced two crustal plates apart.
    Answer: D Source: Section 5-2

50. The Mid-Atlantic Ridge in the Earth's crust is a region where
    A) one tectonic plate is moving below another (subducting).
    B) two tectonic plates are slowly spreading apart.
    C) two tectonic plates are pushing against one another, forcing the ridge upward.
    D) a single hot plume is pushing molten magma or lava through a break in the crust.
    Answer: B Source: Section 5-2

51. The Mid-Atlantic Ridge is being produced by
    A) the weight of the Atlantic Ocean on the thin seabed.
    B) tidal flow of ocean water meeting in mid-Atlantic.
    C) two tectonic plates moving apart because of volcanic upflow.
    D) two tectonic plates pushing together, producing upthrust.
    Answer: C Source: Section 5-2

52. The San Andreas fault in California is an example of
    A) two tectonic plates sliding past each other.
    B) a spreading center, where two tectonic plates are being pushed away from each other.
    C) an upthrust due to a hot spot in the Earth's mantle.
    D) a subduction zone, where one plate is pushed back down into the Earth.
    Answer: A Source: Section 5-2

53. In the modern theory of crustal motion on the Earth's surface, the process of sea floor spreading is described as
    A) the motion of plates away from mid-oceanic ridges and toward continental boundaries.
    B) the rotation of plates around axes that remain stationary on the Earth causing plate-edge collisions.
    C) the motion of plates toward mid-oceanic ridges and away from continental boundaries.
    D) the northward motion of some plates and the southward motion of others, causing earthquakes where they slide past each other.
    Answer: A Source: Section 5-2

54. Approximately how many tectonic plates make up the Earth's surface?
 A) Seven (one under each continent).
 B) Hundreds.
 C) About a dozen.
 D) Two (the eastern and western hemispheres).
 Answer: C Source: Section 5-2

55. What is a typical speed of drift for a continent sliding over the Earth's surface?
 A) A few centimeters per century.
 B) A few centimeters per year.
 C) A few meters per year.
 D) A few centimeters per million years.
 Answer: B Source: Section 5-2

56. The average speed of motion of the plates on the Earth's surface is
 A) a few centimeters per year.
 B) about 10 meters per year.
 C) a few centimeters per century.
 D) very small, less than a millimeter per century.
 Answer: A Source: Section 5-2

57. The process of sea floor spreading and plate tectonic movement on the Earth's surface takes place at a speed of
 A) a few meters per year.
 B) a few centimeters per year.
 C) a few centimeters per century.
 D) less than a millimeter per year.
 Answer: B Source: Section 5-2

58. The African and South American continents are separating at a rate of about 3 cm per year, according to the ideas of plate tectonics. If they are now 5000 km apart, and have moved at a constant speed over this time, how long is it since they were in contact?
 A) 170 million years.
 B) 1.7 million years.
 C) 1.7 billion years.
 D) 170 thousand years.
 Answer: A Source: Section 5-2

59. What is the cause of great mountain ranges on the Earth, such as the Rocky Mountains and the Andes?
    A) Heat from the Earth's interior causing the Earth's crust to expand and then crumple.
    B) A collision of two tectonic plates, where one is folded into mountains while the other is thrust underneath.
    C) Two tectonic plates being pushed apart by molten rock that is being forced up between them.
    D) The carving of continents during ice ages, with the mountains left behind as "islands" in a sea of glaciers.
    Answer: B Source: Section 5-2

60. The Rocky Mountains of North America were formed by
    A) folding of the Earth's crust where two crustal plates collided along a subduction zone.
    B) asteroid impacts during the first billion years of the Earth's history.
    C) shrinking and folding of the Earth's crust as the Earth cooled and contracted.
    D) an ancient mid-ocean ridge now uplifted to become dry land.
    Answer: A Source: Section 5-2

61. The great mountain ranges of the Earth have been produced by
    A) asteroid impacts, since they are just worn-down crater walls.
    B) volcanic eruptions.
    C) collisions between tectonic plates.
    D) wrinkling of the crust as the interior cools and contracts.
    Answer: C Source: Section 5-2

62. To what does the name "Pangaea" refer?
    A) The supercontinent that split into the present North and South America, Europe and Africa.
    B) The outer life-bearing layer of the Earth, including the soil, oceans and atmosphere.
    C) The tectonic plate on which most of North America is riding.
    D) The ocean that covered most of North America millions of years ago.
    Answer: A Source: Section 5-2

63. The supercontinent Pangaea is believed to have been
    A) the original surface of the Earth, before the oceans formed.
    B) the result of a collision of continents, of a type that occurs roughly every 500 million years.
    C) a myth, part of the Atlantis legend.
    D) the first continent that rose from the sea, and broke up 200 million years ago.
    Answer: B Source: Section 5-2

64. In what way are the supercontinents Laurasia, Pangaea, and Gondwanaland related?
    A) Pangaea split into Gondwanaland and Laurasia.
    B) Gondwanaland split into Pangaea and Laurasia.
    C) An earlier, larger supercontinent split into Laurasia, Pangaea, and Gondwanaland.
    D) Laurasia split into Gondwanaland and Pangaea
    Answer: A Source: Section 5-2

65. The motions of large portions of the Earth's surface, the "plates," are caused by
    A) convective flow of material in the Earth's interior.
    B) flexing of the Earth's surface due to solar heating and nighttime cooling.
    C) tidal flow in oceanic waters.
    D) pressure variations in the Earth's atmosphere, both daily and seasonal.
    Answer: A Source: Section 5-2

66. Which of the following statements correctly describes the surface of the Earth?
    A) A thin, deformable crust that allows the continents to slide over it.
    B) A thick, solid crust that holds the continents and ocean floors fixed in position.
    C) Individual and separate solid crustal plates that are pushed around by convection in the underlying mantle.
    D) Individual, solid crustal plates that are pushed around by the Earth's rotation and the tidal forces of the Moon and Sun.
    Answer: C Source: Section 5-2

67. "Continental drift" on the Earth is now thought to be caused by
    A) tidal flexing of the Earth's solid crust by the Moon's gravitational pull.
    B) precession of the Earth's spin axis.
    C) the forces of oceanic tides on the continental shelves around the land-masses.
    D) circulation currents in the deep interior, causing slabs of the Earth's crust to move slowly.
    Answer: D Source: Section 5-2

68. Continental drift on the Earth is a result of
    A) the momentum of impacts from large meteorites hitting the early Earth.
    B) large-scale circulation of partly molten or plastic rock in the Earth's interior.
    C) flexing of the Earth's surface by lunar and solar tides.
    D) water percolating down through geological fault zones, acting as a lubricant.
    Answer: B Source: Section 5-2

69. Deep oceanic trenches on the Earth are locations at which
    A) molten lava oozes out between two tectonic plates that are spreading apart.
    B) volcanic material comes out from the deep interior at volcanic islands, such as Hawaii.
    C) cool surface material on Earth sinks below other material at a tectonic plate boundary.
    D) dense material sank while lighter material rose to the surface during the early geological history of the Earth.
    Answer: C Source: Section 5-2

70. All of the boundaries of the major moving tectonic plates on the Earth's surface are coincident with
    A) the positions of maximum earthquake occurrence.
    B) the edges of the continental shelves around the major continents.
    C) regions where ocean depths are greatest.
    D) the occurrence of major auroral activity.
    Answer: A Source: Section 5-2

71. On the Earth, the majority of earthquakes occur
    A) in the centers of tectonic plates (e.g., North American continent).
    B) along the zone of maximum tidal stress around the equator.
    C) in the arctic and antarctic regions.
    D) along the boundaries of major tectonic plates.
    Answer: D Source: Section 5-2

72. In the modern theory of crustal motion on the Earth's surface, the process of seafloor spreading is described as
    A) the motion of plates toward mid-oceanic ridges and away from continental boundaries.
    B) the upward motions of some plates and the downward motions of others while they remain in fixed positions upon the Earth.
    C) the rotation of plates around axes which remain stationary on the Earth causing plate-edge collisions.
    D) the motion of plates away from mid-oceanic ridges and toward continental boundaries.
    Answer: D Source: Section 5-2

73. Subduction on the Earth is the process by which
    A) volcanic material emerges from the deep interior at a volcanic island such as Hawaii.
    B) cool surface material on Earth sinks below other material at a tectonic plate boundary.
    C) dense material sank while lighter material rose to the surface during the early geological history of the Earth.
    D) molten lava oozes out between two tectonic plates which are spreading apart.
    Answer: B Source: Section 5-2

74. Craters in the abundance seen on the Moon are not apparent on Earth at the present time because
    A) all the potentially damaging interplanetary bodies were stopped by the Earth's atmosphere.
    B) interplanetary objects have avoided the Earth during its history.
    C) plate tectonics has returned cratered surface layers into the Earth's interior and weathering has obliterated the more recent craters.
    D) the Moon protected the Earth from impacts and this resulted in the craters and maria on the Moon.
    Answer: C Source: Section 5-2

75. Where do most of the Earth's earthquakes occur?
    A) Where crustal plates are colliding, separating or sliding past each other.
    B) More or less evenly over the entire Earth's surface.
    C) Only where crustal plates are sliding past each other, as in the San Andreas fault.
    D) Only where crustal plates are colliding head-on.
    Answer: A Source: Section 5-2

76. Which of the following places on Earth experiences frequent earthquakes and volcanic activity because of its location? (See Fig. 5-6, *Discovering the Universe*, Kaufmann & Comins, 5th Ed.)
    A) Australia B) Central Canada C) Iceland D) Brazil
    Answer: C Source: Section 5-2

77. Which two tectonic plates are slowly separating from each other along the Mid-Atlantic Ridge in the south Atlantic, on the Earth's surface? (See Fig. 5-6, *Discovering the Universe*, Kaufmann & Comins, 5th Ed.)
    A) The Pacific and Australia-India plates.
    B) The African and Eurasian plates.
    C) The South American and African plates.
    D) The Nazca and Pacific plates.
    Answer: C Source: Section 5-2

78. Against which tectonic plate is the Australia-India tectonic plate pushing to form the Himalayan mountains? (see Fig. 5-6, *Discovering the Universe*, Kaufmann & Comins, 5th Ed.)
    A) The Eurasian Plate.
    B) The North American Plate.
    C) The Antarctic Plate.
    D) The African Plate.
    Answer: A Source: Section 5-2

79. Which ONE of the following processes was important for melting the entire Earth about 4.6 billion years ago?
    A) Nuclear fusion, or the combining of lighter nuclei to form heavier ones.
    B) Tidal heating of the Earth due to the combined gravitational pulls of the Sun and the Moon.
    C) Heat released by the condensation of water vapor into liquid water, to form the oceans.
    D) Radioactivity, or the spontaneous breaking apart of heavier nuclei into lighter ones.
    Answer: D Source: Section 5-3

80. To what does "planetary differentiation" refer?
    A) The large-scale convection of rock in the mantle of a planet, which on the Earth causes continental drift.
    B) The formation of rocky planets in the hotter, inner solar system and gas giants in the colder, outer regions.
    C) The sinking of heavier elements toward the center of a planet and the floating of lighter elements toward the surface.
    D) The circulation of iron in the core of a planet, resulting in the generation of a magnetic field.
    Answer: C Source: Section 5-3

81. The process by which heavier materials sank into the centers of terrestrial planets, while lighter material rose to the surfaces early in the history of these planets, is known as
    A) subduction.
    B) chemical differentiation.
    C) seafloor spreading.
    D) hot-spot volcanism.
    Answer: B Source: Section 5-3

82. Which of the following mechanisms is most likely to have taken place while the Earth was molten, to form the present structure of the Earth?
    A) Heavy elements sank to the center under gravity while lighter materials rose to the surface and solidified into rocks.
    B) All materials were thoroughly mixed by convection in the molten state, and the Earth remained mixed as it cooled.
    C) Lighter elements sank to the center leaving the heavier material to form the rocky surface after cooling.
    D) Hydrogen and helium became highly compressed by gravity and sank to the core below a layer of heavier rocky material.
    Answer: A Source: Section 5-3

83. Why is it that terrestrial planets are thought to have dense iron cores?
   A) During the formation of the planets, heavy elements accumulated first, then the lighter elements accreted onto them (bimodal accretion).
   B) Magnetism in the iron was sufficiently powerful to pull it to the center as the planet formed (magnetic sorting).
   C) The planets were molten early in their life, and the heavy elements sank and lighter materials floated to the surface (chemical differentiation).
   D) Thermonuclear reactions produced iron as a fusion product in the earlier phases of planetary evolution.
   Answer: C Source: Section 5-3

84. Where is the mantle of the Earth located?
   A) In the outermost atmosphere and beyond, where the planet's magnetic field captures solar wind particles.
   B) Between the core and the crust.
   C) Between the stratosphere and the thermosphere.
   D) Outside of, but in contact with, the crust; i.e., the oceans and atmosphere.
   Answer: B Source: Section 5-3

85. Of what is the mantle of the Earth composed?
   A) Rock B) Iron C) Atmospheric gases D) Water
   Answer: A Source: Section 5-3

86. The Earth's mantle, the semi-molten layer below the crust, is composed largely of what chemical materials?
   A) Minerals rich in iron and magnesium. C) Iron-poor rocks and minerals.
   B) Solid hydrogen and helium. D) Almost pure iron.
   Answer: A Source: Section 5-3

87. The Earth has an average density that is approximately
   A) 2 times that of water. C) more than 10 times that of water.
   B) equal to that of water. D) 5 times that of water.
   Answer: D Source: Section 5-3

88. The core of the Earth occupies what fraction of its radius?
   A) About 1/4. B) Roughly half. C) Less than 10%. D) Almost 80%.
   Answer: B Source: Section 5-3

89. What is the basic structure of the Earth's interior?
    A) Molten iron core, molten rocky mantle, solid rocky crust.
    B) Molten iron inner core, molten rocky outer core, solid rocky mantle, lighter rocky crust.
    C) Solid iron inner core, molten iron outer core, rocky mantle, lighter rocky crust.
    D) Molten iron inner core, solid iron outer core, rocky mantle, lighter rocky crust.
    Answer: C Source: Section 5-3

90. The internal structure of the Earth is
    A) a core of rock and iron, surrounded by a mantle of liquid hydrogen.
    B) a core of solid rock extensively enriched in iron, surrounded by a solid mantle of pure rock.
    C) a large core of iron, partly solid and partly molten, surrounded by a thick, flexible mantle of rock.
    D) a large, solid iron core surrounded by a thick, flexible mantle of rock.
    Answer: C Source: Section 5-3

91. Of what material is the core of the Earth composed?
    A) Roughly half rock and half iron.
    B) Mostly iron.
    C) Rock of similar composition to that in the crust, but much denser.
    D) Titanium and nickel.
    Answer: B Source: Section 5-3

92. The chemical make-up of the central core of the Earth is considered to be
    A) sulfur-rich minerals compressed to a high density.
    B) rocky minerals rich in iron.
    C) almost pure iron.
    D) very close to that at the surface; i.e., silicon-rich rocks and minerals.
    Answer: C Source: Section 5-3

93. What is believed to be the composition of the Earth's core?
    A) About 80% iron and 20% lighter elements.
    B) About half iron and half lighter elements.
    C) Essentially pure iron.
    D) About 1/4 iron and 3/4 lighter elements.
    Answer: C Source: Section 5-3

94. The waves that geologists and geophysicists use to probe the inside of the Earth are
    A) microwaves. B) gravitational waves. C) radio waves. D) seismic waves.
    Answer: D Source: Section 5-3

95. Which scientific approach gives us the most information about the deep interior of the Earth?
   A) Worldwide measurement of low frequency seismic waves produced by earthquakes.
   B) Deep drilling of exploratory holes for science and mineral recovery (e.g., oil).
   C) The study of lava flows from volcanoes.
   D) Measurement of cosmic neutrinos which pass very easily through the Earth.
   Answer: A Source: Section 5-3

96. Which of the following scientific approaches has NOT yet been exploited for the study of the deep interior of Earth?
   A) Study of lavas from volcanoes such as those on Hawaii.
   B) Magnetic field studies.
   C) Manned deep-drill projects into the deep interior via extensive caves beneath the Earth's surface.
   D) Measurements of seismic waves from earthquakes.
   Answer: C Source: Section 5-3

97. Which of the following techniques is the primary one used by geologists and geophysicists to probe the structure of the Earth's core and mantle?
   A) X-ray analysis from satellites.
   B) The study of the deflection of seismic waves from earthquakes.
   C) Direct sampling of interior rock by deep drilling through the ocean floor.
   D) Extrapolation of surface features (e.g., mountain chains) into the deep interior.
   Answer: B Source: Section 5-3

98. What effect does an increase in pressure have on the temperature at which rock melts?
   A) It has no effect on the melting temperature, since the melting temperature is independent of pressure.
   B) It increases the melting temperature.
   C) It lowers the melting temperature.
   D) We do not know yet; the temperature needed to melt rock has not yet been achieved in the laboratory.
   Answer: B Source: Section 5-3

99. The material of the Earth's interior becomes molten at a certain depth below the surface. This depth is
   A) 300 km.
   B) 5200 km.
   C) 2900 km.
   D) This statement is erroneous since nowhere is the interior molten.
   Answer: C Source: Section 5-3

100. The depth from the surface of the Earth to the top of its liquid core is about (see Fig. 5-7, Kaufmann & Comins, *Discovering the Universe*, 5th Ed.)
A) 6400 km. B) 5000 km. C) 30 km. D) 2900 km.
Answer: D Source: Section 5-3

101. The Earth's solid inner core results from the fact that
A) the inner core has a different composition than the outer core, with a higher melting temperature.
B) the temperature of the inner core is lower than the temperature of the outer core, producing a "temperature inversion".
C) the melting temperature of an iron-nickel mixture increases with increasing pressure, and rises above the actual temperature in the inner core.
D) the outer core is heated and melted by friction between the core and the mantle, and this heating does not extend to the inner core.
Answer: C Source: Section 5-3

102. The outer core of the Earth is molten, but the inner core is solid. The reason for this is that
A) the inner core has a different composition than the outer core, with a higher melting point.
B) the melting point of the rocks is raised above the actual temperature by the extremely high pressure.
C) the lower pressure in the inner core allows the material to freeze out of the molten rock.
D) the temperature is lower in the inner core than in the surrounding region.
Answer: B Source: Section 5-3

103. Which of the following statements correctly describes the mantle of the Earth?
A) Solid rock that is hot enough to become semi-molten and flow.
B) Solid, immovable rock upon which the crustal plates can slide.
C) Molten rock.
D) Molten iron.
Answer: A Source: Section 5-3

104. Which of the following statements correctly describes the mantle of the Earth?
   A) Hot, solid rock that flows downward in hotter parts of the mantle and upward in cooler parts.
   B) Hot, solid rock that flows upward in hotter parts of the mantle and downward in cooler parts.
   C) Hot, solid rock that flows upward at the equator and downward at the poles, because of the Earth's rotation.
   D) Hot, solid rock that flows downward at the equator and upward at the poles, because of the Earth's rotation.

   Answer: B Source: Section 5-3

105. What is the relationship between the mantle and the crust of the Earth?
   A) New crust is formed by magma rising from the mantle in some places, and old crust is pushed back down into the mantle in other places.
   B) The crust sits motionless on top of the mantle and does not interact with it.
   C) New layers of crust are formed when magma from the mantle flows out over old crust, and this old crust is pushed back down into the mantle by the weight of the magma on top of it.
   D) Convection in the mantle moves the continents around, but there is no transfer of material from the mantle to the crust or vice-versa.

   Answer: A Source: Section 5-3

106. How is the Earth's magnetic field generated?
   A) By permanent magnetism in the Earth's crustal rocks.
   B) By electric currents in the Earth's core.
   C) By electric currents in the Earth's mantle.
   D) By the flow of electrons and ions in the Earth's magnetosphere.

   Answer: B Source: Section 5-4

107. The magnetic field of the Earth is thought to be caused by
   A) electric currents flowing in the liquid core of the Earth.
   B) localized magnetic anomalies near the Earth's surface.
   C) the flow of solar wind particles around and within the Earth's outer atmospheric region, the magnetosphere.
   D) a magnetized solid iron core in the Earth's interior.

   Answer: A Source: Section 5-4

108. The magnetic field of the Earth is caused by
   A) a solid iron magnet inside the Earth.
   B) the motion of the solar wind blowing by the Earth.
   C) electric currents flowing in the high atmosphere (ionosphere) of the Earth.
   D) electric currents flowing in the molten core.
   Answer: D Source: Section 5-4

109. The Earth's magnetosphere is
   A) the region in the Earth's crust near each magnetic pole.
   B) the atmospheric layer between the stratosphere and thermosphere, where motions are governed by the Earth's magnetic field.
   C) the region beyond the Earth's atmosphere, where the Earth's magnetic field protects us from the solar wind.
   D) the molten core of the Earth, whose motions produce the magnetic field.
   Answer: C Source: Section 5-4

110. What mechanism forms the magnetospheres around several of the planets?
   A) Repulsion of the solar wind by the planetary magnetic field.
   B) The repulsion of solar electromagnetic radiation by the planetary magnetic field.
   C) Heating of the upper atmosphere by solar ultraviolet radiation.
   D) The very rapid spin of these planets, spinning the atmosphere outward.
   Answer: A Source: Section 5-4

111. Where are the Earth's Van Allen radiation belts located?
   A) In the molten iron core.
   B) In the stratosphere.
   C) In the magnetosphere.
   D) In the auroral regions over the north and south magnetic poles.
   Answer: C Source: Section 5-4

112. What are the Van Allen belts?
   A) Regions in the outer solar system beyond Pluto, where comets are thought to originate.
   B) Regions of intense earthquake activity along tectonic plate boundaries on the Earth.
   C) Regions of high-energy charged particles in the Earth's magnetosphere.
   D) Dark regions in Jupiter's atmosphere, circling the planet parallel to the equator.
   Answer: C Source: Section 5-4

113. What are the Van Allen belts?
   A) Two donut-shaped regions of high-energy charged particles in the Earth's magnetosphere.
   B) An oval-shaped region around each of the Earth's magnetic poles where charged particles collide with ions in the Earth's atmosphere.
   C) The inner and outer parts of the asteroid belt.
   D) Regions of high ion concentration in the Earth's upper atmosphere, where radio waves are reflected back toward the surface of the Earth.
   Answer: A Source: Section 5-4

114. What are the names of the two donut-shaped regions where high-energy charged particles are trapped by the Earth's magnetic field in the Earth's magnetosphere?
   A) Van Allen belts.
   B) Auroral ovals.
   C) Ionospheric E and F layers.
   D) Ozone layers.
   Answer: A Source: Section 5-4

115. What causes the auroras?
   A) Charged particles emitting light as they spiral along magnetic field lines in the magnetosphere.
   B) Light from solar flares reflecting from high-altitude clouds in the Earth's atmosphere.
   C) Charged particles from the magnetosphere striking atoms in the upper atmosphere and causing the gases to emit light.
   D) Solar ultraviolet light exciting ozone molecules in the ozone layer.
   Answer: C Source: Section 5-4

116. Auroras on the Earth are caused by
   A) ultraviolet radiation from the Sun, ionizing atoms in the Earth's upper atmosphere.
   B) the reflection of sunlight from the Arctic and Antarctic ice onto the polar skies.
   C) high-energy charged particles from the Earth's magnetic field striking the upper atmosphere.
   D) electrical currents in the ionosphere generated by dynamo action in the Earth's core.
   Answer: C Source: Section 5-4

117. Auroral displays are most often seen
   A) in two bands on either side of the equator, in the tropics.
   B) in circular regions around the N and S geomagnetic poles.
   C) directly above the N and S geomagnetic poles.
   D) at the N and S geographical poles, along the Earth's spin axis.
   Answer: B Source: Section 5-4

118. Where in the northern hemisphere does one most often (almost always!) find the aurora or northern lights?
   A) In an oval around the magnetic north pole between 60 and 70° latitude.
   B) In a small patch at the magnetic north pole.
   C) Only at the midday position over the polar region, when the direct sunlight excites the air to emit light.
   D) Only at the midnight position along a line of longitude from the equator to the north pole.
   Answer: A Source: Section 5-4

119. The auroral display of northern and southern lights, high in the Earth's atmosphere, is caused by
   A) sunlight scattered by very high atmospheric clouds.
   B) fluorescence caused by absorption of solar ultraviolet light by ozone.
   C) reflection of sunlight from the ice in the polar regions.
   D) solar wind electrons hitting the high atmosphere after being accelerated by the magnetosphere.
   Answer: D Source: Section 5-4

120. What is the diameter of the Moon compared to the diameter of the Earth?
   A) About 1/10 of the diameter of the Earth.
   B) Just over 1/2 the diameter of the Earth.
   C) About 1/4 of the diameter of the Earth.
   D) Less than 1/100 of the diameter of the Earth.
   Answer: C Source: Section 5-5

121. Which of the following processes has played the greatest role in shaping the surface of the Moon?
   A) Recent volcanic activity, producing large numbers of crater-like volcanic calderas.
   B) Erosion by wind and atmospheric gases.
   C) Impacts of interplanetary bodies of all sizes.
   D) Motions of tectonic plates, producing mountain ranges wherever they collide.
   Answer: C Source: Section 5-5

122. What is the origin of the majority of lunar craters?
   A) Impacts by space probes.
   B) Volcanic explosions.
   C) Surface collapse after loss of groundwater by evaporation.
   D) Impacts by meteoric material.
   Answer: D Source: Section 5-5

123. Most of the craters on the Moon were formed by
A) volcanic action, the craters being the old calderas of volcanoes.
B) a nuclear war in which the lunar inhabitants wiped themselves off the face of the Moon.
C) bombardment by interplanetary meteoritic material.
D) wind and water erosion of mountains and hills in the distant past.
Answer: C Source: Section 5-5

124. The vast majority of lunar craters were caused by
A) lunar quakes, under gravitational tidal disturbance from the Earth.
B) volcanic eruptions from the Moon's interior.
C) bombardment from space by objects of various sizes.
D) the explosion of rocks under the thermal shock from alternating intense sunlight in daytime to the cold of space at night.
Answer: C Source: Section 5-5

125. Most of the craters on the Moon are thought to have been caused by
A) the intense bombardment by large and small bodies over some specific early period in the Moon's history.
B) the continuous bombardment throughout the Moon's life, including the present and recent past, by large and small asteroids.
C) the collapse of volcanic domes, leaving central peaks in the craters.
D) volcanic activity, leaving behind volcano craters similar to those on Earth.
Answer: A Source: Section 5-5

126. Why do the larger craters on the Moon have central peaks?
A) The incoming projectile was large enough that it was not destroyed, and remained to form the central peak.
B) The crater floor rebounded upward after the initial compression.
C) Debris falling from the crater walls has collected at the center of the crater floor.
D) The impact cracked the crust, and lava flowed into the center of the crater.
Answer: B Source: Section 5-5

127. What are the most common shapes of lunar craters, and why?
A) Random shapes, because mantle convection has deformed the surface and distorted the craters.
B) Round, because most of the craters formed from volcanic explosions, not meteoric impacts.
C) All shapes from round to long and thin, depending on the angle at which the projectile hit the surface.
D) Round, because the incoming projectile vaporized and exploded to form the crater.
Answer: D Source: Section 5-5

128. Why does the ejecta blanket around some lunar craters appear light-colored?
A) The impact punched through the lunar crust into lighter material from the mantle.
B) The crater is relatively young.
C) The crater is relatively old.
D) The incoming projectile was composed of lighter material.
Answer: B Source: Section 5-5

129. What is a mare on the Moon?
A) A sinuous valley, generally caused by the collapse of an old lava tube.
B) A large area of lighter material on the lunar surface.
C) A large crater with a central peak and terracing along the crater walls.
D) A large area of darker material on the lunar surface.
Answer: D Source: Section 5-5

130. Maria are
A) ancient lava flood-plains.
B) heavily cratered highland regions.
C) uplifted regions surrounding large volcanoes.
D) ancient riverbeds, now dry.
Answer: A Source: Section 5-5

131. What is the most likely cause of the smooth and relatively crater-free surfaces of lunar maria?
A) Dust storms which eroded and smoothed the surface.
B) Lava flows in the relatively late geological history of the Moon.
C) Sediments left behind after water had flowed into the basins and evaporated away.
D) Volcanic ash which rained upon the surfaces of the basins in recent geological times.
Answer: B Source: Section 5-5

132. The lunar maria appear smooth because they are
   A) ancient lava flows which occurred soon after the end of an early period of intense bombardment, and which have had relatively few impacts since then.
   B) recent lava flows, occurring within the last billion years, which have obliterated earlier craters.
   C) regions where craters have been obliterated by crustal deformation caused by hot spots and volcanic lava flow from the underlying molten mantle.
   D) ancient sea beds, now dry, dating back to when the Moon had a denser atmosphere and rainfall was abundant.
   Answer: A Source: Section 5-5

133. What is the diameter of Mare Imbrium, the largest lunar "sea," compared to the diameter of the Moon itself?
   A) About 1/3 the diameter of the Moon.
   B) Just over 1/2 the diameter of the Moon.
   C) About 1/100 the diameter of the Moon.
   D) About 1/10 the diameter of the Moon.
   Answer: A Source: Section 5-5

134. Mare Imbrium, the largest of the maria, (the smooth basins on the Earth-facing side of the Moon), has an approximate diameter of (see Fig. 5-17, Kaufmann & Comins, *Discovering the Universe*, 5th Ed., and make a measurement if necessary)
   A) 1100 km. B) 600 km. C) 3500 km. D) 100 km.
   Answer: A Source: Section 5-5

135. What is the name of a long, sinuous crack in a lunar mare?
   A) A regolith. B) A caldera. C) A rift zone. D) A rille.
   Answer: D Source: Section 5-5

136. What is a lunar rille?
   A) A deep trench between two tectonic plates.
   B) A long crack in a mare.
   C) A long, curved mountain range.
   D) A large crater with a central peak.
   Answer: B Source: Section 5-5

137. Lunar rilles are
   A) central peaks of lunar craters.
   B) bright streaks radiating from large impact craters.
   C) large irregular lava basins.
   D) meandering, canyon-like valleys.
   Answer: D Source: Section 5-5

138. A rille is
   A) a shallow winding valley on the Moon.
   B) a meandering canyon on Mars.
   C) a type of geological fault on Venus.
   D) a pattern seen in the tail of a comet.
   Answer: A Source: Section 5-5

139. Where in the universe would you look for a rille?
   A) In the galactic plane, since it is a dark dust band.
   B) On the Moon, since it is a deep winding canyon.
   C) On Mars, since rilles are the "canals" seen by Lowell to extend across the planet surface.
   D) In Saturn's rings, since it is one of the gaps.
   Answer: B Source: Section 5-5

140. How do the near and far sides of the Moon compare to each other?
   A) The near side has several large maria or "seas," whereas the far side has only one small one.
   B) The near side has extensive, heavily cratered regions between and around the smooth maria or "seas," whereas the far side is almost entirely covered with "seas."
   C) Both sides are quite similar to each other: several large maria or "seas" surrounded by heavily cratered terrain.
   D) We don't know; the far side is hidden in perpetual darkness.
   Answer: A Source: Section 5-5

141. The appearance of the entire surface of the Moon could be described as
   A) maria only on the near side, no major maria on the far side.
   B) mostly craters on the near side, extensive maria with few craters on the far side.
   C) uniform distribution of surface features.
   D) dry, waterless terrain in the near side, ice sheets on the dark side.
   Answer: A Source: Section 5-5

142. Examination of the whole surface of the Moon shows us that
   A) there are no differences in surface features around the whole Moon.
   B) the northern hemisphere is distinctly different from the southern hemisphere.
   C) there are two distinctly different sides, that seen from Earth, and that hidden from Earth.
   D) craters exist only on one side of the Moon.
   Answer: C Source: Section 5-5

143. The near and far sides of the Moon are particularly different, as seen by the fact that
   A) the far side is always in darkness.
   B) the number of craters differs markedly, with fewer on the far side.
   C) the far side has no maria.
   D) the average height of the overall terrain is much lower on the far side.
   Answer: C Source: Section 5-5

144. Maria on the Moon exist
   A) uniformly all over the surface of the Moon.
   B) only in a band close to the equator.
   C) only on the Earth-facing side of the Moon.
   D) only in the northern hemisphere.
   Answer: C Source: Section 5-5

145. Which of the following statements is NOT true of the Moon?
   A) There are large, highly cratered regions on both the near and the far sides of the Moon.
   B) There are extensive lava floodplains on both the near and the far sides of the Moon.
   C) Almost all craters on the Moon have been made by meteoroid impacts
   D) There are large basins that were carved out by asteroid impacts.
   Answer: B Source: Section 5-5

146. The mountain ranges on the Moon are
   A) the hard-rock remnants of geological features severely eroded by wind and weather.
   B) lines of extinct volcanoes similar to the Hawaiian Islands, caused by hot-spot vulcanism.
   C) the walls of craters caused by impacts of large objects early in the geological history of the Moon.
   D) the upthrust caused by collisions of moving tectonic plates.
   Answer: C Source: Section 5-5

147. How were the mountain ranges on the Moon formed?
A) They are "spreading centers," where magma from the mantle is rising and pushing tectonic plates apart.
B) They were pushed up by the collisions of tectonic plates on the lunar surface.
C) They were thrust up as crater walls by impacts of large asteroids.
D) They are wrinkles in the crust, created when the Moon cooled and shrunk slightly in size.
Answer: C Source: Section 5-5

148. Most of the mountain ranges on the Moon are
A) the result of the buildup of meteoritic dust from winds on the Moon.
B) the result of plate tectonic movement, similar to that on Earth.
C) the circular edges and rims of large maria, caused by impacts from large objects.
D) the result of water flow and erosion on the Moon's soft surface.
Answer: C Source: Section 5-5

149. The mountains on the Moon were mostly caused by
A) volcanic eruption and buildup of lava.
B) the collisions of crustal plates under tectonic motion.
C) tidal distortion and uplift.
D) impacts from asteroids from outer space.
Answer: D Source: Section 5-5

150. Which of the following processes has played the greatest role in shaping the surface of the Moon?
A) Erosion by wind and atmospheric gases.
B) Impacts of interplanetary bodies of all sizes.
C) Recent volcanic activity, producing large numbers of crater-like volcanic calderas.
D) Motions of tectonic plates, producing mountain ranges wherever they collide.
Answer: B Source: Section 5-5

151. How many times have human beings landed on the Moon?
A) Six times. B) Five times. C) Four times. D) Seven times.
Answer: A Source: Section 5-6

152. Which was the first spacecraft to land human beings on the Moon?
A) Apollo 11. B) Soyuz 5. C) Apollo 10. D) Apollo 13.
Answer: A Source: Section 5-6

153. What is the lunar regolith?
   A) The lower part of the lunar crust that is not extensively cracked by impacts.
   B) The lunar crust and mantle together.
   C) The layer of fine powder covering the lunar surface.
   D) The part of the lunar surface that is not covered with lava flows.
   Answer: C  Source: Section 5-6

154. A regolith is
   A) an extremely large, isolated rock on the surface of a planet or other object.
   B) a lithospheric plate, moved slowly by geologic processes.
   C) a heavily cratered region on a planet or other object.
   D) a layer of pulverized rock on the surface of a planet or other object.
   Answer: D  Source: Section 5-6

155. The texture of the surface of the Moon may be described as
   A) regolith (pulverized rock) in and near craters, and hard bedrock everywhere else.
   B) hard bedrock almost everywhere, since there is very little erosion on the Moon.
   C) regolith (pulverized rock) covering the lunar surface.
   D) eroded basalt held together by subsurface ice (permafrost).
   Answer: C  Source: Section 5-6

156. What name is given to the layer of fine powder and rock fragments that cover the lunar surface?
   A) Regolith  B) Rille  C) Mare  D) Mantle
   Answer: A  Source: Section 5-6

157. What process created the lunar regolith?
   A) Cracking and churning by meteoric bombardment.
   B) Solidification and shrinking of ancient lava flows.
   C) 4.5 billion years of alternating freezing and thawing of subsurface water.
   D) Continuous erosion by charged particles in the solar wind.
   Answer: A  Source: Section 5-6

158. What is a mare on the Moon?
   A) A large area of heavily cratered highland terrain.
   B) The smooth floor of an ancient asteroid impact basin, never subjected to inundation by lava or other volcanic activity.
   C) A very large, ancient lava flow.
   D) A large area smoothed by deformation and tectonic activity.
   Answer: C  Source: Section 5-6

159. Maria are
   A) ancient, sinuous (winding) lava flow channels.
   B) ancient lava floodplains.
   C) heavily cratered highland regions.
   D) uplifted regions surrounding large shield volcanoes.
   Answer: B Source: Section 5-6

160. How old are the lunar maria?
   A) 3.1 to 3.8 billion years old.
   B) 4.0 to 4.3 billion years old.
   C) 1.8 to 2.6 billion years old.
   D) Less than a billion years old.
   Answer: A Source: Section 5-6

161. The smooth, dark maria on the Moon are
   A) immense impact basins that are smooth because they were covered by lava flows after the early, heavy bombardment had ended.
   B) immense impact basins that are smooth because the earlier craters were wiped out by the impact.
   C) areas that, at the time of the early, heavy bombardment, were still molten.
   D) regions that, although as old as the cratered highlands, escaped the early, heavy bombardment.
   Answer: A Source: Section 5-6

162. The lunar maria appear smooth because they are
   A) ancient sea beds, now dry, dating back to when the Moon had a denser atmosphere and rainfall was abundant.
   B) recent lava flows, occurring within the last billion years, and few craters have been formed in this short period.
   C) regions where craters have been obliterated by crustal deformation caused by hot spots in the underlying mantle.
   D) ancient lava flows that occurred soon after the end of an early period of intense bombardment, and have had a few impacts since then.
   Answer: D Source: Section 5-6

163. The smooth surfaces of the lunar maria were most likely caused by
   A) dust storms that smoothed the surface.
   B) lava flows in the early history of the Moon.
   C) volcanic ash that rained upon the surfaces of the basins in ancient times.
   D) water flowing into the basins and allowing sediments to settle over their surfaces.
   Answer: B Source: Section 5-6

164. The type of rock of which the lunar maria are composed is
A) granite. B) limestone. C) anorthosite. D) basalt.
Answer: D Source: Section 5-6

165. The age of Moon rocks has been determined primarily by what method?
A) Careful chemical analysis of the constituents.
B) Counting of the numbers of micrometeoroid craters on the rock surface.
C) Measurements of radioactive elements in the rocks.
D) Careful measurement of magnetic fields "frozen into" the lunar rocks.
Answer: C Source: Section 5-6

166. The age of the moon is determined by
A) measurements of radioactive elements and radioactive dating.
B) radiocarbon dating of remnants of living material such as plant life.
C) careful crater counts over different regions of the Moon and comparison to known meteoroid densities in space.
D) measurements of the relative concentrations of easily melted and evaporated substances, the "volatiles" and the less volatile materials, the "refractory" elements.
Answer: A Source: Section 5-6

167. Most surface rocks on the Earth are younger than a few million years old, whereas ages of lunar rocks have been measured in billions of years. Why is this?
A) Most early surface rocks on Earth have been washed into the sea by weathering and rainwater, and this does not happen on the Moon.
B) The complete surface of the Earth has been covered periodically by younger material from intense volcanic eruptions in the last few million years. No such activity occurred on the Moon.
C) Much of the Earth's surface is continually recycled (created and subducted) by the underlying mantle because of plate tectonic activity, and this does not occur on the Moon.
D) The ages of Earth and Moon are fundamentally different, the Moon being an old object captured from space by a younger Earth.
Answer: C Source: Section 5-6

168. How old are the lunar highlands?
A) Less than a billion years old.
B) 1.8 to 2.6 billion years old.
C) 3.1 to 3.8 billion years old.
D) 4.0 to 4.3 billion years old.
Answer: D Source: Section 5-6

169. The oldest material found on the Moon during manned and unmanned exploration was
A) anorthositic rocks from the highlands.
B) water ice locked into permafrost on the surface of the Moon.
C) smooth, dark glass formed by meteoritic impact.
D) basalt rocks from the mare basins.
Answer: A Source: Section 5-6

170. Taken over the entire surface of the Moon, the older, lighter-colored and heavily cratered highlands (terrae) take up
A) between ½ and ¾ of the lunar surface.
B) between ¼ and ½ of the lunar surface.
C) more than 4/5 of the lunar surface.
D) less than 1/5 of the lunar surface.
Answer: C Source: Section 5-6

171. The smooth, dark maria take up what fraction of the entire surface of the Moon?
A) between ½ and ¾ of the lunar surface.
B) between ¼ and ½ of the lunar surface.
C) less than 1/5 of the lunar surface.
D) more than 4/5 of the lunar surface.
Answer: C Source: Section 5-6

172. The type of rock making up the lunar highlands is
A) limestone. B) basalt. C) granite. D) anorthosite.
Answer: D Source: Section 5-6

173. In what fundamental way do all lunar rocks differ from terrestrial rocks?
A) Lunar rocks are all meteoric in origin.
B) Lunar rocks are all rich in calcium and aluminum.
C) Lunar rocks all derive from ancient lava flows.
D) Lunar rocks and minerals contain no water of crystallization.
Answer: D Source: Section 5-6

174. Water has recently been discovered on the Moon, in the form of
A) fluid water, flowing in deep tunnels, (some of which have collapsed to form rilles on the lunar surface) where it is protected from solar radiation.
B) thin, hazy clouds overlying the dark polar regions, where they are shaded from the Sun's heat.
C) ice, in deep craters at the N and S poles, perpetually shaded from sunlight.
D) permafrost, embedded in the centers of the large, dark maria, the color showing the presence of frost.
Answer: C Source: Section 5-6

175. Which one of the following four theories about the origin of the Moon is now believed to be correct?
   A) The Moon formed separately in a different part of the solar nebula, and was later captured by the Earth.
   B) An object about the size of Mars crashed into the Earth, and debris from the collision formed the Moon.
   C) The Earth was spinning so rapidly while still molten that a piece "spun off" to form the Moon.
   D) The Earth and the Moon formed together, already orbiting each other.
   Answer: B Source: Section 5-7

176. Although we do not yet know precisely how the Moon formed, an important clue is provided by the fact that
   A) the Moon is heavily cratered.
   B) moon rocks resemble material found at or just below the surface of the Earth.
   C) unlike the Earth, the Moon has essentially no atmosphere.
   D) moon rocks and minerals contain essentially no water or other volatile substances with low melting and boiling points.
   Answer: D Source: Section 5-7

177. Which is the most probable heat source that produced extensive, and maybe even total, melting of the Moon at an early stage in its history?
   A) Decay of radioactive elements within it and the impact energy of meteoritic bombardment.
   B) Tidal flexing and distortion caused by its motion around the Earth.
   C) Intense sunlight from the early and very active Sun.
   D) Nuclear fusion reactions occurring in its core.
   Answer: A Source: Section 5-7

178. During what part of the Moon's life were most of the craters on the Moon formed?
   A) Starting about 500 million years after the Moon formed, until about 1.5 billion years after.
   B) During the first 200 million years.
   C) Approximately evenly, from when the Moon first formed until the present day.
   D) During the first 800 million years.
   Answer: D Source: Section 5-7

179. What appears to be the "impact history" of cratering on the Moon?
    A) Heaviest bombardment when the Moon first formed, gradually decreasing to light bombardment today.
    B) Short periods of heavy bombardment alternating with long periods of light bombardment throughout the Moon's life.
    C) An early period of heavy bombardment, then very light bombardment to the present.
    D) More-or-less constant bombardment from the earliest times to the present.
    Answer: C  Source: Section 5-7

180. How many times does the Moon rotate on its axis each time it orbits around the Earth?
    A) Twice.
    B) Zero (the Moon does not rotate on its axis).
    C) Exactly once.
    D) Anywhere from ¾ to 1¼ times, depending on the time of year.
    Answer: C  Source: Section 5-8

181. In its orbit around the Earth, the Moon
    A) rotates once every 24 hours, to keep in step with the Earth.
    B) always keeps the same side toward the Earth.
    C) always keeps the sunlit side toward the Earth.
    D) always keeps the same side toward the Sun.
    Answer: B  Source: Section 5-8

182. To observers on Earth, the Moon shows
    A) its whole surface once per year, as the Earth moves around the Sun.
    B) only the sunlit side at all times.
    C) its whole surface once per month as it orbits the Earth.
    D) the same side to Earth at all times.
    Answer: D  Source: Section 5-8

183. Which of the following statements about the Moon is true?
    A) The Moon does not rotate on its axis.
    B) There is one side of the Moon that is always in darkness.
    C) The Moon can never pass through the Earth's shadow.
    D) There is one side of the Moon from which the Earth can never be seen.
    Answer: D  Source: Section 5-8

184. If astronauts were to set up a permanent settlement at Tranquility Base on the Moon, how many times each year would the Earth rise and set, as seen by a resident of this base?
 A) Never; the Earth would remain essentially motionless in the sky.
 B) Once each year.
 C) 12 times each year.
 D) 13 times each year.
 Answer: A Source: Section 5-8

185. If astronauts were to set up a permanent settlement at Tranquility Base on the Moon, how often would the Sun rise and set, as seen by a resident of this base?
 A) Once each year.
 B) About once per month.
 C) 365 times each year.
 D) Never; the Sun would remain essentially motionless in the sky.
 Answer: B Source: Section 5-8

186. How long is a "lunar day," that is, the time between two successive sunrises and sunsets on the Moon?
 A) About one month.
 B) About one day.
 C) Almost infinite, since the Moon does not rotate about its axis with respect to the Sun.
 D) About one year.
 Answer: A Source: Section 5-8

187. If you were standing on the Moon with the Earth in view, how much time would elapse between two successive "Earthrises"?
 A) About one day.
 B) About one synodic month.
 C) Essentially forever.
 D) About one sidereal month.
 Answer: C Source: Section 5-8

188. Astronauts at a Moonbase that can be seen from Earth will NEVER see
 A) Earthrise or Earthset.
 B) One side of Earth, since the Moon revolves at the same rate as Earth rotates.
 C) The stars moving through their sky since the Moon does not rotate.
 D) Sunrise or sunset, since the Sun will always remain in their sky.
 Answer: A Source: Section 5-8

189. The Moon produces tidal disturbances on the oceans of Earth. In general, there are
   A) one high tide and one low tide per month.
   B) two high and two low tides per day.
   C) one high tide and one low tide per day.
   D) two equal high tides and one low tide per day.
   Answer: B Source: Section 5-8

190. The highest of all tides on the Earth's oceans will occur
   A) at full moon but NOT at new moon, at times when the Sun and Moon are closest to the Earth.
   B) at quarter moons, at times when Sun and Moon are furthest from Earth.
   C) at full or new moon, at times when Moon and Sun are closest to Earth.
   D) at any time, since they are driven by the action of winds in the Earth's atmosphere.
   Answer: C Source: Section 5-8

191. How many "tidal bulges" are there on the Earth, due to the Moon's gravitational pull?
   A) One, on the side of the Earth facing away from the Moon.
   B) One, facing (almost) directly toward the Moon.
   C) Four, one facing (almost) directly toward the Moon and the other three at 90° intervals from this one.
   D) Two, one facing (almost) directly toward the Moon and one (almost) directly away from the Moon.
   Answer: D Source: Section 5-8

192. Sometimes high tides are higher than at other times. What name is given to the highest high tides?
   A) Rip tides. B) Neap tides. C) Yule tides. D) Spring tides.
   Answer: D Source: Section 5-8

193. Sometimes high tides are lower than at other times. What name is given to the lowest high tides?
   A) Spring tides. B) Neap tides. C) Rip tides. D) Pep tides.
   Answer: B Source: Section 5-8

194. What are spring tides?
   A) Any low tide.
   B) High tides that are significantly lower than the average high tide.
   C) High tides that are significantly higher than the average high tide.
   D) Any high tide.
   Answer: C Source: Section 5-8

195. What are neap tides?
   A) High tides that are significantly lower than the average high tide.
   B) Any low tide.
   C) Any high tide.
   D) High tides that are significantly higher than the average high tide.
   Answer: A Source: Section 5-8

196. Spring tides occur
   A) twice a month, at full and new moon.
   B) once a year, in springtime.
   C) once per month, at full moon.
   D) once a day.
   Answer: A Source: Section 5-8

197. When do spring tides occur?
   A) Only when the Moon, Earth and Sun form a straight line, with the Moon on the opposite side of the Earth from the Sun, at full moon.
   B) Whenever the Earth-Moon line makes a 90° angle to the Earth-Sun line.
   C) Only when the Moon and Sun line up on the same side of the Earth, at new moon.
   D) Whenever the Earth, Moon and Sun form a straight line, regardless of which side of the Earth the Moon is on, i.e., at the times of full and new moons.
   Answer: D Source: Section 5-8

198. When do neap tides occur?
   A) Only when the Moon, Earth and Sun form a straight line, with the Moon on the opposite side of the Earth from the Sun.
   B) Whenever the Earth, Moon and Sun form a straight line, regardless of which side of the Earth the Moon is on.
   C) Whenever the Earth-Moon line makes a 90° angle to the Earth-Sun line.
   D) Only when the Moon and Sun line up on the same side of the Earth.
   Answer: C Source: Section 5-8

199. Neap tides occur
   A) once a year, in January.
   B) twice a month, at first and third quarter moon.
   C) twice a month, at full and new moon.
   D) once per month, at new moon.
   Answer: B Source: Section 5-8

200. The water on the side of the Earth that faces away from the Moon
   A) experiences a high tide because the Moon in effect pulls the solid Earth out from under the water on the far side.
   B) experiences no tidal force.
   C) experiences a low tide because all of the Earth's water is pulled toward the side of the Earth that faces the Moon.
   D) experiences either a high tide or a low tide, depending on the angle to the Sun.
   Answer: A Source: Section 5-8

201. The reason for the alignment of the Moon with one face always toward the Earth is that
   A) tides due to gravitational forces from the Earth while the Moon was molten slowed the Moon's rotation.
   B) sporadic (random) impact of meteoroids on the Moon has gradually slowed its rotation.
   C) the Moon started spinning in this way as it was forming and has maintained this rotation.
   D) tides due to gravitational forces from the Sun while the Moon was molten slowed the Moon's rotation.
   Answer: A Source: Section 5-8

202. How quickly is the Moon spiraling away from the Earth?
   A) A few centimeters per year.
   B) A few centimeters per century.
   C) A few centimeters per million years.
   D) A few meters per year.
   Answer: A Source: Section 5-8

203. Which of the following statements is a correct description of the Moon's orbit?
   A) The Moon's distance from the Earth increases and decreases cyclically once every 26,000 years.
   B) The Moon is gradually spiraling toward the Earth.
   C) The Moon's distance from the Earth remains constant from year to year, on average.
   D) The Moon is gradually spiraling away from the Earth.
   Answer: D Source: Section 5-8

204. The Moon, because of the tides on the Earth's oceans, is
   A) moving slowly toward the Earth.
   B) unaffected, and continues to orbit in a constant elliptical path
   C) shrinking.
   D) spiraling outward, away from the Earth.
   Answer: D Source: Section 5-8

205. Which of the following statements is a correct description of the rotation of the Earth?
    A) The average length of a day is gradually getting shorter because the Earth's rate of rotation is speeding up.
    B) The average length of a day is constant from year to year since nothing can change the speed of rotation of the Earth.
    C) The average length of a day varies unpredictably from one year to the next because of the combined effects of solar and lunar tides.
    D) The average length of a day is gradually getting longer because the Earth's rate of rotation is slowing down.
    Answer: D Source: Section 5-8

206. One effect of the tidal drag of the ocean waters on the Earth is
    A) to speed up the Earth in its orbital motion around the Sun.
    B) to slow down the Earth's spin rate.
    C) to tilt its spin axis further and further away from the perpendicular to the ecliptic.
    D) to speed up its rate of spin, thereby gaining energy from the Moon's orbital motion.
    Answer: B Source: Section 5-8

207. The Moon rotates synchronously as it orbits the Earth, always keeping one side pointed toward the Earth, because
    A) of the effect of the magnetic field of the Earth on the magnetic field of the Moon, much like the effect upon a compass needle.
    B) of frictional effects from micro-meteoroids in its orbital plane, especially early in its history.
    C) of the effect of the gravitational pull of the Earth on the tidally induced bulge on the Moon.
    D) it had precisely this rate of spin, equal to its revolution period around the Earth, when it was formed.
    Answer: C Source: Section 5-8

208. The reason that the Moon always keeps one face toward the Earth is because
    A) gravitational forces from the Sun act on the tidal bulge of the Moon.
    B) gravitational forces from the Earth act on the tidal bulge of the Moon.
    C) the impact of asteroids onto the Moon early in its history slowed its rotation rate.
    D) the Moon was spinning in this way when it was formed and it has maintained this rotation.
    Answer: B Source: Section 5-8

# Chapter 6: The Other Terrestrial Planets

1. Which of the following properties do the terrestrial planets, Mercury, Venus, Earth and Mars, have in common?
   A) They all have solid surfaces.
   B) They all have very dense atmospheres.
   C) They all orbit the Sun in a year or less.
   D) They are all cloud-covered.
   Answer: A Source: Chapters 5 and 6

2. The inner planets, in order of increasing planetary radius, are
   A) Mercury, Venus, Earth, Mars.
   B) Mercury, Mars, Venus, Earth.
   C) Mercury, Earth, Venus, Mars.
   D) Mercury, Venus, Earth, Mars.
   Answer: B Source: Chapters 5 and 6

3. Mercury can only be seen easily from Earth
   A) in the winter, when the ecliptic plane is high in the sky at night.
   B) near to the Sun, just after sunset or just before sunrise.
   C) during a lunar eclipse, when the sky is sufficiently dark near to the Moon, since Mercury is always close to the Moon in our sky.
   D) at midnight, when it is high in the sky.
   Answer: B Source: Introduction to Section 6-1

4. Mercury is best seen from Earth with the naked eye at what time of the day?
   A) Close to dawn or dusk, when Mercury is at positions of greatest East and West Elongation.
   B) Near midday, when it is seen through the least amount of the Earth's atmosphere.
   C) At midnight, when Mercury is at opposition and highest in our sky.
   D) During the daytime, when Mercury is at superior conjunction.
   Answer: A Source: Introduction to Section 6-1

5. Why is Mercury difficult to observe from Earth?
   A) It remains close to the Sun in its orbit and is seen in a dark sky only at sunrise or sunset, close to the horizon.
   B) It orbits the Sun very rapidly and moves across our sky very quickly.
   C) It is a very small object which orbits a long way from the Sun, thereby reflecting little light back to Earth.
   D) Its orbit is very elliptical and tilted at a large angle to the ecliptic, and any position on this orbit is only visible at midnight, low in our sky.
   Answer: A Source: Introduction to Section 6-1

6. Why is it relatively difficult to observe Mercury from Earth?
   A) Because its surface is always completely covered in clouds.
   B) Because it is a small object which always appears close to the Sun in the sky.
   C) Because its orbit is such that it is always on the opposite side of the Sun to the Earth.
   D) Because its surface is always glowing from the intense ultraviolet sunlight and from the high temperature of its surface.
   Answer: B Source: Introduction to Section 6-1

7. Why is it relatively difficult to observe details on the surface of Mercury from Earth?
   A) Because its surface is always completely covered in clouds.
   B) Because its orbit always keeps it on the opposite side of the Sun from the Earth.
   C) Because its surface appears very bright from reflected ultraviolet sunlight and emitted infrared infra-red radiation from its very hot surface.
   D) Because it is a small object which always appears close to the Sun in our sky.
   Answer: D Source: Introduction to Section 6-1

8. The Mariner 10 spacecraft was placed in at elliptical orbit with a period of its orbit around the Sun of twice that of Mercury. What is the length of the semimajor axis of Mariner 10's orbit? (Careful with units, and review Kepler's laws, Section 2-3, Kaufmann & Comins, *Discovering the Universe*, 5th Ed.)
   A) 0.61 AU. B) 0.48 AU. C) 0.0125 AU. D) 0.39 AU.
   Answer: A Source: Sections 2-3 and 6-1

9. Which of the planets fits the following description: "A planet with solid, cratered surface which is alternately very hot and very cold, and having no atmosphere"?
   A) Mars. B) Venus. C) Mercury. D) Jupiter.
   Answer: C Source: Section 6-1

10. Which planet in our solar system fits the following description: "A planet with a large iron core, heavily cratered surface, and no atmosphere"?
    A) Mercury. B) Mars. C) Neptune. D) Earth.
    Answer: A Source: Section 6-1

11. How large would the Sun appear to be in the sky (in angular diameter) as seen by someone on Mercury, compared to its angular diameter from Earth?
    A) About 2.5 times larger.
    B) The same size, since it is the same Sun.
    C) About 1.5 times larger.
    D) About 6.25 times larger.
    Answer: A Source: Section 6-1 and Appendix, Table A-1

12. Which planet most resembles the Moon in visible surface features and atmosphere?
A) Venus. B) Mars. C) Mercury. D) Uranus.
Answer: C Source: Section 6-1

13. Approximately how many times in one Earth year will Mercury go completely around the Sun, as viewed from the distant stars? (See the table accompanying Figure 6-1, Kaufmann & Comins, *Discovering the Universe*, 5th Ed.)
A) Twice. B) Three times. C) Four times. D) Once.
Answer: C Source: Section 6-1

14. Approximately how many times in one Earth year will Mercury appear at its point of greatest eastern elongation (e.g., in the western sky at sunset) from Earth? (Review synodic period, Section 2-1 and Table 2-1, Kaufmann & Comins, *Discovering the Universe*, 5th Ed.)
A) Four times. B) Once. C) Three times. D) Twice.
Answer: C Source: Sections 2-1 and 6-1

15. Suppose that, on a certain day, Mercury and Venus are both at greatest elongation in the western sky at sunset. After some time has elapsed, Venus will again be at greatest elongation in the western sky at sunset. During this period, how many times will Mercury have appeared at greatest elongation in the western sky at sunset? (See Section 2-1 and Table 2-1, Kaufmann & Comins, *Discovering the Universe*, 5th Ed., and review synodic period.)
A) Five times. B) Twice. C) Three times. D) Once.
Answer: A Source: Sections 2-1 and 6-1

16. How many times does Mercury rotate around its axis with respect to the distant stars in one orbit round the Sun? (See the table accompanying Figure 6-1, Kaufmann & Comins, *Discovering the Universe*, 5th Ed.)
A) 1.5 B) 3 C) 2/3 D) 2
Answer: A Source: Section 6-1

17. What is the relationship between Mercury's rotation around its spin axis and its revolution about the Sun? (See the table accompanying Figure 6-1, Kaufmann & Comins, *Discovering the Universe*, 5th Ed.)
A) It completes one rotation for every one revolution (synchronous rotation).
B) It completes two rotations for every three revolutions.
C) It completes three rotations for every one revolution.
D) It completes three rotations for every two revolutions.
Answer: D Source: Section 6-1

18. How long is one solar day (e.g., noon to noon) on Mercury, in Earth days? (Review the concept of synodic period in Section 2-1, and see the table accompanying Figure 6-1, Kaufmann & Comins, *Discovering the Universe*, 5th Ed.)
    A) 88 days. B) 1 day. C) 58.7 days. D) 176 days.
    Answer: D Source: Section 6-1

19. Mercury rotates once with respect to the distant stars in 59 days and has a sidereal period of 88 days. If a star is seen to be overhead to an observer on Mercury at midnight on a particular day, how many times will Mercury orbit the Sun before this star is again above the observer's head in a nighttime sky? (Draw a diagram to help in this exercise.)
    A) It will never again reach this specific position. B) 2 C) 1 D) 3
    Answer: B Source: Section 6-1

20. How many times will Mercury rotate with respect to the Sun in one sidereal orbital period? (See the table accompanying Figure 6-1, Kaufmann & Comins, *Discovering the Universe*, 5th Ed.)
    A) 1/2 rotation.
    B) Twice.
    C) Once.
    D) Many times, since Mercury rotates rapidly.
    Answer: B Source: Section 6-1

21. The albedo, or fraction of light reflected from the surface, of Mercury is
    A) very low, because of its dark rocky surface and absence of an atmosphere.
    B) very high, because of the light color of its surface.
    C) variable, and relatively high because of variable cloud cover.
    D) fairly high, about the same as that of the Earth.
    Answer: A Source: Section 6-1

22. Mercury is always much closer to the Sun than is Venus and yet it never appears brighter than Venus even when at maximum brightness. Why is this?
    A) Because Mercury has a high albedo or reflectivity but is very small, and hence appears relatively dark.
    B) We never see more than a small fraction of Mercury's illuminated surface because of its orbital path relative to that of the Earth, hence it always appears dark.
    C) Because Mercury is small, has a dark surface and has no reflecting clouds.
    D) Because Mercury has a thick atmosphere which impedes reflection from its surface, making it appear dark.
    Answer: C Source: Section 6-1

23. Which of the following statements about the similarities of Mercury and our Moon is NOT true?
    A) They both have heavily cratered surfaces.
    B) They have no atmospheres.
    C) They both have very dark surfaces or low reflectivities or albedos.
    D) They both have large, circular and relatively flat basins or maria on parts of their surfaces.
    Answer: D Source: Section 6-1

24. The best estimate for the time of production of most craters on both Mercury and the Moon is
    A) in the first 800 million years after planet formation.
    B) at a relatively constant rate throughout their lives, up to and including the present day.
    C) relatively recently, about 4 billion years after planet formation.
    D) about half-way through their lives, or about 2.3 billion years after planet formation.
    Answer: A Source: Section 6-1

25. What evidence does a planetary geologist use to identify a "young" crater on Mercury?
    A) Observation of few, if any, smaller, overlapping craters within it or on its rim.
    B) Infra-red and photographic results showing evidence of water released recently from the permafrost by the impact.
    C) Observation of a lighter colored central region, since an older crater would have acquired a layer of dark dust over time.
    D) Observation of a sharper rim and central peak compared to older craters. Old craters would have suffered significant erosion by wind and dust.
    Answer: A Source: Section 6-1

26. Mercury, unlike the Moon, has extensive plains between the craters. The reason for this is thought to be that
    A) because of its large iron core, Mercury has a thinner mantle and therefore experienced fewer volcanoes which produce craters.
    B) Mercury is larger than the Moon, and took longer to cool to the point where lava stopped flowing across the surface.
    C) Mercury is closer to the Sun and did not receive as many crater-forming impacts as the Moon.
    D) older craters have been worn down and obliterated by the solar wind.
    Answer: B Source: Section 6-1

27. Mercury appears from spacecraft photographs to resemble the Moon in its surface features but with one important difference, which is
    A) the presence of extensive plains between craters, in contrast to the surface of the Moon.
    B) the presence of retrograde direction of spin compared to most other planets and moons.
    C) evidence of active volcanoes on Mercury.
    D) the presence of clouds and a measurable and significant atmosphere on Mercury.
    Answer: A  Source: Section 6-1

28. Craters on Mercury appear to have been produced
    A) by impacts by objects from space, continuously throughout the planet's history, including very recently in geological time.
    B) by successive expansion and contraction of the planet's surface because of intense heating by the Sun and severe cooling during rotation, since the craters appear to be in irregular lines across the surface.
    C) by impacts from objects from space, early in the planet's history.
    D) by volcanic eruptions, early in the planet's history.
    Answer: C  Source: Section 6-1

29. Where in the solar system would you find the Caloris impact basin?
    A) On our Moon.
    B) On Mars.
    C) On Mercury.
    D) On Jupiter's outer Galilean satellite, Callisto.
    Answer: C  Source: Section 6-1

30. What is the Caloris basin?
    A) An volcanic caldera on Mount Maxwell, on Venus.
    B) A large, lowland area on Mars.
    C) A multi-ringed impact basin on Mercury.
    D) A lunar mare, on the far side of the Moon.
    Answer: C  Source: Section 6-1

31. Caloris Basin, a huge circular region on Mercury surrounded by rings of mountains, appears to have been produced by
    A) wind erosion from huge atmospheric storms, similar to super-hurricanes upon Earth.
    B) the eruption of a large and long-lived volcano which formed the mountains in a similar manner to the formation of the Hawaiian Islands on Earth.
    C) the impact of a massive object in the early phases of the planet's formation, soon after the initial cratering period.
    D) successive expansion and contraction of the planet's surface by intense solar heating and severe cooling as the planet rotated, causing buckling in a similar way to the formation of the North American Rocky Mountains or the South American Andes.
    Answer: C Source: Section 6-1

32. The huge impact of an object which produced the circular Caloris Basin and its ring of mountains on Mercury also produced
    A) a very localized region of intense magnetic field, the only magnetic field possessed by Mercury.
    B) melting of permafrost causing massive floods which produced the deep valleys upon Mercury's surface.
    C) a circular pattern of mountains diametrically opposite to the Basin, caused by the focused seismic waves from the original impact.
    D) a very thick dusty atmosphere which still obscures the planet's surface.
    Answer: C Source: Section 6-1

33. A distinct area of unusually jumbled, hilly terrain has been found on Mercury. How is this terrain believed to have been formed?
    A) By upwelling of magma while Mercury was still hot enough for its mantle to be tectonically active.
    B) By an intense shower of meteoroids near the end of the heavy bombardment era.
    C) By the impact of a large comet.
    D) By seismic waves generated by the Caloris impact.
    Answer: D Source: Section 6-1

34. The huge impact of an object which produced the circular Caloris Basin and its ring of mountains on Mercury also produced
    A) a very thick dusty atmosphere which still obscures the planet's surface.
    B) melting of permafrost causing massive floods which produced deep valleys on Mercury's surface.
    C) a very localized region of magnetic field, the only magnetic field possessed by Mercury.
    D) a pattern of jumbled terrain diametrically opposite to the Basin, caused by the focused seismic waves from the original impact.
    Answer: D Source: Section 6-1

35. What caused the long, meandering scarps (cliffs) which can be seen on Mercury?
    A) Crustal movement similar to continental drift on Earth, where plates have pressed against one another.
    B) Shrinking and folding of the planet's surface as it cooled.
    C) Volcanic eruptions along crustal faults over hot-spots in the mantle.
    D) Large impacts near the end of the early period of heavy bombardment, the scarps being eroded crater walls.
    Answer: B Source: Section 6-1

36. The history of Mercury can be summarized as
    A) volcanic activity and crustal deformation over most of its history, where the only craters to survive have been formed within the last billion years.
    B) rapid cooling to form a thick crust during the early bombardment, with a few isolated lava flows but very dense cratering everywhere else.
    C) early cooling to form a thick crust, with no evidence of volcanism or lava flows at any time in its history.
    D) thin crust formed at first, followed by extensive bombardment, the craters from which were then covered by lava flows which produced extensive lava plains between the remaining craters.
    Answer: D Source: Section 6-1

37. Recent radar observations of Mercury have suggested that
    A) Mercury may have ice at its north and south poles
    B) Mercury may have geologically recent lava flows on the side which was not photographed
    by Mariner 10
    C) Mercury may not be precisely locked into its hypothesized 3-2 spin-orbit coupling
    D) Mercury may have one or more tectonic rift valleys
    Answer: A Source: Section 6-1

38. Which unexpected chemical compound has recently been discovered on Mercury?
    A) Ammonia, escaping from fissures in Caloris Basin.
    B) Cyanide, poisonous to humans, in the troposphere.
    C) Nitrogen dioxide in Mercury's upper atmosphere.
    D) Water ice, at the north and south poles.
    Answer: D Source: Section 6-1

39. Where has water ice been found on Mercury?
    A) Suspended as microscopic crystals in the stratosphere.
    B) In deep fissures on the night side.
    C) In permanently shadowed crater floors at the north and south poles.
    D) In a recently formed crater, probably from a comet impact.
    Answer: C Source: Section 6-1

40. How is water ice able to remain on the surface of Mercury, despite the planet's close proximity to the Sun?
    A) It is continuously replenished by fresh impacts from comets.
    B) It is continuously replenished by condensation of water vapor from volcanoes.
    C) It exists as permafrost below the thermally insulating surface, being exposed only by occasional impacts.
    D) It is permanently shielded from the Sun by crater walls at the north and south poles.
    Answer: D Source: Section 6-1

41. Mercury's atmosphere consists of
    A) sulfur dioxide and hydrogen sulfide from volcanoes.
    B) traces of hydrogen, helium, potassium, sodium, and oxygen.
    C) mostly nitrogen and oxygen, with traces of carbon dioxide and argon.
    D) almost entirely carbon dioxide.
    Answer: B Source: Section 6-1

42. Mercury's atmosphere is
    A) relatively thin, composed of carbon dioxide with small quantities of nitrogen and argon.
    B) almost non-existent.
    C) relatively dense, composed mostly of nitrogen (80%) and oxygen (20%).
    D) very thin, made up of sulfur dioxide and hydrogen sulfide from volcanoes.
    Answer: B Source: Section 6-1

43. What do most astronomers believe is the source of the minute amount of oxygen in Mercury's atmosphere?
 A) Photodissociation of water molecules from ice at the poles.
 B) Bacterial activity in a permafrost layer below the surface.
 C) Ions captured from the solar wind.
 D) Photodissociation of sulfur dioxide ejected from volcanic eruptions.
 Answer: A Source: Section 6-1

44. The Earth's average density (5520 $kg/m^3$) is slightly higher than that of Mercury (5430 $kg/m^3$). This is because
 A) the Earth has a larger iron core in proportion to its size than has Mercury.
 B) the Earth has a denser atmosphere than has Mercury.
 C) the Earth is more massive than Mercury, and has therefore become more compressed, gravitationally.
 D) Mercury's interior has expanded slightly because of radioactive heating, thereby reducing its density since its formation.
 Answer: C Source: Section 6-2

45. Mercury's iron core takes up approximately what fraction of the volume of the planet?
 A) Greater than 95%.
 B) About 17%, very similar to that of Earth.
 C) About 40%.
 D) About 65-70%.
 Answer: C Source: Section 6-2

46. Mercury's average density and the fact that it has a (weak) magnetic field lead to the conclusion that its central core is probably composed of
 A) ices of $H_2O$ and $CH_4$.
 B) solid and/or molten iron.
 C) solid rocks of relatively low density.
 D) molten rock.
 Answer: B Source: Section 6-2

47. Mercury is unique among the inner planets in having
 A) one hemisphere with many craters and few volcanoes and the other hemisphere with few craters and many volcanoes.
 B) no evidence of an iron core.
 C) an iron core which occupies almost half its volume.
 D) extensive volcanic activity and crustal deformation over most of its history, but no evidence of plate tectonics.
 Answer: C Source: Section 6-2

48. The interior of Mercury contains
    A) significant amounts of water as permafrost, since occasional melting has produced deep water-formed valleys which criss-cross the planet's surface.
    B) very little iron, in contrast to Earth, and so the planet has no magnetic field.
    C) mostly ices and very low density material since the planet's average density is close to that of water.
    D) an iron core which occupies a large fraction of the volume of the overall planet and produces a magnetic field.
    Answer: D Source: Section 6-2

49. The internal structure of Mercury is
    A) a dense iron core taking up almost half of the volume of the planet, and a rocky mantle surrounding the core.
    B) a rocky core with a liquid (or perhaps frozen) water mantle and icy surface.
    C) a thick rocky mantle taking up more most of the volume of the planet, overlying a small but dense iron core.
    D) a rocky core, surrounded by liquid metallic hydrogen and a hydrogen-helium atmosphere.
    Answer: A Source: Section 6-2

50. Which physical feature of the planet Mercury causes the deflection of solar wind particles away from its surface?
    A) Its magnetic field, producing a magnetosphere which surrounds the planet.
    B) A dense, ionized atmosphere surrounding the planet.
    C) Mercury's gravitational field.
    D) A stream of ionized atoms, heated by the intense sunlight, which are emitted continuously by the planet.
    Answer: A Source: Section 6-2

51. When the solar wind (mostly protons and electrons, emitted continuously at relatively high speeds by the Sun) approaches Mercury,
    A) it is guided around the planet by its magnetic field.
    B) it is absorbed and stopped by the atmosphere of the planet.
    C) it hits its surface directly, causing heat and a continuous fluorescent glow.
    D) it is repelled by the high electrostatic charge on the planet.
    Answer: A Source: Section 6-2

52. Compared with that of the Earth, Mercury's magnetic field is
    A) of equivalent strength, easily deflecting the solar wind.
    B) weak, but just strong enough to deflect the solar wind.
    C) much more powerful, creating a very large magnetosphere by deflecting the solar wind.
    D) localised and weak, so that it cannot prevent the solar wind from hitting the surface of Mercury.
    Answer: B Source: Section 6-2

53. The magnetic field of Mercury appears to be caused by
    A) a solid magnetized iron core.
    B) electric currents in a molten iron core.
    C) the motion of high-speed solar wind particles around the planet.
    D) electric currents in a region of liquid metallic hydrogen in the core.
    Answer: B Source: Section 6-2

54. According to modern dynamo theory, the production of a planet-wide magnetic field around a terrestrial-type planet requires
    A) rapid rotation and a molten metallic core.
    B) a permanently magnetized solid metallic core.
    C) a molten metallic core, with OR without rotation.
    D) rapid rotation and a metallic core, either solid or molten.
    Answer: A Source: Section 6-2

55. Two conditions appear to be necessary for the generation of a powerful magnetic field in planets which are NOT present simultaneously on Mercury. These conditions are
    A) rapid rotation and a molten iron core.
    B) a molten core and a significant atmosphere.
    C) rapid rotation and a conducting atmosphere.
    D) a solid surface and a significant iron content.
    Answer: A Source: Section 6-2

56. A global magnetic field that deflects the incoming solar wind is found around which terrestrial planet or planets?
    A) Mercury, the Earth, and Mars.
    B) Mercury, Venus, the Earth, and Mars.
    C) Only the Earth.
    D) Mercury and the Earth.
    Answer: D Source: Sections 6-2, 6-5, and 6-11

57. Because Mercury has no appreciable atmosphere, its surface shows extreme temperature changes between night and day, with a range of about
A) 600K. B) 60K. C) 20K. D) 700K.
Answer: A Source: Section 6-2

58. The highest daytime temperature reached on the surface of Mercury is
A) 100°C, high enough to boil water.
B) 0°C, high enough to melt water.
C) 961°C, high enough to melt silver.
D) 430°C, high enough to melt lead and tin.
Answer: D Source: Section 6-2

59. Temperatures on the surface of Mercury are seen to fluctuate between a very cold 100K (-173°C) and an extremely hot 700K (+427°C). What does this measurement indicate about conditions on Mercury?
A) Mercury has a very elliptical orbit and varying distance from the Sun produces these large temperature fluctuations, since intensity varies as the inverse square of the distance from the Sun.
B) Erupting volcanoes occasionally heat the planet's surface to extreme temperatures.
C) The planet is close to the Sun, has no atmosphere to maintain heat from the Sun, and is rotating.
D) The planet has an atmosphere in which the greenhouse effect captures solar radiation to heat the sunlit hemisphere of the planet.
Answer: C Source: Section 6-2

60. Tomorrow's weather forecast for Caloris Basin on Mercury's equator is
A) sunny with scattered cloud, possible afternoon dust storms.
B) overcast, light winds, occasional acid rain.
C) lead and zinc melting by noon; temperatures well below freezing overnight.
D) hot and humid by mid-afternoon; cooling to near freezing overnight.
Answer: C Source: Section 6-2

61. Two conditions appear to be necessary for the generation of a powerful magnetic field in planets that are NOT present simultaneously on Mercury. These conditions are
A) rapid rotation and a molten iron core.
B) a molten core and a significant atmosphere.
C) rapid rotation and a conducting atmosphere.
D) a solid surface and a significant iron content.
Answer: A Source: Section 6-2

62. On the basis of its appearance and general properties, which planetary body could be described as the Earth's twin?
    A) Venus; with about the same mass and diameter, dense atmosphere, and cloud-shrouded.
    B) Pluto; it has a similar size and density, a large moon, and probably life on its surface.
    C) the Moon; somewhat smaller but having the same average density and geology.
    D) Mars; somewhat smaller but with a similar surface, a thin atmosphere, and clouds.
    Answer: A Source: Introduction to Section 6-3

63. Which of the following physical properties of Venus are very similar in value to those of Earth?
    A) Magnetic field and magnetosphere.
    B) Rotation rate around its axis and length of solar day.
    C) Mass and radius and hence average density and surface gravity.
    D) Temperatures of surface and atmosphere.
    Answer: C Source: Introduction to Section 6-3

64. When Venus is at its brightest as seen from the Earth (near greatest elongation) it is
    A) the brightest celestial object in the sky other than the Sun and Moon.
    B) just bright enough to be seen with the unaided eye if one knows exactly where to look for it.
    C) brighter than the full Moon but not as bright as the Sun.
    D) too faint to see without binoculars or a telescope.
    Answer: A Source: Section 6-3

65. Venus appears to be very bright in our skies at certain times because
    A) it is covered by very reflective clouds.
    B) even though its surface is very dark, it is relatively close to the Sun.
    C) it is glowing from the heat of its surface, where the temperature is 750K.
    D) its rocky surface is shiny, like the surface of new volcanic lava.
    Answer: A Source: Section 6-3

66. At greatest elongation, the Earth-Venus line makes an angle with the Sun-Venus line, which is
    A) nearly 180°. B) 90°. C) 0°. D) an acute angle, less than 90°.
    Answer: B Source: Section 6-3

67. The length of a solar "day" on Venus (i.e. time between successive sunrises) is
    A) 243 days. B) 88 days. C) 117 days. D) 224 days.
    Answer: C Source: Section 6-3

68. Compared to the length of a solar day upon Earth (i.e. time between successive sunrises), the length of one solar "day" on Venus is
    A) about half as long, about 10 hours.
    B) about the same.
    C) much shorter about 1 hour.
    D) significantly longer.
    Answer: D Source: Section 6-3

69. Which of the planets fits the following description: "A very hot solid surface, cloud-shrouded, and with a dense $CO_2$ atmosphere"?
    A) Jupiter. B) Mercury. C) Mars. D) Venus.
    Answer: D Source: Section 6-3

70. An Earth-based telescopic view of Venus shows
    A) a crater-covered surface of reddish color.
    B) a smooth, dark surface with few mountain ranges.
    C) evidence of ice-covered polar caps and huge dust storms.
    D) a completely cloud-shrouded planet with high atmospheric wind speeds.
    Answer: D Source: Section 6-3

71. The surface temperature of Venus has been found by radio observations and by remote exploration by spacecraft to be approximately
    A) 750 K. B) 273 K. C) 110 K. D) 300 K.
    Answer: A Source: Section 6-3

72. Which is the hottest planet in the solar system, measured at the surface?
    A) Venus. B) Mercury. C) Earth. D) Mars.
    Answer: A Source: Section 6-3

73. When Venus is seen at inferior conjunction, a ring of light is seen around it, as shown in Fig. 2-8 of Kaufmann and Comins, *Discovering the Universe*, 5th Ed. What does this tell us about Venus?
    A) Venus is very massive, since the gravitational field of Venus is sufficient to bend sunlight around the planet toward us.
    B) the Venus atmosphere is so hot that it glows, even on its dark side.
    C) It has a thick atmosphere, which scatters sunlight toward us.
    D) Venus has auroral displays like those on Earth, which extend over the whole planet.
    Answer: C Source: Sections 2-4 and 6-3

74. The conditions on the surface of Venus are
A) no atmosphere, and hence very variable temperatures under direct sunlight.
B) a dense atmosphere of methane, ammonia and $H_2O$, at a low temperature.
C) a low pressure, low temperature, $CO_2$ atmosphere.
D) a high pressure, high temperature, $CO_2$ atmosphere.
Answer: D Source: Section 6-3

75. The gas that is the major constituent of the atmospheres of Venus and Mars and a minor constituent of the Earth's atmosphere is
A) $CO_2$. B) $O_2$. C) $N_2$. D) $H_2O$.
Answer: A Source: Section 6-3

76. Clouds extend above the surface of Venus to a maximum altitude
A) of thousands of kilometers.
B) of 20 km.
C) of 70 km.
D) with no limit, since there are no clouds on Venus.
Answer: C Source: Section 6-3

77. The clouds in the atmosphere of Venus consist primarily of
A) droplets of $H_2SO_4$ or sulfuric acid.
B) $H_2O$.
C) droplets of $CO_2$.
D) dust particles.
Answer: A Source: Section 6-3

78. The sulfur compounds that are detected in the clouds above the surface of Venus have most probably come from
A) meteoritic material which burned up as it fell into the Venus atmosphere.
B) recent eruptions of volcanoes, spewing gases and dust high into the atmosphere.
C) intense wind storms which have carried surface dust high into the atmosphere.
D) the solar wind, having been captured by the intense magnetic field of the planet.
Answer: B Source: Section 6-3

79. One chemical element which plays a major role in the coloring and chemistry of the Venusian atmosphere and clouds, and probably comes from volcanic eruptions, is
A) sulfur, as dust, sulfur dioxide and sulfuric acid droplets.
B) nitrogen, as gas, nitric oxide and nitric acid droplets.
C) chlorine, as gas, hydrogen chloride and hydrochloric acid droplets.
D) iron, as red dust, iron oxides and iron sulfides.
Answer: A Source: Section 6-3

80. Which particular chemical associated with volcanic emissions has been detected by various techniques in amounts which appear to vary significantly over short time-scales, indicating the presence of active volcanoes on Venus at the present time?
A) Sulfur and sulfur compounds.
B) Ammonia and methane gases.
C) Silicon and silicate dusts.
D) Carbon in $CO_2$ and CO.
Answer: A Source: Section 6-3

81. The cloud structure in the atmosphere of Venus can be described as
A) a clear layer at the surface, a haze layer above it and a high, thick layer of permanent cloud.
B) a permanent and thick cloud layer, extending almost to ground level.
C) isolated clouds, forming and dissipating all the time, with clear sky between them.
D) mostly clear sky with occasional thin, high clouds and dust.
Answer: A Source: Section 6-3

82. The sulfuric acid clouds on Venus
A) are confined to a layer just above the surface about 1 km thick and completely opaque, rendering the surface invisible from above.
B) extend from the surface to a height of 70 km and cover the whole planet.
C) are confined to a narrow layer about 60 km above the planet's surface and cover the whole planet.
D) are in a layer about 70 km above the planet's surface but are patchy, such that the surface can be seen occasionally from above.
Answer: C Source: Section 6-3

83. The lower atmosphere of Venus near to the surface can be described as
A) clear.
B) foggy.
C) very dusty.
D) absent (Venus has almost no atmosphere!).
Answer: A Source: Section 6-3

84. The surface pressure of the atmosphere of Venus compared to that of Earth is
A) about 1/100 atmosphere.
B) about the same as Earth.
C) about 90 atmospheres.
D) extremely small.
Answer: C Source: Section 6-3

85. Tomorrow's weather report for Venus would be
A) cold and clear.
B) snow.
C) overcast and very hot.
D) hot and humid, with clear skies.
Answer: C Source: Section 6-3

86. An observer on the surface of Venus, looking upward during the daytime, would likely see
    A) thin, high clouds near mountains, otherwise clear skies, the Sun moving from west to east.
    B) scattered, billowy clouds, with clear sky between them, the Sun moving from west to east.
    C) a solid cloud deck, with a lighter area toward the Sun which moved from west to east.
    D) a solid cloud deck, the lighter area caused by sunlight moving from east to west.
    Answer: C Source: Section 6-3

87. The severe atmospheric conditions that quickly destroyed spacecraft that soft-landed upon the surface of Venus were
    A) high temperatures, low atmospheric pressure and intense UV radiation from the Sun.
    B) high temperatures, high pressures and corrosive acid clouds and mist.
    C) very low temperatures, a near vacuum and corrosive alkaline clouds and mist.
    D) intense sunlight, including UV, very high pressures and very low temperatures.
    Answer: B Source: Section 6-3

88. Why did the Soviet spacecraft only survive for a few minutes on the Venus surface?
    A) They landed in very rugged terrain and were not able to land upright, and became damaged when they toppled over.
    B) The conditions of extreme pressure, corrosive atmosphere and high temperatures severely damaged it.
    C) They were attacked and destroyed by native inhabitants, but the space agency is not telling the world of this.
    D) They landed very fast because there was insufficient atmosphere to slow down their descent.
    Answer: B Source: Section 6-3

89. The highest temperature in the Venus atmosphere occurs (see Fig. 6-11, Kaufmann & Comins, *Discovering the Universe*, 5th Ed.)
    A) at the planet's surface.
    B) in the clear atmospheric layers below the cloud level, at an altitude of about 30 km.
    C) at the cloud-tops, which are heated by sunlight.
    D) at the height of the thickest clouds, 48 - 52 km, where infrared absorption is highest.
    Answer: A Source: Section 6-3

90. The temperature in the atmosphere of Venus (see Fig. 6-11, Kaufmann and Comins, *Discovering the Universe*, 5th Ed.)
    A) is almost constant with altitude.
    B) decreases smoothly with increasing altitude.
    C) increases smoothly with increasing altitude.
    D) has a complicated structure reaching several maxima and minima at various altitudes.
    Answer: B Source: Section 6-3

91. There are many reasons why a multi-day hiking trip on foot across Aphrodite Terra on Venus would not be advisable, at least not without suitable protection. Which of the following would NOT be a concern?
    A) Water not available.
    B) Very cold night-time temperatures.
    C) Corrosive mists in the atmosphere.
    D) Predominantly carbon dioxide atmosphere.
    Answer: B Source: Section 6-3

92. The mechanism of the "greenhouse effect," which has resulted in very high temperatures on the surface of Venus (and moderate temperatures upon Earth), can be described as
    A) solar infra-red radiation heating the planet surface which then emits visible and UV radiation which is trapped by $CO_2$ in the atmosphere.
    B) solar UV and visible radiation heating the planet surface, the infra-red emissions of which are trapped by $CO_2$ in the atmosphere.
    C) solar UV and visible radiation entering the clouds and triggering chemical reactions in the $CO_2$ and sulfur compounds, the released energy then heating the atmosphere.
    D) solar UV and visible radiation being absorbed by the $CO_2$ of the atmosphere, thereby heating it.
    Answer: B Source: Section 6-4

93. The main reason for the very high temperature (750 K) on the surface of the planet Venus is thought to be
    A) the radiation from hot lava, produced by intense and continuous volcanic action.
    B) the continuous bombardment of the surface by solar wind particles and meteoroids.
    C) the absorption of visible radiation by the Venus surface and the subsequent trapping of infrared radiation emitted by the surface by the atmosphere and clouds.
    D) chemical reactions between constituents of the atmosphere, particularly sulfuric acid.
    Answer: C Source: Section 6-4

94. The component of Venus' atmosphere which is responsible for the excess heating, caused by the greenhouse effect, is
A) $CO_2$. B) $H_2SO_4$ or sulfuric acid droplets. C) $N_2$. D) $H_2O$ vapor.
Answer: A Source: Section 6-4

95. Why has the greenhouse effect been much more effective in raising the surface temperature on Venus than upon the Earth?
A) Because $CO_2$ which traps heat from the planet's surface is the major component in the very dense Venusian atmosphere while it is a only a minor constituent of the Earth's.
B) Because the surface of Venus is much more effective than that of Earth in absorbing solar visible and UV radiation.
C) Because the oceans upon Earth have acted as a thermostat in absorbing much of the heat which would otherwise have raised the Earth's temperature significantly.
D) Because the solar wind, the major cause of heating in the greenhouse effect, is far more intense at Venus' distance from the Sun and Venus has no magnetic field to deflect this solar wind.
Answer: A Source: Section 6-4

96. Why is the surface of Venus hotter than that of Mercury, even though Mercury is much closer to the Sun?
A) Because the thick $CO_2$ atmosphere has prevented re-emission into space of the heat absorbed from sunlight.
B) Because of continuous volcanic activity and the release of hot lava onto the surface.
C) Because chemical reactions within the thick clouds and dense atmosphere are continuously supplying heat to the surface.
D) Because Venus rotates rapidly and this ensures that all of its surface is being heated regularly and uniformly.
Answer: A Source: Section 6-4

97. Spacecraft from which country or countries have landed on the surface of Venus?
A) The former Soviet Union.
B) The United States.
C) The United States and the former Soviet Union.
D) No country has yet successfully landed spacecraft on the surface of Venus.
Answer: A Source: Section 6-4

98. Which spacecraft and technique has given us our most extensive information about the overall surface of Venus?
    A) Radar mapping by the USA Magellan orbiter.
    B) Photography by the four Viking spacecraft (two landers and two orbiters).
    C) Photography by the Venera landers from the former Soviet Union.
    D) Photography from the Space Shuttle at UV wavelengths which can penetrate the Venusian clouds.
    Answer: A   Source: Section 6-5

99. The surface features and overall topology of Venus have been determined primarily by
    A) surface lander vehicles which have explored the surface thoroughly.
    B) radar methods from Venus-orbiting spacecraft, measuring radio echoes from the surface.
    C) balloon-borne spacecraft, launched into the Venus atmosphere by spacecraft.
    D) visible and UV photography from the Space Shuttle.
    Answer: B   Source: Section 6-5

100. The overall geography of the Venus surface has been determined largely by
    A) photography from orbiting spacecraft in long elliptical orbits in order to descend below the clouds during part of their orbits.
    B) visible light photography.
    C) unmanned spacecraft which landed on the surface.
    D) radar techniques from orbiting spacecraft.
    Answer: D   Source: Section 6-5

101. The best images of the overall topology of Venus have been produced by
    A) photography from the Hubble Space Telescope at UV wavelengths to which the Venus atmosphere and clouds are transparent.
    B) visible wavelength images from cameras on board an orbiting spacecraft.
    C) imaging cameras on board two spacecraft which soft-landed upon the surface of Venus.
    D) reflection of microwave and short radio wave radiation from the surface by an orbiting spacecraft.
    Answer: D   Source: Section 6-5

102. A radar mapping satellite orbited which planet in the early 1990s?
    A) Venus.   B) Mercury.   C) The Moon.   D) Mars.
    Answer: A   Source: Section 6-5

103. In the mapping of Venus by the orbiting Magellan spacecraft, what parameter was measured by the sensors in order to produce 3-dimensional maps of the planet's surface?
A) Precise photographs of the extreme limb of the planet, taken at UV wavelengths to which the atmosphere is transparent, which showed detailed profiles of the planet's surface.
B) Pairs of photographs, taken from different angles, which were then combined stereoscopically to produce contour maps.
C) The time delay of the return of reflected radio waves.
D) the wavelength, and hence the Doppler shift, of reflected radio waves.
Answer: C Source: Section 6-5

104. Radar observations are used from Venus-orbiting spacecraft to evaluate mountain heights by measuring the time difference between echos from mountain peaks and from the surrounding plains. In this technique, what would be the time difference for signals reflected from Maxwell Montes, which rises 12 km above the plain? (Hint, think about the geometry and total path length.)
A) 80 ns, or $8 \times 10^{-8}$ s.
B) 80 ms, or $8 \times 10^{-5}$ s.
C) 80 ms, or $8 \times 10^{-2}$ s.
D) 800 ms, or $8 \times 10^{-7}$ s.
Answer: B Source: Section 6-5

105. The geology and geography of the surface of the Venus is best described as
A) heavily cratered, with no major volcanoes or lava flows.
B) mostly volcanic plains, with two continent-sized uplands, and a number of large volcanoes.
C) volcanoes and volcanic uplifts in the northern hemisphere, and cratered plains in the southern hemisphere.
D) colliding surface plates with long mountain chains, rift valleys, and deep subduction trenches.
Answer: B Source: Section 6-5

106. Compared to the surface of Earth, that of Venus is
A) almost completely flat and relatively smooth, apart for two high volcanic mountain ranges.
B) extremely rugged, with deep valleys and many high volcanic mountains.
C) completely covered with innumerable, overlapping craters and old crater walls which constitute mountain ranges.
D) very smooth and flat, with no mountains or structure.
Answer: A Source: Section 6-5

107. How many raised areas that look like continents are seen on Venus?
A) 2 B) 1 C) None. D) Several
Answer: A Source: Section 6-5

108. On a topographical map of Venus, how many large "continents" of high ground rise above the flat plains?
A) 7 B) None C) 1 D) 2
Answer: D Source: Section 6-5

109. What is the highest mountain range on Venus called?
A) Maxwell Montes. B) Alpha Regio. C) Ishtar Terra. D) Olympus Mons.
Answer: A Source: Section 6-5

110. The most common surface features on Venus are
A) impact craters.
B) evidence of plate tectonic motion, including long mountain ranges and subduction troughs.
C) ancient river valleys and huge flood plains.
D) volcanoes and lava flows.
Answer: D Source: Section 6-5

111. The surface of Venus
A) has two mountainous regions on a smooth plain, with no long connected mountain ranges, such as are seen on Earth.
B) has several large individual volcanoes on a smooth plain but no extensive ridged or mountainous regions.
C) is very smooth, with no mountains or mountain ranges.
D) shows long mountain ranges similar to the Rockies/Andes ranges and the mid-oceanic ridges on the Earth.
Answer: A Source: Section 6-5

112. Tectonic activity on Venus differs from that on Earth in that
A) mantle convection appears to be more vigorous and has broken the lithosphere into a multitude of small plates instead of a few large ones.
B) the crust appears to be thicker, and is therefore too rigid to break up into moving plates.
C) the crust appears to be thinner and weaker and cannot support the creation and motion of solid plates.
D) the surface is broken into only two plates, divided by a line at an angle to the equator.
Answer: B Source: Section 6-5

113. The reason why very few impact craters are seen on Venus compared to the Moon is believed to be that
A) the surface of Venus has been subducted back down into the planet several times in its history, thereby removing evidence of impacts.
B) erosion due to rainfall and wind has eroded away all but the most recent craters.
C) lava flows and surface melting have covered all but the most recent craters.
D) Venus formed closer to the Sun than did the Moon, where the cratering rate was much lower.
Answer: C Source: Section 6-5

114. The fact that there are very few impact craters upon the surface of Venus compared to those on Mercury and Mars is thought to be because
A) ocean waters which covered Venus at earlier times washed away any evidence of cratering.
B) wind erosion from its dense atmosphere and chemical action from its corrosive clouds and mist have destroyed most craters.
C) plate tectonic motions have recycled the surface several times since the main bombardment of the planetary system occurred.
D) its thick atmosphere has protected it from most incoming objects and the surface has melted periodically to obliterate old craters.
Answer: D Source: Section 6-5

115. The maximum age of features on the surface of Venus appears to be
A) less than 10 million years, due to wind and chemical erosion.
B) three to four billion years, due to a lack of surface activity since early times.
C) about 700 million years, due to constant resurfacing of the planet by lava flows.
D) about 250 million years, due to recycling of the crust by plate tectonics.
Answer: C Source: Section 6-5

116. The average age of the surface of Venus has been determined primarily from
A) soil analysis by Russian landers.
B) the number of impact craters per unit area of surface.
C) radio-isotope analysis of rocks brought back from Venus by space probes.
D) the amount of weathering of lava flows imaged by the Magellan radar mapper.
Answer: B Source: Section 6-5

117. Venus
    A) has a weak magnetic field, about 1/100 of Earth's field strength.
    B) has a very powerful magnetic field.
    C) has no magnetic field.
    D) has a magnetic field about the strength of that of Earth.
    Answer: C   Source: Section 6-5

118. Spacecraft measurements near to Venus indicate that the planet
    A) has a weak and variable magnetic field.
    B) has a very powerful magnetic field, much stronger than that of Earth.
    C) has no magnetic field.
    D) has a magnetic field which varies in concert with the 11-year solar activity cycle, and is linked to it via the solar wind.
    Answer: C   Source: Section 6-5

119. The one terrestrial planet that rotates in the "wrong" way (opposite to the rotation of most other planets and to the planet's revolutionary direction) is
    A) Earth.   B) Venus.   C) Mercury.   D) Mars.
    Answer: B   Source: Section 6-5

120. Venus rotates
    A) in the same direction as Earth but very slowly.
    B) in the same direction as Earth but very rapidly (in a few hours).
    C) in the opposite direction to Earth but very slowly.
    D) "locked-in" to the Sun, maintaining one side toward it at all times in synchronous rotation.
    Answer: C   Source: Section 6-5

121. Which of the following planets rotates on its axis in a retrograde fashion, opposite to that of most of the planets and opposite to the direction of revolution of the planets?
    A) Saturn   B) Moon   C) Mercury   D) Venus
    Answer: D   Source: Section 6-5

122. Where would the Sun appear to rise on Venus, if you could see through the clouds?
    A) The Sun would not rise or set since Venus rotates synchronously, always keeping one side toward the Sun.
    B) In the east.
    C) In the west.
    D) In the north, since Venus has its spin axis parallel to the plane of its orbit.
    Answer: C   Source: Section 6-5

123. The most likely explanation for the retrograde rotation (in a direction opposite to that of most other planets) of Venus is
   A) the combined gravitational effects of its neighboring planets, Mercury and Earth.
   B) the frictional slowing-down and eventual reversal of Venus' rotation by tidal forces at a time when the planet had deep oceans over its surface.
   C) the frictional drag of its very dense atmosphere on the rotating planet throughout its history.
   D) the impact of a massive object on it early in its history.
   Answer: D Source: Section 6-5

124. What did the Magellan spacecraft tell us about Venus' upper atmosphere, before the spacecraft burned up on entry into the atmosphere?
   A) The upper atmosphere is unexpectedly uniform compared to that of the Earth.
   B) The density of Venus' upper atmosphere varies markedly from one place to another.
   C) The upper atmosphere is almost saturated with water vapour.
   D) The temperature of the upper atmosphere increases unexpectedly to over 1000 K.
   Answer: B Source: Section 6-6

125. Which of the following planets has no moon?
   A) Jupiter B) Pluto C) Venus D) Neptune
   Answer: C Source: Appendix, Table A-3

126. The pressures of the atmospheres at the surfaces of Mercury, Venus, and Mars, in terms of the pressure at the surface of the Earth (known as 1 atmosphere), are
   A) 0.01 atmosphere; almost 100 atmospheres; 1/10 atmosphere.
   B) 0.1 atmosphere; almost 100 atmospheres; 1/10 atmosphere.
   C) 0; almost 100 atmospheres; 0.01 atmosphere.
   D) 0; about 10 atmospheres; 0.1 atmosphere.
   Answer: C Source: Chapter 6

127. Which is the only planet whose surface features can easily be seen through a telescope from Earth?
   A) Jupiter B) Mars C) Venus D) Mercury
   Answer: B Source: Introduction to Section 6-7

128. At opposition, when Mars comes relatively close to Earth, where would it be seen in the sky by an observer in the Earth's northern hemisphere?
   A) High in the north at midnight.
   B) High in the south at midnight.
   C) On the western horizon at midnight.
   D) High in the south at sunset.
   Answer: B Source: Section 2-1

129. On which planet can we see prominent but variable ice caps?
A) The Moon B) Venus C) Mercury D) Mars
Answer: D Source: Introduction to Section 6-7

130. Mars has a period and direction of rotation that are
A) 687 days (1 Martian year), since Mars always turns the same face toward the Sun.
B) a little longer than 24 hours and in the same direction as the Earth.
C) about 240 days and in the opposite direction to Earth (retrograde rotation).
D) just less than 10 hours and in the same direction as Earth.
Answer: B Source: Section 6-7

131. In which of the following physical characteristics are Earth and Mars most similar to each other?
A) Total mass.
B) Length of solar day.
C) Planetary diameter.
D) Number of moons.
Answer: B Source: Sections 6-7 and 6-13

132. In what way did optical illusion mislead earlier visual observers of Mars?
A) Chance alignments of faint dark features looked like manufactured canals and variable dark areas near the equator were interpreted as vegetation.
B) Apparent movement of surface features because of seeing fluctuations of images when viewed through the Earth's atmosphere were interpreted as evidence of movement of life-forms or Martians.
C) Volcano structures were seen as eye-shaped images, and were interpreted as having been made by intelligent beings to indicate their presence on Mars.
D) Moving areas of obscured detail on the planet were interpreted as massive flash floods rather than dust storms.
Answer: A Source: Section 6-7 and Insight 6-2

133. Observers in the 19th century reported seeing many straight-line features criss-crossing the surface of Mars, and interpreted these to be canals constructed by intelligent beings. What is the most likely present-day explanation for these observations?
A) They were lines of dark volcanoes similar to those of the Hawaiian Islands on Earth.
B) They were rifts in the Martian surface at the boundaries of geological tectonic plates.
C) They were linear, stationary cloud formations (mountain lee wave clouds) and weather fronts moving around with the planet.
D) They were optical illusions caused by vague shadings on the planet surface.
Answer: D Source: Section 6-7

134. What observations of the Martian surface led Lowell to the conclusion that intelligent life-forms existed upon Mars?
   A) Melting icecaps, a network of linear features that look like canals, and varying dark surface marking, assumed to be vegetation.
   B) Geometrical structures and patterns which are apparently the remains of buildings and cities.
   C) Sculpted mountains in the shape of humanoid heads, obviously carved to announce the presence of intelligent life to distant observers.
   D) Lakes and rivers of water flowing from polar icecaps and detected by strong specular reflection of sunlight.
   Answer: A Source: Section 6-7

135. The so-called "canals," which Schiaparelli reported seeing upon the surface of Mars, were actually
   A) river valleys, caused by massive floods early in Mars' history.
   B) lines of volcanoes along faults in the Martian surface.
   C) the remnants of the walls of ancient craters which have been eroded by winds and dust over Mars' history.
   D) an optical illusion.
   Answer: D Source: Section 6-7

136. What is the "face on Mars"?
   A) A sculpted hill carved either by a long-dead Martian civilisation or by visiting aliens.
   B) A natural arrangement of volcanoes and lava plains which gives the appearance of a face when viewed from Earth.
   C) A small carved portrait placed on the Martian surface by one of the Viking landers.
   D) A natural rock formation which looks like a face in the right lighting.
   Answer: D Source: Section 6-8

137. What did the 1998 spacecraft Mars Orbiter find when it looked at the "face on Mars"?
   A) It found that the face had two noses, showing that it had been carved by a non-human sculptor.
   B) It showed the face to be a natural rock formation.
   C) It found that the shifting sands had uncovered two other, smaller faces on either side of it.
   D) Nothing; the face and the rocks around it were either gone or buried by sand.
   Answer: B Source: Section 6-8

138. Which of the following features were NOT found on Mars when spacecraft finally visited the planet?
A) Craters. B) Straight canals. C) Deep, winding canyons. D) Extinct volcanoes.
Answer: B Source: Sections 6-8 and 6-9

139. The dark markings near the equator of Mars show seasonal variations because of
A) variations in the dust coverage on the surface, caused by winds.
B) changes in the growth of vegetation.
C) changes in the flow of water, released by sunlight from permafrost.
D) changes in coverage of rocks by $CO_2$ ice as the temperature varies from above to below the freezing point of $CO_2$.
Answer: A Source: Section 6-9

140. The dark, seasonal markings on Mars, which intensify in the spring and fade in the autumn, are probably
A) dust being blown by winds, alternately covering and uncovering dark rocks.
B) lighting effects which vary as the angle of the Sun changes through the seasons.
C) plant life.
D) microbial activity in the soil.
Answer: A Source: Section 6-9

141. Which of the planets fits the following description: "A cool, solid surface, with a thin $CO_2$ atmosphere and occasional dust clouds"?
A) Mercury B) Jupiter C) Venus D) Mars
Answer: D Source: Sections 6-9 and 6-11

142. Craters on Mars have flatter bottoms and their rims appear to be more worn down than those on either the Earth's Moon or Mercury. What has caused this?
A) Erosion by wind and infilling by dust storms.
B) Erosion by rainfall, now and in the past.
C) Erosion by the seasonal fall and subsequent melting of carbon dioxide snow.
D) Breakdown of the rocks by the extreme temperature variations between day and night.
Answer: A Source: Section 6-9

143. The "snow" that occasionally falls upon Mars and covers the bottoms of craters is most probably made of
A) water ice.
B) carbon dioxide ice.
C) frozen sulfuric acid droplets.
D) very fine white dust, disturbed occasionally by fierce wind storms.
Answer: B Source: Section 6-9 and Figure 6-19

144. Which of the following features have NOT been seen or detected on Mars?
A) Active volcanoes.
B) Dust storms.
C) Advancing and receding polar icecaps.
D) Valley fog.
Answer: A Source: Section 6-9

145. Which of the following statements best characterizes the surface of the planet Mars?
A) 80% of the surface is rolling plains, with a number of major volcanoes and only two continent-sized uplands.
B) There are several moving lithospheric plates rimmed by long mountain chains, deep subduction trenches, and several large rift valleys.
C) All of the big volcanoes are in the northern hemisphere, and most of the craters are in the southern hemisphere.
D) The surface is continuously resurfaced by ongoing volcanic activity; consequently there are almost no visible impact craters.
Answer: C Source: Section 6-9

146. The overall geography of Mars can be best summarized as
A) smooth plains where continuous resurfacing by ongoing volcanic activity has hidden older impact craters and other details.
B) mostly rolling plains, with several volcanoes on top of two continent-sized uplands.
C) major volcanoes in the northern hemisphere, extensively cratered plains in the southern hemisphere, separated by one major valley system.
D) moving lithospheric plates whose motions have produced long folded mountain chains, deep subduction trenches, and several large rift valleys.
Answer: C Source: Section 6-9

147. Where are most of the extinct volcanoes located on Mars?
A) Around the southern polar cap.
B) Along the bottom of the deep valley, Valles Marineris, which was originally formed by enormous geological stresses.
C) In a line along the equator, the line of maximum tidal stress on the planet.
D) In the northern hemisphere.
Answer: D Source: Section 6-9

148. Where in the planetary system is the massive extinct supervolcano, Olympus Mons?
   A) On Io, one of the moons of Jupiter.
   B) On the mountain range, Ishtar Terra, on Venus.
   C) On Mars.
   D) On Earth.
   Answer: C Source: Section 6-9

149. Olympus Mons is
   A) a volcano on Mars.
   B) a mountain on Venus.
   C) a long-lived anticyclone or spot on Jupiter.
   D) a valley on the Moon.
   Answer: A Source: Section 6-9

150. The major volcanoes on Mars have formed
   A) in mountain belts where the planet's surface is being stressed as it is bent and subducted back into the mantle.
   B) over individual stationary "hot-spots" in the underlying molten mantle.
   C) on long, interconnected ridges where magma, rising from the mantle, is pushing the crust apart.
   D) where shrinkage of the crust during cooling early in the planet's history has wrinkled the surface.
   Answer: B Source: Section 6-9

151. Which of the following statements appears to be TRUE for Mars?
   A) The oppressive heat on Mars has kept the planet's crust thin and prevented the onset of plate tectonics.
   B) Plate tectonics is only just beginning and will be important in the distant future, because geologic processes happen more slowly on a small planet.
   C) Plate tectonics has not been important, because Mars is small and has cooled more rapidly than the Earth to form a thick solid crust.
   D) Plate tectonics is the dominant process creating large-scale surface features such as mountain ranges and volcanoes.
   Answer: C Source: Section 6-9

152. Two observed features on Mars appear to lead to contrary geological conclusions-the appearance of massive solitary volcanoes, apparently caused by "hot spots" under the planet's surface and the discovery of a long, deep, rift valley across the planet. Why is this a contradiction?
   A) Massive volcanoes should have covered the planet's surface, including the valley, with dust and ash, but the valley remains as a major feature.
   B) Volcanoes suggest a molten inner planet, while a rift valley suggests the presence of a solid core.
   C) Single massive volcanoes indicate "hot-spot" volcanism with no plate tectonic motion, while a rift valley suggests otherwise.
   D) Ancient volcanoes show no sign of erosion by water, whereas the valley suggests prolonged rainfall and erosion.
   Answer: C   Source: Section 6-9

153. On which object would you find the great valley system, the Valles Marineris?
   A) On the far side of the Moon
   B) Venus
   C) Mars
   D) Callisto (one of the large satellites of Jupiter)
   Answer: C   Source: Section 6-9

154. What are the Valles Marineris?
   A) A long scarp-and-trough system in the lava plains of Mercury.
   B) A system of tectonic faults on the Jovian satellite Ganymede.
   C) A system of deep trenches bordering Aphrodite Terra on Venus.
   D) A large rift valley system associated with the great volcanoes on Mars.
   Answer: D   Source: Section 6-9

155. The Martian valley Valles Marineris is shown in Fig. 6-22, Kaufmann and Comins, *Discovering the Universe,* 5th Ed. How long is this valley in terms of Earth-bound distances?
   A) The full width of the North American continent at mid-latitudes, a few thousand kilometers.
   B) About 50 km, a significant distance on this small planet.
   C) A few hundred kilometers; New York to Washington, D.C., USA.
   D) Half the length of the Earth's equator, about 20,000 km.
   Answer: A   Source: Section 6-9

156. We know that water exists on Mars, but where and in what state does it NOT exist on this planet?
   A) In permafrost, below the surface.
   B) Flowing in river valleys.
   C) As water vapor in the atmosphere and as clouds.
   D) In polar icecaps.
   Answer: B Source: Section 6-10

157. Where has water been detected upon Mars?
   A) Only as atmospheric water vapor, never condensing out as clouds, liquid water or solid ice.
   B) As a liquid in under-surface lakes and rivers.
   C) As a liquid flowing along the numerous flood valleys and meandering stream beds.
   D) In permafrost and polar icecaps, and as water vapor in the atmosphere.
   Answer: D Source: Section 6-10

158. Which of the following statements is true for Mars?
   A) There is no evidence of there ever being water on Mars.
   B) There are linear canals, apparently built to carry water for irrigation, which criss-cross the planet.
   C) Water can be seen in the bottoms of river valleys on spacecraft photographs.
   D) Mars has dry riverbeds but no liquid water on its surface at the present time.
   Answer: D Source: Section 6-10

159. On which planetary body can distinct evidence be seen for the flow of water at an earlier time?
   A) Venus. B) The Earth's Moon. C) Mars. D) Titan, the moon of Saturn.
   Answer: C Source: Section 6-10

160. What significant evidence is there for the idea that large quantities of water once flowed on the planet Mars?
   A) Frozen, dust-covered lakes inside ancient craters.
   B) Deep, winding canyons and flood plains.
   C) A network of relatively straight canals linking polar and equatorial regions.
   D) Clouds and frost which formed above and around the Viking spacecraft.
   Answer: B Source: Section 6-10

161. Which of the following signs of water is NOT seen on Mars?
A) Occasional clouds around the large volcanoes.
B) Evidence of permafrost under the Martian surface.
C) Meltwater pools at the edges of the polar caps.
D) Water ice (as opposed to $CO_2$ ice) in the polar caps.
Answer: C Source: Section 6-10

162. The polar caps on Mars are most likely made up of
A) light-colored dust, blown there by the Martian dust storms.
B) sulfur dioxide and sulfur compounds.
C) volcanic outflows of light-colored lava and dust similar to that produced by Earth-based volcanoes (such as Mt. St. Helens).
D) water and $CO_2$ ices.
Answer: D Source: Section 6-10

163. The polar caps on Mars are now known to consist of
A) only $H_2O$ ice, which does not melt easily and survives the summer heat.
B) methane ($CH_4$) ammonia ($NH_3$) and water ($H_2O$) ices, whose relative abundances vary with the seasons.
C) only $CO_2$ ice, which is very volatile and melts easily.
D) $CO_2$ ice, which evaporates easily, overlying thicker and more long-lived $H_2O$ ice.
Answer: D Source: Section 6-10

164. The very rapid recession of the edge of the white polar cap region toward the poles in springtime on Mars is caused by
A) the evaporation of a thin layer of $H_2O$ ice crystals (frost or snow).
B) the rapid evaporation of $CO_2$ ice.
C) the increased growth of vegetation from mid-latitudes toward the poles.
D) the melting of a thick layer of solid $H_2O$ ice and the subsequent run-off of water.
Answer: B Source: Section 6-10

165. The pressure of atmospheric carbon dioxide ($CO_2$) at the surface of Mars, compared to the atmospheric pressure at the Earth's surface, is
A) about the same.
B) extremely small (less than 1 millionth).
C) about 90 times greater.
D) about 1/100.
Answer: D Source: Section 6-10

166. There are many reasons why a multi-day hiking trip on foot through the Valles Marineris on Mars would not be advisable, at least not without suitable protection. Which of the following conditions would NOT be a concern?
  A) The possibility of dust storms.
  B) The predominantly carbon dioxide atmosphere.
  C) High levels of ultraviolet radiation.
  D) Oppressively high atmospheric pressure.
  Answer: D Source: Section 6-10

167. How have we obtained samples of Martian rocks?
  A) We have not yet been able to obtain any Martian rocks.
  B) Rocks blasted off Mars by impacts, and landed on Earth as meteorites.
  C) Rocks collected and returned to Earth by astronauts.
  D) Sample return missions, where a robotic rover collected rocks for return by the lander.
  Answer: B Source: Section 6-10

168. The SNC meteorites are
  A) basaltic rocks from the Moon, made up of solidified lava.
  B) carbonaceous chondrites from the asteroid belt, containing naturally occurring amino acids.
  C) rocks from Mars, containing water-soaked clay.
  D) granitic rocks from Venus, formed by solidification of a molten crust.
  Answer: C Source: Section 6-10

169. Scientists have now examined rocks that came from Mars in their laboratories on Earth because
  A) several meteorites found in Antarctica are now thought to have been knocked off Mars and fell to Earth.
  B) a large impact which occurred on the Martian surface early in its history created a vast dust cloud through which Earth passes once per year and samples of this dust have now been collected by the Space Shuttle.
  C) spacecraft have been to the surface of Mars and have returned with samples of Martian rocks and soil.
  D) an orbiting spacecraft scooped up Martian dust during its passage by Mars and then returned to Earth.
  Answer: A Source: Section 6-10

170. In view of the present surface and atmospheric conditions on Mars, why would there be no liquid water on its surface?
   A) The very low atmospheric pressure would allow the water to boil and evaporate rapidly.
   B) The UV radiation from the Sun would have dissociated the molecules into hydrogen (which would leave the planet) and oxygen, which is still present.
   C) The water would all be frozen, given the prevailing day and night temperatures.
   D) It would have reacted chemically with the surface rocks.
   Answer: A Source: Section 6-10

171. Which of the following would NOT be a weather report from Mars, in view of the present atmospheric conditions on that planet?
   A) Clear and cold.
   B) Dry and windy with blowing and drifting dust.
   C) Rain, tapering to intermittent showers by noon.
   D) Dry-ice snow.
   Answer: C Source: Section 6-10

172. Tomorrow's weather forecast for Mars is
   A) sunny and cold; Mars has no atmosphere to retain heat.
   B) overcast, possible light rain.
   C) possible frost overnight, then sunny.
   D) continuous rain, possible flooding in low-lying areas.
   Answer: C Source: Sections 6-10 and 6-11

173. Tomorrow's (and most day's) weather forecast for Mars is likely to be
   A) sunny and clear; Mars has no atmosphere in which clouds can form.
   B) continuous rain, possible flooding in low-lying areas.
   C) sunny, possible thin high clouds, windy.
   D) heavy overcast, possible light acid rain and mist.
   Answer: C Source: Sections 6-10 and 6-11

174. What is the major constituent of the atmosphere of Mars?
   A) $CO_2$ (carbon dioxide).
   B) $H_2$ (hydrogen).
   C) $H_2O$ (water vapor).
   D) $CH_4$ (methane or natural gas).
   Answer: A Source: Section 6-11

175. The tilt of the equator of Mars to its orbital plane
   A) is very similar to that of Earth, about 25°.
   B) is 90°.
   C) is 0°.
   D) varies periodically through the Martian year..
   Answer: A Source: Section 6-11

176. The equator of Mars is tilted with respect to its orbital plane, and so Mars
   A) shows no seasonal variation at all.
   B) experiences very long (20 year) seasonal variations.
   C) shows similar seasons to Earth, each season lasting about twice the length of seasons on Earth.
   D) occasionally experiences small seasonal variations..
   Answer: C Source: Section 6-11

177. Mars experiences similar seasonal changes to those on Earth because
   A) the length of the year is very close to that of Earth.
   B) its has about the same shape of elliptical orbit as the Earth, and this produces similar changes in solar radiation intensity as the planet orbits the Sun.
   C) the length of its day is very close to an Earth day.
   D) its spin axis is tilted at about the same angle to its orbital plane as is the Earth's axis.
   Answer: D Source: Section 6-11

178. Earth and Mars possess two properties in which they are very similar. What are they?
   A) Planet diameter and inclination of equator to the ecliptic plane.
   B) Orbital period and length of solar day.
   C) Length of solar day and diameter.
   D) Length of solar day and inclination of equator to the ecliptic.
   Answer: D Source: Sections 6-7 and 6-11

179. A major feature of the atmosphere of Mars is
   A) very dense clouds shrouding most of the planet.
   B) occasional strong winds and dust storms.
   C) very high temperatures and pressures.
   D) a chemical mixture very similar to that of Earth.
   Answer: B Source: Section 6-11

180. The distinctive red color of Mars is probably caused by
   A) iron oxides or rust in the soil.
   B) the scattering of blue sunlight out of the optical beam by dust in the atmosphere, similar to sunsets upon Earth.
   C) red dust which is suspended high above the surface by winds and filters the sunlight.
   D) progressive reddening of sunlight as it traverses the interplanetary dust between the Sun and Mars and then Mars and Earth.
   Answer: A Source: Section 6-11

181. The reddish color of Mars is probably due to
   A) iron oxides such as rust.
   B) vegetation turning red in the Martian autumn.
   C) sulfur compounds thrown out by active volcanoes.
   D) the glow from the very high temperature surface on the sunlit parts of Mars.
   Answer: A Source: Section 6-11

182. The dominant component of the soil on Mars is probably
   A) iron oxides.
   B) basaltic lava pulverized by meteoritic bombardment.
   C) volcanic ash from eruptions in recent geological times.
   D) sedimentary rocks laid down by massive floods early in Mars' history.
   Answer: A Source: Section 6-11

183. What material produces the distinct red color of Mars?
   A) Red-colored vegetation, which seems to fluctuate seasonally, particularly near the equator.
   B) $CO_2$, since it absorbs blue and green light preferentially.
   C) Rust, or iron oxides.
   D) Scattered sunlight from very fine dust, similar to sunset effects upon Earth.
   Answer: C Source: Section 6-11

184. The Martian magnetic field is
   A) comparable to Earth's magnetic field.
   B) non-existent-no spacecraft has ever detected a magnetic field on Mars.
   C) weak and localized, not at all like the global magnetic field of Earth.
   D) much stronger and more extensive than the Earth's magnetic field.
   Answer: C Source: Section 6-11

185. The Martian magnetic field is apparently created by
 A) electric currents in the molten, metallic core.
 B) permanent magnetism in the solidified core.
 C) permanent magnetism in the crust.
 D) the interaction of the solar wind with the Martian ionosphere.
 Answer: C Source: Section 6-11

186. The origin of the Martian magnetic field appears to be
 A) convection of molten, iron-rich lava in the Martian mantle.
 B) remnant magnetism from an earlier global magnetic field when Mars was young.
 C) magnetic material deposited by impacts of iron meteorites.
 D) an electric dynamo still operating in the small Martian core.
 Answer: B Source: Section 6-11

187. Where in the solar system did the Viking spacecraft land?
 A) On Europa, one of the moons of Jupiter.
 B) On Mars.
 C) On the Earth's Moon
 D) On Venus.
 Answer: B Source: Section 6-12

188. Viking 1 and 2, the planet-exploring spacecraft, were sent to which planet?
 A) Venus B) Jupiter C) Mars D) Mercury
 Answer: C Source: Section 6-12

189. The only experiments carried out so far to detect life on another world were performed by
 A) the Apollo astronauts.
 B) the Viking landers.
 C) the Venera landers.
 D) Pathfinder and Sojourner.
 Answer: B Source: Section 6-12

190. On which of the following planets have experiments been carried out to search for life-forms or evidence of life?
 A) Mercury B) Mars C) Venus D) Jupiter
 Answer: B Source: Section 6-12

191. The exploratory life-sciences experiments on board the Viking spacecraft landers found evidence
   A) of very reactive chemistry in the Martian surface rocks but no evidence of life or remnants of life-forms.
   B) that primitive life-forms had existed upon Mars earlier in its history but that they had not survived.
   C) of primitive life-forms such as elementary bacteria which should not be a hazard when Man explores Mars.
   D) of a very sterile environment in which life could not have existed and a very chemically inert soil, reacting with almost no reageants.
   Answer: A Source: Section 6-12

192. What were the results of the life-sciences experiments on board the Viking landers?
   A) A slight fizzing due to released carbon dioxide, but very little else.
   B) Nothing at all; the Martian soil is completely chemically inert and biologically sterile.
   C) Strong reactions releasing oxygen, from apparent biological causes (e.g., bacteria).
   D) Strong reactions releasing oxygen, but from non-biological causes.
   Answer: D Source: Section 6-12

193. Several components of the atmosphere and the environment on Mars render it sterile and antiseptic and would destroy life on the planet. Which of the follow is NOT one of these factors?
   A) Hydrogen peroxide in the soil.
   B) Sulfuric acid droplets in the atmosphere.
   C) Solar UV radiation, which is not absorbed by the thin atmosphere.
   D) Ozone in the atmosphere, produced by solar UV light.
   Answer: B Source: Section 6-12

194. Where have some scientists hypothesized that they have found direct evidence for life, either contemporary or ancient, beyond the Earth (although the hypothesis has been strongly disputed by many other scientists)?
   A) In "orange soil" found on the Moon.
   B) In a meteorite composed of ancient Martian rock.
   C) In spectra of the dark deposits along fissures in the ice of Jupiter's satellite, Europa.
   D) In the "soil" (regolith) at the Viking 2 landing site on Mars.
   Answer: B Source: Section 6-12

195. What evidence have some scientists claimed for ancient life on Mars (although the hypothesis has been strongly disputed by many other scientists)?
 A) Calcium carbonate deposits similar to seashells in a sample of Martian limestone.
 B) Possible fossilized bacteria in a Martian meteorite.
 C) An apparent thumbprint on a rock at the Mars Pathfinder landing site.
 D) Peroxides in the "soil" (regolith), found by experiments on the Viking landers.
 Answer: B Source: Section 6-12

196. Phobos and Deimos are moons of which planet?
 A) Uranus B) Venus C) Mars D) Jupiter
 Answer: C Source: Section 6-13

197. When were the two moons of Mars, Phobos, and Deimos, discovered?
 A) 1610 B) 1930 C) 1877 D) 1846
 Answer: C Source: Section 6-13

198. The moons of Mars are
 A) spherical and smooth-surfaced (no visible craters or volcanoes).
 B) irregularly shaped, cratered, and grooved.
 C) spherical, and have active volcanoes on their surfaces.
 D) ice-covered, approximately spherical but flattened by rapid rotation.
 Answer: B Source: Section 6-13

199. The shapes and sizes of the two moons of Mars are
 A) irregular but quite large compared to the planet, between 500 and 1500 km across.
 B) spherical and quite large compared to the planet, about 1000 km in diameter, similar to the largest asteroid.
 C) irregular in shape and very small, only several tens of km across.
 D) almost spherical but very small, only 10 and 30 km in diameter, respectively.
 Answer: C Source: Section 6-13

200. The Martian moon, Phobos, orbits near the equatorial plane of Mars in just over 7.5 hours, in the same direction as the planet rotation. How then would you see Phobos move across the Martian sky from the surface of the planet?
  A) It would rise in the west, move rapidly across the sky and set in the east, and appear several times per Martian day.
  B) It would remain almost stationary in the sky, in almost synchronous orbit, since its period is close to Mars' rotation period.
  C) It would rise in the east, move rapidly across the sky and set in the west, and appear several times per Martian day.
  D) It would rise in the west, move rapidly across the sky and set in the east, and would do this only once per Martian day.
  Answer: A Source: Section 6-13

# Chapter 7: The Outer Planets

1. When is the best time to observe Jupiter from the Earth?
   A) When it is high above our southern horizon at sunrise or sunset.
   B) When it is at opposition and hence is closest to Earth.
   C) When it is at maximum eastern or western elongation, and is then furthest away from the Sun in our sky at sunrise or sunset.
   D) When it is at conjunction, when it appears closest to the Sun and is then at its brightest.
   Answer: B Source: Sections 2-1 and 7-1

2. At what point in its orbit would Jupiter appear to be faintest when viewed from Earth?
   A) Conjunction.
   B) Opposition.
   C) When the line from the Earth to Jupiter is at a right angle to the line from Jupiter to the Sun.
   D) Its apparent brightness does not vary with orbital position since it has an almost circular orbit.
   Answer: A Source: Sections 2-1 and 7-1

3. At what point in its orbit would Jupiter appear to be brightest when viewed from the Earth?
   A) When the Earth-Jupiter line is at right angles to the Sun-Jupiter line
   B) Conjunction.
   C) Opposition.
   D) Its brightness will not vary with orbital position because of its almost circular orbit.
   Answer: C Source: Sections 2-1 and 7-1

4. Which of the following spacecraft did NOT visit Jupiter?
   A) Galileo B) Pioneer 10 C) Voyager 1 D) Viking
   Answer: D Source: Section 7-1

5. Which spacecraft was the most recent to reach Jupiter?
   A) Galileo B) Voyager 2 C) Soyuz 15 D) Pioneer 2
   Answer: A Source: Section 7-1

6. Which spacecraft recently dropped a probe into the atmosphere of Jupiter?
   A) Galileo B) Cassini C) Voyager D) Ulysses
   Answer: A Source: Section 7-1

7. Which is the biggest planet in the solar system?
   A) Neptune B) Jupiter C) Saturn D) Earth
   Answer: B Source: Section 7-1

8. What is the mass of Jupiter compared to other objects in the solar system?
   A) Twice the mass of all other planets combined.
   B) As large as the mass of Saturn and Neptune combined.
   C) Half the mass of the Sun.
   D) Twenty times the mass of all other planets combined.
   Answer: A Source: Section 7-1

9. How many times more massive would Jupiter have to be to be a star (generate energy by nuclear fusion)?
   A) 1000 times more massive.
   B) 10 times more massive.
   C) 75 times more massive.
   D) 2 times more massive.
   Answer: C Source: Section 7-1

10. The fraction of the mass of the planetary system (excluding the Sun) which is concentrated in the planet Jupiter is about (see Section 7-1, Kaufmann & Comins, *Discovering the Universe*, 5th Ed.)
    A) 50% B) 98% C) 70% D) 10%
    Answer: C Source: Section 7-1

11. Compared to that of Earth, the mass of Jupiter is
    A) about 300 times larger.
    B) about 1/300, because of Jupiter's low density.
    C) several thousand times larger.
    D) about 11 times as large.
    Answer: A Source: Section 7-1

12. The intensity of sunlight reaching Jupiter is approximately what fraction of that at Earth's orbital distance? (Hint: Think about the inverse-square law, Foundations III.)
    A) 1/100 B) About the same. C) 1/25 D) 1/5
    Answer: C Source: Section 7-1 and Foundation III

13. Which object or group of objects below rotates the most quickly around their own axes?
    A) The terrestrial planets.
    B) The Jovian planets.
    C) The Sun.
    D) The Earth's Moon.
    Answer: B Source: Section 7-1

14. The apparent angular diameter of Jupiter when viewed from Earth appears to vary with time because
    A) the distance between Jupiter and Earth varies.
    B) the absorption of radiation by interplanetary dust between Jupiter and Earth varies with time.
    C) the amount of sunlight reaching Jupiter varies with time.
    D) Jupiter is a fluid planet and pulsates with a long natural period of oscillation.
    Answer: A Source: Sections 7-1 and 1-1, Toolbox 1

15. What are the characteristic features on the visible surface of Jupiter?
    A) Light and dark bands of clouds parallel to the equator.
    B) Large volcanoes and a long, deep rift valley.
    C) A Bluish-green, almost featureless, cloud layer.
    D) A bluish tint with high, white clouds and dark storms.
    Answer: A Source: Section 7-1

16. What is the physical appearance of Jupiter, as seen from the Earth or a spacecraft?
    A) Solid ice surface, some craters but highly modified by ice flow.
    B) Densely cratered surface with one large impact basin.
    C) A series of dark belts and light zones parallel to the equator.
    D) A uniform bluish color with a high-level haze.
    Answer: C Source: Section 7-1

17. Tomorrow's weather forecast for someone standing on the surface of Jupiter is
    A) sunny and clear, since Jupiter has no atmosphere in which clouds can form.
    B) overcast, possible rain with snow at higher elevations.
    C) sunny, possible thin, high cloud.
    D) the question is meaningless, since there is no solid surface upon which to stand.
    Answer: D Source: Section 7-1

18. Evidence of volcanism (lava outflow, etc.) either active or ancient, is NOT found on
    A) Mars. B) Earth's Moon. C) Earth. D) Jupiter.
    Answer: D Source: Section 7-1

19. What are the dark, reddish bands on Jupiter called?
    A) Voids B) Belts C) Zones D) Streams
    Answer: B Source: Section 7-1

20. What are the light-colored bands on Jupiter called?
    A) Streams B) Belts C) Zones D) Plumes
    Answer: C Source: Section 7-1

21. On which planet is the Great Red Spot found?
A) Neptune B) Jupiter C) Saturn D) Mars
Answer: B Source: Section 7-1

22. The existence of the Great Red Spot of Jupiter has been known since
A) the first fly-by of a spacecraft, Pioneer 10, in December, 1973.
B) first light at the 200 inch telescope on Mt Palomar, in 1948.
C) the 1600s.
D) the arrival at Jupiter of Voyager 1, with its imaging cameras, in 1979.
Answer: C Source: Section 7-1

23. What is the Great Red Spot?
A) A large and long-lived, possibly permanent, storm on Jupiter.
B) A dark polar hood in the clouds of Titan, a satellite of Saturn.
C) A lava lake on Io, a satellite of Jupiter.
D) A large, anticyclonic storm on Neptune, discovered by the Voyager spacecraft.
Answer: A Source: Section 7-1

24. What is the Great Red Spot on Jupiter?
A) The point where charged particles from the satellite Io collide with Jupiter's cloud tops.
B) The summit of a large mountain that rises above the upper cloud level.
C) A region over the South Pole of Jupiter where ammonia compounds have condensed in the colder atmosphere.
D) A large, long-lived, anticyclonic storm that is maintained by the circulation in Jupiter's atmosphere.
Answer: D Source: Section 7-1

25. The Great Red Spot is
A) the colored polar cap of Jupiter.
B) a large, long-lived storm system in Jupiter's atmosphere.
C) the top of a massive mountain penetrating through Jupiter's clouds.
D) a temporary storm in Jupiter's atmosphere, lasting a few months.
Answer: B Source: Section 7-1

26. The Great Red Spot is
A) a large, red crater on Mars.
B) a hot-spot on Venus, detected by Russian landers and the U.S. Magellan orbiter.
C) a large region at the north pole of Saturn, rotating rapidly.
D) a large, stable, circulating storm on the surface of Jupiter.
Answer: D Source: Section 7-1

27. One distinctive feature that is visible on the "surface" of Jupiter through a telescope from Earth is
    A) the Cassini Division.
    B) the Great Red Spot.
    C) Maxwell Montes.
    D) Olympus Mons.
    Answer: B Source: Section 7-1

28. The lifetime of the Great Red Spot of Jupiter appears to be
    A) about 2 to 4 weeks between successive appearances, similar to sunspots.
    B) at least 300 years, from visual records.
    C) about one Jupiter orbital period between successive appearances, since the spot is produced by tidal effects.
    D) well over 2000 years, from ancient Greek records.
    Answer: B Source: Section 7-1

29. What is the dominant circulation pattern in Jupiter's atmosphere (i.e., at the visible "surface")?
    A) Alternating bands of eastward and westward flow parallel to the equator, with light and dark ovals between these flows.
    B) Uniform eastward flow of the entire atmosphere, with occasional dark storms and turbulent swirls.
    C) Air rising at the equator, flowing north and south toward the poles, then sinking and returning to the equator at a lower level
    D) Isolated storms and turbulent swirls, with little overall flow pattern in any particular direction.
    Answer: A Source: Section 7-1

30. The rotation periods of Jupiter and Saturn are
    A) very short, on the order of 1 hour.
    B) relatively short, on the order of 10 hours.
    C) long, on the order of several days.
    D) very long, several weeks, because of their great size and mass.
    Answer: B Source: Section 7-1

31. The rotation periods for the Jovian planets Jupiter, Saturn, Uranus, and Neptune are
    A) reasonably long, on the order of several Earth days.
    B) much shorter than Earth's, between 1 and 2 hours.
    C) very long, on the order of years, because of the sizes of these planets.
    D) somewhat shorter than Earth's, between 10 to 20 hours.
    Answer: D Source: Section 7-1

32. Detailed observations of Jupiter's rotation suggest that
    A) it is slowing down noticeably at the present time.
    B) it is not a rigid object, equatorial regions rotating faster than polar regions.
    C) it rotates in two separate parts, equatorial regions rotating in a direction opposite to polar regions.
    D) it rotates as a solid body, equatorial and polar regions showing the same rotational period.
    Answer: B Source: Section 7-1

33. The interesting feature of Jupiter's rotation is the fact that
    A) its rotation rate has slowed down significantly since it was first observed through telescopes in the 1600s.
    B) it rotates in a direction opposite to that of most of the planets and opposite to its direction of revolution around the Sun.
    C) its axis of rotation lies almost in the plane of its orbit.
    D) regions at different latitudes appear to rotate at different rates.
    Answer: D Source: Section 7-1

34. What is the overall composition of Jupiter, excluding its rocky core?
    A) 75% hydrogen, 15% helium, 10% everything else.
    B) Pure hydrogen.
    C) Hydrogen and helium only, nothing else.
    D) 86% hydrogen, 13% helium, 1% everything else.
    Answer: D Source: Section 7-1

35. What is the composition of Jupiter's clouds?
    A) Water ice crystals.
    B) Ice crystals of water and carbon dioxide.
    C) Ice crystals of methane, ammonia and water.
    D) Liquid droplets of water and ammonia.
    Answer: C Source: Section 7-1

36. The composition of the clouds which we see on Jupiter is
    A) similar to those of the Earth (water droplets and crystals of frozen water) in the lower levels, but very different (e.g., ammonia crystals and other chemicals) in the higher levels.
    B) similar to Earth clouds (water droplets and crystals of frozen water) in the higher levels, but very different (e.g., ammonia crystals) in the lower levels.
    C) similar to Earth clouds through the whole atmosphere- water droplets and crystals of frozen water.
    D) very different from Earth clouds, composed almost entirely of ammonia and ammonium hydrosulfide crystals with almost no water.
    Answer: A Source: Section 7-1

37. Which of the following chemicals is the least abundant in the outer atmosphere of Jupiter?
    A) $CH_4$, methane.
    B) $NH_3$, ammonia.
    C) $CO_2$, carbon dioxide.
    D) $H_2O$, water vapor.
    Answer: C Source: Section 7-1

38. The major components of the visible surfaces of the Jupiter and Saturn atmospheres, other than hydrogen and helium, have been found to be
    A) $H_2O$ (water), $CO_2$ (carbon dioxide).
    B) $N_2$ (nitrogen), $O_2$ (oxygen), $CO_2$ (carbon dioxide).
    C) dust and iron oxides.
    D) $CH_4$ (methane), $NH_3$ (ammonia), $H_2O$ (water).
    Answer: D Source: Section 7-1

39. Biological activity on Jupiter
    A) is impossible because it would be destroyed by the intense solar ultraviolet light.
    B) is impossible because it would be destroyed by Jupiter's predominantly hydrogen atmosphere.
    C) might be possible because the warm temperature below the clouds allows liquid-water oceans to exist on Jupiter's surface.
    D) might be possible suspended in the clouds, since lightning in methane and ammonia gases and water vapor can generate organic compounds.
    Answer: D Source: Sections 7-1 and 18-1

40. How thick is Jupiter's gaseous atmosphere?
    A) 10 km B) 150 km C) 65,000 km D) 2000 km
    Answer: B Source: Section 7-1

41. What causes the banded structure on Jupiter's visible "surface" as seen from the Earth?
    A) An underlying north-south flow pattern stretched into bands by Jupiter's rapid rotation.
    B) An underlying rising and falling convection pattern, stretched into bands by Jupiter's rapid rotation.
    C) The "sweeping" of Jupiter's clouds through magnetic field lines from Jupiter's magnetosphere.
    D) The breaking up of strong eastward flow due to Jupiter's rapid rotation, by underlying mountain ranges.
    Answer: B Source: Section 7-1

42. The light-colored zones on Jupiter are
    A) regions of warmer, ascending gas, which stretch all the way around the planet.
    B) volcanic plumes stretched around the planet by Jupiter's high speed of rotation.
    C) high-speed jet-streams that have been warmed by solar ultraviolet radiation.
    D) regions of cooler, descending gas, which stretch all the way around the planet.
    Answer: A Source: Section 7-1

43. The dark belts of Jupiter are
    A) regions of warmer, ascending gas, which stretch all the way around the planet.
    B) ancient stream channels, now dry, dating back to when Jupiter had abundant rainfall.
    C) dark clouds formed when ammonia condenses at high altitudes in regions of up-welling gas.
    D) regions of cooler, descending gas, which stretch all the way around the planet.
    Answer: D Source: Section 7-1

44. The high-speed winds observed on Jupiter occur mainly
    A) near the centers of the dark belts.
    B) at the boundaries between the dark belts and the light zones.
    C) in a north-south direction from the dark belts toward the light zones.
    D) near the centers of the light zones.
    Answer: B Source: Section 7-1

45. The large-scale atmospheric circulation pattern on Jupiter is characterized predominantly by
    A) strong winds blowing westward at all latitudes, so that the entire atmosphere rotates more slowly than the planet.
    B) strong winds blowing eastward at all latitudes, so that the entire atmosphere rotates faster than the planet.
    C) strong winds blowing parallel to the equator, but in opposite directions at different latitudes.
    D) isolated cyclones (low-pressure areas) and anticyclones (high-pressure areas), as on Earth.
    Answer: C Source: Section 7-1

46. How thick are the cloud layers on Jupiter's visible "surface"?
    A) About 100 km.
    B) Greater than 10,000 km, as seen through the dark ovals, holes in the cloud layers.
    C) Very thin, only about 10 km.
    D) About 1000 km.
    Answer: A Source: Section 7-1

47. What was the most surprising result that the *Galileo Probe* found when it penetrated the atmosphere of Jupiter?
    A) It measured only traces of the $NH_3$, $NH_4SH$, and water vapor cloud layers, which are easily measured spectroscopically from Earth.
    B) It found no evidence of hydrogen at all in the atmosphere of Jupiter.
    C) It hit a hard surface just under the cloud layers, contrary to our understanding of Jupiter's atmospheric structure.
    D) It measured vast quantities of invisible water vapor above the cloud layers
    Answer: A Source: Section 7-1

48. What is the basic structure of the planet Jupiter?
    A) Rocky core, liquid hydrogen mantle, gaseous atmosphere.
    B) Rocky core, liquid methane and water mantle, gaseous atmosphere.
    C) Entirely liquid hydrogen, except for a thin, gaseous atmosphere.
    D) Rocky core, frozen water mantle, thin methane atmosphere.
    Answer: A Source: Section 7-2

49. What is the basic structure of the planet Jupiter?
    A) A thick mantle of liquid hydrogen with a rocky core and a relatively thin, gaseous atmosphere.
    B) A thick, gaseous atmosphere over a thin mantle of liquid hydrogen and a rocky core.
    C) A large, rocky core with a thin mantle of liquid hydrogen and a thin, gaseous atmosphere.
    D) A rocky core, liquid hydrogen mantle, and a gaseous atmosphere, all of about the same thickness.
    Answer: A Source: Section 7-2

50. What are the fractional proportions of the components which make up the mass of Jupiter?
    A) 96% as hydrogen, helium and a small fraction of other gases and 4% rock.
    B) 75% hydrogen and helium, very little as other gases and 25% rock.
    C) Almost pure hydrogen with maybe 1% rock in the central core.
    D) 50% hydrogen, 49% helium and 1% rock in the core of the planet.
    Answer: A Source: Section 7-2

51. The internal structure of the two largest Jovian planets (from the center outward) is
    A) rocky core, liquid methane mantle, gaseous methane atmosphere.
    B) iron-nickel core, rocky mantle, solid crust.
    C) rocky core, liquid molecular hydrogen layer, liquid metallic hydrogen layer.
    D) rocky core, liquid metallic hydrogen layer, liquid molecular hydrogen layer.
    Answer: D Source: Section 7-2

52. The overall interior structure of the Jovian planets is expected to be
    A) three-layered; a rocky core, a liquid mantle of hydrogen and an extensive gaseous atmosphere.
    B) three-layered; an inner molten iron core, a semi-fluid mantle and a solid crust.
    C) two-layered; a large, solid, rocky core surrounded by an extensive gaseous atmosphere.
    D) four-layered; a solid inner core, a liquid iron outer core, a semi-fluid mantle and a solid crust.
    Answer: A Source: Section 7-2

53. Where in the solar system would you look for liquid hydrogen?
    A) Nowhere, since there is nowhere cold enough in the solar system to liquefy hydrogen.
    B) In the deep interiors of Jupiter and Saturn.
    C) On the surface of Venus, beneath the clouds.
    D) On the polar caps of Mars.
    Answer: B Source: Section 7-2

54. In order of increasing density, the Jovian planets would be listed as
A) Neptune, Uranus, Saturn, Jupiter.
B) Jupiter, Saturn, Uranus, Neptune.
C) Neptune, Saturn, Uranus, Jupiter.
D) Saturn, Uranus, Jupiter, Neptune.
Answer: D Source: Section 7-2

55. The low average density of Jupiter (about 1300 kg/$m^3$ compared with that of water, 1000 kg/$m^3$) indicates that this planet is composed mainly of
A) helium as gas and liquid only, since low temperatures and great pressures are needed to form solid helium.
B) water, compressed somewhat by gravity, maybe in the form of ice.
C) hydrogen, in liquid or gaseous form.
D) methane, ammonia and water, from spectroscopic observation of its atmosphere.
Answer: C Source: Section 7-2

56. The Jovian planets have high masses and hence generate powerful gravitational fields, and yet have low average densities. What does this indicate about their interiors?
A) They are composed almost entirely of water.
B) They are composed almost entirely of hydrogen and helium.
C) The interiors are hot and gaseous, like those of cool stars.
D) They are composed almost entirely of ice crystals and dust.
Answer: B Source: Section 7-2

57. The deepest central sections of the interiors of Jupiter and Saturn are thought to be composed of
A) methane, ammonia, and water vapor.
B) magnetized iron cores.
C) rocky cores.
D) liquid metallic hydrogen.
Answer: C Source: Section 7-2

58. The magnetic field strength at Jupiter's equator exceeds that of Earth by a factor of about
A) 14. B) a million. C) 20,000. D) 2.
Answer: A Source: Section 7-2

59. Jupiter has a magnetic field that is
A) almost nonexistent.
B) much more powerful than that of Earth.
C) about the same strength and extent as that of Earth.
D) weak and variable, sometimes existing only at the Great Red Spot.
Answer: B Source: Section 7-2

60. What is the source of Jupiter's intense magnetic field?
   A) Electric currents in Jupiter's liquid hydrogen mantle.
   B) Electric currents in Jupiter's molten rocky core.
   C) A permanently magnetized iron core.
   D) Electric currents in ionized layers of Jupiter's atmosphere.
   Answer: A Source: Section 7-2

61. The material in the interiors of Jupiter and Saturn that is thought to be responsible for their powerful magnetic fields is
   A) molten iron and nickel.
   B) liquid metallic hydrogen.
   C) solid magnetic iron.
   D) liquid "ices": $NH_3$ (ammonia), $CH_4$(methane), $H_2O$ (water).
   Answer: B Source: Section 7-2

62. The requirements for the generation of a powerful magnetic field in a Jovian planet (e.g., Jupiter, Saturn) appear to be
   A) solid interior throughout the planet, and slow rotation.
   B) liquid "metal" interior and relatively rapid rotation.
   C) solid iron core forming a permanent magnet.
   D) liquid "metal" core and interior, and very slow rotation.
   Answer: B Source: Section 7-2

63. What conditions are considered to be necessary for a planet to be able to generate an intense magnetic field?
   A) Electrically conducting material in its interior and slow rotation, since rapid rotation will destroy a magnetic field.
   B) Relatively rapid rotation and electrically conducting material in its interior.
   C) A solid iron core into which a magnetic field was induced early in the planet's history.
   D) An ionized and electrically conducting layer in its atmosphere.
   Answer: B Source: Section 7-2

64. The magnetosphere of Jupiter is
   A) a large region outside Jupiter occupied by its magnetic field and filled with high-energy charged particles.
   B) a narrow layer in Jupiter's atmosphere, just above the cloud tops, in which intense electric currents flow and generate the planet's magnetic field.
   C) the magnetized hydrogen in the inner regions of Jupiter just outside the solid core, where the planet's magnetic field is produced.
   D) a region of charged particles extending along the orbit of the satellite Io, forming a ring around Jupiter.
   Answer: A Source: Section 7-2

65. The magnetosphere of Jupiter is
   A) a doughnut-shaped region similar to the Van Allen belts around the Earth containing high-speed protons and electrons whose motions produce the planet's magnetic field.
   B) the inner regions of Jupiter just outside the solid core which contain liquid metallic hydrogen, in which electric currents flow to produce the planet's magnetic field.
   C) a narrow layer in which intense electric currents flow, just above the cloud tops in the planet's atmosphere, generating the planet's magnetic field.
   D) a large cavity created and maintained within the solar wind stream by the planet's magnetic field, filled with extremely hot ionized plasma.
   Answer: D Source: Section 7-2

66. The shape and dimensions of the magnetosphere surrounding Jupiter are controlled by
   A) the outward motion of the atoms of Jupiter's outer atmosphere as it rotates rapidly within the planet's powerful gravitational field.
   B) the pressure of the ionized gas of the solar wind against the planet's magnetic field.
   C) solar radiation pressure pushing against the planet's outer atmosphere.
   D) the pressure of the solar wind against the outer atmosphere of Jupiter.
   Answer: B Source: Section 7-2

67. Compared to the stability of the size and shape of the Earth's magnetosphere against varying solar wind pressure, that of Jupiter
   A) is much more stable, because of Jupiter's much stronger magnetic field and its greater distance from the Sun and hence weaker solar wind pressure.
   B) is extremely weak, the magnetosphere disappearing whenever the solar wind pressure increases beyond a certain limit.
   C) is very weak, varying wildly in size and shape.
   D) about as stable, varying a little under strong solar wind changes.
   Answer: C Source: Section 7-2

68. If we could see the full extent of the magnetosphere of Jupiter from Earth, how big would it appear in our sky?
    A) About as large as the full Moon.
    B) About 16 times larger than Jupiter itself.
    C) 16 times larger than the full Moon.
    D) About twice as large as Jupiter itself.
    Answer: C Source: Section 7-2

69. If we could see Jupiter's magnetosphere with our eyes, how big would it appear?
    A) It would cover essentially our entire sky whenever Jupiter was above the horizon.
    B) It would be easily visible, several times larger than the Full Moon.
    C) It would be just big enough to see, almost at the limit of naked-eye visibility.
    D) It would be too small to be visible without a telescope.
    Answer: B Source: Section 7-2

70. How does the magnetosphere of Jupiter compare to that of the Earth?
    A) Much larger, because of Jupiter's large size and rapid rotation.
    B) Jupiter has no detectable magnetic field.
    C) Very similar to that of the Earth.
    D) Much smaller, because Jupiter is made up mostly of hydrogen.
    Answer: A Source: Section 7-2

71. How did Comet Shoemaker-Levy 9 achieve lasting fame?
    A) It just missed the Earth, passing between the Moon and the Earth.
    B) It exploded after passing too close to the Sun.
    C) It became the first comet to be visited by spacecraft.
    D) It crashed into Jupiter.
    Answer: D Source: Section 7-3

72. What were the results of the impacts of the fragments of Comet Shoemaker-Levy 9 into Jupiter?
    A) Fireballs hotter than the Sun and dark splotches that lasted for months.
    B) Holes punched through the clouds and craters gouged into Jupiter's surface.
    C) Essentially no visible effects, since the fragments plummeted deep below the cloud layers before being destroyed.
    D) Rings of debris flung into orbit around Jupiter's equator.
    Answer: A Source: Section 7-3

73. What effect did Comet Shoemaker-Levy 9 have on the atmosphere of Jupiter?
   A) It created dark patches that lasted for months.
   B) It had no lasting effect beyond the initial burst of hot gas.
   C) It created white clouds of ammonia and water ice crystals that lasted for several days.
   D) It disrupted the atmospheric circulation of light and dark zones for almost a month.
   Answer: A Source: Section 7-3

74. How many moons are known to orbit Jupiter?
   A) Eleven.
   B) At least sixteen.
   C) None, just like Mercury and Venus.
   D) Four, discovered by Galileo.
   Answer: B Source: Section 7-4

75. How many large, spherical moons are in orbit around Jupiter?
   A) None. B) Four. C) Eleven. D) Sixteen.
   Answer: B Source: Section 7-4

76. The four giant moons of Jupiter were discovered by.
   A) Newton. B) Galileo. C) the Voyager spacecraft. D) Ptolemy.
   Answer: B Source: Section 7-4

77. How many moons of Jupiter were seen by Galileo?
   A) Twelve.
   B) Four.
   C) None, he was unable to see them with the naked eye.
   D) Only one.
   Answer: B Source: Section 7-4

78. The four major moons of Jupiter are collectively named after which early astronomer?
   A) Ptolemy. B) Galileo. C) Newton. D) Copernicus.
   Answer: B Source: Section 7-4

79. For how long has it been known that the planet Jupiter has moons?
   A) Since ancient times.
   B) Since the Voyager spacecraft flybys in 1979.
   C) Since Galileo turned his telescope to the sky, in 1610.
   D) Since 1948, when the Mt. Palomar telescope was built.
   Answer: C Source: Section 7-4

80. Which of the following early telescope observations probably convinced Galileo that the Copernican heliocentric model provided a better explanation for the solar system than the Greek geocentric model?
    A) Sunspots on the Sun and apparent solar rotation.
    B) Moons seen to be orbiting another planet, Jupiter, rather than the Earth.
    C) Venus changing in angular size with time.
    D) Apparent change in angular size of the Moon as it orbits the Earth.
    Answer: B Source: Section 7-4

81. Which of the following motions is seen to be characteristic of the four Galilean moons of Jupiter?
    A) They each keep the same face toward the planet at all times.
    B) They orbit the planet in the opposite direction to its rotation.
    C) They each keep the same face toward the Sun at all times.
    D) They orbit the planet in a plane carrying them over both the north and south poles of Jupiter.
    Answer: A Source: Section 7-4

82. The moons of Jupiter behave in which way, in terms of the relationship between the period P and radius a of their orbits around the planet?
    A) They follow a Keplerian relationship, $P^2 = ka^3$, where k is a constant which is different from that in the relation governing planetary motion around the Sun.
    B) The moons do not obey the Keplerian relation $P^2 = ka^3$, because their motion about Jupiter is affected by Jupiter's motion about the Sun.
    C) The moons do not obey the Keplerian relation $P^2 = ka^3$, since they orbit Jupiter and not the Sun.
    D) They obey the Keplerian relation, $P^2 = ka^3$, in which k is the same as that for planetary motion around the Sun, since it is a universal constant.
    Answer: A Source: Section 7-4

83. What is significant about the rotational and revolutional motions of the Galilean moons of Jupiter and of our own Moon?
    A) Nonsynchronous rotation, with independent periods of rotation and revolution, and spin axes perpendicular to their orbit.
    B) Synchronous rotation, with one face always pointed toward the planet.
    C) Nonsynchronous rotation, with axes of rotation in any direction with respect to their orbital plane.
    D) Synchronous rotation, with one face always pointed toward the Sun.
    Answer: B Source: Section 7-4

84. Brightness variations of Jupiter's moons as they orbit the planet indicate that the relation between the spin around their axes and their orbital motions is that the
   A) moons rotate exactly once per orbital period.
   B) moons' rotation is controlled by the gravitational influence of the Sun, and they always keep one face toward it, producing the observed brightness variations.
   C) moons rotate on their axes independently of their orbital motion.
   D) moons do not rotate at all while orbiting the planet..
   Answer: A Source: Section 7-4

85. If the orbital period of Jupiter's moon, Ganymede, is 7.15 days, what would be the view from a point on its surface at the equator?
   A) Both the Sun and Jupiter would remain fixed in the sky above Ganymede.
   B) The Sun, if visible, would always be at the same point in the sky, but Jupiter would rise and set with a period of about 7.15 days.
   C) Jupiter, if visible, would always remain at the same point in the sky but the Sun would rise and set with a period of about 7.15 days.
   D) Both Jupiter and the Sun would rise and set with a period of about 7.15 days.
   Answer: C Source: Section 7-4

86. The general shape of the Galilean moons of Jupiter is
   A) irregular, rounded surfaces.
   B) approximately spherical, but with significant flattening of the polar axes, (oblate shape).
   C) approximately spherical, but with significantly shorter equatorial diameters than polar diameters (prolate shape).
   D) almost perfectly spherical.
   Answer: D Source: Section 7-4

87. Which of the following lists of planets and satellites of planets are in the correct order of increasing size (with Europa being one of the major moons of Jupiter)?
   A) Europa, Moon, Mercury, Mars.
   B) Europa, Mercury, Moon, Mars.
   C) Moon, Europa, Mars, Mercury.
   D) Mercury, Moon, Europa, Mars.
   Answer: A Source: Section 7-4

88. The average densities of the Galilean moons of Jupiter follow which pattern with increasing distance from the planet?
    A) Average density increases with distance from the planet.
    B) Average density is the same for all moons, since they were made from the same material.
    C) Average density shows NO pattern with distance, the highest density moon being Ganymede, the largest moon.
    D) Average density decreases with distance from the planet.
    Answer: D Source: Sections 7-4, 5, 6, and 7

89. The inner two Galilean satellites of Jupiter differ from the outer two by having
    A) much lower average densities.
    B) almost the same average density, but much older, more heavily cratered surfaces.
    C) almost the same average density, but much younger, less-cratered surfaces.
    D) much higher average densities.
    Answer: D Source: Sections 7-4, 5, 6, and 7

90. Which satellite of Jupiter is volcanically active?
    A) Europa. B) Io. C) Ganymede. D) Callisto.
    Answer: B Source: Section 7-4

91. What characteristic of Jupiter's satellite, Io, makes it different from any other known satellite in the solar system?
    A) It has a permanent, dense atmosphere.
    B) It is volcanically active.
    C) It has geyser-like plumes of nitrogen gas.
    D) Its surface is broken into heavily cratered and lightly cratered regions in a pattern similar to plate tectonics.
    Answer: B Source: Section 7-4

92. Jupiter's satellite Io has numerous black spots on its surface. What are these spots believed to represent?
    A) Collapse features in an icy crust.
    B) Volcanic vents.
    C) Impact craters.
    D) Debris from comets.
    Answer: B Source: Section 7-4

93. What percentage of Io's surface is covered with lava flows?
    A) Zero; the surface is perpetually ice-covered. B) 40%. C) 100%. D) 5%.
    Answer: D Source: Section 7-4

94. The most geologically active object in the planetary system at the present time is
   A) Triton, a moon of Neptune.
   B) Mars.
   C) Io, a moon of Jupiter.
   D) the Earth's Moon.
   Answer: C Source: Section 7-4

95. The eruptions observed on Io are thought to most clearly resemble
   A) geysers, where material is shot upward by the pressure of gas produced below the surface.
   B) explosions, where material is thrown upward by a single burst and then falls back to the surface.
   C) terrestrial mid-ocean ridges, where upwelling molten rock pushes the crust apart.
   D) volcanoes, producing lava flows and columns of erupting silicate ash.
   Answer: A Source: Section 7-4

96. Which of the following is NOT seen on Jupiter's satellite, Io?
   A) Impact craters. B) Lava flows. C) Volcanic plumes. D) Sulfur dioxide frost.
   Answer: A Source: Section 7-4

97. Why would you expect to see no craters, such as those on the Moon or Mars, on Io, the innermost Galilean moon of Jupiter?
   A) Because volcanoes are continuously depositing new material onto the surface.
   B) Because the surface is always re-entering the planet's interior by subduction in rapid plate tectonic motion, similar to but faster than that upon Earth.
   C) Because the continuous $H_2O$ rainfall will quickly erode and wash away all trace of craters.
   D) Because the liquid surface cannot maintain a crater, just as the Earth's oceans cannot do so.
   Answer: A Source: Section 7-4

98. How would "Interplanetary Travel" advertise a holiday on Jupiter's satellite, Io?
   A) The largest number of volcanoes for your travel dollar anywhere in the Solar System!
   B) Exquisite ethane lakes, hydrocarbons beyond your wildest dreams!
   C) Glaciers galore for your hiking pleasure under star-studded skies!
   D) Greatest deep-sea diving beneath thick ice layers!
   Answer: A Source: Section 7-4

99. The major chemical constituent of the layers of material continuously being deposited on the surface of Io, the Moon of Jupiter, by "volcanic" action is
   A) silicate dust and rock. B) water. C) sulfur. D) hydrogen.
   Answer: C Source: Section 7-4

100. The "snow" which falls continuously on the surface of Io, the innermost Galilean moon of Jupiter, is composed of
A) water crystals.
B) ammonia crystals.
C) sulfur and sulfur dioxide.
D) methane crystals.
Answer: C Source: Section 7-4

101. Which chemical or chemicals appear to play a prominent role in the "volcanoes" of Io?
A) Water and steam.
B) Sulfur and sulfur dioxide.
C) Molten lava.
D) Methane and ammonia.
Answer: B Source: Section 7-4

102. Which gas is thought to be responsible for the tremendous eruptions of material from Jupiter's innermost Galilean moon, Io, by providing the propulsive forces?
A) Carbon dioxide, $CO_2$.
B) Steam, or heated water vapor.
C) Sulfur in vapor form.
D) Sulfur dioxide, $SO_2$.
Answer: D Source: Section 7-4

103. The "volcanoes" of Io, the innermost Galilean moon of Jupiter, most resemble in behavior which Earth-bound phenomenon?
A) Undersea volcanoes, where hot lava produces explosive boiling of the water.
B) Lava-flow volcanoes occurring on the tops of mountains, such as those on Hawaii.
C) Regularly spouting geysers in relatively flat hot-springs areas.
D) Mountain volcanoes such as Mount St. Helens, exhibiting rare but devastating explosions that throw material high into the atmosphere.
Answer: C Source: Section 7-4

104. What peculiar feature accompanies Io in its orbit around Jupiter?
A) An auroral storm in the magnetosphere which always keeps pace with Io.
B) A torus or ring of ionized sulfur and oxygen atoms and electrons.
C) A comet-like tail of rocks and dust, shining by reflected sunlight.
D) A narrow ring of rocks and dust, Jupiter's ring, at about Io's orbital distance.
Answer: B Source: Section 7-4

105. What is believed to be the source of the many colors seen on the surface of Io, the inner large satellite of Jupiter?
A) Sulfur and sulfur compounds.
B) Organic (carbon) compounds discolored by solar ultraviolet light.
C) Impurities in a water-ice crust.
D) Phosphorous compounds.
Answer: A Source: Section 7-4

106. The required heating of the large Galilean moon, Io, of Jupiter, in order to produce volcanic activity is probably caused by
   A) radioactive elements in its surface.
   B) nuclear fission within its interior.
   C) tidal distortion by Jupiter and its other moons.
   D) its original heat of formation.
   Answer: C Source: Section 7-4

107. The source of intense heating in the interior of Jupiter's moon Io, causing continuous and intense volcanic activity, is
   A) frictional heating as the solar wind impacts on the moon surface.
   B) tidal flexing and distortion, caused by Jupiter and the other large moons.
   C) solar UV and visible radiation.
   D) heat released by continuous shrinkage after creation, transforming potential gravitational energy to heat.
   Answer: B Source: Section 7-4

108. One of the most important sources of heat in the interiors of moons that orbit close to giant planets is
   A) reflection of sunlight from the planet's surface.
   B) thermal energy received from the planet, since Jupiter radiates more energy than it receives from the Sun.
   C) decay of radioactive elements within the moons.
   D) continuous tidal distortion from the planet and other moons.
   Answer: D Source: Section 7-4

109. The heating of the interior of Io, the innermost Galilean moon of Jupiter, is caused by
   A) thermal heating from Jupiter, since Jupiter emits more radiation than it receives from the Sun
   B) continuous tidal distortion by Jupiter and the other moons
   C) continual bombardment by meteoroids, attracted by Jupiter's enormous gravitational pull
   D) the fact that Io moves inside Jupiter's magnetosphere, where the temperature is as high as anywhere in the solar system
   Answer: B Source: Section 7-4

110. Io, one of the major Jupiter moons, is undergoing extensive volcanic activity associated with interior heating that is caused by
A) the effect of Jupiter's Van Allen particles.
B) tidal distortion and internal friction because of flexing.
C) solar radiation and heat falling on the surface
D) original heat, caused by gravitational condensation at the moon's formation.
Answer: B Source: Section 7-4

111. What is the heat source which causes the extensive volcanic activity observed on Jupiter's satellite, Io?
A) Tidal stresses from Jupiter and Io's companion moons.
B) "Primordial" heat remaining from when Io first formed
C) Heat released by radioactivity in Io's core and mantle.
D) Frictional heating by mantle convection and crustal tectonics.
Answer: A Source: Section 7-4

112. As a result of measurements by the Galileo spacecraft, what is now believed to be the internal structure of Io?
A) A thick mantle of rock over a core of iron and iron sulfide.
B) A thick mantle of ice over a core of rock.
C) Entirely or almost entirely rock, with no evidence of ice or a differentiated metallic core.
D) A thick crust of ice over a rocky mantle, and a small, possibly molten, iron core.
Answer: A Source: Section 7-4

113. Which of the Moons of Jupiter is characterized by an exceptionally smooth, icy surface, few craters, and many streaks and cracks?
A) Io. B) Europa. C) Callisto. D) Ganymede.
Answer: B Source: Section 7-5

114. The outer three Galilean moons of Jupiter differ from Io, the innermost such moon, by having surfaces of
A) water ice. B) carbon dioxide. C) sulfur. D) smoothly polished rock.
Answer: A Source: Sections 7-4, 5, 6, and 7

115. The surface of Europa, one of the Galilean moons of Jupiter, appears to be covered with
A) many ancient craters and maria.
B) dark areas of older crust separated by lighter, grooved terrain.
C) rugged mountain ranges and ancient volcanoes.
D) a smooth layer of ice, crossed by many cracks.
Answer: D Source: Section 7-5

116. What appears to have caused the extensive cracking and streaking of the surface of Europa?
  A) Tidal flexing by Jupiter.
  B) Shrinking of the satellite as it cooled.
  C) Expansion of the surface as the ice froze.
  D) Impacts by cometary debris.
  Answer: A Source: Section 7-5

117. Jupiter's satellite Europa is believed to be made up of
  A) ice and rock mixed throughout, in roughly equal proportions.
  B) a large, rocky core covered with an "ocean" of ice.
  C) ice, with possibly a small, rocky core.
  D) rock only.
  Answer: B Source: Section 7-5

118. What physical mechanism most probably caused the very long cracks and streaks that crisscross the surface of Europa, the moon of Jupiter?
  A) They are frozen rivers that, in warmer times, were flowing across the moon's surface.
  B) They are the tops of gigantic greenhouses built by inhabitants of Europa to protect their cucumber crops.
  C) Volcanic eruptions caused lava flows that then froze in place.
  D) Tidal flexing by Jupiter cracked the surface, and subsurface fluids gushed upward and froze.
  Answer: D Source: Section 7-5

119. Europa, one of the Galilean satellites of Jupiter, has a surface consisting of
  A) a relatively young, icy crust covered with a network of streaks and cracks, and only a few impact craters.
  B) rock, heavily cratered like the highlands of our Moon.
  C) an icy crust showing two interlocking types of terrain, one ancient and heavily cratered, the other younger with systems of parallel grooves.
  D) an ancient, icy crust covered with numerous craters; no surface cracks or groove belts which would indicate internal activity.
  Answer: A Source: Section 7-5

120. What does the surface of Europa, a satellite of Jupiter, look like?
   A) Entirely ice, with fractures and ridges going in every direction, and no features more than about a hundred meters high.
   B) Entirely ice, with light and dark areas containing many impact craters, and parallel ridges up to about a kilometer high.
   C) Rocky and densely cratered, except for a few large, dark, relatively uncratered plains.
   D) We don't know; the surface is hidden below a dense layer of clouds.
   Answer: A Source: Section 7-5

121. On which other world in the solar system do we find evidence of ice rafts (now apparently frozen), similar in many respects to ice rafts in the Earth's Arctic Ocean?
   A) Pluto, in the equatorial region facing most directly toward the Sun.
   B) Mars, at the edges of the polar caps.
   C) The icy surface of Jupiter's satellite, Europa.
   D) In the tropical regions of Saturn's satellite, Titan.
   Answer: C Source: Section 7-5

122. What appears to be the relationship between the surface of Europa and its interior?
   A) A solid ice crust extends from the surface to the rocky mantle.
   B) A frozen ocean with one small, irregular "continent" and several mountain peaks rises above the ice.
   C) A light, rocky surface and crust "floats" on a denser but partially molten layer of rock.
   D) A solid surface of ice overlies a layer of water or slush.
   Answer: D Source: Section 7-5

123. Which of the following features have NOT been found on Europa, one of the Galilean satellites of Jupiter?
   A) Mountains. B) Ice. C) Cracks running along the surface. D) Craters.
   Answer: A Source: Section 7-5

124. Based on its average density of 2,970 kg per cubic meter, what is the internal structure of Jupiter's moon, Europa, thought to be?
   A) Half rock and half water and ice.
   B) Rock with a 100-km thick ocean of water and ice.
   C) Rock with a small iron core.
   D) Almost entirely ice.
   Answer: B Source: Section 7-5

125. Where in the universe would you look for Ganymede?
A) Near to Jupiter, since it is one of Jupiter's moons.
B) In the asteroid belt.
C) Around the planet Mars, since it is one of the moons of Mars.
D) On the Moon, since it is one of the largest maria.
Answer: A Source: Section 7-6

126. Which is the largest planetary satellite in the solar system?
A) Jupiter's satellite, Ganymede.
B) Saturn's satellite, Titan.
C) Our Moon.
D) Neptune's satellite, Triton.
Answer: A Source: Section 7-6

127. The most massive planetary satellite in the solar system is
A) Jupiter's satellite, Ganymede.
B) the Earth's Moon.
C) Jupiter's satellite, Europa.
D) Saturn's satellite, Titan.
Answer: A Source: Section 7-6

128. What features dominate the surface of Ganymede, one of Jupiter's satellites?
A) Craters densely spread over the entire surface.
B) Old, dark, highly cratered polygons separated by younger, lighter, grooved terrain.
C) Volcanoes, lava lakes, and sulfur-dioxide frost.
D) Many cracks and streaks and very few craters in an otherwise smooth, icy surface.
Answer: B Source: Section 7-6

129. Ganymede, one of the Galilean satellites of Jupiter, has a surface consisting of
A) an ancient, icy crust covered with numerous craters; no surface cracks or groove belts which would indicate internal activity.
B) rock, heavily cratered like the highlands of our Moon.
C) an icy crust showing two interlocking types of terrain, one ancient and heavily cratered, the other younger with systems of parallel grooves.
D) a relatively young, icy crust covered with a network of streaks and cracks, and only a few impact craters.
Answer: C Source: Section 7-6

130. The surface of Ganymede, a Galilean satellite of Jupiter, is characterized by
   A) multi-colored deposits of sulfur and sulfur compounds, with several lava lakes and active volcanic plumes.
   B) bright, eroded, ancient terrain broken by darker, younger areas which show signs of tectonic activity in the past, such as water or slush volcanoes and faulting along subduction zones.
   C) dark areas of ancient terrain and bright, younger but still very old, areas with many folded ridges indicating tectonic activity on Ganymede in the distant past.
   D) dark areas of ancient terrain, and bright areas with almost no craters and many signs of recent tectonic activity such as fracturing, creating ice rafts and eruptions of icy lava.

   Answer: C Source: Section 7-6

131. The structure of Ganymede, a Galilean satellite of Jupiter, is thought to be
   A) about half rock and half ice, with the rock at the center and the ice outside.
   B) a thick crust of ice, a rocky mantle, and metallic core.
   C) mostly or entirely rock with no firm evidence of an iron core, and active volcanoes on its surface.
   D) a large rocky core with a thin layer (100 km thick) of ice and water over it.

   Answer: B Source: Section 7-6

132. How are the relative ages of the different types of terrain on Ganymede estimated?
   A) The darker the coloration the older the terrain.
   B) The denser the cratering the older the terrain.
   C) The smoother the surface the older the terrain.
   D) The colder the surface the older the terrain.

   Answer: B Source: Section 7-6

133. What appears to be the origin of the light, grooved terrain on the surface of Ganymede?
   A) Impacts by cometary debris.
   B) Expansion and cracking of the surface as the ice froze.
   C) Cracking of the surface due to shrinkage of the satellite as it cooled.
   D) Tidal flexing by Jupiter.

   Answer: B Source: Section 7-6

134. Ganymede, one of the large Galilean satellites of Jupiter, is very different from the Earth, but in which one of the following ways does Ganymede resemble the Earth?
 A) Ganymede appears to have had plate tectonics at some time in its past.
 B) Based on its density, Ganymede appears to have a thick, rocky mantle and large iron core under its icy crust.
 C) Ganymede has active volcanoes.
 D) Unlike all other moons in the solar system, Ganymede has a dense atmosphere.
 Answer: A Source: Section 7-6

135. What features dominate the surface of Callisto, one of Jupiter's satellites?
 A) Craters densely spread over the entire surface.
 B) Old, dark, highly cratered polygons separated by younger, lighter, grooved terrain.
 C) Volcanoes, lava lakes, and sulfur-dioxide frost.
 D) Many cracks and streaks and very few craters in an otherwise smooth, icy surface.
 Answer: A Source: Section 7-7

136. The structure of Callisto, the outer Galilean satellite of Jupiter, is thought to be
 A) a thick ice crust and a slushy mantle of ice and water over a core of rock.
 B) mostly or entirely rock with no firm evidence of an iron core, and active volcanoes on its surface.
 C) a large rocky core with a thin layer (100 km thick) of ice and water over it.
 D) a jumbled mixture of about half rock and half ice.
 Answer: A Source: Section 7-7

137. Which of the following statements describes the surface of Callisto, the outer Galilean moon of Jupiter?
 A) Ice-covered surface with dark coating, maybe of meteoritic dust or hydrocarbon material, making the moon appear dark.
 B) Ammonia and methane ice surface, with methane clouds above it, making the moon appear very bright.
 C) Sulfur-coated surface, with molten sulfur lakes and active volcanoes.
 D) Very dark, rocky surface, with a thin ice frost which occasionally disappears by evaporation in sunlight.
 Answer: A Source: Section 7-7

138. The surface of Callisto, the outer Galilean moon of Jupiter, can best be described as
   A) rough and mountainous, with evidence of either volcanoes or geyser-like activity.
   B) relatively smooth with many small craters, but no evidence of large impacts.
   C) relatively smooth, with a great many small impact craters, and one large basin created by an asteroid impact.
   D) very smooth, with a network of streaks and cracks, and very few craters.
   Answer: C Source: Section 7-7

139. Callisto, the outer Galilean moon of Jupiter, has a surface consisting of
   A) an ancient, icy crust covered with numerous craters; no surface cracks or groove belts which would indicate internal activity.
   B) an icy crust showing two interlocking types of terrain, one ancient and heavily cratered, the other younger with systems of parallel grooves.
   C) rock, heavily cratered like the highlands of our Moon.
   D) a relatively young, icy crust covered with a network of streaks and cracks, and only a few impact craters.
   Answer: A Source: Section 7-7

140. How many of the large, Galilean satellites are in synchronous rotation around Jupiter (always turning the same face toward Jupiter)?
   A) The inner two. B) All four. C) Only Io. D) The inner three.
   Answer: B Source: Sections 7-4, 5, 6, and 7

141. Jupiter's outermost eight satellites (moons) appear to be
   A) about the same size as our Moon or Mercury.
   B) asteroids captured by Jupiter.
   C) rocky or icy debris left over from the formation of the Jovian satellite system.
   D) dust grains.
   Answer: B Source: Section 7-8

142. What are the physical characteristics of the "other" (non-Galilean) moons of Jupiter?
   A) 500 to 1000 km in diameter and irregular in shape.
   B) 100 to 400 km in diameter and spherical in shape.
   C) Less than 200 km diameter and irregular in shape.
   D) All sizes up to 1500 km diameter, with the smaller ones irregular and the larger ones spherical.
   Answer: C Source: Section 7-8

143. Jupiter's ring was discovered by
A) direct ground-based photography.
B) visual observations by Galileo.
C) momentary occultation of starlight as the planet and the rings moved in front of a star.
D) spacecraft photography.
Answer: D Source: Section 7-8

144. Rings of dust and icy particles are found around which planets?
A) All four of the Jovian planets.
B) All planets which have moons associated with them.
C) All four of the terrestrial planets.
D) Only Saturn.
Answer: A Source: Section 7-8

145. What kind of ring or rings does Jupiter have?
A) One thin, dark ring made up of rocky particles about 1/1000 mm in size.
B) Three wide, bright rings, composed of icy particles about 1 cm to 1 m in size.
C) Several narrow, dark rings composed of icy particles less than 1 mm in size, darkened by ultraviolet radiation.
D) Only a ring of ions at Io's orbital distance, but no rings of solid particles.
Answer: A Source: Section 7-8

146. The faint ring around Jupiter is
A) in the planet's equatorial plane, tilted 3° to the ecliptic.
B) in a plane containing the north and south poles of the planet.
C) in the planet's orbital plane, tilted 1.3° to the ecliptic.
D) in the ecliptic plane.
Answer: A Source: Section 7-8

147. Does Jupiter have rings orbiting the planet?
A) Yes, several thin, dark rings of particles made up of (or else coated with) methane ice.
B) Yes, three very wide, bright rings and several faint, thin ones.
C) Yes, three very thin rings composed of fine dust particles.
D) No.
Answer: C Source: Section 7-8

148. Which planet has wide, bright rings that are easily visible from Earth?
A) Saturn. B) Neptune. C) Venus. D) Jupiter.
Answer: A Source: Section 7-9

149. What observing technique was being used when Saturn's rings were originally discovered?
A) Spacecraft photography.
B) Ground-based photography.
C) Momentary occultation of starlight as the planet moved in front of a star.
D) Visual observations through a telescope.
Answer: D Source: Section 7-9

150. "Markings" on the surface of Saturn are
A) similar in appearance to those on Jupiter but much less distinct.
B) non-existent, since Saturn shows completely uniform cloud-tops.
C) similar to those on Jupiter, but much more pronounced.
D) of a completely different pattern from those on Jupiter, and more distinct.
Answer: A Source: Section 7-9

151. Saturn's atmosphere does not show the same colorful contrast that we see in Jupiter's atmosphere. This is because
A) Saturn's features are obscured by an upper-cloud deck of methane ice crystals, whereas Jupiter has too high a temperature for methane ice.
B) Saturn has counter-flowing eastward and westward winds like Jupiter, but lacks the three differently colored cloud levels.
C) Saturn has a similar circulation pattern to Jupiter, but it is obscured by a thick, hazy atmosphere.
D) Saturn's clouds and circulation pattern resemble those of the Earth (individual cyclones and anticyclones) rather than those of Jupiter.
Answer: C Source: Section 7-9

152. Spectroscopy of the upper layers of Saturn has revealed an atmosphere containing large quantities of which of the following gases, in addition to the hydrogen and helium?
A) Nitrogen ($N_2$), oxygen ($O_2$) and water vapor ($H_2O$).
B) Clouds of sulfuric acid droplets, ($H_2SO_4$).
C) Carbon dioxide ($CO_2$) and traces of water vapor ($H_2O$.)
D) Methane ($CH_4$), ammonia ($NH_3$) and water vapor ($H_2O$).
Answer: D Source: Section 7-9

153. What is the mass of Saturn compared to the mass of Jupiter?
A) About 1.5 times the mass of Jupiter.
B) Almost the same mass.
C) About 1/10 the mass of Jupiter.
D) About 1/3 the mass of Jupiter.
Answer: D Source: Section 7-9

154. Other than the rings, how does the appearance of Saturn differ from that of Jupiter?
   A) Saturn has many more belts and zones than Jupiter, with large storms distorting their shapes.
   B) Saturn's visible surface is basically featureless, with no hint of the belts and zones of Jupiter.
   C) There are belts and zones on Saturn, but they are very faint and hazy compared to Jupiter.
   D) Saturn shows an ever-changing system of dark storms and light eddies, without the belts and zones of Jupiter.
   Answer: C Source: Section 7-9

155. The major constituent of Saturn is
   A) rock. B) nitrogen. C) hydrogen. D) carbon dioxide.
   Answer: C Source: Section 7-9

156. What is the interior structure of Saturn?
   A) A thick mantle of liquid hydrogen with a rocky core and a relatively thin, gaseous atmosphere.
   B) A thick, gaseous atmosphere over a thin mantle of liquid hydrogen and a rocky core.
   C) A large, rocky core with a very thin mantle of liquid hydrogen and a thin, gaseous atmosphere.
   D) A large liquid hydrogen core overlain by a thin, gaseous atmosphere.
   Answer: A Source: Section 7-9

157. Saturn is less massive than Jupiter, but has almost the same size. Why is this?
   A) Saturn is composed of lighter material than is Jupiter.
   B) The smaller mass exerts less gravitational force and is unable to compress the mass as much as in Jupiter.
   C) Saturn's interior is hotter than that of Jupiter.
   D) Saturn is rotating faster than Jupiter, and the increased centrifugal force results in a larger size.
   Answer: B Source: Section 7-9

158. Which is the least dense planet in the solar system?
   A) Uranus. B) Jupiter. C) Pluto. D) Saturn.
   Answer: D Source: Section 7-9

159. The rings of Saturn are in which plane with respect to the planetary system?
  A) A plane inclined at an angle to the both the orbital and equatorial planes of the planet and to the ecliptic plane, which is why we can easily see the rings face-on from Earth.
  B) The ecliptic plane.
  C) The equatorial plane of Saturn.
  D) The orbital plane of Saturn around the Sun.
  Answer: C Source: Section 7-10

160. The rings of Saturn alternately appear very distinct and then almost disappear when viewed from Earth over a period of a few years because
  A) the ice crystals from which they are made melt and re-freeze as the planet approaches and recedes from the Sun.
  B) the plane of the rings is tilted with respect to the ecliptic plane and thus appear edge-on at times.
  C) the Earth is very much closer to Saturn at opposition than at conjunction, hence the rings are more easily seen at this time.
  D) the solar wind occasionally blows away the ring particles when the Sun is particularly active.
  Answer: B Source: Section 7-10

161. The physical structure of Saturn's rings is
  A) hot, ionized gas from the planet's magnetosphere.
  B) a sequence of many hundred separate rings made of ice and rock particles.
  C) a thin solid ring, structured from ice and rock.
  D) a thin but extensive gas cloud over the equator.
  Answer: B Source: Section 7-10

162. The size distribution of particles in the rings of the Jovian planets is
  A) from pebble-sized fragments to objects a few meters in diameter, with significant amounts of fine dust particles in some rings.
  B) a mixture of gas (mostly hydrogen) and dust grains, none larger than about 1 mm across.
  C) only dust and smoke-like particles.
  D) individual bodies varying in size from 10 m to 1 km across, with no smaller components.
  Answer: A Source: Section 7-10

163. The particles in Saturn's rings are composed of
   A) rocks with the reflectivity of dark asphalt.
   B) water ice or rock coated with water ice.
   C) a mixture of iron and nickel.
   D) ammonia and methane ice, possibly with rocky centers.
   Answer: B Source: Section 7-10

164. What is the typical thickness of Saturn's rings?
   A) 15 km. B) 300 m. C) 2 km. D) 150 km.
   Answer: C Source: Section 7-10

165. Approximately how thick (top to bottom) are Saturn's rings, compared to the total width from inside to outside?
   A) About 1/1,000,000 of the width.
   B) About 1/50,000 of the width.
   C) About 1/10 of the width.
   D) About 1/2000 of the width.
   Answer: B Source: Section 7-10

166. The rings of Saturn are seen by
   A) fluorescence, a glow produced by photochemistry when material is irradiated by solar UV light and/or high-speed cosmic particles.
   B) emitted light from the molecules of the material of the rings, such as methane, ammonia etc..
   C) reflected and scattered sunlight.
   D) reflected light from Saturn, since we can see them at night.
   Answer: C Source: Section 7-10

167. What basic optical process makes the rings of Saturn visible to observers upon Earth?
   A) Absorption of sunlight, making the rings appear dark against the bright sunlit planet.
   B) Scattering of sunlight.
   C) Refraction of sunlight by ice crystals.
   D) Self-emission of light from atoms of the ring material under UV excitation.
   Answer: B Source: Section 7-10

168. The particles in Saturn's rings
   A) move in circular Keplerian orbits, the inner particles moving fastest.
   B) all move as if they are one solid disk.
   C) move in circular orbits, with the outer particles moving fastest because they are furthest from the planet.
   D) revolve in different directions depending upon the distance from the planet.
   Answer: A Source: Section 7-10

169. How does the orbital period P of particles moving round in the rings of Saturn depend upon their distance R from the center of the planet?
   A) P is constant for all R, since the rings move as a solid disk.
   B) P is proportional to $1/R^2$
   C) P increases directly with R.
   D) $P^2$ is proportional to $R^3$.
   Answer: D Source: Section 7-10

170. Which of the following describes the motions of the particles in the rings of Saturn?
   A) They move in zigzag patterns within the ring system because of interaction with the major moons.
   B) They move in randomly oriented elliptical orbits and collide frequently with each other.
   C) They all move in concert as if they were a solid sheet, because of electrostatic interaction.
   D) They each move in almost circular Keplerian orbits around the planet.
   Answer: D Source: Section 7-10

171. In the rings of Saturn, there are two main rings, the outer A ring and the inner B ring. How are the particles in these rings seen to move when examined (by Doppler Effect of reflected sunlight) from Earth?
   A) Speeds of the particles vary because of attraction by the Sun, such that A particles sometimes move faster, while at other times B particles move faster.
   B) Particles in the A ring move more slowly than those in the B ring.
   C) Particles in the A ring move faster than those in the B ring, since they are further from the planet.
   D) Speeds of the particles in all rings are the same, since mutual gravitational forces between them ensure that they move as a solid disk.
   Answer: B Source: Section 7-10

172. A manned space mission to Saturn some time in the future places an observer in a spacecraft within its ring system. What motions would this observer see between the ring particles and the spacecraft?
   A) The particles would be moving in ellipses and would therefore sometimes move ahead of and at other times fall behind the spacecraft.
   B) Particles closer to the planet would overtake the spacecraft, while particles further from the planet would be overtaken by the spacecraft, following Kepler's laws.
   C) All particles would be moving with the same orbital speed and would travel around the planet together with the spacecraft, since the motions of particles and spacecraft are all governed by gravity.
   D) Particles orbiting further from the planet would be traveling faster and would overtake the spacecraft, while particles closer to the planet would be moving slower and would be overtaken by the spacecraft.
   Answer: B Source: Section 7-10

173. What is the Cassini division?
   A) The layer of relatively clear air separating Saturn's upper cloud deck from the middle cloud deck.
   B) The boundary between the bright B ring and the faint C ring in Saturn's rings.
   C) A wide, dark gap in Saturn's rings.
   D) The division between the two dark equatorial belts of Jupiter.
   Answer: C Source: Section 7-10

174. Where is the Cassini division found in our solar system?
   A) Between the terrestrial and the Jovian planets.
   B) Between two major groups of asteroids in the asteroid belt.
   C) Between two major ring systems around Saturn.
   D) Between two band systems on the visible "surface" of Jupiter.
   Answer: C Source: Section 7-10

175. The Cassini division is
   A) a major division in the rings of Saturn, seen from Earth.
   B) a gap between two groups of asteroids in the asteroid belt.
   C) a gap between two mountain ranges on the Moon.
   D) the division between terrestrial and Jovian planets.
   Answer: A Source: Section 7-10

176. How was the Cassini division created in Saturn's rings?
   A) One of Saturn's satellites exerts a resonant pull on particles in the division, clearing a gap.
   B) A small moon orbits within the division, clearing particles from the gap.
   C) An intense region of high-energy electrons in Saturn's magnetosphere at that distance has eroded the particles from the gap.
   D) The rings simply formed that way in the ancient past.
   Answer: A Source: Section 7-10

177. The major gaps in the rings of Saturn are most likely caused by
   A) combined gravitational forces from Saturn and its moons, which deflect the paths of particles that stray into the gaps.
   B) mutual gravitational interactions between the multitude of particles in the rings.
   C) streams of charged particles, similar to the Earth's Van Allen belts, in Saturn's magnetosphere.
   D) the intervention of a massive body, that moved through the rings in their early history, leaving the gaps.
   Answer: A Source: Section 7-10

178. What is the composition of Saturn's rings?
   A) Water ice or ice-coated rock.
   B) Particles of methane and ammonia ice.
   C) Sodium and sulfur ions
   D) Small grains of rock.
   Answer: A Source: Section 7-10

179. What size are the particles making up Saturn's rings?
   A) Dust grains a few micrometers in diameter.
   B) Up to about 10 cm diameter.
   C) Up to about 10 m diameter.
   D) "Snowflakes" up to about a millimeter in diameter.
   Answer: C Source: Section 7-10

180. Why is the F ring much narrower than the main rings?
   A) There are not enough particles available to make a wider ring.
   B) The ring is in a stronger part of Saturn's gravitational field and cannot spread out any further.
   C) Two "shepherd" satellites focus the particles into a narrow ring.
   D) The ring is constrained by Saturn's strong magnetic field.
   Answer: C Source: Section 7-10

181. Two tiny but significant satellites that follow nearly identical orbits around Saturn are called shepherd moons since they
   A) trail streams of gases behind them, like comet tails.
   B) clear particles from the Cassini division.
   C) concentrate particles into the narrow F ring of Saturn.
   D) lift dust particles up into "spokes" above the rings.
   Answer: C Source: Section 7-10

182. The gravitational effect that confines the particles of the F ring of Saturn to a narrow orbit is (see Figs. 7-23 and 7-24, Kaufmann & Comins, *Discovering the Universe*, 5th Ed.)
   A) the gravitational effects of the major moons of Saturn, such as Mimas and Enceladus.
   B) the gravitational influence of two small shepherding satellites in orbits adjacent to the ring.
   C) major gravitational distortion caused by the neighboring planet, Jupiter.
   D) the pressure of the solar wind upon these particles.
   Answer: B Source: Section 7-10

183. The physical mechanism that is thought to control the motion and position of material in the narrow F ring around Saturn is
   A) the effect of the planet's intense magnetic field upon the material.
   B) the effect of sunlight focused on this material by the planet's atmosphere.
   C) the confining gravitational interactions between this material and two shepherd satellites.
   D) the complex gravitational interactions between the major moons of Saturn and the F ring material.
   Answer: C Source: Section 7-10

184. The main gravitational effect that organizes the particles in the rings of Saturn and Uranus into specific narrow orbits is
   A) collisions between the ring particles and the major moons of the planets.
   B) perturbations by orbiting satellites.
   C) perturbations by neighboring planets, such as Jupiter and Neptune.
   D) the complex gravitational effects of the planet and the Sun on the ring particles.
   Answer: B Source: Section 7-10

185. Around which planet does the satellite Titan orbit?
   A) Saturn. B) Neptune. C) Pluto. D) Jupiter.
   Answer: A Source: Section 7-11

186. Saturn's moon Titan has a size
   A) about the same size as the moons of Mars.
   B) intermediate between Mercury and the Earth's moon.
   C) intermediate between Mars and Earth.
   D) intermediate between Mercury and Mars.
   Answer: D Source: Section 7-11

187. Saturn's moon, Titan, is different from all other moons of planets because
   A) it possesses a thick atmosphere.
   B) continuously erupting volcanoes are observed upon it.
   C) its orbit carries it directly over both poles of the planet.
   D) lakes of water with floating icebergs are seen upon its surface.
   Answer: A Source: Section 7-11

188. What characteristic of Saturn's satellite, Titan, makes it different from any other known satellite in the solar system?
   A) It is volcanically active.
   B) Its surface is broken into heavily cratered and lightly cratered regions in a pattern similar to plate tectonics.
   C) It has a permanent, dense atmosphere.
   D) It has geyser-like plumes of nitrogen gas.
   Answer: C Source: Section 7-11

189. Which is the only satellite in the solar system known to possess a permanent, dense, atmosphere?
   A) Titan. B) Callisto. C) Charon. D) Ganymede.
   Answer: A Source: Section 7-11

190. Which of the following satellites of planets in our solar system has a significant and dense atmosphere?
   A) Triton, a moon of Neptune.
   B) Titan, a moon of Saturn.
   C) Io, a moon of Jupiter.
   D) the Moon, of Earth.
   Answer: B Source: Section 7-11

191. Which of the moons of the giant planets is known to have a significant atmosphere?
   A) Titan, a moon of Saturn.
   B) Triton, a moon of Neptune.
   C) Callisto, a moon of Jupiter.
   D) Europa, a moon of Jupiter.
   Answer: A Source: Section 7-11

192. What is the appearance of Saturn's satellite, Titan, when viewed from space?
A) Dark islands of ice in a bright ethane sea, occasional clouds.
B) Volcanoes, lava flows and sulfur dioxide frost.
C) Ice criss-crossed with innumerable streaks and cracks.
D) Surface not visible under a featureless cloud cover.
Answer: D Source: Section 7-11

193. The major constituent of the atmosphere of Titan (the largest moon of Saturn) is
A) methane, $CH_4$.
B) water vapor, $H_2O$.
C) nitrogen, $N_2$.
D) carbon dioxide, $CO_2$.
Answer: C Source: Section 7-11

194. What are the main constituents of the atmosphere of Titan, a satellite of Saturn?
A) Carbon dioxide.
B) Ammonia and methane.
C) Nitrogen and oxygen.
D) Nitrogen and methane.
Answer: D Source: Section 7-11

195. How was methane, $CH_4$, first discovered on Titan, the giant moon of Saturn?
A) By noting the formation on its surface of colored "ice," characteristic of methane ice.
B) Spectroscopically, by noting specific absorptions in reflected sunlight.
C) By chemical "sniffers," carried by Voyager 2 when it passed very close to Titan.
D) By detecting the light of burning methane spectroscopically, similar to that seen from oil or gas well flares.
Answer: B Source: Section 7-11

196. How is nitrogen thought to have become the dominant gas in the atmosphere of the moon Titan?
A) From the break-up of ammonia into nitrogen and hydrogen, with the hydrogen being lost to space.
B) From outgassing through volcanic vents.
C) It has been present from primordial times, when Titan first formed.
D) From the collision of comets with the surface of Titan.
Answer: A Source: Section 7-11

197. Which chemical in the atmosphere of Titan (a moon of Saturn) plays the same role that water plays on Earth, by producing "rain," "snow," and "ice" at the temperature encountered on that moon?
A) Nitrogen, $N_2$ B) Ethane, $C_2H_6$ C) Oxygen, $O_2$ D) Ammonia, $NH_3$
Answer: B Source: Section 7-11

198. What would be a typical weather forecast on Titan, Saturn's largest moon?
A) Occasional sulfur clouds and sulfur dioxide fog, from volcanic eruptions.
B) Hydrocarbon fog and ethane rain in a nitrogen atmosphere.
C) Turbulent winds in an ammonia, methane, and water vapor atmosphere, with dense clouds of ammonia compounds and water ice.
D) Dust storms and high winds in a thin $CO_2$ atmosphere.
Answer: B Source: Section 7-11

199. The surface temperature of Titan, the largest moon of Saturn, is
A) 178 K (195°C). B) 368 K (+95°C). C) 95 K (-178°C). D) 273 K (0°C).
Answer: C Source: Section 7-11

200. How would "Interplanetary Travel" advertise a holiday on Titan, one of the satellites of Saturn?
A) Hot and dry-never rains-beautiful sulfurous skies!
B) Glaciers galore for your hiking pleasure under star-studded skies!
C) Exquisite ethane lakes, hydrocarbons beyond your wildest dreams!
D) The largest number of volcanoes for your travel dollar anywhere in the Solar System!
Answer: C Source: Section 7-11

201. Based on their average densities, the six moderate-sized satellites of Saturn are believed to be composed
A) mostly of ice, with perhaps a small, rocky core.
B) of about half rock and half ice.
C) mostly of rock, with only a thin mantle of ice.
D) mostly of solid hydrogen.
Answer: A Source: Section 7-11

202. What is the visual appearance of Uranus from space?
A) Blue-green with white, high-altitude clouds and dark storms.
B) Reddish belts and light zones parallel to the equator.
C) Blue-green and featureless.
D) Perpetually covered with yellowish, sulfur-rich clouds.
Answer: C Source: Section 7-12

203. The discoverer of the planet Uranus was
A) Clyde Tombaugh.
B) Edmund Halley.
C) William Herschel.
D) Galileo Galilei.
Answer: C Source: Section 7-12

204. How was Uranus discovered?
    A) By accident, by an astronomer who was conducting a sky survey.
    B) By an astronomer studying old photographs of the sky, several years after they were taken.
    C) By a careful search in the 1930s by an astronomer who was convinced it must be there.
    D) By careful application of Newton's laws to the motion of other planets.
    Answer: A Source: Section 7-12

205. The smallest angle in the sky that can be resolved from the Earth's surface, using the best telescopes under the best observing conditions, is approximately 0.5 arc sec. (You may wish to review the small angle formula in Toolbox 1-1 of Kaufmann & Comins, *Universe*, 5th Ed.) Consequently the smallest feature which can be resolved at the distance of Uranus is about
    A) 1/8 of the diameter of Uranus.
    B) 1/20 of the diameter of Uranus.
    C) 1/3 of the diameter of Uranus.
    D) 1.5 times the diameter of Uranus.
    Answer: A Source: Section 7-12 and Toolbox 1-1

206. Which of the following planets was discovered by accident, by an astronomer who was surveying the sky with his telescope but had no idea that he would discover a planet?
    A) Neptune. B) Pluto. C) Saturn. D) Uranus.
    Answer: D Source: Section 7-12

207. What are the most abundant gases in the atmosphere of Uranus?
    A) Carbon dioxide and nitrogen.
    B) Hydrogen and helium.
    C) Nitrogen and ammonia.
    D) Methane and water.
    Answer: B Source: Section 7-12

208. What gives Uranus its blue-green coloration?
    A) Absorption of red light by methane gas.
    B) Absorption of blue and green light by ammonia gas.
    C) Emission by blue and green spectral lines of ethane and propane.
    D) Impurities of phosphorous and sulfur in the otherwise white ice crystals of water, methane, and ammonia.
    Answer: A Source: Section 7-12

209. As photographed by Voyager 2, the atmosphere of Uranus shows
    A) dark, whirlpool-like features and high, white clouds of methane ice crystals.
    B) a pinkish haze due to scattering from silicate dust particles.
    C) light zones and dark belts circling the planet parallel to the equator.
    D) a faint haze over the north pole, but is otherwise featureless.
    Answer: D Source: Section 7-12

210. Which planet has its rotation axis tilted at 98 degrees from the vertical to its orbit?
    A) Venus. B) Uranus. C) Neptune. D) Earth.
    Answer: B Source: Section 7-12

211. The major planet whose spin axis lies almost in its orbital plane is
    A) Uranus. B) Mercury. C) Neptune. D) Mars.
    Answer: A Source: Section 7-12

212. By what angle is the rotation axis of Uranus tilted from the vertical to its orbit?
    A) Almost 180 degrees.
    B) Less than 2 degrees.
    C) 98 degrees.
    D) 25 degrees (almost the same as the Earth's tilt).
    Answer: C Source: Section 7-12

213. As seen by an observer floating above the clouds on Uranus, where would the Sun be located in midsummer in the northern hemisphere?
    A) Almost directly above the Equator.
    B) Directly over a line of latitude 25 degrees north of the Equator.
    C) Almost directly above the North Pole.
    D) Directly over a line of latitude 25 degrees south of the Equator.
    Answer: C Source: Section 7-12

214. The expected seasonal changes on Uranus because of its orbital and spin-axis alignments, compared to those on Earth, will be
    A) very much exaggerated.
    B) absent, because of the alignment of the spin axis.
    C) much less.
    D) the same.
    Answer: A Source: Section 7-12

215. Seasonal variations at a particular point on Uranus during a Uranian year would be
   A) extreme, since its spin axis is nearly in its orbital plane.
   B) almost nonexistent, since Uranus moves in an almost perfectly circular orbit and its distance from the Sun remains constant.
   C) not present at any point on the planet, since dense clouds shield it from climate changes.
   D) nonexistent, since such variations at any point on the planet will be smoothed out during its long "year" by the planet's rapid rotation.
   Answer: A Source: Section 7-12

216. Which of the following effects is now thought to be the most likely cause for the inclinations of spin axis of several of the planets such as Uranus (and even Earth) to their orbital planes?
   A) Tidal distortion and deflection caused by neighboring planets.
   B) Major collision with another planet-like body.
   C) Steady force on one hemisphere of the planet from the highly directional solar wind.
   D) Out-of-balance force on the spin axes of planets from their moons, some of which are very massive.
   Answer: B Source: Section 7-12

217. Which planets rotate about their axis in a retrograde direction?
   A) Venus, Uranus, and Pluto.
   B) Mercury, Mars, and Uranus.
   C) Venus, Uranus, and Neptune.
   D) Mercury, Venus, and Neptune.
   Answer: A Source: Section 7-12

218. What is believed to be the basic structure of the interior of Uranus?
   A) Rocky core, thick layer of liquid hydrogen, thin gaseous atmosphere.
   B) Iron core, thick layer of rock, thin gaseous atmosphere.
   C) Rocky core, thick layer of water, thin gaseous atmosphere.
   D) Rocky core, thick layer of water, thick layer of liquid hydrogen, thin gaseous atmosphere.
   Answer: D Source: Section 7-12

219. The surprising fact about the magnetic field of Uranus compared to that of the Earth or Jupiter is
   A) that it is much more intense than that of any other planet.
   B) that its axis is precisely aligned with the spin axis of the planet.
   C) that its axis makes a larger angle to the planet's spin axis than any other planet.
   D) that it is extremely small or absent.
   Answer: C Source: Section 7-12

220. What are the characteristics of the magnetic field of Uranus?
   A) It is aligned almost exactly along the planet's axis of rotation, through the center of Uranus.
   B) It passes through the center of Uranus but is tilted almost 60 degrees from the axis of rotation.
   C) It is aligned almost parallel to the planet's axis of rotation, but is offset from the planet's center.
   D) It is tilted almost 60 degrees to the axis of planetary rotation and is offset from the planet's center.
   Answer: D Source: Section 7-12

221. Compared to Earth, Jupiter and Saturn, the magnetic fields of Uranus and Neptune
   A) are much more powerful, dominating the motions of clouds on these planets.
   B) are non-existent, since Uranus and Neptune have no magnetic fields.
   C) have axes tilted a long way from the spin axis of the planets.
   D) have axes which are aligned almost exactly with the planet's spin axis.
   Answer: C Source: Section 7-12

222. How were the rings of Uranus discovered?
   A) By the Hubble Space Telescope, observing in infrared light.
   B) From Earth, when the each ring momentarily blocked off light from a background star.
   C) By Voyager 2 during its pass through the Uranian system.
   D) By the Ulysses spacecraft when observing from above the Sun's north pole.
   Answer: B Source: Section 7-13

223. The ring system around Uranus was originally discovered by what observing technique?
   A) Infrared observation from the IRAS spacecraft in Earth orbit.
   B) Observation by the cameras on board the Voyager 1 spacecraft.
   C) Radar reflection from the particles in the rings.
   D) Occultation of light from a star as Uranus (and the rings) passed in front of it.
   Answer: D Source: Section 7-13

224. What scientific method was used to discover the rings around Uranus?
   A) Occultation of a star as the planet and rings moved in front of it.
   B) X-ray photography from the Einstein satellite.
   C) Direct photography from Earth.
   D) Spacecraft exploration of the planet.
   Answer: A Source: Section 7-13

225. The rings of Uranus are
   A) wide, dense, and very bright (70% reflectivity).
   B) intrinsically very bright (70% reflectivity) but hard to detect because they are very narrow.
   C) broad, diffuse (almost transparent) bands made up of almost nothing bigger than dust particles.
   D) narrow and very dark (1% reflectivity).
   Answer: D Source: Section 7-13

226. The material in the Uranus ring system differs from that in the Saturn ring system in what important way?
   A) The particle sizes are larger, those of the Saturnian system being only snowflake or dust grain sizes.
   B) It reflects much more sunlight than that of the Saturn ring material.
   C) It reflects much less sunlight than does the material of Saturn's rings.
   D) The individual particles move as a solid ring for some reason, not in Keplerian orbits as do the Saturnian ring particles.
   Answer: C Source: Section 7-13

227. The satellites of Uranus orbit the planet
   A) in the plane of the planet's equator, and therefore approximately in the plane of the ecliptic.
   B) in the plane of the ecliptic, and therefore almost at right angles to the plane of Uranus' equator.
   C) in the plane of the planet's equator, and therefore in a plane almost at right angles to the ecliptic.
   D) in various directions, each one reflecting the satellite's trajectory at the time it was captured by Uranus.
   Answer: C Source: Section 7-13

228. Miranda is a satellite of
   A) Neptune. B) Jupiter. C) Uranus. D) Pluto.
   Answer: C Source: Section 7-13

229. Which one of the following statements correctly describes Miranda, one of the satellites of Uranus?
   A) Heavily cratered, icy surface with deformation and flowing due to heating in some regions.
   B) Icy surface with chaotically varied terrain.
   C) Rocky surface with active volcanoes.
   D) Icy, frost-covered surface with vents of rising gas visible in the northern hemisphere.
   Answer: B Source: Section 7-13

230. Miranda, a satellite of Uranus,
   A) shows an ancient surface covered with impact craters.
   B) has a thick atmosphere containing methane, nitrogen and hydrocarbons.
   C) shows active geysers and resurfacing due to water flows.
   D) appears to have reassembled from separate parts after being shattered by an impact.
   Answer: D Source: Section 7-13

231. Which planetary satellite shows strong evidence of having once been disrupted by an impact and subsequently re-assembled by gravity?
   A) Triton. B) Callisto. C) Miranda. D) Io.
   Answer: C Source: Section 7-13

232. How was Neptune discovered?
   A) By a careful search with the Hubble Space Telescope.
   B) By a careful application of Newton's laws to the motion of Uranus.
   C) Accidentally by an astronomer studying the sky visually through a telescope.
   D) Accidentally, by an astronomer taking photographs for an all-sky survey.
   Answer: B Source: Section 7-14

233. Which planet was discovered by applying Newton's laws to the motion of another planet?
   A) Neptune. B) Mars. C) Uranus. D) Mercury.
   Answer: A Source: Section 7-14

234. The spacecraft that successfully photographed Uranus on a 1986 fly-by mission was
   A) Viking B) Magellan C) Galileo D) Voyager
   Answer: D Source: Section 7-14

235. Which is the eighth planet from the Sun (in order of increasing mean distance or semimajor axis)?
   A) Saturn. B) Uranus. C) Pluto. D) Neptune.
   Answer: D Source: Section 7-14

236. What is the visual appearance of Neptune from space?
   A) Reddish belts and light zones parallel to the equator.
   B) Blue-green with white, high-altitude clouds and dark storms.
   C) Blue-green and featureless.
   D) Perpetually covered with yellowish, sulfur-rich clouds.
   Answer: B Source: Section 7-14

237. Which planet is characterized by a blue-green appearance with dark storms and white, high-altitude methane clouds?
   A) Uranus. B) Neptune. C) Saturn. D) Pluto.
   Answer: B Source: Section 7-14

238. The Great Dark Spot (which is not the same as the Great Red Spot) was found on which planet or moon?
   A) Jupiter. B) Mars. C) Neptune. D) Miranda, a moon of Uranus.
   Answer: C Source: Section 7-14

239. The Great Dark Spot on Neptune, photographed by Voyager 2 during its flyby of the planet, was
   A) a region of up-welling gas in Neptune's atmosphere, above a hot-spot on its surface, possibly a volcano.
   B) a high-pressure system (anticyclone).
   C) a volcanic caldera.
   D) a low-pressure system (cyclone).
   Answer: B Source: Section 7-14

240. The Great Dark Spot on Neptune
   A) came into existence sometime between the Voyager flyby in 1989 and when the Hubble Space Telescope photographed Neptune in 1994.
   B) has been visible through telescopes since at least as far back as 1665.
   C) disappeared sometime between the Voyager flyby in 1989 and when the Hubble Space Telescope photographed Neptune in 1994.
   D) was the short-lived result of a comet crash in 1995.
   Answer: C Source: Section 7-14

241. The atmospheres of Uranus and Neptune consist of
  A) mostly carbon dioxide, with a small amount of nitrogen.
  B) mostly methane, with small quantities of hydrogen and helium.
  C) mostly hydrogen and helium, with significant amounts of methane, ammonia and water vapor.
  D) mostly hydrogen and helium, with significant amounts of methane and water vapor but very little ammonia.
  Answer: D Source: Section 7-14

242. Neptune's predominantly blue appearance is caused by
  A) scattering of the blue end of the solar spectrum preferentially by air, similar to the process in the Earth's atmosphere.
  B) the fact that solar light has lost much of its red light by scattering in the interplanetary medium at the distance of Neptune.
  C) absorption of the red end of the spectrum of reflected sunlight by methane in its atmosphere.
  D) auroral emissions caused by solar wind particles exciting the atoms and molecules in Neptune's high atmosphere.
  Answer: C Source: Section 7-14

243. Neptune's high cirrus clouds consist of
  A) methane ice crystals.
  B) droplets of sulfuric acid.
  C) crystals of water ice.
  D) ammonia ice crystals.
  Answer: A Source: Section 7-14

244. At the time of the Voyager flybys of Neptune, the patterns and motions of its atmosphere most closely resembled those of which other planet?
  A) Uranus, with no very distinct features.
  B) Jupiter, with clouds and rotating spots.
  C) Saturn, with indistinct cloud bands.
  D) Venus, with rapidly moving clouds of sulfuric acid droplets, and a thick $CO_2$ atmosphere.
  Answer: B Source: Section 7-14

245. One distinct difference between the two otherwise similar planets, Uranus and Neptune, is
   A) the lack of moons around Neptune.
   B) the almost featureless visible image of Uranus, compared to storms and clouds on Neptune.
   C) belts and zones on Uranus, visible through haze, compared to the totally cloud-enshrouded Neptune.
   D) the absence of an internal magnetic field and surrounding magnetosphere at Neptune.
   Answer: B Source: Section 7-14

246. Which ONE of the following four statements applies to ALL FOUR of the Jovian planets?
   A) They are almost entirely hydrogen and helium; only about 1% of the planet's mass is made up of heavier elements.
   B) They are thought to have substantial rocky cores.
   C) Their spin axes are all approximately perpendicular to their orbital planes.
   D) Liquid metallic hydrogen makes up a large part of their interiors.
   Answer: B Source: Sections 7-2, 9, 12, and 14

247. Which planets are believed to have a thick layer water mixed with ammonia and methane in their interiors?
   A) Saturn and Neptune.
   B) Saturn, Uranus and Neptune.
   C) Uranus and Neptune.
   D) Saturn and Uranus.
   Answer: C Source: Sections 7-12 and 14

248. Uranus and Neptune resemble each other in many ways. However, in which of the following ways do Uranus and Neptune NOT resemble each other (i.e., which statement is incorrect)?
   A) Both planets (probably) have three layers: rocky core, watery mantle, and thick hydrogen-helium atmosphere.
   B) Both planets appear basically featureless in photographs from Voyager 2.
   C) Both planets have a system of rings circling the planet.
   D) Both planets have much stronger magnetic fields than does the Earth.
   Answer: B Source: Sections 7-12 and 14

249. Which of the following statements correctly describes how Neptune differs from Uranus?
A) Neptune is circled by a system of narrow, dark rings, whereas those of Uranus are wide like those of Saturn.
B) Neptune has a number of vortices and ice clouds visible in its atmosphere, whereas Uranus' atmosphere appears almost featureless.
C) Neptune's magnetic axis is closely aligned with its rotation axis, whereas for Uranus these two axes are at a large angle to each other.
D) Neptune rotates quite rapidly (16 hours), whereas Uranus rotates very slowly (243 days).
Answer: B Source: Sections 7-12 and 14

250. Does Neptune have rings orbiting the planet?
A) Yes, a system of thin, dark rings of particles of methane ice.
B) Yes, three very thin rings composed of fine dust particles.
C) No.
D) Yes, three very wide, bright rings and several faint, thin ones.
Answer: A Source: Section 7-15

251. Why are the rings of Neptune dark?
A) The particles are composed of methane ice that has been darkened by radiation damage.
B) The particles have highly reflective surfaces, but the rings are far from the Sun where there is little light.
C) The rings are composed of small, dark rock particles.
D) Although the particles have highly reflective surfaces, there are very few particles in the rings.
Answer: A Source: Section 7-15

252. Which planet has a system of smaller moons with highly elliptical orbits and one large moon that orbits in a retrograde direction?
A) Uranus. B) Neptune. C) Saturn. D) Pluto.
Answer: B Source: Section 7-15

253. Triton, the giant moon of Neptune, differs from all other major moons of planets because
A) it orbits in a retrograde direction, opposite to the planet's rotation.
B) it orbits faster than the rotation of the planet.
C) its orbit takes it over the planet's poles.
D) its orbit is very elliptical.
Answer: A Source: Section 7-15

254. Which property of Triton, the moon of Neptune, makes it significantly different from all other major moons in the solar system?
  A) It has an extremely dark, smooth surface and consequently was not discovered until very recently.
  B) It has an atmosphere of nitrogen and oxygen, and $H_2O$ clouds.
  C) Its orbit is at right angles to the equator of its mother planet.
  D) It orbits in a direction opposite to the rotation and revolution of its mother planet.
  Answer: D Source: Section 7-15

255. What features characterize the visible surface of Triton, Neptune's largest moon?
  A) Dense atmosphere permanently shrouded in clouds.
  B) Wrinkled surface, frozen lakes, and plumes of nitrogen gas.
  C) Ice heavily cratered by ancient impacts.
  D) Lava flows, volcanoes, and sulfur dioxide frost.
  Answer: B Source: Section 7-15

256. Triton, the largest satellite of Neptune, has
  A) a surface of ice in which ancient, densely cratered regions are surrounded by interconnecting systems of parallel ridges.
  B) a surface of ice with frozen lakes, plumes of escaping gas, and few craters.
  C) a thick atmosphere that hides the surface from view.
  D) a densely cratered surface of ice with at least one ringed structure indicating an ancient asteroid impact.
  Answer: B Source: Section 7-15

257. Triton, the largest satellite of Neptune, has
  A) a very thin atmosphere of nitrogen gas.
  B) a dense atmosphere of nitrogen and methane.
  C) a dense atmosphere of methane and ammonia.
  D) no known atmosphere at all.
  Answer: A Source: Section 7-15

258. Which planetary satellites are known to have plumes of gas escaping through their surfaces?
  A) Io and Triton.
  B) Ganymede and Charon.
  C) Our Moon, Europa, and Miranda.
  D) Io and Titan.
  Answer: A Source: Section 7-15

259. The plumes which were seen rising from the surface of Triton, Neptune's largest satellite, are believed to be
A) sulfur dioxide from geysers heated by tidal stresses.
B) nitrogen gas evaporated by heat from the Sun.
C) volcanic ash from eruptions similar to, but much smaller than, an Earth-bound eruption.
D) water, methane and ammonia ice crystals above volcanic vents.
Answer: B Source: Section 7-15

260. A striking characteristic on Triton, the largest satellite of Neptune, is
A) a brightness variation, with one hemisphere being as bright as ice and the other hemisphere as dark as asphalt.
B) an interconnecting network of parallel grooves, indicating tectonic activity in geologically recent times.
C) plumes of nitrogen gas rising from the icy surface, possibly as a result of solar heating.
D) a crater so large that the impact which created it must have come close to shattering the satellite.
Answer: C Source: Section 7-15

261. How would "Interplanetary Travel" advertise a holiday to Neptune's satellite, Triton?
A) Hot and dry -- never rains -- beautiful sulfurous skies!
B) Exquisite ethane lakes, hydrocarbons beyond your wildest dreams!
C) Skate on frozen nitrogen lakes all morning, bask beside nitrogen geysers in the afternoon!
D) The largest number of volcanoes for your travel dollar anywhere in the Solar System!
Answer: C Source: Section 7-15

262. Which planetary satellites are known to have plumes of gas escaping through their surfaces?
A) Io and Triton.
B) Io and Titan.
C) Ganymede and Charon.
D) Europa and Miranda.
Answer: A Source: Section 7-15

263. What future awaits Triton, the largest satellite of Neptune?
A) Probable destruction from impact by Pluto.
B) Escape from Neptune after billions of years, as Triton gradually spirals outward.
C) Gravitational capture by Pluto.
D) Tidal break-up as it spirals closer to Neptune.
Answer: D Source: Section 7-15

264. Which planet was discovered in this century?
A) Ceres. B) Uranus. C) Pluto. D) Neptune.
Answer: C Source: Section 7-16

265. The planet Pluto was discovered in
A) 1781. B) 1609. C) 1930. D) 1846.
Answer: C Source: Section 7-16

266. How was the planet Pluto discovered?
A) By searching photographs of the sky for an object that moved day by day.
B) By Voyager spacecraft cameras, which were used between planetary encounters to survey the planetary system.
C) By the infrared cameras on the IRAS spacecraft.
D) By prediction using Newton's laws, to account for the deviations from uniform orbits of Uranus and Neptune.
Answer: A Source: Section 7-16

267. Which was the eighth planet from the Sun in 1990?
A) Charon. B) Pluto. C) Uranus. D) Neptune.
Answer: B Source: Section 7-16

268. Which planet in our solar system has the moon with the largest diameter compared to the diameter of the planet?
A) Earth. B) Neptune. C) Pluto. D) Saturn.
Answer: C Source: Section 7-16

269. Of the nine major planets in our solar system, which one has the greatest orbital inclination (orbit at the greatest angle to that of the Earth)?
A) Earth. B) Mars. C) Pluto. D) Mercury.
Answer: C Source: Section 7-16 and Appendix, Table 1

270. Of the nine major planets in our solar system, which one has the greatest orbital eccentricity (most elliptical orbit)?
A) Earth. B) Pluto. C) Mercury. D) Mars.
Answer: B Source: Section 7-16 and Appendix, Table 1

271. What is the moon of the planet Pluto called?
A) Triton. B) Charon. C) Chiron. D) Callisto.
Answer: B Source: Section 7-16

272. If you were standing on Pluto, how often would you see the satellite Charon rise above your horizon each day?
   A) Never; Charon is a synchronous satellite like the communications satellites we put in orbit around the Earth.
   B) Once each six-hour day as Pluto rotates on its axis.
   C) Twice each six-hour day, since Charon is in a retrograde orbit.
   D) Once every two days, since Charon orbits in the same direction that Pluto rotates, but more slowly.
   Answer: A Source: Section 7-16

273. What method was recently used to determine the diameters of Pluto and its moon, Charon?
   A) Radio interferometry, with arrays of telescopes producing extremely high resolution images.
   B) Photography from a visiting spacecraft.
   C) Observation of the mutual eclipses of planet and moon.
   D) Observation of the disappearance of the planet and then the moon behind our Moon in an occultation.
   Answer: C Source: Section 7-16

274. What technique has allowed scientists to map surface brightness distribution on Pluto's surface?
   A) Occultation of this planet by the Earth's Moon during the 1980s.
   B) Imaging by the Voyager spacecraft in 1989.
   C) Eclipses of the planet's surface by its moon, Charon, during 1985-1990.
   D) Direct photography using adaptive-optics telescopes on the Earth in 1995.
   Answer: C Source: Section 7-16

275. What is unique about the Pluto-Charon system, compared to all other planets in the solar system?
   A) Charon is an icy moon, but is in orbit around a giant planet made mostly of liquid hydrogen.
   B) Both Pluto and Charon are in synchronous rotation, so each object maintains the same face toward the other object at all times.
   C) Pluto has only one satellite.
   D) Both Pluto and Charon are volcanically active, with lava flows and vents of sulfur-dioxide gas.
   Answer: B Source: Section 7-16

276. The Pluto-Charon system moves in which way in its mutual motion?
    A) There is no relationship between rotation period of Pluto and orbital period of Charon.
    B) Charon orbits Pluto with exactly Pluto's rotation period.
    C) Charon orbits Pluto once while Pluto rotates twice.
    D) Charon orbits Pluto twice while Pluto rotates once.
    Answer: B Source: Section 7-16

277. What is the probable composition of both Pluto and its satellite, Charon, based on their average densities?
    A) About half rock and half ice.
    B) Rock with perhaps a 100-km thick ice mantle.
    C) Mostly ice, with perhaps a small rocky core.
    D) Liquid hydrogen, with a small rocky core and a gaseous atmosphere.
    Answer: A Source: Section 7-16

# Chapter 8: Vagabonds of the Solar System

1. Which solar system object was found on January 1, 1801, located between the orbits of Mars and Jupiter?
   A) Halley's Comet.
   B) The Kuiper belt object 1993 SC.
   C) The asteroid Ceres.
   D) The asteroid Gaspra.
   Answer: C Source: Section 8-1

2. A chunk of rock and metal 10 kilometers in diameter orbiting the Sun would be called
   A) an asteroid. B) a meteoroid. C) a comet. D) a moon.
   Answer: A Source: Section 8-1

3. An asteroid is
   A) another name for the nucleus of a comet, a volatile object which moves around the Sun in a long, elliptical orbit.
   B) a meteorite before it enters the atmosphere and plunges to Earth.
   C) a small, easily recognizable group of stars within a constellation.
   D) a planetesimal moving in an orbit around the Sun.
   Answer: D Source: Section 8-1

4. The asteroid belt exists between the orbits of the planets
   A) Earth and Mars.
   B) Mars and Jupiter.
   C) Venus and Earth.
   D) Jupiter and Saturn.
   Answer: B Source: Section 8-1

5. Most of the asteroids of our solar system move around the Sun between the orbits of which planets?
   A) Venus and Earth.
   B) Mars and Jupiter.
   C) Earth and Mars.
   D) Jupiter and Saturn.
   Answer: B Source: Section 8-1

6. What is the largest known asteroid in our solar system?
   A) Pallas. B) Phobos. C) Ceres. D) Gaspra.
   Answer: C Source: Section 8-1

7. The size of the largest asteroid, Ceres, compared to the largest mare or impact basin on the Moon, Mare Imbrium, is
   A) not comparable, since all asteroids are very small objects, (1 km diameter), while most maria are large (100-1000 km diameter).
   B) much smaller, only about 1/3 the size.
   C) very similar, about 1000 km across.
   D) much larger, by a factor of more than 2.
   Answer: C Source: Sections 5-5 and 8-1

8. When compared to the size of our Moon, the biggest asteroids have diameters
   A) about the same size.
   B) between 1/10 and 1/3 as large.
   C) very much larger, by a factor of at least 5.
   D) very much smaller (less than 1/10).
   Answer: B Source: Section 8-1

9. Which of the following statements is **NOT** true for asteroids?
   A) Only a minority of all asteroids are in the asteroid belt.
   B) Some asteroids have orbits which carry them inside the Earth's orbit.
   C) Some asteroids occupy the same orbit as Jupiter.
   D) The total mass of all asteroids is much smaller than the mass of the Earth.
   Answer: A Source: Section 8-1

10. The total number of asteroids bright enough to be visible on photographs taken from Earth is estimated to be about
    A) 30. B) 100,000. C) 3000. D) one million.
    Answer: B Source: Section 8-1

11. The total number of asteroids orbiting the Sun among the planets is estimated to be
    A) hundreds of billions.
    B) several thousands.
    C) more than a million.
    D) a few hundreds.
    Answer: C Source: Section 8-1

12. How would a typical asteroid appear on a time exposure photograph of the sky as it orbited the Sun, if the camera were tracking the background stars?
    A) It would produce a short trail as it moved slowly against the background stars.
    B) It would produce a flash of light as it crossed the field of view of the camera.
    C) It would look like any other star, a small extra dot not shown on star charts of this area of the sky.
    D) It would look like a small, diffuse patch against the sharp images of stars because of the dust and gas surrounding it.
    Answer: A Source: Section 8-1

13. How would we be able to detect a large asteroid if it were heading straight for the Earth?
    A) It would appear as a short trail against the background stars on a sky-tracked long-exposure photograph and its spectrum would show no Doppler shift.
    B) It would appear as a slowly brightening point of light where no star had previously been charted, and the spectrum of sunlight reflected from it would be blue-shifted by the Doppler effect.
    C) It would appear as a slowly brightening star-like object where no star was previously charted, with a red-shifted solar spectrum of reflected light.
    D) It would appear as a slowly brightening and growing diffuse sphere of light where no star was charted, because of light scattered from the dust and gas surrounding it, and show a blue-shifted spectrum.
    Answer: B Source: Section 8-1

14. Most asteroids
    A) are dark, irregular in shape and heavily cratered.
    B) are dark and spherical in shape, with many craters on their surfaces.
    C) are spherical and ice-coated, and hence are light-colored and shiny.
    D) have irregular shape, are covered with very light-colored dust, reflecting sunlight well.
    Answer: A Source: Section 8-1

15. The general shape or form of most asteroids is thought to be
    A) double, since most asteroids are binary systems, orbiting the Sun together.
    B) loose collections of very small particles, loosely held together by mutual gravity.
    C) perfectly spherical.
    D) irregular.
    Answer: D Source: Figures 8-5 and 8-6

16. What is the diameter of the largest asteroid in our solar system, Ceres, compared to the diameter of the Earth, which is about 12,800 km?
    A) About 1/4. B) About 1/2. C) Less than 1/10. D) Smaller than 1/100.
    Answer: C Source: Section 8-1

17. If all the material in the asteroid belt were to be combined to produce a planet, how big would it be?
    A) About 1500 km in diameter, significantly smaller than the Moon.
    B) Only a few km in diameter, similar to an average mountain on Earth.
    C) About the size of Earth with a diameter of about 13,000 km.
    D) About the size of Mercury, with a diameter of about 5000 km.
    Answer: A Source: Section 8-1

18. Only a few of the largest asteroids appear to be spherical. Why do you think this is?
    A) Self-gravity was sufficient to pull them into this shape during their early history.
    B) They solidified from spherical gas clouds in their early history and retained this shape.
    C) The visible outer atmospheres of these large asteroids are spherical even though the underlying surfaces are irregular.
    D) Repeated collisions with other asteroids have worn them down to spheres.
    Answer: A Source: Section 8-1

19. The average sidereal period for an asteroid moving around the Sun in the asteroid belt, according to Kepler's law, is
    A) 1.99 years. B) 2.8 years. C) 46.8 years. D) 4.68 years.
    Answer: D Source: Sections 2-3 and 8-1

20. If an asteroid were to be moving in a circular orbit around the Sun with an orbital period of 1/5 of that of Jupiter, what would be the radius of its orbit?
    A) 3.65 AU. B) 1.78 AU. C) 1.04 AU. D) 15.2 AU.
    Answer: B Source: Sections 2-3 and 8-1

21. The asteroid belt is believed by most astronomers to be composed of
    A) the remnants of a gaseous planet, disrupted by a massive impact.
    B) genuine leather.
    C) rather dirty ice-balls similar to the nuclei of comets.
    D) rocky debris left over from the formation of the solar system.
    Answer: D Source: Section 8-1

22. Computer simulations of the formation of the solar system show that the material in the vicinity of the asteroid belt is not in the form of one large planet because
    A) most of the material originally in the asteroid belt crashed into Mars, creating the heavily cratered terrain we see there.
    B) a violent collision destroyed two protoplanets, the debris from which became the asteroid belt.
    C) Jupiter's gravitational pull flung most of the material in this region out of the solar system and prevented coalescence of the rest.
    D) this region is where the gravitational field of the Sun is balanced by that of Jupiter, and this prevented coalescence of matter into a planet.
    Answer: C Source: Section 8-1

23. The Kirkwood gaps (see Fig. 8-4, Kaufmann & Comins, *Discovering the Universe*, 4th Ed.) are in the
    A) equatorial region of the Sun, where no sunspots are found.
    B) asteroid belt, where there are very few asteroids.
    C) rings of Saturn, where there is less material than at other radii.
    D) spectrum of hydrogen gas, where light has been absorbed by molecules first identified by Kirkwood.
    Answer: B Source: Section 8-2

24. The Kirkwood gaps are caused by
    A) the gravitational pull of Jupiter, which nudges asteroids into new orbits.
    B) large asteroids on the outer fringe of the Asteroid Belt, which gravitationally affect the paths of smaller objects within the Belt.
    C) large asteroids moving in circular orbits within the Asteroid Belt, which sweep out and collect smaller objects in their path.
    D) large asteroids whose orbits carry them periodically through the Asteroid Belt, where they sweep out a path and leave it devoid of asteroids.
    Answer: A Source: Section 8-2

25. What effect does Jupiter have upon asteroids in the asteroid belt?
    A) It will disturb the orbits of all the asteroids in the belt, slowing them down and causing them to spiral slowly in towards the Sun.
    B) It will only disturb the orbits of those asteroids whose orbital distances (or semi-major axes) are a simple fraction (e.g., 1/2, 1/3, 2/3, 2/7) of the radius of Jupiter's orbit.
    C) It will perturb only the orbits of those asteroids whose orbital periods are a simple fraction (e.g., 1/2, 1/3, 2/3, 2/7) of its orbital period.
    D) It will have no effect whatsoever upon such small objects since they are a long way away from Jupiter and Jupiter's gravitational influence varies as the inverse square of distance, by Newton's law.
    Answer: C Source: Section 8-2

26. What is the relationship between the Kirkwood Gaps in the Asteroid Belt and the Cassini division in the rings of Saturn?
    A) Both are caused by disruptions of the orbits of small objects by larger planets or moons. In both cases, the periods of the small objects are simple fractions of those of the larger disturbing object.
    B) Both are caused by large objects passing through swarms of smaller objects, sweeping out gaps in the swarms.
    C) Both were discovered by observers from the same group; Kirkwood and Cassini both worked at the same observatory.
    D) Both are caused by disruptions of the orbits of small objects by larger planets or moons. In both cases, the orbital distance of small objects in the gaps is related by simple fractions to the orbital distance to the disturbing object.
    Answer: A Source: Section 8-2

27. The asteroid belt has a gap where few objects are found because of repeated gravitational disturbances from Jupiter. At what distance from the Sun will this gap be found if objects in the gap have a period of one-third of that of Jupiter? (Hint: Use Kepler's Law.)
    A) 2.5 AU. B) 3.28 AU. C) 7.86 AU. D) 2.6 AU.
    Answer: A Source: Sections 2-3 and 8-2

28. What is unusual about the asteroid Mathilde, which was studied and photographed by the NEAR spacecraft (Fig. 8-6, Kaufmann & Comins, *Discovering the Universe*, 5th Ed.)?
    A) It is not much denser than water.
    B) It has a much higher mass for its volume than any other known asteroid.
    C) It has a very bright surface, possibly caused by fresh material thrown out by impacts.
    D) It has almost no craters visible anywhere on its surface.
    Answer: A Source: Section 8-2

29. The surfaces of Ida and Mathilde, asteroids photographed by the Galileo spacecraft, (Figs. 8-5 and 8-6, Kaufmann & Comins, *Discovering the Universe*, 5th Ed.) show
    A) icy surfaces criss-crossed with cracks and systems of parallel grooves.
    B) ancient surfaces densely covered with large and small overlapping craters, like the surface of the highland areas of the Moon.
    C) irregular, somewhat rounded and moderately cratered surfaces.
    D) young, sharp, jagged surfaces due to fragmentation by collision with other asteroids, and few craters.
    Answer: C Source: Section 8-2

30. Which two spacecraft have taken detailed photographs of the surface of an asteroid?
    A) Magellan and Cassini.
    B) Voyager 1 and Galileo.
    C) Galileo and NEAR.
    D) NEAR and Cassini
    Answer: C Source: Section 8-2

31. Asteroids which orbit the Sun at the same distance as Jupiter are known as
    A) Jupitoids. B) Trojans. C) Apollo asteroids. D) Adenoids.
    Answer: B Source: Section 8-3

32. Where do the Trojan asteroids orbit the Sun?
    A) Within the asteroid belt, so-named because they are large.
    B) In circular orbits at the same orbital distance as Jupiter.
    C) In circular orbits at the same orbital distance as Earth.
    D) In elliptical orbits which cross the orbit of Earth.
    Answer: B Source: Section 8-3

33. What kind of orbit is traced by the Trojan asteroids as they move around the Sun?
    A) Circular orbits at about 2.8 AU from the Sun, within the asteroid belt.
    B) Long, elliptical orbits which cross the orbit of Earth.
    C) Circular orbits at Jupiter's orbital distance, at angles of $\pm 60°$ away from the planet.
    D) Elliptical orbits which carry them from outside Neptune's orbit to inside that of Jupiter.
    Answer: C Source: Section 8-3

34. What will be the orbital sidereal period of a Trojan asteroid? (See Fig. 8-7 and Table A-1, Appendix, Kaufmann & Comins, *Discovering the Universe* 5th Ed.)
   A) 11.86 years.
   B) It is difficult to be specific, because they all have different orbital periods, depending upon their masses.
   C) 1.88 years.
   D) 5.9 years, the same as most asteroids in the asteroid belt.
   Answer: A Source: Section 8-3

35. If an asteroid is found to be orbiting in a circular path around the Sun at the same distance as Jupiter, (5.2 AU), what will be its orbital period compared to that of Jupiter, which is 11.86 years?
   A) Exactly 1/2 or 5.93 years, since it will be in a synchronous orbit with Jupiter.
   B) The same as Jupiter, 11.86 years.
   C) About 10 times as long, or 118.6 years, since the Sun's gravitational force is much smaller on such a small object.
   D) About 1/10 of Jupiter's period, since it is a much smaller object.
   Answer: B Source: Sections 2-3 and 8-3

36. The two Lagrangian points, $L_4$ and $L_5$, in the Jupiter-Sun planetary system are
   A) positions in space at Jupiter's orbital distance from the Sun where the combined gravitational forces from the Sun and Jupiter produce a minimum in which asteroids can become trapped.
   B) an area between the Sun and Jupiter where the gravitational forces upon an object from these massive bodies are equal and opposite.
   C) areas in the asteroid belt where gravitational interaction of Jupiter with asteroids disturbs their orbits and causes a Kirkwood gap.
   D) points at high latitudes on Jupiter where auroras (called Lagrangian auroras on Jupiter) occur most frequently.
   Answer: A Source: Section 8-3

37. The stable Lagrangian points near to two objects in space are points where
   A) the total gravitational force is zero.
   B) the combined gravitational field is flat (maximum or minimum).
   C) the combined gravitational field is a minimum.
   D) the combined gravitational field is a maximum.
   Answer: C Source: Section 8-3

38. One significant feature of the Lagrangian points $L_4$ and $L_5$ produced by the Sun and Jupiter is that
   A) they are points of maximum gravitational field near to Jupiter where the major moons Io and Europa are held.
   B) gravitational fields combine to produce a minimum in the field, thereby trapping asteroids at these points.
   C) gravitational fields combine to enhance the overall force on particles passing through them, accelerating them out of the solar system.
   D) they are regions of reduced gravitational field within the asteroid belt from which asteroids can escape, producing the gaps within the belt.
   Answer: B Source: Section 8-3

39. Asteroids should be able to remain stably trapped at the $L_4$ and $L_5$ Lagrangian points of the Earth-Sun system (see Section 8-3 and Fig. 8-7, Kaufmann & Comins, *Discovering the Universe*, 5th Ed.), provided that they are NOT subjected to other, stronger, gravitational influences. As a test of this stability, calculate how strong Jupiter's gravity is at the Earth's $L_4$ and $L_5$ points (when Jupiter is closest to them) compared to the strength of Earth's gravity at the same points. (You will need to use Newton's law of gravitation (see Toolbox 2-2); it will also help to draw a diagram.) Compared to the Earth's gravitational force at these points, Jupiter's gravity is
   A) 18 times stronger.
   B) 1/18 as strong.
   C) 76 times stronger.
   D) 12 times stronger.
   Answer: A Source: Section 8-3 and Toolbox 2-2

40. Asteroids whose orbits carry them across the Earth's orbit are known as
   A) Amor asteroids.
   B) Kirkwood asteroids.
   C) Apollo asteroids.
   D) Trojan asteroids.
   Answer: C Source: Section 8-3

41. What is the difference between an Apollo asteroid and an Amor asteroid?
   A) Apollo asteroids approach the orbit of Mercury, whereas Amor asteroids pass only within the orbit of the Earth.
   B) Apollo asteroids orbit entirely inside the Earth's orbit and therefore do not cross it, whereas Amor asteroids cross the Earth's orbit and might therefore hit the Earth.
   C) Apollo asteroids pass within the orbit of the Earth, whereas Amor asteroids pass only within the orbit of Mars.
   D) Apollo asteroids pass near terrestrial planets, whereas Amor asteroids remain in the same orbit as Jupiter.
   Answer: C Source: Section 8-3

42. A useful technique for estimating the shapes of many asteroids has been
    A) measurement by many closely spaced observers of occultations of stars by asteroids.
    B) the measurement of brightness variations caused by asteroid rotation.
    C) photography from spacecraft, such as Galileo.
    D) direct photography of the asteroid's shape from Earth.
    Answer: B Source: Section 8-3

43. The major difference between the orbital paths of comets and those of the asteroids in the asteroid belt is that
    A) comet orbits are mostly circular and in the ecliptic plane, whereas the asteroids have elliptical orbits inclined at random to the ecliptic plane.
    B) cometary orbits are highly elliptical and at random inclinations to the ecliptic plane compared to the circular orbits of asteroids in the ecliptic plane.
    C) asteroids orbit the Sun continuously whereas all comets approach the Sun's vicinity only once before leaving the solar system.
    D) comets never approach closer to the Sun than approximately Jupiter's orbit whereas some asteroids approach very close to the Sun.
    Answer: B Source: Sections 8-1 and 8-4

44. The most likely origin of the "dirty snowballs," which become comets when they are deflected into orbits that bring them closer to the Sun, is
    A) the asteroid belt, since most asteroids are actually comet nuclei.
    B) the Kuiper and Oort belts surrounding the solar system.
    C) the icy surfaces of the moons of Jupiter and Saturn.
    D) dust and gas clouds in the Milky Way galaxy.
    Answer: B Source: Section 8-4

45. The Kuiper belt is
    A) a random distribution of short-period comets extending from inside the orbit of Jupiter to approximately the orbit of Neptune.
    B) a spherical distribution of distant comets around the Sun, extending out about 50,000 AU.
    C) a flat or donut-shaped distribution of distant comets around the Sun, extending out about 500 AU.
    D) another name for the asteroid belt.
    Answer: C Source: Section 8-4

46. The Oort cloud is
    A) another name for the asteroid belt.
    B) a random distribution of short-period comets extending from inside the orbit of Jupiter to approximately the orbit of Neptune.
    C) a spherical distribution of distant comets around the Sun, extending out about 50,000 AU.
    D) a flat or donut-shaped distribution of distant comets around the Sun, extending out about 500 AU.
    Answer: C Source: Section 8-4

47. What is the approximate orbital period of a comet nucleus that orbits the Sun on the outer fringes of the Oort cloud? (See Section 8-4, Kaufmann & Comins, *Discovering the Universe*, 5th Ed.)
    A) 10,000 years. B) 100,000 years. C) 10 million years. D) 1000 years.
    Answer: C Source: Section 8-4

48. Most comet nuclei are believed to be
    A) carbonaceous chondrite meteorites, carbon material, ignited by sunlight and trailing long smoke trails.
    B) pieces of dirty ice, ejected from the surface of the icy satellites of the outer planets by asteroid impacts.
    C) pieces of rock or iron chipped from asteroids by impacts.
    D) pieces of dusty ice, left over from the formation of the solar system.
    Answer: D Source: Section 8-4

49. The nucleus of Comet Halley, as seen on close-up photographs taken by the Giotto spacecraft, is
    A) roughly spherical, moderately cratered, and covered with dark dust.
    B) potato-shaped and darker than coal.
    C) roughly spherical, light-colored, icy, and covered with many cracks and grooves.
    D) oblong, with a bright ice surface.
    Answer: B Source: Section 8-4

50. The measured diameter of the core of Halley's comet is about
    A) $10^7$ km. B) 10 km. C) 100 m. D) $10^6$ km.
    Answer: B Source: Section 8-4

51. Compared to the coma or the visible, fuzzy ball of a comet, $10^6$ km in diameter, the diameter of the actual nucleus of the comet is
    A) about the size of the largest asteroid, 1/1000, or 1000 km.
    B) very small, about $1/10^5$, or 10 km.
    C) extremely small, $1/10^8$, or 10 m.
    D) about 1/100, or 10,000 km.
    Answer: B Source: Section 8-4

52. The nucleus of a typical comet is
    A) irregular in shape, with a bright and very reflective, icy surface.
    B) spherical, smooth and very light-colored, being composed mostly of ice.
    C) irregular in shape, with a very dark and cratered surface.
    D) spherical, with a very smooth, dark surface.
    Answer: C Source: Section 8-4

53. The typical size of the coma or gas cloud surrounding the comet nucleus as it reaches its closest point to the Sun is
    A) about 10 km. B) about $10^7$ km. C) about $10^6$ km. D) about $10^5$ km.
    Answer: C Source: Section 8-4

54. How large does the coma, or gas cloud surrounding a comet nucleus, become when it reaches its closest point to the Sun?
    A) About $10^6$ km, as big as the Sun.
    B) About $10^7$ km, close to 1/10 of Mercury's orbit.
    C) Only 10 km, but it glows brightly in sunlight.
    D) About $10^8$ km, close to the size of Mercury's orbit.
    Answer: A Source: Section 8-4

55. The huge hydrogen cloud which surrounds the nucleus of a comet, discovered by its UV emission, has a typical diameter of about
    A) $2 \times 10^7$ km. B) 1/2 AU. C) 2 A.U. D) $10^6$ km.
    Answer: A Source: Section 8-4

56. In a single photograph of a comet and its tail, the only direction that one can determine with certainty is
    A) the direction away from the Sun, since the tail is pushed in this direction by the solar wind.
    B) the direction towards the Sun indicated by the tail direction, since gas and dust in the tail is attracted towards the Sun by its gravity.
    C) the direction towards Jupiter, since the gravity of this giant planet pulls the tail material towards it.
    D) the direction in which the comet is moving, from the trailing tail.
    Answer: A Source: Section 8-4

57. Comet tails are the result of
    A) interplanetary material streaming into the comet because of its gravity.
    B) interplanetary dust, collected by the comet as it moves in its orbit.
    C) sunlight glinting upon the central icy comet core.
    D) melting and evaporation of ices from the comet core.
    Answer: D Source: Section 8-4

58. The tail of a comet
    A) is longest when the comet is closest to the Sun.
    B) remains constant in length throughout its complete orbital path.
    C) is longest when the comet is furthest from the Sun, since it is then unaffected by sunlight.
    D) is longest when the comet is closest to Jupiter, being pulled out by its gravity.
    Answer: A Source: Section 8-4

59. A comet's tail
    A) points towards the Sun, because it is caused by jets of gases evaporated from the comet's nucleus on the side heated by the Sun.
    B) always points towards the nearest planet, being attracted by its gravity field as the comet passes by it.
    C) always points away from the Sun, regardless of the motion of the comet.
    D) always trails behind the comet in its orbit, and so points away from the Sun only while the comet is approaching the Sun.
    Answer: C Source: Section 8-4

60. The gas and ion tail of a comet
    A) is always blown away from the comet in the anti-Sun direction by the solar wind.
    B) lies between the comet and the Sun, because of gravitational attraction.
    C) always trails along the orbital path, being left behind by the comet.
    D) always lies in the ecliptic plane, since a comet is a part of the solar system.
    Answer: A Source: Section 8-4

61. The ionized gas tail of a comet is always aligned
    A) in the comet-Sun line.
    B) along the comet's direction of motion.
    C) along the line between the comet and the nearest planet to it in its orbital motion.
    D) along the celestial equator.
    Answer: A Source: Section 8-4

62. A comet's tail always points from the comet head
    A) towards the Sun, because of gravitational attraction.
    B) in a direction along its orbital path, always behind the comet.
    C) away from the Sun.
    D) towards the nearest planet, because of gravitational attraction for the tail material.
    Answer: C Source: Section 8-4

63. The particular feature of a comet that exhibits the most structure and always points away from the Sun is
    A) its hydrogen envelope.
    B) its coma, or gas cloud.
    C) its ion or gas tail.
    D) its dust tail.
    Answer: C Source: Section 8-4

64. What mechanism controls the direction in which a comet's ion tail is aligned in space?
    A) Its direction of motion, since the tail simply trails behind the comet in its orbit.
    B) The flow of solar wind past the comet's head.
    C) The gravitational attraction of the tail material towards the giant planet, Jupiter.
    D) The gravitational attraction of the tail material towards the Sun.
    Answer: B Source: Section 8-4

65. The dust tail of a comet has which of the following characteristics?
    A) Long, straight, structured and pointed directly away from the Sun.
    B) Narrow and straight and pointed directly at the Sun at all times.
    C) Spherical, very large and of low brightness, centered upon the comet nucleus, showing up only on UV photographs.
    D) Curved, wide and without structure, but thin and transparent to starlight.
    Answer: D Source: Section 8-4

66. Dust grains released by the melting of ice in a comet nucleus
    A) become a straight, highly structured and very variable tail, blown away from the comet by the solar wind.
    B) become a cloud around the nucleus, the coma, scattering sunlight very efficiently at blue wavelengths.
    C) become a uniform, curved tail, moving away from the comet under radiation pressure from sunlight.
    D) drift away from the Sun along magnetic field lines, outlining the structure of this field.
    Answer: C   Source: Section 8-4

67. The aphelia (furthest distances from the Sun) of comet's orbits
    A) are mostly confined to the region between Mars and Jupiter, although some comets have orbits which reach beyond the orbit of Pluto.
    B) are all located far beyond the orbit of Pluto.
    C) are always closer to the Sun than the orbit of Pluto.
    D) can be located anywhere from inside the orbit of Pluto to as far as 100,000 AU from the Sun.
    Answer: D   Source: Section 8-4

68. The orbits of comets
    A) can extend far beyond the orbit of Pluto and are primarily in the ecliptic plane.
    B) are confined to distances closer to the Sun than the orbit of Pluto and are mostly in the ecliptic plane.
    C) are confined to distances closer to the Sun than the orbit of Pluto and are oriented randomly in the solar system.
    D) can extend far out beyond the orbit of Pluto and can be oriented in any direction in the solar system.
    Answer: D   Source: Section 8-4

69. The orbits of comets are
    A) primarily in the plane of the ecliptic and confined to distances closer to the Sun than approximately the orbit of Pluto.
    B) randomly oriented in the solar system and confined to distances closer to the Sun than approximately the orbit of Pluto.
    C) randomly oriented in the solar system and can extend far beyond the orbit of Pluto.
    D) primarily in the plane of the ecliptic and can extend far out beyond the orbit of Pluto.
    Answer: C   Source: Section 8-4

70. How far will Halley's comet be from the Sun when it reaches its furthest point from the Sun, or aphelion, if its sidereal period is 76 years? (Caution: This calculation needs Kepler's law and a little care. Assume that the comet's perihelion distance from the Sun is negligible.)
A) About 1324 AU, well beyond the orbit of Pluto.
B) About 36 AU, between the orbits of Neptune and Pluto.
C) About 18 AU, between the orbits of Saturn and Uranus.
D) About 9 AU, between the orbits of Jupiter and Saturn.
Answer: B Source: Sections 2-3 and 8-4

71. A typical comet in an elliptical orbit around the Sun will lose what fraction of its mass by melting each time it passes close to the Sun (i.e., at each perihelion passage)?
A) 1/100. B) Very small, less than 1/10,000. C) 1/10. D) 1/1,000.
Answer: A Source: Section 8-5

72. The number of times that a typical comet can pass close to the Sun (i.e., the number orbits which the comet can complete) before it is completely vaporized or destroyed is about
A) millions. B) 100. C) once. D) 1000.
Answer: B Source: Section 8-5

73. A shooting star is
A) a small particle of interplanetary dust, burning up and glowing as it enters the Earth's atmosphere.
B) the leading scorer on a basketball team.
C) a violently erupting star, ejecting matter rapidly away from it into interstellar space.
D) a near-neighbor star, moving rapidly across our field of view.
Answer: A Source: Section 8-6

74. The luminous trails of small dust particles that are completely vaporized in the Earth's atmosphere are known as
A) meteorites. B) meteoroids. C) auroral flashes. D) meteors.
Answer: D Source: Section 8-6

75. A piece of rock from outer space that reaches the Earth's surface after surviving a fiery passage through the Earth's atmosphere is known as
A) a meteoroid. B) a meteorite. C) a meteor. D) an asteroid.
Answer: B Source: Section 8-8

76. A small particle of rock orbiting the Sun would be called
A) a meteorite. B) a micrometer. C) a meteor. D) a meteoroid.
Answer: D Source: Section 8-6

77. A meteoroid is the name used to describe a solid particle that
    A) is drifting around in space.
    B) has fallen to Earth from space.
    C) originated on the Moon, but was knocked onto the Earth by a massive impact.
    D) burns up as it falls through the Earth's atmosphere.
    Answer: A Source: Section 8-6

78. The Barringer meteorite crater is located in
    A) Australia.
    B) Quebec, Canada.
    C) Arizona, U.S.A.
    D) the Yucatan Peninsula, Mexico.
    Answer: C Source: Section 8-7

79. The estimated impact energy of the object which produced the Barringer Crater in Arizona, in terms of explosive power (tons of TNT equivalent, a somewhat dubious and frightening scale!) is
    A) similar to the first nuclear weapons, less than 10 kilotons.
    B) about 1 megaton.
    C) about 20 megatons, similar to the most powerful hydrogen bombs.
    D) greater than 1000 megatons.
    Answer: C Source: Section 8-7

80. The cause of the "meteor showers," seen at regular times each year upon Earth, is most probably
    A) the Earth moving through the remnant dust and rock fragments of an old comet which are orbiting the Sun in the comet's old orbit.
    B) sunspot activity and the resultant geomagnetic disturbances.
    C) unstable weather conditions on Earth.
    D) the Earth running into material within the spiral arm structure of the Milky Way.
    Answer: A Source: Section 8-7

81. A meteor shower occurs when
    A) the Earth passes through a swarm of dust particles in space.
    B) a meteor is about to get married.
    C) the head of a comet hits the Earth's atmosphere.
    D) the Earth passes through the asteroid belt.
    Answer: A Source: Section 8-7

82. A meteor shower results from
    A) the Earth passing through debris of an old comet.
    B) material re-entering the Earth's atmosphere after being ejected into space by violent volcanic eruptions on Earth.
    C) a small piece of rock fragmenting as it passes through the Earth's atmosphere.
    D) material ejected by a massive impact upon the Moon, the Earth's gravity attracting it towards Earth.
    Answer: A Source: Section 8-7

83. A meteor shower, or the appearance of many more "shooting stars" at a particular time in the year from a specific sky direction, is related to which astronomical phenomenon?
    A) The passage of Earth through the remnants of an old comet.
    B) The Earth's passage through part of the asteroid belt.
    C) The passage of Earth through intense streams of solar wind.
    D) The Earth's passage through different parts of the spiral arms of the Galaxy.
    Answer: A Source: Section 8-7

84. Why do the meteors that are seen in the sky in a particular meteor shower appear to come from one specific direction in the sky?
    A) This is the direction along which objects will pass upon being attracted to the Sun from outer space by its gravity.
    B) This specific direction is always along the ecliptic plane because meteor showers occur when Earth catches up with a collection of particles moving in its orbit.
    C) Because this is the direction of the orbit of the comet which disintegrated to produce the shower.
    D) The meteors only appear to come from a specific direction because of the Earth's orbital motion.
    Answer: C Source: Section 8-7

85. Meteoritic material that has been found recently in Antarctica is believed to have come from another planet. Which planet is this?
    A) Mars. B) Mercury. C) Venus. D) Jupiter.
    Answer: A Source: Section 8-8

86. The estimated total infall of meteoritic and extraterrestrial material from space per day upon the Earth is
    A) about 30 tons. B) about 1 million tons. C) about 300 tons. D) less than 1 ton.
    Answer: C Source: Section 8-8

87. Interplanetary material
    A) hits the Earth at specific times of the year in the form of small particles which produce meteor showers, but does not fall upon Earth at other times.
    B) falls upon Earth only very rarely in the form of single large objects, but these individual impacts can devastate parts of the Earth and threaten life upon Earth.
    C) falls upon Earth at the rate of several hundred tons per day, mostly as micrometeoroids.
    D) occasionally hits the Earth in the form of fairly large objects which form craters, but there is no continuous stream of incoming matter.
    Answer: C Source: Section 8-8

88. Which are the most common types of meteoroids in space?
    A) Irons. B) Carbonaceous chondrites. C) Stones. D) Stony-irons.
    Answer: C Source: Section 8-8

89. The most common meteorites to hit the Earth are
    A) the carbonaceous chondrites.
    B) the iron meteorites.
    C) the stony meteorites.
    D) the stony-iron meteorites.
    Answer: C Source: Section 8-8

90. Stony meteorites
    A) are very much like ordinary silicate rocks.
    B) contain large quantities of carbon and $H_2O$, and even hydrocarbons and amino acids.
    C) have solid iron cores surrounded by rocky silicate shells.
    D) are made of solid iron with small quantities of nickel and cobalt.
    Answer: A Source: Section 8-8

91. What fraction of the material arriving on the Earth from outer space is in the form of iron meteorites?
    A) 10 to 20%. B) 95%. C) 50%. D) A few %.
    Answer: D Source: Section 8-8

92. What fraction of the material arriving on the Earth from outer space is in the form of stony meteorites?
    A) A few %. B) 10 to 20%. C) 95%. D) 50%.
    Answer: C Source: Section 8-8

93. With which of the following descriptions would you identify a rock as a fallen meteorite?
    A) A rock containing large transparent crystals of common salt, embedded in sandstone.
    B) A layered rock, consisting mainly of limestone.
    C) An irregular and very heavy solid iron rock, with a distinctive crystal structure throughout its interior.
    D) A basalt rock full of hollow spaces, shiny and dark colored, with fine details upon its surface.
    Answer: C Source: Section 8-8

94. The Widmanstatten pattern found in many iron meteorites consists of
    A) a myriad of small iron crystals, formed when the rock cooled quickly in the vacuum of space.
    B) fracture lines created by the original impact which knocked the meteoroid off the parent asteroid.
    C) large crystals formed as the iron cooled slowly over many millions of years.
    D) small holes created by the heating and partial melting of the meteorite as it entered the Earth's atmosphere.
    Answer: C Source: Section 8-8

95. The Widmanstatten pattern of crystals found in many iron meteorites is evidence that the meteorites
    A) cooled slowly over many millions of years.
    B) cooled quickly, in only a few years at most.
    C) have suffered highly energetic collisions with asteroids or other meteorites.
    D) have been subject to periodic heating by close approaches to the Sun.
    Answer: A Source: Section 8-8

96. Stony-iron meteorites are believed to
    A) originate from undifferentiated asteroids (same composition throughout).
    B) have been ejected by volcanoes on Mars.
    C) be pieces of primordial solar system material, unaltered since the solar system formed.
    D) originate from differentiated asteroids (in which iron sank to the center).
    Answer: D Source: Section 8-8

97. The existence of distinct and separate types of meteorites, stony, stony-iron, and iron, is probably a result of
    A) their formation in different parts of the early solar nebula, with stones condensing closer to the Sun and the irons further away from the Sun.
    B) differentiation of material in molten asteroid interiors, with the iron sinking to the core, followed by fragmentation and separation of iron core and rocky shell.
    C) preferential accretion of iron particles to other iron particles because of their magnetic properties, leaving stony particles to accrete separately.
    D) different amounts of heating and "erosion" of the outer layers of meteorites as they pass through the Earth's atmosphere, the irons losing all their outer rocky shells.
    Answer: B Source: Section 8-8

98. Which of the following biochemical materials have been found and identified in carbonaceous chondrites that have hit the Earth?
    A) Amino acids, or proteins.
    B) Living single-cell organisms.
    C) Living (?) viruses.
    D) Lichens and mosses.
    Answer: A Source: Section 8-8

99. Perhaps the most interesting and puzzling material to be found inside rocks that have come from outer space is
    A) water. B) pure iron. C) amino acids, or proteins. D) radioactive material.
    Answer: C Source: Section 8-8

100. Recent calculations show that the Tunguska explosion in Siberia in 1908 was probably caused by
    A) a small stony asteroid about 80 m across, exploding well above the ground.
    B) a small comet nucleus about 1 km across, suddenly vaporizing in the atmosphere.
    C) a large stony meteorite about 1 m across, striking the ground at a high speed.
    D) a small nuclear explosion.
    Answer: A Source: Section 8-9

101. What has been discovered in some meteorites that suggests that the formation of our Sun and solar system might have been triggered by a supernova explosion?
    A) Super-pure iron, only produced in nuclear reactions.
    B) Fusion crusts around most meteorites, indicating that they had been heated intensely at some time in their history.
    C) Amino acids, which can only be produced in very specific conditions.
    D) Decay products of short-lived radioactive elements, which could only have been produced in intense nuclear reactions.
    Answer: D Source: Section 8-9

102. A meteorite that recently fell to Earth was found to contain evidence of the existence earlier in its history of a short-lived radioactive isotope of aluminum, $^{26}Al$, which can only be produced under extremely energetic conditions and in nuclear reactions. What conclusion can be drawn from this observation?
   A) Radioactivity occurs spontaneously in normal matter, and so this finding is not surprising.
   B) The meteorite had probably passed through radioactive clouds in space before hitting Earth and collected the $^{26}Al$ at that time.
   C) An energetic nuclear event, a supernova, occurred near the Sun, producing the $^{26}Al$ at about the time that the meteorite was formed.
   D) The meteorite became so hot on its descent through the Earth's atmosphere that it became radioactive and $^{26}Al$ was formed at that time.
   Answer: C Source: Section 8-9

103. The metal that is relatively abundant in meteorites but that is generally rare upon Earth has been found in specific layers of clay throughout the world, and its study is helping to evaluate the effect of meteor impacts upon the Earth's climate and its inhabitants in recent geological time (e.g., dinosaurs). This metal is
   A) iridium. B) nickel. C) iron. D) cobalt.
   Answer: A Source: Section 8-10

104. The astronomical event that is now thought to have occurred some 65 million years ago resulting in the death of a large fraction of all living species and leaving a layer of clay containing an enhanced concentration of a rare metal, iridium, in the geological record in rocks throughout the Earth was
   A) a very large volcanic eruption on Earth.
   B) a supernova which exploded relatively close to the solar system.
   C) the impact upon the Earth of an asteroid.
   D) an extraordinary solar eruption or flare.
   Answer: C Source: Section 8-10

105. On the basis of the recent interpretation of geological evidence, the impact of a 10-km diameter asteroid on the surface of the Earth would very likely
   A) disrupt the global ecology and cause the extinction of a large percentage of all species living on the Earth.
   B) create significant damage near to the impact site but have relatively little lasting world-wide effect.
   C) shatter the Earth into fragments.
   D) completely destroy life on Earth.
   Answer: A Source: Section 8-10

106. An impact which took place at about the time of the extinction of the dinosaurs (and may in fact have caused their extinction) is believed to have created
   A) the Manicouagan Crater in Quebec.
   B) the Barringer Crater in Arizona.
   C) the Hudson Bay in northern Canada.
   D) the Chicxulub Crater in the Yucatan Peninsula.
   Answer: D Source: Section 8-10

107. The Chicxulub Crater, believed by many scientists to have been created by the impact that caused the extinction of the dinosaurs, is
   A) a large, round valley with a central peak, high in the Andes Mountains.
   B) hidden under the Yucatan Peninsula and adjacent Carribean Sea.
   C) a prominent landmark in northern Arizona.
   D) an eroded, water-filled depression in the Canadian Shield area of northern Quebec.
   Answer: B Source: Section 8-10

108. What is the specific mechanism by which the Chicxulub impact is believed to have caused the extinction of the dinosaurs and many other species, 65 million years ago?
   A) Dust thrown up by the impact blotted out the Sun and disrupted the food chain over a period of several years.
   B) Iridium poisoning due to fallout of meteoritic material destroyed a large fraction of the creatures within several days of the impact.
   C) Shock waves and tsunamis (tidal waves) killed the majority of living creatures within a few minutes of the impact.
   D) The incoming asteroid contained alien biological material that destroyed susceptible species within a few weeks of the impact.
   Answer: A Source: Section 8-10

# Chapter 9: Our Star, the Sun

1. What is the photosphere of the Sun?
   A) The core of the Sun, where nuclear energy is generated.
   B) The visible "surface" of the Sun.
   C) The middle layer of the Sun's atmosphere.
   D) The region of convecting gases below the visible surface of the Sun.
   Answer: B Source: Section 9-1

2. What is the Sun's photosphere?
   A) The upper layer of the Sun's atmosphere.
   B) The envelope of convective mass motion in the outer interior of the Sun.
   C) The lowest layer of the Sun's atmosphere.
   D) The middle layer of the Sun's atmosphere.
   Answer: C Source: Section 9-1

3. What name is given to the visible "surface" of the Sun?
   A) Corona. B) Chromosphere. C) Photosphere. D) Prominence.
   Answer: C Source: Section 9-1

4. The thickness of the photosphere, or the visible "surface" of the Sun, is
   A) several times 10,000 km.
   B) about 4000 km.
   C) about 400 km.
   D) only about 1 km.
   Answer: C Source: Section 9-1

5. The temperature of the Sun's photosphere is
   A) 5800 K. B) about 10,000 K. C) 4300 K D) close to 1 million K.
   Answer: A Source: Section 9-1

6. The approximate temperature of the visible surface of the Sun is
   A) 2000 K. B) 10,000 K. C) 4300 K. D) 5800 K.
   Answer: D Source: Section 9-1

7. How does the average density of the Sun compare to that of the planet Jupiter?
   A) The Sun has approximately the same average density as Jupiter.
   B) The Sun is considerably less dense than Jupiter.
   C) It is not possible to specify an average density for an object as large as the Sun.
   D) The Sun is many times denser than Jupiter.
   Answer: A Source: Section 9-1, Table 9-1, Append Table A-2

8. If the temperature of the solar surface is 5,800 K, and Wien's law for the peak wavelength of the spectrum of the Sun, assumed to be a blackbody, is given by $l_{max}$ T = $2.9 \times 10^6$, with T in kelvins and l in nanometers (nm), what is the expected dominant wavelength of the Sun?
   A) 50 nm B) 600 nm C) 500 nm D) 300 nm
   Answer: C Source: Sections 9-1 and 4-1

9. The surface of the Sun is divided into light-colored areas with dark boundaries in a cellular pattern. What are these cells called?
   A) Sunspots. B) Granules. C) Filaments. D) Spicules.
   Answer: B Source: Section 9-1

10. What is the cellular granulation pattern seen on the visible surface of the Sun?
    A) The cells are regions of nuclear energy generation in the Sun's photosphere.
    B) The cells are the base of a circulation pattern that extends from the photosphere to the outer corona.
    C) Each cell is a region of stronger magnetic field, which compresses and heats the gas within it.
    D) The cells are the tops of rising blobs of hot gas in the Sun's convective interior.
    Answer: D Source: Section 9-1

11. The granular appearance of the surface of the Sun is evidence of what phenomenon occurring in or on the Sun?
    A) Rapid rotation of the surface layers.
    B) Thermonuclear fusion in its interior.
    C) The regular impact of meteoroids and comets onto the solar surface.
    D) Convective motion under the solar surface.
    Answer: D Source: Section 9-1

12. Granulation on the surface of the Sun is caused by
    A) magnetic field disturbances above the solar surface.
    B) convective currents carrying heat from beneath the surface.
    C) nuclear fusion processes occurring just below the surface.
    D) differential rotation of the Sun.
    Answer: B Source: Section 9-1

13. The granulation pattern seen on the surface of the Sun results from
    A) strong magnetic fields cooling the gas in certain regions.
    B) convection of gas in the region under the photosphere.
    C) heating of the photosphere by solar flares.
    D) prominences and filaments above the solar surface.
    Answer: B  Source: Section 9-1

14. Granulation or the mottled appearance of the whole solar surface is an indication of what physical process at work in the Sun?
    A) The outflow of neutrinos from the interior.
    B) Thermonuclear fusion of hydrogen in the Sun's surface layers.
    C) Rapid rotation of the Sun.
    D) Convective motion of gases in the upper portion of the Sun's interior.
    Answer: D  Source: Section 9-1

15. A typical granule on the surface of the Sun
    A) is about 1000 km across and lasts for a few minutes.
    B) is a few thousand kilometers across and lasts for about two solar rotations.
    C) is about 30,000 km across and lasts for several hours.
    D) arches quietly for several days over a sunspot group.
    Answer: A  Source: Section 9-1

16. The centers of the granular cells on the surface of the Sun are brighter than the edges of the cells because
    A) the centers are composed of different gases than the edges.
    B) the centers are hotter than the edges.
    C) the centers are denser than the edges.
    D) the centers are cooler than the edges.
    Answer: B  Source: Section 9-1

17. If granulation on the Sun's surface is a result of convective motion below it, and the centers of granular cells are where material is upwelling from below and returning in the regions between, what is the expected temperature distribution across a granular cell?
    A) The center of the cell will be hotter than the edges.
    B) The center of the cell will be cooler than the edges.
    C) Alternate cell centers will be hot and cold, with the edges at an intermediate temperature.
    D) The temperature will be uniform across the cell.
    Answer: A  Source: Section 9-1

18. Spectral lines observed in the granules seen at the center of the Sun's disk are
    A) split by the Zeeman effect due to the strong magnetic fields in the granule.
    B) blueshifted near the center of the granule and redshifted near the edge of the granule.
    C) redshifted near the center of the granule and blueshifted near the edge of the granule.
    D) always redshifted, because granules are caused by gas descending into the Sun.
    Answer: B Source: Section 9-1

19. What type of intensity distribution does one see at the edge or limb of the Sun at visible wavelengths?
    A) Limb brightening.
    B) An image of uniform brightness right to the limb.
    C) An image with uniform distribution except where active regions occur.
    D) Limb darkening.
    Answer: D Source: Section 9-1

20. When we view the Sun's disk in visible light, we see less deeply into the Sun near the limb than at the center of the disk because of interaction of the light with atoms of the gas. What conclusion can be drawn from the observation that the Sun appears less bright near the limb than it does at the disk center?
    A) The temperature of the gas increases with increasing height in the solar atmosphere.
    B) The light from the solar limb is redshifted because of solar rotation, and this gives the appearance of cooler gas at the limb.
    C) The light has to travel through more of the solar corona from the limb, hence it is reduced in intensity and appears cooler.
    D) The temperature of the gas falls with increasing height in the solar atmosphere.
    Answer: D Source: Section 9-1

21. What is the reason why the edge of the Sun's visible disk is darker than the center?
    A) We cannot see as deeply into the Sun near the edge and the gas is cooler there, so it emits less light.
    B) We cannot see as deeply into the Sun near the edge and the gas is less dense there, so it emits less light.
    C) The gases near the edge are in regions of stronger magnetic fields.
    D) We see more deeply into the Sun near its edge and the gas is cooler at these deeper levels.
    Answer: A Source: Section 9-1

22. The center of the disk of the visible Sun appears brighter than the edges because
    A) we see into deeper and cooler layers at the center of the solar disk.
    B) we see a greater contribution from the corona of the Sun at the center of the disk.
    C) we see into deeper and hotter layers at the center of the disk.
    D) cooler sunspots are more visible at the Sun's edge than they are at the center of the disk.
    Answer: C Source: Section 9-1

23. The Sun is about a thousand times more massive than the planet Jupiter. Why, then, does it have about the same average density as Jupiter?
    A) The Sun has a much weaker gravitational field than Jupiter.
    B) The Sun is composed of lighter elements than Jupiter.
    C) The Sun is rotating much faster than Jupiter, and is supported by the resulting centrifugal force.
    D) The Sun is much hotter than Jupiter, and is supported by the resulting high pressure.
    Answer: D Source: Section 9-8

24. What is the name of the layer of the Sun's atmosphere that appears as a pinkish ring just outside the visible disk of the Sun during a total solar eclipse?
    A) The convective zone.
    B) The corona.
    C) The chromosphere.
    D) The photosphere.
    Answer: C Source: Section 9-2

25. The word chromosphere refers to
    A) a light-emitting region just outside the event horizon of a black hole.
    B) a dense, spherical interstellar cloud of glowing gas.
    C) a layer in the Sun's atmosphere.
    D) a layer in the Earth's atmosphere, just below the ionosphere.
    Answer: C Source: Section 9-2

26. Where is the chromosphere on the Sun?
    A) It is the layer above the visible surface of the Sun.
    B) It is the visible surface of the Sun.
    C) It is the layer below the visible surface of the Sun.
    D) It is the outermost part of the Sun's atmosphere.
    Answer: A Source: Section 9-2

27. The visible light coming from the solar chromosphere is dominated by light at what wavelength(s)? (See Fig. 9-3, Kaufmann & Comins, *Discovering the Universe*, 5th Ed.)
   A) Continuous spectrum over all wavelengths from blue to red, scattered by chromospheric material.
   B) The green emission line from iron atoms that have lost 13 electrons, Fe XIV.
   C) The red hydrogen Balmer Ha emission line.
   D) Continuous spectrum over all wavelengths, crossed by numerous dark absorption lines.
   Answer: C Source: Section 9-2

28. What are the names of the three layers in the Sun's atmosphere, in order from lowest to highest?
   A) Photosphere, corona, chromosphere.
   B) Photosphere, chromosphere, corona.
   C) Chromosphere, photosphere, corona.
   D) Corona, chromosphere, photosphere.
   Answer: B Source: Sections 9-1, 2, and 3

29. What are the names of the three layers in the Sun's atmosphere, in order from highest to lowest?
   A) Photosphere, corona, chromosphere.
   B) Photosphere, chromosphere, corona.
   C) Chromosphere, photosphere, corona.
   D) Corona, chromosphere, photosphere.
   Answer: D Source: Sections 9-1, 2, and 3

30. What is a spicule on the Sun?
   A) A bright arc of gas suspended above the edge of the visible disk of the Sun.
   B) A long, thin, curved line of bright gas in the corona.
   C) A jet of rising gas in the chromosphere.
   D) A small, bright cell in the photosphere.
   Answer: C Source: Section 9-2

31. Spicules on the solar surface are
   A) intense eruptions from sunspot groups and active regions, associated with solar flares.
   B) jets of gas surging out of the photosphere of the Sun into the chromosphere, usually at supergranule boundaries.
   C) curtain-like structures hanging over sunspot regions.
   D) streams of solar coronal material, usually seen only during a total solar eclipse.
   Answer: B Source: Section 9-2

32. Where would you expect to find spicules?
    A) In binary star systems in which one star is a neutron star attracting and collecting mass from the other star.
    B) In the atmosphere of the Sun.
    C) In supernova remnants.
    D) In interstellar clouds heated by hot, massive stars.
    Answer: B Source: Section 9-2

33. Where do spicules tend to occur on the Sun?
    A) Randomly over the surface of the Sun.
    B) At the boundaries of granules.
    C) At the boundaries of supergranules.
    D) In the vicinity of sunspot groups.
    Answer: C Source: Section 9-2

34. What is the name of a jet of rising gas in the chromosphere of the Sun?
    A) A spicule B) A granule C) A prominence D) A flare
    Answer: A Source: Section 9-2

35. Material in solar spicules travels outwards in the solar atmosphere at typical speeds of 20 km/s. What would be the observed wavelength of the Balmer Ha hydrogen spectral line emitted by this gas compared to that from stationary solar material?
    A) 0.044 nm shorter than the Ha from stationary solar material.
    B) 0.000067 nm shorter than the Ha from the stationary solar material.
    C) 0.044 nm longer than the Ha from stationary solar material.
    D) There will be no shift since the light is emitted by hydrogen gas in both the spicule and the stationary solar material.
    Answer: A Source: Sections 9-2 and 4-7

36. Supergranules on the Sun are
    A) large areas in which the rapid convection of the gas destroys all granules that would otherwise form in that area.
    B) very large but otherwise ordinary granules.
    C) another name for large, long-lived sunspot groups.
    D) large areas of rising and falling gas containing hundreds of ordinary granules.
    Answer: D Source: Section 9-2

37. Where is the coolest region in the Sun?
    A) In the lower corona.
    B) In the photosphere.
    C) In the convective zone.
    D) In the lower chromosphere.
    Answer: D Source: Section 9-2

38. What name is given to the outer atmosphere of the Sun?
A) The convective zone.
B) The radiative zone.
C) The chromosphere.
D) The corona.
Answer: D Source: Section 9-3

39. What is the corona on the Sun?
A) The region above the Sun's north and south poles.
B) The Sun's inner atmosphere, just above the photosphere.
C) The large region beyond (outside of) the Sun's atmosphere.
D) The Sun's outer atmosphere.
Answer: D Source: Section 9-3

40. The total light emitted by the solar corona, which is seen most effectively during a total solar eclipse, is equivalent in brightness to
A) the average brightness of the Milky Way.
B) about one thousandth of that of the solar photosphere.
C) the average brightness of the night sky at a dark site.
D) the brightness of the full moon, about one millionth as bright as the solar photosphere.
Answer: D Source: Section 9-3

41. The visible corona of the Sun is most effectively photographed
A) in spring and fall seasons, because of the tilt of the spin axis of the Sun.
B) during lunar eclipses, when the sky is darker.
C) during solar eclipses.
D) at solar maximum periods, over a period of a few years.
Answer: C Source: Section 9-3

42. How did astronomers first detect the high temperatures in the corona of the Sun?
A) By observing emission lines of highly ionized elements, like iron.
B) By direct measurements using space probes.
C) By observing the effects the high temperature has on Mercury and Venus.
D) By measuring the brightness of the corona in visible (white) light.
Answer: A Source: Section 9-3

43. Which of the following features appear in the spectrum of the solar corona, and indicate very high gas temperatures?
A) Bright emission from the hydrogen Balmer line, H¯a, at the red end of the spectrum
B) Intense emission lines from highly ionized atoms, such as iron.
C) Dark absorption lines from H, Ca, and Fe on a continuous bright spectrum.
D) Intense continuous emission in the infrared part of the spectrum.
Answer: B Source: Section 9-3

44. Observations of the spectrum of the solar corona reveal emission lines that originate in atoms from which many electrons have been stripped. What conclusion can be drawn from this result?
   A) The magnetic field intensity is high enough to drag electrons from the atoms.
   B) The pressure of the gas is sufficient to squeeze the electrons from the atoms.
   C) The solar rotation speed at coronal height is such as to reduce the ability of atoms to retain electrons.
   D) The atomic collision energies and hence the gas temperatures are extremely high.
   Answer: D Source: Section 9-3

45. To what does the symbol Fe XIV refer?
   A) A compound of iron, xenon, iodine and vanadium.
   B) Iron atoms that have lost 14 electrons.
   C) Iron atoms that have lost 13 electrons.
   D) Iron atoms that have lost 15 electrons.
   Answer: C Source: Section 9-3

46. The temperature of the corona of the Sun
   A) is very hot, about $10^6$ K.
   B) is very cool, since it is furthest from the heat source.
   C) is about the same as that of the photosphere, 5800 K.
   D) is about twice as hot as the photosphere, 12,000 K.
   Answer: A Source: Section 9-3

47. What is the temperature of the solar corona?
   A) 5800 K.
   B) 2000 to 3000 K.
   C) 1,000,000 to 2,000,000 K.
   D) 50,000 to 100,000 K.
   Answer: C Source: Section 9-3

48. The corona of the Sun has a temperature that is
   A) about 1 to 2 million K.
   B) about the same as the photosphere, about 6000 K.
   C) noticeably less than the photosphere, about 1000 to 2000 K.
   D) about 10 K, since it merges with cold interstellar space.
   Answer: A Source: Section 9-3

49. One particular feature of the solar corona is
    A) its very cold temperature.
    B) its very uniform density and structure.
    C) its variation with time over periods of a few minutes.
    D) its very high temperature.
    Answer: D Source: Section 9-3

50. Why is the solar corona so much hotter than the photosphere?
    A) Energy is carried upward through the chromosphere by magnetic fields.
    B) The corona absorbs part of the light passing through it from the photosphere.
    C) Energy is carried upward through the chromosphere by convective gas motion.
    D) The corona is so hot simply because it is so much closer to the Sun than we are.
    Answer: A Source: Section 9-3

51. How much mass will the Sun lose to space during its lifetime, through the solar wind?
    A) Up to 25% of its total mass.
    B) A few thousandths of its total mass.
    C) Well over one-half of its total mass.
    D) A few millionths of its total mass.
    Answer: A Source: Section 9-3

52. What is the solar wind?
    A) The circulation of gases between the equator and the poles of the Sun.
    B) The constant flux of photons from the Sun's visible surface.
    C) The Sun's outer atmosphere streaming out into space.
    D) The storm of waves and vortices on the Sun's surface generated by a solar flare.
    Answer: C Source: Section 9-3

53. Sunspots are
    A) cooler, darker regions on the Sun's surface.
    B) hotter, deeper regions in the Sun's atmosphere.
    C) cooler regions of the Sun's high corona.
    D) the shadows of cool, dark curtains of matter, hanging above the solar surface.
    Answer: A Source: Section 9-4

54. Which one of the following is NOT considered to be a feature of the quiet Sun?
    A) A sunspot. B) A granule. C) A spicule. D) The solar wind.
    Answer: A Source: Section 9-4

55. Which of the following statements is NOT true for sunspots?
    A) They are cooler than the surrounding photosphere of the Sun.
    B) They increase and decrease in number, relatively regularly.
    C) They often occur in pairs of opposite magnetic polarity.
    D) They occur in regions of lower-than-average magnetic fields.
    Answer: D Source: Section 9-4

56. The major feature which distinguishes a sunspot from other regions on the Sun is
    A) much brighter emission of light from it.
    B) its very powerful magnetic field.
    C) a coronal hole existing above it.
    D) faster rotation around the Sun's axis than neighboring regions.
    Answer: B Source: Section 9-4

57. What is the structure of a typical large sunspot?
    A) A dark center surrounded by a less dark area.
    B) A roughly circular, dark region with a lighter central area.
    C) An irregular dark area of uniform darkness.
    D) Usually round and of uniform darkness.
    Answer: A Source: Section 9-4

58. What is the average length of time from one maximum in the number of sunspots on the Sun to the next maximum?
    A) 4 1/2 years. B) 11 years. C) 22 years. D) 7 years.
    Answer: B Source: Section 9-4

59. How does the number of sunspots on the Sun vary with time?
    A) Relatively regularly, with a period of about 11 years.
    B) They increase for about 11 years and then decrease again over the next 11 years.
    C) They vary irregularly, with no periodicity.
    D) They increase and decrease every year as Earth revolves around the Sun.
    Answer: A Source: Section 9-4

60. The 11 year sunspot cycle on the Sun is
    A) an extremely regular build-up and decay of the number of sunspots, with a precise period of 11.3 years.
    B) the regular movement of a relatively constant number of sunspots from the poles to the equator of the Sun over an 11-year period.
    C) a somewhat irregular but always present cycle of build-up and decay of sunspot numbers.
    D) an irregular cycle averaging about 11 years, which sometimes disappears entirely.
    Answer: D Source: Section 9-4

61. What is the lifetime of a typical sunspot?
    A) From a few hours to a few months.
    B) From a few years to a few decades.
    C) 11 years.
    D) Here today, gone tomorrow!
    Answer: A Source: Section 9-4

62. How does the temperature inside the umbra of a sunspot compare to the solar photosphere outside the sunspot?
    A) They are about the same temperature.
    B) The umbra is about 800 K hotter.
    C) The umbra is about 4000 K cooler
    D) The umbra is about 1500 K cooler.
    Answer: D Source: Section 9-4

63. The umbra of a sunspot is about 1500 K cooler than the surrounding solar photosphere. How will the light from the umbra compare to the light from the rest of the photosphere?
    A) The light from both will be the same color, but the umbra will emit less light per square meter.
    B) The light from the umbra will be redder.
    C) The light from the umbra will be bluer.
    D) Both will emit light of the same color and the same intensity (per square meter).
    Answer: B Source: Section 9-4

64. Sunspots are seen to be cooler than the rest of the Sun's surface, sometimes by as much as 1500 K. What would be the peak wavelength of the radiation from the sunspot when compared to that from the rest of the Sun?
    A) It would be at a shorter or longer wavelength, depending on the position of the spot.
    B) It would be at a longer wavelength.
    C) It would be the same, since the light still originates at the Sun.
    D) It would be at a shorter wavelength.
    Answer: B Source: Section 9-4

65. If sunspots are cooler than the photosphere (by at least 1000 K), what will be the peak wavelength in a sunspot spectrum compared with the peak wavelength of the photospheric spectrum?
   A) The same, since the chemical composition is the same (hydrogen and helium).
   B) Either shorter or longer, depending on the magnetic field strength in the sunspot.
   C) Shorter
   D) Longer
   Answer: D Source: Section 9-4

66. The average sunspot group on the solar surface will last for about
   A) one-half rotation of the Sun.
   C) one rotation of the Sun.
   B) two rotations of the Sun.
   D) one day.
   Answer: B Source: Section 9-4

67. Who, to our knowledge, first measured the rotation period of the Sun?
   A) Galileo B) Roemer C) Aristotle D) Ptolemy
   Answer: A Source: Section 9-4

68. Galileo observed the phenomenon of solar rotation in the early 1600s by
   A) measuring the Doppler shift of hydrogen spectral lines from the E and W limbs of the Sun.
   B) watching bright regions of hydrogen gas drift across the Sun.
   C) noting the periodic (monthly) variation of auroral disturbances or northern lights.
   D) measuring the motion of sunspots across the solar surface.
   Answer: D Source: Section 9-4

69. The rotation of the Sun is
   A) fastest at the equator, slowest at mid-latitudes, rising to intermediate speeds near the poles.
   B) slowest at the equator, faster at mid-latitudes, and fastest near the poles.
   C) fastest at the equator, slower at mid-latitudes, and slowest near the poles.
   D) fastest at mid-latitudes, slower at the equator, and slowest near the poles.
   Answer: C Source: Section 9-4

70. How can we characterize the rotation of the Sun?
   A) Differential rotation, with the equator rotating more slowly than the poles.
   B) Like a solid body (all parts rotating equally).
   C) In a banded pattern, with alternating bands of fast and slow rotation.
   D) Differential rotation, with the equator rotating faster than the poles.
   Answer: D Source: Section 9-4

71. What is the rotation period of the Sun?
    A) About one rotation per day.
    B) About two rotations per year.
    C) About one rotation per month.
    D) About four rotations per month.
    Answer: C Source: Section 9-4

72. The equatorial regions of the Sun are seen to rotate with an approximate period of
    A) about 15 d ays. B) about 1 year. C) about 33 days. D) about 25 days.
    Answer: D Source: Section 9-4

73. Why should you NEVER look directly at the Sun?
    A) It can lead to baldness.
    B) It is bad for the complexion.
    C) It uses up valuable time.
    D) Looking directly at the sun causes blindness.
    Answer: D Source: Section 9-4

74. What is the character of the sunspot cycle?
    A) Starting at sunspot minimum, spots first appear far from the equator, new spots appearing closer to the equator as they increase in number, and then die out close to the equator.
    B) Starting at sunspot minimum, new spots appear close to the equator, appear farther from the equator as they increase in number, then die out at high latitudes.
    C) Sunspots increase and decrease in number over 11 years, with no discernible dependence on latitude on the Sun.
    D) Starting at sunspot minimum, new spots appear uniformly over the Sun, then gradually become concentrated at mid-latitudes as they increase and then decrease in number.
    Answer: A Source: Section 9-4

75. Over the course of a sunspot cycle of about 11 years, the regions of sunspot occurrence on the Sun move
    A) equatorward, moving from 30° to 10° latitude.
    B) poleward, moving from 10° to 30° latitude.
    C) from the northern to the southern hemisphere or vice-versa, across the equator, within the range of ±30° latitude.
    D) equatorward, moving from the poles to the equator.
    Answer: A Source: Section 9-4

76. What is the Zeeman effect?
    A) The generation of magnetic fields by the motion of charged particles.
    B) The splitting of spectral lines from a light source in a magnetic field.
    C) The inability of charged particles to move perpendicular to magnetic field lines.
    D) The attraction of two areas of opposite magnetic polarity toward each other.
    Answer: B Source: Section 9-5

77. What is the Zeeman effect?
    A) When the temperature of a light source is increased, the wavelength of maximum emission decreases.
    B) When light is shone onto a metal surface, electrons are ejected from the metal only if the wavelength of the light is shorter than some critical wavelength.
    C) When a light source is located in a magnetic field, the spectral lines it emits are split into two or more components.
    D) When a light source is moving relative to an observer, the wavelengths of its spectral lines are shifted to longer or shorter wavelengths.
    Answer: C Source: Section 9-5

78. The Zeeman effect describes the shift in wavelength of light caused by
    A) relative motion of the source and observer.
    B) magnetic fields acting on the radiating atoms.
    C) the extreme mass and gravitational field of the source.
    D) the light passing through a transparent medium.
    Answer: B Source: Section 9-5

79. The Zeeman effect describes what change in spectral lines?
    A) Splitting of lines because the atoms are within an intense magnetic field.
    B) Their shift because of the movement of the source.
    C) Broadening associated with high temperatures.
    D) The change in relative intensity of different lines from sources of different temperature.
    Answer: A Source: Section 9-5

80. The Zeeman effect refers to
    A) the splitting of spectral lines when magnetic fields are applied to atoms.
    B) the drift of sunspots across the solar disk.
    C) the brightening of sunlight near sunspots.
    D) the shift of spectral lines because of solar rotation.
    Answer: A Source: Section 9-5

81. Intense magnetic fields have been found to exist in sunspots by the observation of what specific physical effect?
   A) The Doppler shift of light from sunspots.
   B) The Zeeman effect, the splitting of spectral absorption lines.
   C) The measurement of relative strengths of spectral absorption lines of gases in the sunspot.
   D) The observation of ionized atoms in the region of the sunspots.
   Answer: B  Source: Section 9-5

82. The strength of the magnetic field in a sunspot is estimated from Earth by
   A) measuring the shape of structures seen in the corona above sunspots during solar eclipses.
   B) measuring the size and brightness of the sunspot.
   C) observing the wavelength splitting of atomic spectral lines by the Zeeman effect.
   D) measuring the Doppler shift of spectral lines of light emitted from above the sunspot.
   Answer: C  Source: Section 9-5

83. In 1908, Hale noticed that the spectral lines from sunspots are split into closely spaced components, and concluded that the magnetic fields in the sunspots must be very strong. What is the name of the effect Hale was observing?
   A) The Zeeman effect.
   B) The photoelectric effect.
   C) The Wilson-Bappu effect.
   D) The Doppler effect.
   Answer: A  Source: Section 9-5

84. An astronomer observing certain regions of the Sun through a spectroscope notices that the spectral lines emitted from these regions are split into two or more components. From this observation, what can the astronomer conclude about these regions?
   A) They contain fast-moving gas.
   B) They contain strong magnetic fields.
   C) They are very hot.
   D) They contain strong gravitational fields.
   Answer: B  Source: Section 9-5

85. What is a typical magnetic field strength inside a sunspot?
   A) A million times stronger than the Earth's magnetic field.
   B) A few times stronger than the Earth's magnetic field.
   C) A few thousand times stronger than the Earth's magnetic field.
   D) 1/100 of the strength of the Earth's magnetic field.
   Answer: C  Source: Section 9-5

86. Which of the following statements is NOT true for sunspots?
    A) They increase and decrease in number, relatively regularly.
    B) They occur in regions of lower-than-average magnetic fields.
    C) They often occur in pairs of opposite magnetic polarity.
    D) They are cooler than the surrounding photosphere of the Sun.
    Answer: B Source: Section 9-5

87. The major feature that distinguishes a sunspot from other regions on the Sun is
    A) the much brighter emission of light from it.
    B) its very powerful magnetic field.
    C) a coronal hole existing above it.
    D) faster rotation around the Sun's axis than neighboring regions.
    Answer: B Source: Section 9-5

88. What is the average length of a complete solar cycle of sunspots and magnetic fields?
    A) 4.5 years. B) 22 years. C) 7 years. D) 11 years.
    Answer: B Source: Section 9-5

89. Solar magnetic activity at the present time in history seems to vary almost periodically with a time-scale of
    A) 100 years. B) 22 years. C) 5 minutes. D) 2 years.
    Answer: B Source: Section 9-5

90. The Maunder minimum refers to
    A) a period of 70 years during which essentially no sunspots were seen.
    B) the minimum in the number of sunspots which occurs roughly every 11 years.
    C) the region of lowest temperature in the photosphere, between the decreasing temperature in the solar interior and the increasing temperature in the chromosphere.
    D) the unexpectedly low number of neutrinos observed from the Sun.
    Answer: A Source: Section 9-5

91. Mean temperatures in Europe during the Maunder minimum, when virtually no sunspots were seen between 1645 and 1715, were
    A) unchanged, within statistical uncertainty.
    B) lower than average.
    C) more strongly fluctuating above and below average compared to the period from 1715 to present.
    D) higher than average.
    Answer: B Source: Section 9-5

92. Which recently discovered fact about the Sun might have some bearing on climate changes and the weather on Earth?
    A) The Sun's overall energy output depends upon the 11-year sunspot cycle.
    B) Solar wind seems to originate in cooler regions of the corona, the coronal holes.
    C) There are far less neutrinos emitted from the Sun than are predicted.
    D) The Sun's surface is oscillating up and down every 5 minutes.
    Answer: A Source: Section 9-5

93. How does the Sun's overall magnetic field behave?
    A) The northern and southern hemispheres have the same magnetic polarity, and this polarity reverses every 11 years.
    B) The northern and southern hemispheres have opposite magnetic polarity, and this polarity reverses every 11 years.
    C) The poles of the Sun have the opposite magnetic polarity from the equator, and this polarity reverses every 11 years.
    D) Magnetic polarity is randomly distributed over the Sun, while the strength of the magnetic field increases and decreases with an 11 year cycle.
    Answer: B Source: Section 9-5

94. What causes sunspots?
    A) Differential rotation on the Sun creates vortices, or eddies, which are cooler and darker than the rest of the solar surface.
    B) Solar flares cause the photosphere to expand and cool in the vicinity of the flare.
    C) Magnetic fields breaking through the photosphere inhibit gas motion where the field is strong.
    D) Magnetic fields below the photosphere pull gas down, creating holes in the photosphere.
    Answer: C Source: Section 9-5

95. Sunspots are caused by
    A) the impact of meteoroids and comets on the solar surface.
    B) coronal holes, darkening the surface.
    C) differential rotation and its effect on weak magnetic fields just under the solar surface.
    D) dark clouds hanging over the surface, above the magnetic field regions.
    Answer: C Source: Section 9-5

96. What is the cause of the sunspot cycle on the Sun?
    A) Differential rotation on the Sun creates eddies, or vortices, which cool the photosphere and create sunspots, then the eddies gradually cancel each other out and the cycle starts over.
    B) Comets crashing into the Sun cool the photosphere and create sunspots. The 11-year sunspot cycle is the result of an 11-year periodicity in the flux of comets.
    C) Subsurface magnetic fields are twisted by the Sun's differential rotation and break through the surface as sunspots, then gradually cancel each other and return below the surface.
    D) Subsurface magnetic fields are concentrated by the Sun's differential rotation. These fields remain under the surface and prevent heat from reaching the photosphere, creating sunspots. The concentrated fields gradually cancel each other out, and the cycle starts over.
    Answer: C Source: Section 9-5

97. What is a plage?
    A) A bright area on the photosphere.
    B) A region in the corona that looks bright against the darkness of space but dark against the brighter photosphere.
    C) A bright area in the chromosphere.
    D) A sudden eruption on the photosphere, in the vicinity of sunspot groups.
    Answer: C Source: Section 9-6

98. What causes plages?
    A) Compression and heating of chromospheric gas by magnetic fields.
    B) Blobs of convecting gas rising through the photosphere and depositing their energy in the chromosphere.
    C) Matter descending along loops of magnetic field in the corona.
    D) Sunspots preventing heat from reaching the chromospheric gases above them.
    Answer: A Source: Section 9-6

99. What name is given to a brighter region in the chromosphere, often in association with a sunspot?
    A) Plage. B) Granule. C) Filament. D) Prominence.
    Answer: A Source: Section 9-6

100. What name is given to a dark line or streak often seen in photographs of the solar photosphere, and often observed in association with sunspots?
    A) A spicule. B) A filament. C) A granule. D) A plage.
    Answer: B Source: Section 9-6

101. What is the name of a large loop of bright gas extending outwards from the edge of the Sun (often seen during total solar eclipses)?
A) A spicule. B) A filament. C) A prominence. D) A plage.
Answer: C Source: Section 9-6

102. Which of the following phenomena on the Sun do NOT appear to be sources of particles traveling out into the solar system from the Sun?
A) Spicules. B) Coronal holes. C) Flares. D) Eruptive prominences.
Answer: A Source: Section 9-6

103. An arching column of gas suspended over a sunspot group is called a
A) spicule. B) coronal hole. C) flare. D) prominence.
Answer: D Source: Section 9-6

104. What is a filament on the Sun?
A) A sunspot which has been stretched by solar differential rotation.
B) A spicule seen in profile near the edge of the Sun's limb.
C) A prominence seen in silhouette against the photosphere.
D) A plage near the end of its life, when it is fading away.
Answer: C Source: Section 9-6

105. What is a prominence on the Sun?
A) A loop of gas supported by magnetic fields.
B) A jet of gas shot upward from an emerging sunspot.
C) A shock wave created by the eruption of a solar flare.
D) Another name for a plage.
Answer: A Source: Section 9-6

106. What is the name of a sudden eruptive surge on the surface of the Sun?
A) A sunspot. B) A prominence. C) A plage. D) A flare.
Answer: D Source: Section 9-6

107. How long does a typical flare last on the Sun?
A) 5 days. B) 20 minutes. C) 40 seconds. D) 2 months.
Answer: C Source: Section 9-6

108. Solar flares occur at what positions on the solar disk?
 A) Only within sunspot groups.
 B) Only in a narrow band along the solar equator.
 C) Only in coronal holes.
 D) Only at the polar regions.
 Answer: A Source: Section 9-6

109. Solar flares, the violent eruptive events on the Sun, occur most frequently
 A) in or above complex sunspot groups.
 B) over single isolated but large sunspots.
 C) along the solar equator at positions aligned with Jupiter's position, being caused by tidal disturbance on the Sun.
 D) within solar coronal holes, from that the solar wind originates.
 Answer: A Source: Section 9-6

110. What is the source of the x-ray emitted by the solar corona?
 A) High-energy charged particles spiraling along the coronal magnetic fields.
 B) Radioactivity in the coronal gases.
 C) X-ray from the solar photosphere scattered from ions in the corona.
 D) The high temperature gas of the corona.
 Answer: D Source: Section 9-6

111. The extremely high gas temperatures in the solar corona mean that this region is best observed at wavelengths of
 A) visible light.
 B) infrared light.
 C) x-ray.
 D) Balmer Ha light from hydrogen gas.
 Answer: C Source: Section 9-6

112. The bright x-ray image which one obtains of the solar corona when the Sun is photographed at this wavelength indicates that the gas temperature at these heights is
 A) extremely low, much cooler than the photosphere.
 B) extremely high, above $10^6$ K.
 C) about the same temperature as the photosphere.
 D) about twice that of the photosphere.
 Answer: B Source: Section 9-6

113. How do the physical conditions within a coronal hole compare with those of the rest of the solar corona?
A) Lower temperature, but very high density.
B) High temperature, but very low density.
C) Higher temperature and higher density.
D) Lower temperature and lower density.
Answer: D Source: Section 9-6

114. A coronal hole shows up most prominently on what kind of photograph?
A) x-ray.
B) any wavelength of visible light.
C) red light of the first Balmer line of hydrogen, H¯a, at 656.3 nm.
D) radio wavelengths.
Answer: A Source: Section 9-6

115. What is the source of the solar wind?
A) Gas escaping through coronal holes.
B) Gas flung out from solar flares.
C) Gas flung out from the Sun's equatorial region by the centrifugal force due to the Sun's rotation.
D) Gas escaping from x-ray-bright regions of the solar corona.
Answer: A Source: Section 9-6

116. Coronal holes are thought to be the source of
A) fluctuation in human behavior (e.g., astrology).
B) the solar wind.
C) powerful loops of magnetic field, linked to active regions.
D) dust released from the Sun.
Answer: B Source: Section 9-6

117. The solar wind is
A) the name for the electromagnetic radiation coming from the Sun.
B) a violent explosive expansion of specific regions of the Sun's atmosphere at certain times.
C) a gentle outflow of solar material, mostly protons and electrons, which is always moving outward from the Sun.
D) the inflow of matter onto the Sun under gravitational attraction.
Answer: C Source: Section 9-6

118. The solar wind is made up primarily of
   A) hydrogen nuclei and electrons, some He nuclei.
   B) equal numbers of all light elements, up to oxygen, no electrons.
   C) electrons and He nuclei (the "ash" of nuclear fusion).
   D) hydrogen nuclei, with very few electrons.
   Answer: A Source: Section 9-6

119. The solar wind appears to originate mainly from which regions of the Sun?
   A) Coronal holes
   B) Granulation cells
   C) Sunspots
   D) All over the surface, with no preferred location
   Answer: A Source: Section 9-6

120. The solar wind originates primarily
   A) in flare explosions.
   B) in sunspots.
   C) near the solar equator, where solar spin reduces the gravitational field.
   D) in coronal holes, which are cooler regions in the corona.
   Answer: D Source: Section 9-6

121. The solar wind is ionized gas flowing outward
   A) only from solar flares.
   B) from sunspots.
   C) primarily through coronal holes.
   D) more or less uniformly from the entire solar surface.
   Answer: C Source: Section 9-6

122. What is nuclear fusion?
   A) A nucleus changing to become the nucleus of a different element by emitting an electron and a neutrino.
   B) Two nuclei sticking together to form a new, heavier nucleus.
   C) A heavy nucleus splitting apart to form two lighter nuclei.
   D) The removal of electrons from atoms to form ions.
   Answer: B Source: Section 9-7

123. Nuclear fusion is
   A) the splitting of heavier nuclei to produce lighter nuclei and energy.
   B) the combining together of hydrogen atoms to produce hydrogen molecules, $H_2$, and energy.
   C) the combining of electrons with nuclei to produce atoms and release energy.
   D) the combining together of light nuclei (e.g., hydrogen) to produce heavier nuclei (e.g., helium) and energy.
   Answer: D Source: Section 9-7

124. What is the energy source for the Sun?
   A) Primordial heat left over from when the Sun first formed.
   B) Radioactivity.
   C) Thermonuclear fusion in the core.
   D) Heat released by gravitational contraction.
   Answer: C Source: Section 9-7

125. What process provides the power for the Sun?
   A) The fission of uranium to form lead.
   B) The fusion of helium into carbon.
   C) The fusion of hydrogen into helium.
   D) The emission of neutrinos.
   Answer: C Source: Section 9-7

126. The Sun's source of energy at the present time is thought to be
   A) gravitational contraction.
   B) thermonuclear fusion (combination) of hydrogen atoms.
   C) the chemical burning of hydrogen gas with oxygen.
   D) thermonuclear fission (splitting) of heavy elements into hydrogen.
   Answer: B Source: Section 9-7

127. By how much does the mass of the Sun decrease each second because of the energy radiated from it (its luminosity)? (See the discussion in Section 9-7 of Kaufmann & Comins, *Discovering the Universe*, 5th Ed.)
   A) $3.9 \times 10^{26}$ kg. B) $6.0 \times 10^{11}$ kg. C) $4.2 \times 10^{9}$ kg. D) $2.0 \times 10^{7}$ kg.
   Answer: C Source: Section 9-7

128. At the present time, the energy of the Sun is generated
   A) in its central core only, by fusion of hydrogen nuclei.
   B) in its central core only, by fission of heavy nuclei.
   C) throughout the Sun, by gravitational contraction.
   D) in the central core by fusion of helium nuclei, and in an outer shell by fusion of hydrogen nuclei.
   Answer: A Source: Section 9-7

129. The dominant energy source that powers the Sun at the present time is
A) release of energy trapped in strong magnetic fields.
B) thermonuclear fusion of helium into heavier elements in the core.
C) release of gravitational energy as the Sun slowly contracts.
D) thermonuclear fusion of hydrogen into helium in the core.
Answer: D Source: Section 9-7

130. From which fusion reaction does the Sun derive its power?
A) 2H Þ He. B) 4H Þ He. C) 4He Þ O. D) 3He Þ C.
Answer: B Source: Section 9-7

131. In the thermonuclear process that is thought to heat the Sun, the nuclei of which chemical elements are converted to other nuclei to produce the requisite energy?
A) Hydrogen converted to helium.
B) Iron, in a chain reaction, leading eventually to hydrogen.
C) Helium converted to hydrogen.
D) Uranium converted to lead.
Answer: A Source: Section 9-7

132. Hydrogen "burning" by fusion reactions occurs only in the deep interior of the Sun (and other stars), because
A) only in the core is the temperature low enough, and the density high enough, to allow fusion to occur.
B) this is the only place in the Sun where there is sufficient hydrogen.
C) the requisite conditions of high temperature and high density only occur there.
D) the density is sufficiently low there that hydrogen atoms can collide with each other often enough.
Answer: C Source: Section 9-7

133. A hydrogen nucleus (a proton) has a charge of +1 and a helium nucleus has a charge of +2. Why, then, does it require four protons to form helium in the core of the Sun?
A) Two of the protons become neutrons.
B) Two helium nuclei are formed from the four protons.
C) Two of the protons are ejected back into the solar material.
D) Two of the protons are converted into neutrinos.
Answer: A Source: Section 9-7

134. What is a positron?
A) The nucleus of a hydrogen atom.
B) The nucleus of a helium atom.
C) A positive electron.
D) A chargeless, massless particle.
Answer: C Source: Section 9-7

135. A positron is
   A) a hydrogen nucleus.
   B) a positively charged electron.
   C) a charged neutron.
   D) a positively charged neutrino.
   Answer: B Source: Section 9-7

136. A positron is
   A) an anti-electron, the same as a normal electron except for its positive charge.
   B) another name for a proton, or a hydrogen nucleus.
   C) a charged neutron.
   D) a positively charged neutrino, having positive charge and very small or zero mass.
   Answer: A Source: Section 9-7

137. What happens to the positrons produced by the nuclear reactions in the core of the Sun?
   A) They combine with neutrons to form protons.
   B) They collide and stick together to form helium.
   C) They collide with electrons, producing energy.
   D) They escape from the Sun into space.
   Answer: C Source: Section 9-7

138. When four protons collide to form helium, what fraction of the original mass of the protons is converted to energy?
   A) 3/4 of a percent. B) 4%. C) 1/20 of a percent. D) 100%.
   Answer: A Source: Section 9-7

139. Which of the following physical products are not produced by the Sun during the thermonuclear processes in which hydrogen nuclei are combined together in its core?
   A) Neutrinos.
   B) Positive electrons, or positrons.
   C) Helium nuclei.
   D) Heavy nuclei such as uranium.
   Answer: D Source: Section 9-7

140. Apart from the helium nuclei and energy that are produced in thermonuclear reactions between protons in the center of the Sun, what are the other by-products?
   A) Positive electrons (positrons), gamma rays and neutrinos.
   B) Gamma rays and neutrinos.
   C) Protons, neutrinos and negative electrons.
   D) Gamma rays, negative electrons and neutrinos.
   Answer: A Source: Section 9-7

141. Thermonuclear fusion reactions in the core of the Sun convert four hydrogen atoms into one helium atom. The helium atom has
    A) more mass than the four hydrogen atoms, because energy is produced in the reaction, and the energy adds mass by $E = mc^2$.
    B) an undetermined amount of mass, depending on the temperature at which the reaction occurs.
    C) the same mass as the four hydrogen atoms, because the mass of any product has to equal the sum of the mass of its parts.
    D) less mass than the four hydrogen atoms, because the energy produced is lost from the atom, and energy is equivalent to mass by $E = mc^2$.
    Answer: D  Source: Section 9-7

142. The total time that the Sun will spend converting hydrogen to helium in its core is
    A) about 4.5 million years.
    B) about 1 million years.
    C) about 10 billion years ($10^{10}$ years).
    D) at least 200 billion years ($2 \times 10^{11}$) years.
    Answer: C  Source: Section 9-7

143. How much longer can the Sun continue to generate energy by nuclear reactions in its core?
    A) About 5 billion years.
    B) About 5 million years.
    C) About 50 billion years.
    D) About 500,000 years.
    Answer: A  Source: Section 9-7

144. Approximately where is the Sun in terms of its total lifetime?
    A) It is about 1/4 of the way through its life.
    B) It is about 1/10 of the way through its life.
    C) It is about 3/4 of the way through its life.
    D) It is about half way through its life.
    Answer: D  Source: Section 9-7

145. How much matter is converted to energy in the Sun each second?
    A) 600 tons.  B) 6 billion tons.  C) 6 tons.  D) 600 million tons.
    Answer: D  Source: Section 9-7

146. Who first postulated that hydrogen fusion reactions are responsible for generating the energy produced by the Sun?
    A) George Gamow
    B) Arthur Eddington
    C) Albert Einstein
    D) Stephen Hawking
    Answer: B  Source: Section 9-7

147. What stops the Sun from collapsing under the force of its own gravity?
   A) Ions and electrons are pushed apart by the electric forces between their charges.
   B) Neutrinos from the core collide with gas atoms and prevent them from falling inwards.
   C) It is held up by gas pressure due to the very high temperature inside it.
   D) The interior of the Sun is under such high pressure that it is a liquid, and liquids are incompressible.
   Answer: C Source: Section 9-8

148. To what do the words "hydrostatic equilibrium" in the Sun refer?
   A) The balance of gravity inward and gas pressure outward.
   B) The balance of gas pressure inward and heat outward.
   C) The balance of gas pressure outward and magnetic forces inward.
   D) The creation of one helium nucleus for the "destruction" of every four hydrogen nuclei.
   Answer: A Source: Section 9-8

149. The Sun has existed for a very long time without change in its appearance or behavior. This means that it must be in hydrostatic equilibrium. Under these conditions, which two parameters must be in balance within the Sun?
   A) Hydrogen gas pressure and helium gas pressure.
   B) The magnetic field and the force of gravity.
   C) The force of gravity and outward gas pressure.
   D) The numbers of hydrogen and helium nuclei.
   Answer: C Source: Section 9-8

150. Any object will collapse under its own weight unless something stops it. In an ordinary star like the Sun, this collapse is prevented by
   A) the rotation of the star, generating centrifugal force.
   B) turbulence and up-welling in the atmosphere of the star.
   C) gas pressure inside the star.
   D) the star's solid core.
   Answer: C Source: Section 9-8

151. The energy transfer process that operates in the Sun via mass motion is known as
   A) conduction. B) convection. C) radiation. D) thermonuclear fusion.
   Answer: B Source: Section 9-8

152. What is the dominant mechanism by which energy is transported through the core of the Sun?
 A) By neutrinos streaming outwards through the Sun's material (particle transport of energy).
 B) By collisions of faster-moving particles with slower-moving particles (conductive transport of energy).
 C) By hotter gas rising and cooler gas falling (convective transport of energy).
 D) By photons (radiative transport of energy).
 Answer: D Source: Section 9-8

153. What is the dominant mechanism by which energy is transported through the outer regions of the solar interior?
 A) By hotter gas rising and cooler gas falling (convective transport of energy).
 B) By neutrinos streaming outwards through the Sun's material (particle transport of energy).
 C) By photons (radiative transport of energy).
 D) By collisions of faster-moving particles with slower-moving particles (conductive transport of energy).
 Answer: A Source: Section 9-8

154. Of the three ways in which heat energy is transmitted from one place to another (radiation, conduction, convection), which is (or are) important in the solar interior below the surface of the Sun?
 A) Radiation alone, by photons of energy.
 B) Conduction, convection and radiation.
 C) Radiation and convection.
 D) Conduction and convection.
 Answer: C Source: Section 9-8

155. Of the three ways in which energy is transported in nature, which two are important in the Sun?
 A) Radiation and convection.
 B) Convection and conduction.
 C) The statement is wrong; all three are equally important in the Sun.
 D) Radiation and conduction.
 Answer: A Source: Section 9-8

156. The mechanism at work when energy is transmitted by convection is
 A) the successive exchange of radiant energy between atoms.
 B) the passage of radiation through a gas at the speed of light.
 C) the mass motion of hot gases.
 D) the fusion of hydrogen nuclei into helium nuclei.
 Answer: C Source: Sections 9-8

157. The average time taken for energy generated by thermonuclear fusion in the center of the Sun to reach the surface layers and escape is calculated to be
A) about a hundred thousand years.
B) about 1 year.
C) only a few seconds.
D) about ten million years.
Answer: A Source: Section 9-8

158. Energy is transported from the center of the Sun to the surface by
A) mostly convection; radiation only in the outer layers.
B) convection in the thermonuclear core; radiation everywhere else.
C) mostly radiation; convection only in the outer layers.
D) radiation in the thermonuclear core; convection everywhere else.
Answer: C Source: Section 9-8

159. Convection currents in the Sun's interior occupy what fraction of the Sun's radius?
A) Only the inner 10% of the radius, at the core.
B) The inner 80% of the radius.
C) The outer 20% of the radius.
D) The whole radius.
Answer: C Source: Section 9-8

160. The order of the layers or parts of the Sun, as radius increases, is
A) radiative zone, convection zone, corona, chromosphere, photosphere.
B) corona, chromosphere, convection zone, photosphere, radiative zone.
C) radiative zone, convection zone, chromosphere, photosphere, corona.
D) radiative zone, convection zone, photosphere, chromosphere, corona.
Answer: D Source: Section 9-8

161. The temperature at the center of the Sun, where thermonuclear processes take place, is approximately
A) 1.5 million K. B) 4500 K, as shown by sunspots. C) 6000 K. D) $1.5 \times 10^7$ K.
Answer: D Source: Section 9-8

162. The temperature of the Sun throughout its radius and including its atmosphere
A) decreases outward from the center, but then increases again in the atmosphere.
B) is almost constant from the center to the surface, but falls abruptly above the visible surface.
C) increases and decreases several times between the center and the surface, then decreases through the atmosphere.
D) decreases smoothly outward from the center, gradually merging into the cold of the interplanetary medium.
Answer: A Source: Section 9-8

163. Where is most of the mass of the Sun concentrated?
   A) In the convective zone.
   B) In the photosphere.
   C) It is spread uniformly through the Sun.
   D) In the inner core.
   Answer: D Source: Section 9-8

164. One method that has been used successfully in recent times to investigate the deep interior of the Sun has been to observe
   A) the progress of a solar-impacting comet.
   B) regular 5-minute oscillations and fluctuations of the surface.
   C) the deep atmospheric conditions, as detected by a spacecraft that entered the solar atmosphere.
   D) the spectrum and behavior of a sunspot.
   Answer: B Source: Section 9-8

165. What is a neutrino?
   A) A positive electron.
   B) An uncharged electron.
   C) A chargeless, massless particle.
   D) An uncharged proton.
   Answer: C Source: Section 9-8

166. The neutrino is
   A) an elusive, subatomic particle having little or no mass, and difficult to detect.
   B) a heavy, uncharged nuclear particle, easily detected.
   C) a very small asteroid-like body orbiting the Sun.
   D) another name for an electron that carries a positive charge instead of a negative charge.
   Answer: A Source: Section 9-8

167. The neutrino is
   A) a tiny particle that interacts very weakly with matter, with extremely low or zero mass and no charge.
   B) another name for the neutron, a component of almost all atomic nuclei, with a mass close to that of the proton and no charge.
   C) another name for a photon of very high energy, i.e., short wavelength electromagnetic radiation with great penetrating power.
   D) a massive but very elusive nuclear particle that carries most of the energy generated in the core of the Sun to the surface, but that then decays to release electromagnetic radiation (i.e., light).
   Answer: A Source: Section 9-8

168. What happens to the neutrinos produced by the nuclear reactions in the core of the Sun?
   A) They collide and stick together with protons to form helium nuclei.
   B) They combine with protons to form neutrons.
   C) They collide with electrons, producing energy.
   D) They escape from the Sun into space.
   Answer: D Source: Section 9-8

169. A neutrino produced in the nuclear furnace in the core of the Sun
   A) can penetrate easily through the gas of the Sun's interior and through the solid Earth.
   B) can penetrate easily through the Sun's gaseous interior but will be stopped by the surface of the solid Earth.
   C) can penetrate easily through the Sun's interior but will be deflected away from the Earth by its magnetic field.
   D) can penetrate easily through both the gaseous Sun's interior and the solid Earth, but will be easily stopped by chemicals containing chlorine.
   Answer: A Source: Section 9-8

170. The time taken for neutrinos generated in the thermonuclear reactions at the center of the Sun to escape from its surface is
   A) about 1 million years.
   B) about 1 year.
   C) instantaneous, since they travel faster than the speed of light.
   D) a few seconds.
   Answer: D Source: Section 9-8

171. How many neutrinos from the Sun pass through each square inch of your body every second?
   A) 100 million. B) 100 billion. C) a few hundreds. D) 100 trillion.
   Answer: B Source: Section 9-8

172. Which of the following particles or types of radiation will provide the most direct information on the processes of nuclear fusion that are occurring in the solar core?
   A) X-rays from the solar corona.
   B) Visible light from the photosphere.
   C) Protons in the solar wind and from solar flares.
   D) Neutrinos.
   Answer: D Source: Section 9-8

173. Which technique has been used for the past 20 years to attempt to measure solar neutrinos?
   A) The interactions of neutrinos with radioactive argon nuclei to produce chlorine nuclei that are then measured chemically.
   B) Production of radioactive argon nuclei by neutrino interaction with chlorine nuclei in deep underground tanks.
   C) The proton-proton reaction in which neutrinos play an intermediate role.
   D) The interaction of neutrinos with water molecules in huge underground tanks.
   Answer: B Source: Section 9-8

174. The solar neutrino experiment designed by Raymond Davis has detected a rate of solar neutrinos arriving at the Earth which is
   A) less than 1% of the predicted rate.
   B) almost exactly equal to the predicted rate.
   C) almost double the predicted rate.
   D) about 1/3 of the predicted rate.
   Answer: D Source: Section 9-8

175. What problem have observers of solar neutrinos run into?
   A) The neutrinos are of the wrong type (muon neutrinos, instead of the predicted electron neutrinos).
   B) The neutrinos are about twice as energetic on average than is predicted by theoretical models of the Sun.
   C) Only about 1/3 of the expected number of neutrinos is observed, compared to theoretical models of the Sun.
   D) About six times as many neutrinos are observed than expected from theoretical models of the Sun
   Answer: C Source: Section 9-8

176. Which of the following lines of research is NOT used in the study of the solar interior?
   A) Visible light spectroscopy observing the hydrogen Balmer series.
   B) Observations of surface oscillations on the Sun.
   C) Theoretical studies, using physics.
   D) Neutrino astronomy.
   Answer: A Source: Section 9-8

# Build Your Foundation III: The Stars

1. How far away is the nearest star beyond the Sun?
   A) Between 1 and 2 light years away.
   B) About 4 light years away.
   C) About 1/4 light year away
   D) About 1/10 light year away.
   Answer: B Source: Foundations III-Introduction

2. How far away is the nearest star beyond the Sun?
   A) Between 1/2 and 1 parsec away.
   B) About 12 parsecs away.
   C) Between 1 and 2 parsecs away.
   D) About 4 parsecs away.
   Answer: C Source: Foundations III-Introduction

3. What is parallax?
   A) The shift in position of an object as it moves.
   B) The apparent shift in position of an object as we move.
   C) The angle taken up by the size (e.g., diameter) of an object, as seen by us.
   D) The distance to an object, measured in parsecs.
   Answer: B Source: Foundations III-1

4. Stellar parallax is
   A) the circular or elliptical motion of a star in a binary system, as the two stars orbit around each other.
   B) the apparent shift that we see in the position of a nearby star as we orbit the Sun.
   C) the difference between the apparent magnitude and the absolute magnitude of a star.
   D) the apparent change in the distance to a star if its light is dimmed by passing through interstellar clouds.
   Answer: B Source: Foundations III-1

5. As you drive along a road, trees in the middle distance seem to shift in position relative to far-away hills. What name is given to this phenomenon?
   A) The inverse-square law. B) The Doppler effect. C) Parallax. D) Perspective.
   Answer: C Source: Foundations III-1

6. How many stars (other than the Sun) have an angle of parallax greater than one second of arc?
   A) Over a hundred. B) Eight. C) None. D) Only one.
   Answer: C Source: Foundations III-1

7. How is stellar parallax defined?
   A) The angle subtended by the diameter of the Earth's orbit as seen from the star.
   B) The angle through which a star moves in our sky in one year, due to the motion of both the star and the Earth.
   C) The angle subtended by the radius of the Earth's orbit as seen from the star.
   D) The angle taken up by the diameter of a star as seen from the Earth.
   Answer: C Source: Foundations III-1

8. Parallax of a nearby star is used to estimate its
   A) apparent magnitude.
   B) physical size or diameter.
   C) distance from Earth.
   D) surface temperature.
   Answer: C Source: Foundations III-1

9. Which of the following properties of nearby stars is determined by a measurement of stellar parallax?
   A) Rotation periods.
   B) Spectral type and surface temperature.
   C) Distance from Earth.
   D) Apparent magnitude.
   Answer: C Source: Foundations III-1

10. The most straightforward way to determine the distance to a nearby star involves the measurement of
    A) stellar parallax.
    B) magnitude differences.
    C) the Zeeman effect.
    D) the star's spectrum.
    Answer: A Source: Foundations III-1

11. How can we tell that some stars are relatively close to us in the sky?
    A) Because they appear to move periodically back and forth against the background stars because of the Earth's movement around the Sun.
    B) Because they appear to be extremely bright and must therefore be very close to us.
    C) Because they are occasionally occulted or eclipsed by our Moon, hence they must be close.
    D) Because the light from these stars shows only a very small redshift caused by the universal expansion of the universe, so they must be close.
    Answer: A Source: Foundations III-1

12. Why is there such a thing as stellar parallax?
    A) Because stars move in space.
    B) Because stars have size (they are not really just points of light).
    C) Because the Earth rotates about its own axis.
    D) Because the Earth moves in space.
    Answer: D Source: Foundations III-1

13. The apparent motion against the background sky, as a result of Earth's motion through 1 AU, of a star whose distance from the Sun is 80 pc, (its stellar parallax), is
A) 0.0125 arc second.
B) 0.0125 radian or 0.72°.
C) 80 arc seconds.
D) 0.0125 arc minute.
Answer: A Source: Foundations III-1

14. What is the relationship between stellar parallax (p) measured in seconds of arc and distance (d) measured in parsecs?
A) $d = 1/p^2$ B) $d = p^2$ C) $d = p$ D) $d = 1/p$
Answer: D Source: Foundations III-1

15. A particular star has an angle of parallax of 0.2 arcsec. What is the distance to this star?
A) 5 pc B) 0.2 pc C) 2 pc D) 50 pc
Answer: A Source: Foundations III-1

16. A particular star has an angle of parallax of 0.1 arcsec. What is the distance to this star?
A) About 10 ly. B) About 33 ly. C) About 3.3 ly. D) About 0.1 ly.
Answer: B Source: Foundations III-1

17. A particular star is 20 pc away from the Earth. What is the angle of parallax for this star?
A) 6 arcsec. B) 20 arcsec. C) 0.02 arcsec. D) 0.05 arcsec.
Answer: D Source: Foundations III-1

18. If stellar parallax can be measured to a precision of about 0.01 arcsec using telescopes on the Earth to observe stars, to what distance does this correspond in space?
A) 500 pc. B) 200 pc. C) 0.01 pc. D) 100 pc.
Answer: D Source: Foundations III-1

19. A particular star is about 30 ly from the Earth. What is the angle of parallax for this star?
A) 0.01 arcsec. B) 0.1 arcsec. C) 0.03 arcsec. D) 30 arcsec.
Answer: B Source: Foundations III-1

20. If a nearby star shows a parallax of 0.5 arc seconds (when the Earth moves through 1 AU, by definition) at what is its distance from Earth, in light-years?
A) 6.52 ly. B) 3.26 ly. C) 2 ly. D) 1.83 ly.
Answer: A Source: Foundations III-1

21. The triple star system a Centauri has a parallax (the largest known) of 0.75 arc second. How far is this star system from the Sun in light-years? (Careful with units)
A) 4.35 ly. B) 0.75 ly. C) 1.33 ly. D) 0.41 ly.
Answer: A Source: Foundations III-1

22. If the Hipparchos satellite measures the parallax motion of a star against the background stars and concludes that the star has a parallax of 0.004 arc seconds, how far is that star from us?
    A) 0.004 pc or 0.013 light-years.
    B) 400pc or 1300 light-years.
    C) 250 pc or 815 light-years.
    D) 25pc or 81.5 light-years.
    Answer: C Source: Foundations III-1

23. A star at a distance of 80 pc will have a parallax (apparent motion in the sky as a result of Earth's motion through 1 AU) of
    A) 0.0125 arc minute.
    B) 80 arc seconds.
    C) 0.0125 arc second.
    D) 0.001125 arc second.
    Answer: C Source: Foundations III-1

24. Using telescopes on the Earth, how far out into space can stellar parallax provide a measure of the distances to stars with reasonable accuracy?
    A) 10 pc B) 2000 pc C) 100 pc D) 500 pc
    Answer: C Source: Foundations III-1

25. What does apparent magnitude tell us about a star?
    A) The total output of electromagnetic energy emitted at all wavelengths from the star.
    B) The brightness the star would appear to have if it were exactly 10 pc from the Earth.
    C) The brightness of a star as it appears in our sky.
    D) The intrinsic brightness of a star (the total light actually emitted by the star).
    Answer: C Source: Foundations III-2

26. Apparent magnitude is a measure of
    A) the intrinsic brightness (actual light output) of a star.
    B) the size (diameter) of a star.
    C) the temperature of a star.
    D) the brightness of a star, as seen from the Earth.
    Answer: D Source: Foundations III-2

27. The relative brightness of stars as we see them in our sky are represented by their
    A) apparent magnitudes.
    B) luminosities.
    C) surface temperatures.
    D) absolute magnitudes.
    Answer: A Source: Foundations III-2

28. If two stars differ by one magnitude, what is the ratio of their brightnesses?
    A) 100 B) 2.5 C) 2 D) 10
    Answer: B Source: Foundations III-2

29. Two stars that differ from each other by five magnitudes have a ratio of brightnesses of
A) 25 B) 10 C) 100 D) 2.5
Answer: C Source: Foundations III-2

30. A star of apparent magnitude +1 appears
A) fainter than a star of apparent magnitude +2.
B) farther away than a star of apparent magnitude +2.
C) either brighter or fainter than a star of apparent magnitude +2, depending on the distance to the stars.
D) brighter than a star of apparent magnitude +2.
Answer: D Source: Foundations III-2

31. A star of apparent magnitude +5 appears
A) brighter than a star of apparent magnitude +3.
B) farther away than a star of apparent magnitude +3.
C) either brighter or fainter than a star of apparent magnitude +3, depending on the distance to the stars.
D) fainter than a star of apparent magnitude +3.
Answer: D Source: Foundations III-2

32. A star of apparent magnitude +4.7 appears
A) brighter than a star of apparent magnitude +4.8.
B) either brighter or fainter than a star of apparent magnitude +4.8, depending on the distance to the stars.
C) fainter than a star of apparent magnitude +4.8.
D) farther away than a star of apparent magnitude +4.8.
Answer: A Source: Foundations III-2

33. A star of apparent magnitude +3.5 appears
A) brighter than a star of apparent magnitude +3.3.
B) farther away than a star of apparent magnitude +3.3.
C) either brighter or fainter than a star of apparent magnitude +3.3, depending on the distance to the stars.
D) fainter than a star of apparent magnitude +3.3.
Answer: D Source: Foundations III-2

34. A star of apparent magnitude-3 appears
    A) either brighter or fainter than a star of apparent magnitude-2, depending on the distance to the stars.
    B) farther away than a star of apparent magnitude-2.
    C) fainter than a star of apparent magnitude-2.
    D) brighter than a star of apparent magnitude-2.
    Answer: D Source: Foundations III-2

35. A star of apparent magnitude-2 appears
    A) fainter than a star of apparent magnitude-3.
    B) brighter than a star of apparent magnitude-3.
    C) farther away than a star of apparent magnitude-3.
    D) either brighter or fainter than a star of apparent magnitude-3, depending on the distance to the stars.
    Answer: A Source: Foundations III-2

36. A star of apparent magnitude-1.5 appears
    A) brighter than a star of apparent magnitude +2.0.
    B) farther away than a star of apparent magnitude +2.0.
    C) fainter than a star of apparent magnitude +2.0.
    D) either brighter or fainter than a star of apparent magnitude +2.0, depending on the distance to the stars.
    Answer: A Source: Foundations III-2

37. A star of apparent magnitude +2.1 appears
    A) either brighter or fainter than a star of apparent magnitude-1.2, depending on the distance to the stars.
    B) farther away than a star of apparent magnitude-1.2.
    C) fainter than a star of apparent magnitude-1.2.
    D) brighter than a star of apparent magnitude-1.2.
    Answer: C Source: Foundations III-2

38. How many times brighter than a magnitude +4.0 star is a magnitude +3.0 star?
    A) Twice as bright
    B) 100 times brighter
    C) 2.512 times brighter
    D) 4/3 (1.333) times brighter
    Answer: C Source: Foundations III-2

39. How many times brighter than a magnitude +6.0 star is a magnitude +1.0 star?
    A) 2.512 times brighter
    B) 100 times brighter
    C) 6 times brighter
    D) 5 times brighter
    Answer: B Source: Foundations III-2

40. To match the light intensity from a first-magnitude star would require the equivalent light from how many second-magnitude stars?
A) About 10 B) About 0.4, or 1/2.5 C) 2 D) About 2.5
Answer: D Source: Foundations III-2

41. How many stars of 6th magnitude in a small cluster would it take for the cluster to appear as bright as a single 1st magnitude star?
A) 5 B) 6 C) $10^5$ D) 100
Answer: D Source: Foundations III-2

42. Sirius, visually the brightest star in our sky, has an apparent magnitude of about -1.5 (see Appendix, Kaufmann & Comins, *Discovering the Universe*, 5th Ed.) while the Andromeda galaxy has an apparent magnitude of about +3.5. What is the ratio of their brightness, as seen by Earth-bound observers?
A) The Andromeda galaxy is 100 times brighter than Sirius.
B) The Andromeda galaxy is 5 times brighter than Sirius.
C) The Andromeda galaxy is 100 times fainter than Sirius.
D) The Andromeda galaxy is 2 times fainter than Sirius.
Answer: C Source: Foundations III-2

43. The statement that the apparent magnitude of a variable star has increased indicates that
A) its brightness has increased.
B) its surface temperature has decreased.
C) its brightness has decreased.
D) its surface temperature has increased.
Answer: C Source: Foundations III-2

44. By approximately how many magnitudes is the star Sirius fainter than the full Moon in our sky? (See Fig. III-2. Kaufmann & Comins, *Discovering the Universe*, 5th Ed.)
A) 11 B) 5 C) 2 D) 15
Answer: A Source: Foundations III-2

45. A star that has an apparent magnitude of 0 (see Fig. III-2, Kaufmann & Comins, *Discovering the Universe*, 5th Ed.)
A) would have infinite brightness, since 1/0 = infinity.
B) would be fainter than Spica (alpha Virginis), which has an apparent magnitude of +1.0.
C) would not be emitting any light, and therefore could not be seen from the Earth.
D) would be brighter than Deneb (alpha Cygni), which has an apparent magnitude of +1.2.
Answer: D Source: Foundations III-2

46. Light leaving a point source spreads out so that the apparent brightness, $I$, of light per unit area varies with distance $d$ according to which following law (μ means "proportional to")?
A) $I$ μ $1/d$. B) $I$ μ $d^2$. C) $I$ μ $1/d^2$. D) $I$ = constant.
Answer: C Source: Foundations III-2

47. Suppose that at night a light bulb is moved to a distance twice as far away as it was at the start. How bright will it appear compared to when it was at its earlier distance?
A) 1/4 as bright. B) 1/2 as bright. C) 1/8 as bright. D) 1/16 as bright.
Answer: A Source: Foundations III-3

48. Suppose that two identical stars (having the same total light output) are located such that star A is at a distance of 5 pc and star B is at a distance of 25 pc. How will star B appear, compared to star A?
A) Star B will be 1/2.2 as bright as star A.
B) Star B will be 1/20 as bright as star A.
C) Star B will be 1/25 as bright as star A.
D) Star B will be 1/5 as bright as star A.
Answer: C Source: Foundations III-3

49. The intensity of sunlight per square meter reaching Jupiter is approximately what fraction of that at the Earth's orbital distance? (See Fig. III-3, Kaufmann & Comins, *Discovering the Universe*, 5th Ed.)
A) 1/5. B) 25 times. C) About the same. D) 1/25.
Answer: D Source: Foundations III-3

50. The intensity of sunlight per square meter reaching Saturn is approximately what fraction of that at the Earth's orbital distance? (See Fig. III-3, Kaufmann & Comins, *Discovering the Universe*, 5th Ed.)
A) 1/9.5 B) 1/90 C) 90 times. D) About the same.
Answer: B Source: Foundations III-3

51. How much more light falls on a unit area of Mercury's surface than on an equivalent area of the Moon if Mercury is at 0.4 AU and the Earth's Moon is at 1.0 AU from the Sun, and the inverse-square law holds?
A) 0.4 B) 2.5 C) 6.25 D) 0.16
Answer: C Source: Foundations III-3

52. The star a Centauri C and the star Groombridge 34 B have the same apparent magnitude, but a Centauri C is 1.3 pc away from the Earth and Groombridge 34 B is 3.5 pc away. What is the luminosity of Groombridge 34 B, compared to a Centauri C? (See Fig. III-3, Kaufmann & Comins, *Discovering the Universe*, 5th Ed.)
   A) 2.7 times brighter.
   B) 7.2 times brighter.
   C) 2.7 times fainter.
   D) 7.2 times fainter.
   Answer: B Source: Foundations III-3

53. How is absolute magnitude defined?
   A) It is the apparent magnitude a star would have if all of the energy from the star were concentrated in the visual region.
   B) It is the apparent magnitude a star would have if the star were located at exactly 10 pc from the Earth.
   C) It is the apparent magnitude a star would have if the star were located at exactly 10 AU from the Earth.
   D) It is the apparent magnitude a star would have if the star were located at exactly 10 ly from the Earth.
   Answer: B Source: Foundations III-3

54. The absolute magnitude of a star is the brightness the star would appear to have if it were placed at what distance from Earth?
   A) 32.6 light-years.
   B) 1 astronomical unit.
   C) 10 light-years.
   D) The distance to the galactic center.
   Answer: A Source: Foundations III-3

55. At what distance are stars assumed to be from the Earth when they are represented by their absolute (as opposed to their apparent) magnitude?
   A) 1 light-years. B) 10 parsecs. C) 10 light-years. D) 10 astronomical units.
   Answer: B Source: Foundations III-3

56. A particular star is at a distance of 20 pc from the Earth. For this star, the apparent magnitude will be
   A) the same as the absolute magnitude, since magnitude is independent of distance.
   B) a smaller number than the absolute magnitude.
   C) either larger or smaller than the absolute magnitude, depending on the temperature and diameter of the star.
   D) a larger number than the absolute magnitude.
   Answer: D Source: Foundations III-3

57. A particular star is at a distance of 5 pc from the Earth. For this star, the apparent magnitude will be
    A) either larger or smaller than the absolute magnitude, depending on the temperature and diameter of the star.
    B) a larger number than the absolute magnitude.
    C) the same as the absolute magnitude, since magnitude is independent of distance.
    D) a smaller number than the absolute magnitude.
    Answer: D  Source: Foundations III-3

58. The star Alphard has an apparent magnitude of 2.0, and the star Megrez has an apparent magnitude of 3.3. The only thing that can be said with certainty about Alphard is that
    A) it is brighter than Megrez, as seen in our sky.
    B) it has a greater luminosity than Megrez.
    C) it is fainter than Megrez, as seen in our sky.
    D) it is closer than Megrez.
    Answer: A  Source: Foundations III-3

59. The star Alderamin has an apparent magnitude of 2.4 and an absolute magnitude of 1.4. From this information (assuming that the starlight has not been dimmed by interstellar clouds) we can say for sure that
    A) Alderamin is less than 10 light-years away.
    B) Alderamin is more than 10 light-years away.
    C) Alderamin is more than 10 parsecs away.
    D) Alderamin is less than 10 parsecs away.
    Answer: C  Source: Foundations III-3

60. The star Fomalhaut has an apparent magnitude of 1.15 and an absolute magnitude of 2.0. From this information (assuming that the star has not been dimmed by interstellar clouds) we can say for sure that Fomalhaut is
    A) more than 10 light-years away.
    B) less than 10 parsecs away.
    C) more than 10 parsecs away.
    D) less than 10 light-years away.
    Answer: B  Source: Foundations III-3

61. The star Alderamin has an apparent magnitude of 2.4 and an absolute magnitude of 1.4. The star Merak has an apparent magnitude of 2.4 and absolute magnitude of 0.5. Assuming that neither star has been dimmed by interstellar clouds, we can say for sure that
    A) Merak is closer to us than is Alderamin.
    B) Merak and Alderamin are the same distance from us.
    C) Merak is an intrinsically fainter star than is Alderamin.
    D) Merak is farther away from us than is Alderamin.
    Answer: D  Source: Foundations III-3

62. The star g Phoenicis has an apparent magnitude of 3.4 and an absolute magnitude of -4.6. The North Star (Polaris) has an apparent magnitude of 2.0 and an absolute magnitude of -4.6. Assuming that no light has been absorbed or scattered by interstellar dust, we can say for sure that
A) Polaris is farther away from us than g Phoenicis.
B) both stars are the same distance away from us.
C) Polaris is closer to us than g Phoenicis.
D) Polaris appears fainter in our sky than g Phoenicis.
Answer: C Source: Foundations III-3

63. A particular star has an apparent magnitude of +12 and an absolute magnitude of +5. According to the equation in Astronomer's Toolbox III-2, Kaufmann and Comins, *Discovering the Universe*, 5th Ed., what is the distance to this star?
A) 125 ly B) 125 pc C) 250 pc D) 250 ly
Answer: C Source: Foundations III-3

64. A star whose absolute magnitude M is +2.2 is seen to have an apparent magnitude when viewed from Earth of m = +5.2. According to the equation in Astronomer's Toolbox III-2, Kaufmann and Comins, *Discovering the Universe*, 5th Ed., how far away is the star?
A) 3/5 or 0.6 pc B) 130 pc C) 40 pc D) 12.3 pc
Answer: C Source: Foundations III-3

65. A star whose distance from Earth is 100 pc has an apparent magnitude of m = +2.5. According to the equation in Astronomer's Toolbox III-2, Kaufmann and Comins, *Discovering the Universe*, 5th Ed., what is its absolute magnitude, M?
A) -47.5 B) -7.5 C) -2.5 D) +7.5
Answer: C Source: Foundations III-3

66. The star Arietis has an apparent magnitude of +2.7 and a distance of 52 light-years. According to the equation in Astronomer's Toolbox III-2, what is its absolute magnitude, M?
A) +3.7 B) +1.7 C) -0.9 D) +6.2
Answer: B Source: Foundations III-3

67. What is a star's luminosity?
    A) The total energy emitted by the star per second, measured in watts.
    B) The apparent magnitude the star would have if it were located exactly 10 ly from the Earth.
    C) The amount of energy received on one square meter of a planet's surface exactly 1 AU from the star, per second.
    D) The apparent magnitude the star would have if it were located exactly 10 pc from the Earth.
    Answer: A Source: Foundations III-3

68. The luminosity of a star is
    A) another name for its color or surface temperature.
    B) its brightness as seen by people on Earth.
    C) its brightness if it were to be at a distance of 10 parsecs (32.6 light-years) from Earth.
    D) its total energy output into all space, over all wavelengths.
    Answer: D Source: Foundations III-3

69. The luminosity of a star is a unique measure of its
    A) velocity of recession away from us.
    B) total energy output.
    C) physical size.
    D) temperature.
    Answer: B Source: Foundations III-3

70. The luminosity of a star is
    A) its apparent magnitude.
    B) the light emitted by the star within the sensitive range of the eye.
    C) the total energy emitted at all wavelengths into all space, from its whole surface.
    D) the total energy emitted at all wavelengths toward the Earth.
    Answer: C Source: Foundations III-3

71. The luminosity of a star is
    A) its brightness when measured from a distance of 10 parsecs or 32.6 light-years.
    B) its total energy output emitted at all wavelengths into all space.
    C) its brightness when measured from Earth.
    D) the energy output of 1 square meter of its surface into all space at all wavelengths.
    Answer: B Source: Foundations III-3

72. The Sun's absolute magnitude is about +5. The brightest stars in our sky have absolute magnitudes of about -10. What is the luminosity of these stars compared to that of the Sun, assuming that they have similar spectral light distributions?
A) 5 times less.
B) 1 million times less.
C) 15 times greater.
D) 1 million times greater.
Answer: D Source: Foundations III-3

73. Two stars in our sky have the same apparent brightness. If neither of them is hidden behind gas or dust clouds, then we know that
A) they must have the same temperature.
B) they must be at the same distance away from us.
C) they may be at different distances, in which case the farther one must have the greater luminosity.
D) they may be at different distances, in which case the nearer one must have the greater luminosity.
Answer: C Source: Foundations III-3

74. How bright (in absolute magnitude) are the intrinsically brightest stars in the universe?
A) -10 B) 0 C) +1 D) +17
Answer: A Source: Foundations III-3

75. How bright (in absolute magnitude) are the intrinsically faintest stars in the universe?
A) +1 B) -10 C) 0 D) +17
Answer: D Source: Foundations III-3

76. How bright (in terms of total energy output per second) are the brightest stars in the universe, compared to the Sun?
A) About 1,000,000 times brighter.
B) About 1000 times brighter.
C) About 1,000,000,000 times brighter.
D) About 100 times brighter.
Answer: A Source: Foundations III-3

77. How bright (in terms of total energy output per second) are the faintest stars in the universe, compared to the Sun ?
A) About 1/10,000 as bright.
B) About 1/1000 as bright.
C) About 1/100,000 as bright.
D) About 1/100 as bright.
Answer: C Source: Foundations III-3

# Chapter 10: The Nature of Stars

1. The technique called photometry in stellar astronomy is the measurement of
   A) the arrival times of photons from variable and pulsating stars, in order to determine accurately the pulsation or rotation periods of these stars.
   B) the precise positions and relative motions of stars in the galaxy, from which galactic structure and overall rotation can be determined.
   C) the intensity of light from stars through several limited-bandpass filters from which surface temperature, variability, luminosity etc. of stars can be determined.
   D) the relative absorption of light by different atoms and molecules in high resolution spectra of starlight, from which stellar temperatures can be estimated.
   Answer: C Source: Section 10-1

2. When observed through a set of photometric filters, the brightness of a distant star is seen to be brightest through the ultraviolet filter, less bright through the blue filter, and faintest in through the yellow filter. What conclusion can be drawn from this information, assuming no absorption of light between the star and Earth?
   A) There is insufficient information to draw a conclusion about star surface temperature.
   B) The star has a very low surface temperature.
   C) The star has a very high surface temperature.
   D) The star has an intermediate temperature, close to that of the Sun.
   Answer: C Source: Section 10-1

3. Optical glass filters are used to select certain portions of a star's light for photometry. Which of the following filters would most closely match the sensitivity of our eyes, which have a peak sensitivity at about 550 nm wavelength? (See Fig. 3-1 of Kaufmann & Comins, *Discovering the Universe*, 5th Ed.)
   A) Yellow. B) Blue. C) Red. D) Ultraviolet.
   Answer: A Source: Section 10-1 and Figure 3-1

4. The difference in the brightness of a star as seen through two different colored filters, for example blue and yellow, is directly related to which stellar property?
   A) Radius. B) Distance from Earth. C) Luminosity. D) Surface temperature.
   Answer: D Source: Section 10-1

5. A particular star is brighter as seen through a blue filter than through a yellow filter. Which of the following surface temperatures is possible for this star? (See Fig. 10-1, Kaufmann & Comins, *Discovering the Universe*, 5th Ed.)
   A) 4,500 K. B) 12,000 K. C) 6,000 K. D) 3,000 K.
   Answer: B Source: Section 10-1

6. A particular star is fainter as seen through a blue filter than through a yellow filter. Which of the following surface temperatures is possible for this star? (See Fig. 10-1, Kaufmann & Comins, *Discovering the Universe*, 5th Ed.)
   A) 12,500 K. B) 38,000 K. C) 8,800 K. D) 3800 K.
   Answer: D Source: Section 10-1

7. A particular star is approximately equally bright when viewed through a blue filter and through a yellow filter. What is the approximate surface temperature of this star? (See Fig. 10-1, Kaufmann & Comins, *Discovering the Universe*, 5th Ed.)
   A) 6,000K.
   B) 12,000 K.
   C) 3,000K.
   D) It is not possible for a star to be equally bright at two different wavelengths.
   Answer: A Source: Section 10-1

8. The star Rigel, in the constellation Orion, appears brighter through a blue filter than it does through a yellow filter. Suppose that a second star is found which has the same brightness as Rigel through the yellow filter, but is brighter than Rigel through the blue filter. From this information, we can say conclusively that the second star has
   A) the same temperature but a lower luminosity.
   B) a higher temperature.
   C) the the same temperature but a higher luminosity.
   D) a lower temperature.
   Answer: B Source: Section 10-1

9. The star Regulus, in the constellation Leo, appears brighter through a blue filter than it does through a yellow filter. Suppose that a second star is found which has the same brightness as Regulus through the blue filter, but is brighter than Regulus through the yellow filter. From this information, we can say conclusively that the second star has
   A) the same temperature but a lower luminosity.
   B) a higher temperature.
   C) the same temperature but a higher luminosity.
   D) a lower temperature.
   Answer: D Source: Section 10-1

10. In order for Balmer series lines to show up strongly in absorption in stellar spectra, significant numbers of hydrogen atoms have to have electrons in the $n = 2$ energy level. What then does the appearance of such lines in a stellar spectrum tell us about the temperature of the star surface?
    A) It must be high enough to ionize the hydrogen atoms by collision in order that they can absorb from this level.
    B) It must be reasonably high to excite the electrons to this level by collisions, but not high enough to ionize the atoms.
    C) It must be reasonably low so that no atoms will have electrons excited beyond this energy level (e.g., to n = 3).
    D) It tells us very little about the temperature, since hydrogen gas will show significant Balmer absorption, whatever the surface temperature.
    Answer: B Source: Section 10-2

11. Why is there a limited range of stellar surface temperatures around 10,000 K around which neutral hydrogen gas will absorb visible light in the Balmer series?
    A) Because electrons in hydrogen have to be at the n = 2 energy level in order to produce absorption in this series. If the gas is too cold, most atoms are in the n = 1 state and if it is too hot, most atoms are ionized.
    B) Because there must be electrons at the n = 3 energy level in order for Balmer absorption to occur. If the gas is too cold, electrons are only in the n = 1 and 2 levels while if the gas is too hot, the gas is ionized and no electrons are left in the hydrogen atoms.
    C) Because electrons must be in the ground state n = 1 in order to undergo Balmer absorption. If the gas is too cold, electrons cannot be excited from this level while, if it is too hot, there are no electrons left in the n = 1 level.
    D) Because there must be sufficient continuum radiation from the stellar surface in the visible region to be absorbed by the hydrogen gas.
    Answer: A Source: Section 10-2

12. Why are Balmer absorption lines very weak in the spectra of stars with low surface temperatures, significantly below 10,000K, for example?
    A) Because hydrogen atoms have no electrons in any energy levels at these temperatures.
    B) Because atoms need electrons which have been excited by high-temperature collisions to the n = 2 level in order to undergo Balmer absorption.
    C) Because there is no emitted continuum radiation at Balmer-line wavelengths when the gas is so cool, and so absorption will not be seen.
    D) Because the hydrogen atoms have to be hot enough to be ionized in order to show Balmer absorption.
    Answer: B Source: Section 10-2

13. In order for absorption lines in the Paschen series of hydrogen to be seen in the IR spectrum of a star (see Sec. 4-6 of Kaufmann & Comins, *Discovering the Universe*, 5th Ed.), the temperature of its surface must be high enough to excite electrons by collision to the
    A) n = 3 energy level.
    B) n = 2 energy level.
    C) n = 4 energy level.
    D) ionization level.
    Answer: A Source: Sections 10-2 and 4-6

14. Spectral classification of stars into the lettered categories, O,B,A,F,G,K,M, is carried out by
    A) finding the wavelength of peak emission in the continuum spectrum of the star.
    B) determining their relative masses by the study of binary star motions, in order to place them into their proper mass classification.
    C) examining the relative depths of absorption lines from various neutral and ionized atoms in a stellar spectrum.
    D) determining the total energy emitted at all wavelengths by stars, taking account of the full spread of wavelengths and their distances, in order to place the star into its luminosity class.
    Answer: C Source: Section 10-3

15. The spectrum of an ordinary main-sequence star is a
    A) series of emission lines, mostly from hydrogen, the major constituent of stellar surfaces, which occasionally overlap to produce sections of continuous color.
    B) continuum of colors crossed by dark absorption lines, caused by absorption by cooler atoms and molecules at the star's surface.
    C) smooth continuum of color, peaking at a specific wavelength whose position is dependent upon the star's surface temperature.
    D) continuum of colors, crossed by brighter lines caused by emission from the hot atoms and molecules on the star's surface.
    Answer: B Source: Section 10-3

16. The chemical makeup of the Sun's surface can be determined
    A) by taking a sample of the star's surface with a space probe.
    B) by examining the chemicals present in a meteorite, since it is part of the solar system.
    C) by solar spectroscopy.
    D) by measuring the components of the solar wind with Earth-orbiting spacecraft.
    Answer: C Source: Section 10-3

17. Surface temperature of a nearby star can be determined most precisely by measuring what parameters?
    A) The Doppler shift of the star's spectral lines.
    B) The star's position on the HR diagram.
    C) The relative strengths of absorption lines from different atoms (e.g., H, Ca) and molecules (e.g., TiO) in the star's spectrum.
    D) The relative strengths of emission lines from different atoms and ions in the star's spectrum.
    Answer: C   Source: Section 10-3

18. Spectral types of stars (e.g., O, B, A, F, G, K, M) define uniquely their
    A) surface temperatures.
    B) luminosities.
    C) sizes or radii.
    D) absolute magnitudes.
    Answer: A   Source: Section 10-3

19. The surface temperature of a nearby star can best be determined from spectral classification by examining
    A) the pattern of spectral absorption lines from various atoms.
    B) the relative intensities of light measured through different photometric filters.
    C) the peak wavelength of the star's continuum blackbody spectrum.
    D) the pattern of emission lines which are on the star's spectrum.
    Answer: A   Source: Section 10-3

20. Which of the following sequences of stellar spectral classifications is in correct order of increasing temperature?
    A) K,M,G,F,A,B,O.
    B) M,K,G,F,A,B,O.
    C) O,B,A,F,G,K,M.
    D) A,B,F,G,K,M,O.
    Answer: B   Source: Section 10-3

21. The sequence of letters which is used to classify star surface temperatures, as determined by relative spectral absorption line strengths, in order of decreasing temperature, is
    A) A,B,F,G,K,M,O.
    B) O,B,A,F,G,K,M.
    C) O,F,M,G,A,B,K.
    D) M,O,F,K,G,A,B.
    Answer: B   Source: Section 10-3

22. The Sun's classification in terms of its surface temperature, as determined from absorption lines in its spectrum, is
    A) M9.   B) G2.   C) B2.   D) O1.
    Answer: B   Source: Section 10-3

23. Which of the following letters representing spectral classification signifies the hottest stellar surface temperature?
A) G. B) B. C) K. D) A.
Answer: B Source: Section 10-3

24. Which of the following four spectral classifications represents the coolest stellar surface temperature?
A) G. B) B. C) K. D) A.
Answer: C Source: Section 10-3

25. The spectral class of the star Enif is K2 while that of the Sun is G2. Which of the following conclusions can be drawn about Enif from this information?
A) It is hotter than the Sun.
B) It is cooler than the Sun.
C) It is intrinsically brighter than the Sun.
D) It is intrinsically fainter than the Sun.
Answer: B Source: Section 10-3

26. Absorption line strengths are used in the spectral classification of stars and the determination of surface temperatures. Which of the following atomic or molecular constituents will exhibit strong absorption lines in spectra from stars with very high surface temperature?
A) H. B) He II. C) Ca II. D) TiO.
Answer: B Source: Section 10-3

27. If the surface temperature of a star is very low, which of the following atomic or molecular constituents will produce the most prominent absorption lines in its spectrum?
A) Fe II. B) TiO. C) Mg II. D) He II.
Answer: B Source: Section 10-3

28. Which of the following molecules produces the strong absorption bands in the spectrum of a cool M-type star, as can be seen in Fig. 10-3 of Kaufmann and Comins, *Discovering the Universe*, 5th Ed.?
A) HCl, hydrogen chloride.
B) TiO, titanium oxide.
C) CaI, calcium iodide
D) $H_2O$, water vapor.
Answer: B Source: Section 10-3

29. The symbol He II refers to a
A) helium atom which has lost two electrons.
B) neutral helium atom (atomic number = 2) which has lost no electrons.
C) helium molecule, which contains two helium atoms.
D) helium atom which has lost one electron.
Answer: D Source: Sections 10-3 and 4-5

30. Fe XII is an ionized iron atom with
    A) 1 electron removed.
    B) 12 electrons removed.
    C) 11 electrons removed.
    D) 13 electrons removed.
    Answer: C Source: Sections 10-3 and 4-5

31. How many electrons are missing from the ionized silicon atom Si IV?
    A) 4. B) 1. C) 5. D) 3.
    Answer: D Source: Sections 10-3 and 4-5

32. The reason why a doubly ionized helium gas, He III, will not produce absorption lines in a stellar spectrum is
    A) that the gas temperature on any star is always so high that electrons are excited beyond energy levels in these atoms from which visible photons can be absorbed.
    B) because doubly ionized helium is meaningless, since neutral helium only contains one electron.
    C) that a doubly ionized helium atom is simply a helium nucleus stripped of all its electrons, and so cannot absorb visible photons.
    D) that the gas temperature on star surfaces can never be hot enough to excite electrons in this atom to levels from which they will absorb visible photons.
    Answer: C Source: Section 10-3

33. Which of the following atoms or ions will produce the strongest absorption lines in the spectra of stars with the highest surface temperatures?
    A) H I, neutral hydrogen.
    B) Fe I, neutral iron.
    C) Ca II, singly ionized calcium.
    D) He II, ionized helium.
    Answer: D Source: Section 10-3

34. Which of the following atoms or ions will produce strong absorption lines in the spectra of stars with the relatively cool surface temperatures?
    A) TiO, molecules of titanium oxide.
    B) He I, neutral helium.
    C) Ca II, ionized calcium.
    D) Mg II, ionized magnesium.
    Answer: A Source: Section 10-3

35. The spectrum of a star shows the following absorption line characteristics: "Very strong H, weaker Mg II and Si II, and no He I or Ca II lines". Classify the spectrum of this star, using the graph of absorption line strengths in the spectra of stars shown in Fig. 10-3 of Kaufmann and Comins, *Discovering the Universe*, 5th Ed.
    A) A. B) K. C) B. D) G.
    Answer: A Source: Section 10-3

36. Use the graph of absorption line strengths in Fig. 10-3 of Kaufmann and Comins, *Discovering the Universe*, 5th Ed., to determine the spectral class of a star with the following absorption lines in its spectrum: "Very strong He I lines; weaker H lines; weak Si III lines; no Ca II, Fe II, or He II lines".
A) B. B) M. C) A. D) F.
Answer: A Source: Section 10-3

37. What will be the spectral class of a star whose spectrum shows the following absorption lines: "Very strong Ca II lines; weaker Fe I lines; equal strength, but weaker Fe II and Ca I lines; no He I, He II, or TiO lines"? (Hint: Use the Spectral Type - Temperature graph in Fig. 10-3 of Kaufmann & Comins, *Discovering the Universe*, 5th Ed.)
A) K. B) A. C) F. D) O.
Answer: A Source: Section 10-3

38. The spectrum of a very distant star shows spectral absorption lines of ionized helium, He II, AND molecular absorption bands from titanium oxide, TiO. What would be your conclusion about this star?
A) There must be cool, interstellar gas containing TiO between the star and Earth.
B) It is obviously the spectrum of a binary system, two stars close together, a hot star and a cooler companion, unresolved as separate stars from our distance but contributing separate spectra.
C) There must be a very hot atmosphere containing helium gas, overlying a much cooler stellar surface.
D) The star must have a thick, cool atmosphere overlying a hot stellar atmosphere.
Answer: B Source: Section 10-3

39. From its position in the Hertzsprung-Russell diagram of Fig. 10-4 on p. 237 of Kaufmann and Comins, *Discovering the Universe*, 5th Ed., what can you conclude about the star Mira compared to the Sun?
A) Mira is cooler, redder and intrinsically fainter than the Sun.
B) ira is cooler and redder but intrinsically brighter than the Sun.
C) Mira is hotter than the Sun and intrinsically brighter.
D) Mira is hotter and more blue, but intrinsically fainter than the Sun.
Answer: B Source: Section 10-4

40. The Hertzsprung-Russell diagram is a plot of
A) apparent brightness against intrinsic brightness of a group of stars.
B) apparent brightness against distance for stars near to the Sun.
C) absolute magnitude (or intrinsic brightness) against temperature of a group of stars.
D) luminosity against mass of a group of stars.
Answer: C Source: Section 10-4

41. What are the two physical parameters of stars which are plotted in the Hertzsprung-Russell diagram?
 A) luminosity and mass.
 B) radius and mass.
 C) mass and surface temperature.
 D) luminosity and surface temperature.
 Answer: D Source: Section 10-4

42. Which two physical parameters of stars are plotted on the Hertzsprung-Russell diagram to show the systematics of a group of stars (e.g., a cluster)?
 A) Surface temperature and mass.
 B) Luminosity and surface temperature.
 C) Luminosity and radius.
 D) Mass and apparent magnitude.
 Answer: B Source: Section 10-4

43. In the Hertzsprung-Russell diagram of Fig. 10-4 of Kaufmann and Comins, *Discovering the Universe*, 5th Ed., which of the following is the correct sequence of stars in order of increasing intrinsic brightness?
 A) Sirius B, Deneb, Procyon B, the Sun.
 B) The Sun, Procyon B, Deneb, Sirius B.
 C) Procyon B, Sirius B, the Sun, Deneb.
 D) Deneb, the Sun, Sirius B, Procyon B.
 Answer: C Source: Section 10-4

44. In the Hertzsprung-Russell diagram Fig. 10-4 of Kaufmann and Comins, *Discovering the Universe*, 5th Ed., which of the following is the correct sequence of stars in order of increasing temperature?
 A) Deneb, the Sun, Sirius B, Procyon B.
 B) Procyon B, Sirius B, the Sun, Deneb.
 C) The Sun, Procyon B, Deneb, Sirius B.
 D) Sirius B, Deneb, Procyon B, the Sun.
 Answer: C Source: Section 10-4

45. Using Figs. 10-4 and 10-5, determine which of the following is the correct sequence of stars in order of increasing size or stellar radius?
 A) Mira, Betelguese, the Sun, Sirius B.
 B) Sirius B, the Sun, Mira, Betelguese.
 C) Sirius B, the Sun, Betelguese, Mira.
 D) Betelguese, Mira, the Sun, Sirius B.
 Answer: B Source: Section 10-4

46. A star in the lower left part of the Hertzsprung-Russell diagram, compared to a star in the middle of the diagram, is
 A) smaller. B) cooler. C) larger. D) brighter.
 Answer: A Source: Section 10-4

47. Compared to a star in the middle of the Hertzsprung-Russell diagram, a star in the upper right part of the diagram is
    A) fainter.
    B) hotter.
    C) larger.
    D) non-existent, since there are no stars which appear in the upper right part of the diagram.
    Answer: C Source: Section 10-4

48. Where on the Hertzsprung-Russell diagram do most local stars in our Universe congregate?
    A) On the main sequence, where stars are generating energy by fusion reactions.
    B) In the supergiant area, where the most massive stars spend a significant time.
    C) In the giants area, where most stars spend the longest time of their lives.
    D) In the white dwarf area, the "graveyard" of stars.
    Answer: A Source: Section 10-4

49. What fraction of the stars surrounding the Sun are main sequence stars?
    A) Very few of them, about 20%.
    B) Almost all of them, about 90%.
    C) Roughly half of them, about 55%.
    D) There are no main sequence stars close to the Sun.
    Answer: B Source: Section 10-4

50. What is a dwarf star?
    A) A star of about the same size (diameter) as the Earth.
    B) A main sequence star.
    C) A large, planetary object, such as Jupiter.
    D) Any star which is significantly smaller than a giant or supergiant star.
    Answer: B Source: Section 10-4

51. What is a white dwarf star?
    A) A main sequence star with a surface temperature near 12,000 K.
    B) A star of about the same size (diameter) as the Earth.
    C) Any star which is significantly smaller than a giant or supergiant star.
    D) A large, planetary object, such as Jupiter.
    Answer: B Source: Section 10-4

52. If the surface temperatures of white dwarf stars are 4 times that of the Sun and energy output per unit area of a star depends upon the 4th power of the temperature by the Stefan-Boltzmann relation, why then are white dwarfs intrinsically so faint?
    A) Because they are shrouded in very thick atmospheres.
    B) Because they are moving rapidly away from the Sun and their spectra are extremely red-shifted, hence they appear faint at visible wavelengths.
    C) Because they have very thin atmospheres which do not emit continuum radiation but only line emissions, like a low density gas.
    D) Because they are very small.
    Answer: D Source: Section 10-4

53. Measurements indicate that a certain star has a very high intrinsic brightness (100,000 times as bright as our Sun) and yet is relatively cool (3500 K). How can this be?
    A) The star must be quite small.
    B) The star must be very large.
    C) The star must be a in the upper part of the main sequence.
    D) There must be an error in observation, since no star can have these properties.
    Answer: B Source: Section 10-4

54. The following are parameters of stars which astronomers obtained from their measurements. Which of these conclusions is obviously erroneous, on the basis of the positions of these alleged stars on the Hertzsprung-Russell diagram in Fig. 10-5 of Kaufmann and Comins, *Discovering the Universe*, 5th Ed.? ($L_S$, $R_S$, are the luminosity and radius of the Sun, respectively)
    A) Luminosity = $10^4$ $L_S$, Radius = 100 $R_S$, Temperature = 5,000 K; Conclusion: A red giant star.
    B) Luminosity = $L_S$, Radius = $R_S$, Temperature = 6,000 K; Conclusion: Main sequence star.
    C) Luminosity =1/100 $L_S$, Radius =1/100 $R_S$, Temperature =20,000 K; Conclusion: A white dwarf star.
    D) Luminosity = $L_S$, Radius = 1/10 $R_S$, Temperature = 20,000 K; Conclusion: A white dwarf star.
    Answer: D Source: Section 10-4

55. Using the Hertzsprung-Russell diagram, (Fig. 10-5 of Kaufmann & Comins, *Discovering the Universe*, 5th Ed.), determine which type of star has the following characteristics: "Surface temperature of 40,000 K and luminosity 100,000 times that of the Sun".
    A) Red giant.
    B) Cool, red, main sequence star.
    C) White dwarf.
    D) Hot, blue, main sequence star.
    Answer: D Source: Section 10-4

56. Using the Hertzsprung-Russell diagram (Fig. 10-5 of Kaufmann & Comins, *Discovering the Universe*, 5th Ed.), determine which type of star has the following characteristics: "A star with surface temperature 10,000 K and luminosity 1/100 times that of the Sun".
A) Main sequence. B) Red giant. C) Red supergiant. D) White dwarf.
Answer: D Source: Section 10-4

57. A red supergiant star is found to have a surface temperature of 2500 K and a luminosity 100,000 times that of the Sun. Use the Hertzsprung-Russell diagram , Fig. 10-5 of Kaufmann and Comins, *Discovering the Universe*, 5th Ed. to determine its approximate radius, compared to that of the Sun.
A) About 1000 times larger.
B) About 100 times larger.
C) Almost the same.
D) About 10 times larger.
Answer: A Source: Section 10-4

58. What will be the intrinsic brightness or luminosity of a white dwarf star with the same temperature as the Sun? (see Fig. 10-5, Kaufmann & Comins, *Discovering the Universe*, 5th Ed.)
A) 4 time the Sun's luminosity.
B) $10^{-4}$ of the Sun's luminosity.
C) Since it has the same surface temperature, it will have the same brightness.
D) $10^{-2}$ of the Sun's luminosity.
Answer: B Source: Section 10-4

59. A white dwarf star whose temperature is the same as that of the Sun will have a radius which is (see Fig. 10-5, Kaufmann & Comins, *Discovering the Universe*, 5th Ed.)
A) 2 times smaller than the Sun's radius.
B) 10 times smaller than that of the Sun.
C) the same size as the Sun, since the white dwarf has the same temperature.
D) 100 times smaller than that of the Sun.
Answer: D Source: Section 10-4

60. By what standard technique do astronomers find the luminosity class (I, II, III, IV, or V) of a star?
A) By studying the absorption lines in the star's spectrum.
B) By timing how long it takes for the star to be eclipsed by a companion in an eclipsing binary star system.
C) By observing the diameter of the star on a photographic plate or CCD image.
D) By combining the apparent magnitude with the measured distance to the star.
Answer: A Source: Section 10-5

61. What is the physical reason why astronomers can find the luminosity class (I, II, III, IV, or V) of a star using the star's spectrum?
A) The absorption lines in the spectrum are affected by the star's surface temperature.
B) The relative amounts of hydrogen, helium, and other elements are different for stars of different luminosity classes.
C) The absorption lines in the spectrum are affected by the density and pressure of the star's atmosphere.
D) The wavelength of maximum emission (given by Wien's law) is affected by the size of the star.
Answer: C Source: Section 10-5

62. The luminosity class of a star (I, II, III, IV, or V) is most closely related to which physical characteristic of the star?
A) Surface temperature.
B) Radius.
C) Chemical composition (amount of hydrogen, helium, etc.).
D) Radial velocity.
Answer: B Source: Section 10-5

63. A star with a surface temperature of 5000 K and a luminosity of greater than $10^4$ times that of the Sun is a member of which luminosity class? (See Fig. 10-6, Kaufmann & Comins, *Discovering the Universe*, 5th Ed.)
A) I, Supergiant. B) V, Main sequence. C) III, Giant. D) II, Bright giant.
Answer: A Source: Section 10-5

64. A star with a surface temperature of 4000 K and a luminosity of about $10^{-2}$ times that of the Sun is a member of which luminosity class? (See Fig. 10-6, Kaufmann & Comins, *Discovering the Universe*, 5th Ed.)
A) III, Giant. B) II, Bright giant. C) I, Supergiant. D) V, Main sequence.
Answer: D Source: Section 10-5

65. The star Hadar is classified as B1 II, which means that it is
A) a cool giant. B) a hot, bright giant. C) a cool supergiant. D) a hot supergiant.
Answer: B Source: Sections 10-4 and 10-5

66. The star Arcturus is classified as K2 III, which means that it is
A) a cool giant.
B) hot giant.
C) a cool main sequence star.
D) a cool supergiant.
Answer: A Source: Sections 10-4 and 10-5

67. The star Spica is classified as B1 V, which means that it is
   A) a hot supergiant.
   B) a cool main sequence star.
   C) a cool giant.
   D) a hot, main sequence star.
   Answer: D Source: Sections 10-4 and 10-5

68. Barnard's star, one of our near neighbors, is classified as M5 V. This means that it is
   A) a cool giant.
   B) a cool main sequence star, a red dwarf.
   C) a hot, main sequence star.
   D) a cool supergiant, a huge star.
   Answer: B Source: Sections 10-4 and 10-5

69. Which of the following spectral-luminosity classes corresponds to a red supergiant?
   A) M3 V. B) M2 I. C) G2 III. D) B7 I.
   Answer: B Source: Sections 10-4 and 10-5

70. The star Elnath is classified as B7 III while the star Al Na'ir is classified as B7 IV. Compared to Al Na'ir, Elnath has
   A) about the same intrinsic brightness, but is considerably cooler.
   B) about the same surface temperature, but is intrinsically much brighter.
   C) about the same intrinsic brightness, but is considerably hotter.
   D) about the same surface temperature, but is intrinsically much fainter.
   Answer: B Source: Sections 10-4 and 10-5

71. The spectral-luminosity class of the star a Arae is B2 V, and that of p Herculi is K3 II. Using ONLY this information, and without referring to tables or diagrams, we can tell
   A) only that a Arae is cooler than p Herculi.
   B) only that a Arae is hotter than p Herculi.
   C) that a Arae is cooler and intrinsically fainter than p Herculi.
   D) that a Arae is hotter but intrinsically fainter than p Herculi.
   Answer: B Source: Sections 10-4 and 10-5

72. The spectral-luminosity class of the star Spica is B1 V, and that of the star t Ceti is G8 V. From this, we know that
   A) t Ceti is cooler and has a lower luminosity than Spica.
   B) t Ceti is hotter but has the same luminosity as Spica.
   C) t Ceti is hotter and has a lower luminosity than Spica.
   D) t Ceti is cooler but has the same luminosity as Spica.
   Answer: A Source: Sections 10-4 and 10-5

73. The star z Canis Majoris has an absolute magnitude of -2.4 and a spectral class of B2. Using Figures 10-4 and 10-6 of Kaufmann and Comins, *Discovering the Universe*, 5th Ed., how would z Canis Majoris be classified?
A) B2 I. B) B-type white dwarf. C) B2 V. D) B2 III.
Answer: C Source: Sections 10-4 and 10-5

74. A particular star has an absolute magnitude of 0 and a spectral class similar to that of the Sun. Using Figures 10-4 and 10-6 of Kaufmann and Comins, *Discovering the Universe*, 5th Ed., how would this star be classified?
A) G2 I. B) G2 III. C) G-type white dwarf. D) G2 V.
Answer: B Source: Sections 10-4 and 10-5

75. A particular star has an absolute magnitude of +12 and a spectral class of A5. Using Figures 10-4 and 10-6 of Kaufmann and Comins, *Discovering the Universe*, 5th Ed., how would this star be classified?
A) A5 III. B) A5 V. C) A5 I. D) A-type white dwarf.
Answer: D Source: Sections 10-4 and 10-5

76. What is spectroscopic parallax?
A) The apparent change in position of a nearby star compared to distant background stars, due to the motion of the Earth around the Sun.
B) The change in position of the absorption lines in a star's spectrum due to the Doppler shift, caused by the star's motion around the center of mass in a binary star system.
C) The distance to a star, measured using the spectral-luminosity class of the star and the inverse square law.
D) The apparent change in position of the absorption lines in a star's spectrum due to the Doppler shift, caused by the Earth's motion around the Sun.
Answer: A Source: Section 10-6

77. Two stars, one classified A4 V and the other A4 III, have the same apparent magnitude. There is no significant amount of absorption of starlight by interstellar material. From this information we know that
A) the A4 V star is cooler than the A4 III star.
B) the A4 V star is hotter than the A4 III star.
C) the A4 V star is further from the Sun than the A4 III star.
D) the A4 V star is closer to the Sun than the A4 III star.
Answer: D Source: Section 10-6

78. Two stars, one classified A4 V and the other F8 V, have the same apparent magnitude. There is no significant amount of absorption of starlight by interstellar material. From this information we know that
A) both stars are at the same distance from the Sun.
B) the A4 V star is smaller than the F8 V star.
C) the A4 V star is closer to the Sun than the F8 V star.
D) the A4 V star is further from the Sun than the F8 V star.
Answer: D Source: Section 10-6

79. What proportion of visible stars in our night-time sky are multiple-star systems, such as binary stars?
A) Less than 1%
B) Nearly 100%
C) Only about 1/4 or 25%
D) 1/2 or about 50%
Answer: D Source: Introduction to Section 10-7

80. Which one of the following statements is correct for an isolated star (i.e., a star which is not in a binary star system)?
A) There are several ways to measure its mass accurately.
B) It is not possible to measure the star's mass accurately.
C) Its mass can be measured accurately only if its luminosity and temperature can be measured.
D) Its mass can be measured accurately only if its distance can be found.
Answer: B Source: Introduction to Section 10-7

81. What is the only way to measure the mass of a star accurately?
A) Measure its spectral type and luminosity class, then use the HR diagram.
B) Measure its distance using trigonometric parallax and its brightness using photometry.
C) Measure its gravitational effect on another object.
D) It is not possible to measure the mass of a star.
Answer: C Source: Introduction to Section 10-7

82. How do astronomers measure the masses of stars?
A) By observing the star's brightness at different wavelengths (colors).
B) By measuring the star's brightness, and obtaining its radius using the HR diagram.
C) By measuring the star's brightness, temperature, and distance.
D) By observing the motion of two stars in a binary star system.
Answer: D Source: Introduction to Section 10-7

83. Which important stellar parameter can be derived from the study of binary stars mutually bound to each other by gravitational forces?
    A) Surface temperatures of the stars.
    B) The distance of the stars from Earth.
    C) Stellar masses.
    D) The age of the stars.
    Answer: C Source: Introduction to Section 10-7

84. Which important stellar parameter can be best determined by observations of binary stars?
    A) Pulsation period.
    B) Surface temperature.
    C) Distance from Earth.
    D) Stellar mass.
    Answer: D Source: Introduction to Section 10-7

85. One important aspect of the study of binary star systems, as distinct from single stars, is that it provides
    A) a verification of the Doppler equation for wavelength shift of light from moving objects.
    B) a measurement of the surface temperatures of stars.
    C) a measurement of the masses of stars.
    D) a measurement of the composition (abundances of elements) inside stars.
    Answer: C Source: Introduction to Section 10-7

86. How do two unequal-mass stars move around each other, in general, in a binary star system?
    A) The low-mass star moves in a circular orbit around high-mass star which remains stationary.
    B) In a single circular orbit around the same center, and always on opposite sides from each other.
    C) In elliptical orbits, about a common "center of mass".
    D) In straight lines, back and forth past each other.
    Answer: C Source: Section 10-7

87. What is the difference between an optical double star and a visual binary star?
    A) The stars in an optical double star are actually orbiting each other, whereas a visual binary is an illusion: the stars are at vast distances from each other and are not actually orbiting each other.
    B) An optical double is an illusion. The stars are at vast distances from each other and are not actually orbiting each other, whereas in a visual binary the stars are actually orbiting each other.
    C) Optical double stars can be seen as separate stars only through a telescope, whereas visual binaries can be seen with the unaided eye (e.g., the star Mizar in the Big Dipper's handle).
    D) There is no difference; they are two names for the same thing.
    Answer: B Source: Section 10-7

88. In a particular binary star system, only one star is visible because the other star is too faint to see at that distance. An astronomer measures the size (semimajor axis) and period of the orbit of the visible star. From this information the astronomer
    A) can calculate the sum of the masses of the two stars, but not the mass of each star separately.
    B) can calculate the mass of the visible star, but not that of the unseen star.
    C) cannot calculate anything about the mass; both stars have to be visible to do this.
    D) can calculate the mass of each star.
    Answer: A Source: Section 10-7

89. A particular star in a binary star system orbits the other in an elliptical orbit with a semimajor axis of 3 AU and a period of 5 years. What is the sum of the masses of the two stars in the system?
    A) 1.1 $M_{\odot}$. B) 0.07 $M_{\odot}$. C) 13.9 $M_{\odot}$. D) 0.9 $M_{\odot}$.
    Answer: A Source: Section 10-7 and Toolbox 10-1

90. Two stars in a binary system orbit around a common point which is
    A) always exactly midway between the two stars.
    B) closer to the less massive star.
    C) always inside one of the stars.
    D) closer to the more massive star.
    Answer: D Source: Section 10-7

91. When two stars of unequal mass orbit each other under their mutual gravitational attraction, where is the center of mass of the system located?
   A) At a point halfway between the centers of the stars.
   B) At the center of the more massive star.
   C) At a point between the two stars, closer to the more massive star.
   D) At a point between the stars, closer to the less massive star.
   Answer: C Source: Section 10-7

92. The relationship between mass and luminosity of stars on the main sequence is that
   A) luminosity is independent of the stellar mass.
   B) the luminosity of stars reaches a peak at around 1 solar mass, and decreases as mass increases and decreases beyond this limit.
   C) the greater the stellar mass, the less the luminosity.
   D) the larger the stellar mass, the larger the luminosity.
   Answer: D Source: Section 10-8

93. Where are the most massive stars to be found in the main sequence of a Hertzsprung-Russell diagram?
   A) Main sequence stars all have approximately the same mass, by definition.
   B) The lower, right end.
   C) In the center section near to the Sun's position, with lower mass stars on either side.
   D) The upper, left end.
   Answer: D Source: Section 10-8

94. What will be the mass of a main-sequence star which has a luminosity 1000 times greater than that of the Sun? (See Fig. 10-9, Kaufmann & Comins, *Discovering the Universe*, 5th Ed.)
   A) $10^5$ solar masses.
   B) 1000 solar masses.
   C) 0.1 solar mass.
   D) 5 solar masses.
   Answer: D Source: Section 10-8

95. Using the Hertzsprung-Russell diagram and the mass-luminosity relationship for main sequence stars (Fig. 10-9 of Kaufmann & Comins, *Discovering the Universe*, 5th Ed.), determine which of the following type of main sequence stars will have the highest mass.
   A) G. B) A. C) M. D) O.
   Answer: D Source: Section 10-8

96. Using the Hertzsprung-Russell diagram in Fig. 10-4 of Kaufmann and Comins, *Discovering the Universe*, 5th Ed., and the mass-luminosity relationship for main sequence stars shown in Fig. 10-9, which of the following is the correct sequence of stars in increasing order of mass?
   A) Barnard's Star, the Sun, Altair, Regulus.
   B) Regulus, Barnard's Star, the Sun, Altair.
   C) Barnard's Star, Altair, the Sun, Regulus.
   D) Regulus, Altair, the Sun, Barnard's Star.
   Answer: A Source: Section 10-8

97. Use the Hertzsprung-Russell diagram (Fig. 10-5) and the mass-luminosity relation (Fig. 10-9) in Kaufmann and Comins, *Discovering the Universe*, 5th Ed. to estimate the mass of Vega, which is an AO V main-sequence star with a surface temperature of about 10,000K.
   A) Between 3.0 and 5.0 solar masses.
   B) About 1.0 solar masses.
   C) Between 1.5 and 2.0 solar masses.
   D) About 10 solar masses.
   Answer: C Source: Section 10-8

98. The radial-velocity curve of a star in a binary star system is a plot against time of
   A) the variation of Doppler shift of its spectral lines, and hence of its speed towards or away from us.
   B) the speed of the star in a direction perpendicular to the line of sight to the star.
   C) the position of the star in celestial coordinates.
   D) the temperature of the star as determined from the movement of the peak wavelength of its spectrum.
   Answer: A Source: Section 10-9

99. Absorption lines in the spectra of some binary stars are seen to change periodically from single to double lines and back again. Why is this?
   A) Motion toward and away from Earth during their orbital motion results in Doppler shift of light from these stars at times and no shift when the stars are moving perpendicular to the line of sight.
   B) Oscillations on the surfaces of the stars leads to Doppler-shifted lines.
   C) The effect of the gravitational field of one star on the atoms of the second star produces spectral line shifts periodically.
   D) The magnetic field of one star produces Zeeman splitting of spectral lines in atoms of the second star, periodically.
   Answer: A Source: Section 10-9

100. An eclipsing binary system consists of
   A) two stars orbiting each other, in which periodic spectral line shifts because of Doppler shift are measured.
   B) two stars which are clearly resolved as separate but orbit each other and are obviously gravitationally bound to each other.
   C) two stars whose combined light output towards Earth varies regularly as one star moves in front of the other periodically.
   D) a star which is periodically eclipsed by the Moon.
   Answer: C Source: Section 10-10

101. An eclipsing binary system consists of
   A) two mutually orbiting and gravitationally bound stars which are close enough to be resolved when viewed from Earth.
   B) two stars which periodically eclipse each other, as seen from Earth.
   C) two stars in which spectral lines move back and forth periodically because of Doppler shift, indicating mutually orbiting stars.
   D) a star which is periodically eclipsed by the Moon.
   Answer: B Source: Section 10-10

102. The light intensity from a particular star remains essentially constant, except for short and regular decreases by a fixed amount. What is the explanation for this phenomenon?
   A) One star is regularly eclipsing its companion as they move in mutual orbits whose plane is close to the line of sight.
   B) Shells of absorbing gas and dust are being periodically ejected from the star's surface and are subsequently dispersing into space.
   C) A variable star is pulsating in size, temperature, and intensity.
   D) A massive, dark planet is periodically passing in front of the star's visible surface.
   Answer: A Source: Section 10-10

103. Which of the following observations would **NOT** be an indication of a binary star system?
   A) The "star" appears to move in a straight line against a background field of stars.
   B) A "star" appears to become periodically dimmer for a few hours at a time.
   C) A "star" image separates into two distinct images periodically and then blends again, periodically.
   D) The "star" appears to wiggle in its path across our sky against the background stars.
   Answer: A Source: Sections 10-9 and 10-10

104. What condition is necessary in order for us to see eclipses of stars in binary star systems?
   A) The line of sight from Earth to the star system must be in, or very close to, the orbital plane of the stars.
   B) The line of sight from Earth to the star system must be very close to the perpendicular to the orbital plane of the stars.
   C) One of the stars must be much bigger than the other in order to hide its smaller companion when the orbital plane is at a large angle to the line of sight.
   D) The stars must have very similar surface temperatures whatever the inclination of their orbital plane to the line of sight, in order for us to see a significant eclipse.
   Answer: A Source: Section 10-10

105. Which of the following major perturbations can occur to a close binary system and radically alter the evolution and behaviour of the two individual stars?
   A) The eclipsing of the light from one star by the other, when viewed from Earth.
   B) The gravitational disturbance of one star's motion by its companion, to force it to move in an orbit.
   C) The heating of the localized areas of the atmosphere of one star by its companion.
   D) The transfer of matter from one star to its companion.
   Answer: D Source: Section 10-10

106. What particular and very important phenomenon frequently occurs in binary star systems where the stars are very close together?
   A) The less massive star spirals slowly into its more massive companion because of tidal interactions.
   B) The radiation from the hotter star will slowly heat and evaporate away the cooler star.
   C) The less massive star, in its elliptical orbit, will repeatedly pass through the thin, extended atmosphere of the second star, producing periodic rises and falls in light output from the star system.
   D) Mass lost from one star is deposited upon its companion.
   Answer: D Source: Section 10-10

107. The components of a binary star, particularly if they are close, can influence each other in various ways. Which of the following is **NOT** likely to be an effect of one star upon its companion?
   A) Mass can be transferred from one star to its companion.
   B) A very hot star can heat part of its cooler companion to produce a hot spot.
   C) Intense radiation from a hot star can produce nuclear reactions on the surface of a cooler companion and initiate a nova explosion.
   D) The gravitational force of one star will make its companion move in an orbit, rather than remaining stationary.
   Answer: C Source: Section 10-10

# Chapter 11: The Life Cycles of Stars

1. The space between stars is known to contain
   A) variable amounts of gas but no dust, which only forms in planetary systems near stars.
   B) dust and gas, both atomic and molecular.
   C) a perfect vacuum.
   D) large quantities of dust that absorb light, but no gas, either atomic or molecular.
   Answer: B Source: Section 11-1

2. Observations of similar star clusters at different distances from the Sun show that they appear to be fainter than expected on the basis of distance alone. This is because
   A) the photons of light become "tired" and appear less bright as they travel.
   B) the cosmological redshift has moved some of the light into the infrared spectral region.
   C) light is scattered and absorbed by interstellar dust and gas between distant clusters and the Earth.
   D) star clusters are systematically smaller and hence less bright, the further they are from the galactic center and hence from the Sun.
   Answer: C Source: Section 11-1

3. How much of the visible mass of our galaxy is in the form of gas and dust spread out between the stars?
   A) 90% B) Less than 1%. C) 0% D) 50%
   Answer: C Source: Section 11-1

4. What is the most abundant element in the universe?
   A) Helium. B) Oxygen. C) Hydrogen. D) Carbon.
   Answer: C Source: Section 11-1

5. Which is the second most abundant element in the universe (after hydrogen)?
   A) Helium. B) Nitrogen. C) Carbon. D) Iron.
   Answer: A Source: Section 11-1

6. Which are the two most abundant elements in the universe?
   A) Nitrogen and oxygen.
   B) Hydrogen and helium.
   C) Hydrogen and oxygen.
   D) Hydrogen and carbon.
   Answer: B Source: Section 11-1

7. Which of the following molecules is likely to be the most common in interstellar space?
   A) OH, hydroxyl B) $H_2$ C) $H_2O$, water D) CO, carbon monoxide
   Answer: B Source: Section 11-1

8. How have molecules such as formaldehyde ($H_2CO$) been detected in interstellar clouds?
   A) By observing the chemical reactions in which they are created.
   B) Only by theoretical modeling, knowing that the component elements (H, C, O) are present.
   C) By direct sampling by space probes.
   D) By molecular emission lines.
   Answer: D Source: Section 11-1

9. In which part of the electromagnetic spectrum are molecules most easily detected?
   A) Visible light. B) Ultraviolet light. C) Radio waves. D) X-ray.
   Answer: C Source: Section 11-1

10. At what wavelengths have astronomers mapped and studied the distribution of the giant molecular clouds in space?
    A) Millimeter wavelengths, using radio telescopes.
    B) Long radio wavelengths, greater than 20 cm, since molecules are very efficient radio emitters.
    C) UV, since molecules are efficient UV emitters and the clouds are hot.
    D) Visible light, using photography.
    Answer: A Source: Section 11-1

11. Which of the following easily observed molecular species is used as a tracer for the fundamental but difficult-to-observe $H_2$ molecules in giant molecular clouds?
    A) $CO_2$ B) $H_2O$ C) CO D) OH
    Answer: C Source: Section 11-1

12. Hydrogen in molecular form, $H_2$, is thought to be very abundant in gas clouds in space, but these molecules emit radiation relatively inefficiently, since they are symmetrical molecules. Which other molecule occurs in close association with $H_2$ and is used as a probe for molecular clouds?
    A) Methane, $CH_4$ B) CO C) $CO_2$ D) $H_2O$
    Answer: B Source: Section 11-1

13. In star-forming regions in interstellar space, which molecule is the easiest to detect?
    A) Formaldehyde ($H_2CO$)
    B) Ammonia ($NH_3$)
    C) Hydrogen ($H_2$)
    D) Carbon monoxide (CO)
    Answer: D Source: Section 11-1

14. An astronomer observes that in a particular interstellar cloud there are 500 molecules of carbon monoxide (CO) per cubic meter. What abundance of hydrogen gas ($H_2$) does the astronomer infer from this observation?
    A) 50,000 molecules of $H_2$ per cubic meter.
    B) 5,000,000 molecules of $H_2$ per cubic meter.
    C) 500,000,000 molecules of $H_2$ per cubic meter.
    D) 500 molecules of $H_2$ per cubic meter.
    Answer: B Source: Section 11-1

15. How is gas distributed in interstellar space?
    A) Uniformly distributed through space.
    B) Concentrated in river-like streams of gas that extend across the galaxy.
    C) In clumps, concentrated in interstellar clouds.
    D) Concentrated around existing stars, because of the stars' gravitational pull.
    Answer: C Source: Section 11-1

16. What is the typical mass of a giant molecular cloud?
    A) 10 to 100 solar masses.
    B) 1000 to 10,000 solar masses.
    C) 100,000 to 1,000,000 solar masses.
    D) 10 million to 1 billion solar masses.
    Answer: C Source: Section 11-1

17. What is a typical size for a giant molecular cloud?
    A) 1000 light years across.
    B) Anything up to about a light year across.
    C) 100 light years across.
    D) 10 light years across.
    Answer: C Source: Section 11-1

18. Giant molecular clouds, which are major sites of star formation, can be up to
    A) 10 pc across and contain a few thousand solar masses of material.
    B) 1000 pc across and contain 100 million solar masses of material.
    C) 100 pc across and contain 2 million solar masses of material.
    D) 10 times the size of the solar system and contain 2 to 3 solar masses of material.
    Answer: C Source: Section 11-1

19. What fraction of the mass of the universe is hydrogen (as represented by the abundance in giant molecular clouds)?
    A) 98% B) Almost 50% C) 74% D) 2%
    Answer: C Source: Section 11-1

20. What fraction of the mass of the universe is helium (as represented by the abundance in giant molecular clouds)?
    A) 25% B) 2% C) 10% D) 50%
    Answer: A Source: Section 11-1

21. A typical dark nebula of gas and dust from which a star might form has which of the following dimensions?
    A) About 10 solar masses inside a diameter of about 10 parsecs.
    B) A few thousand solar masses inside a diameter of about 10 parsecs.
    C) 1 or 2 solar masses inside a diameter of about 100 AU, about the size of the solar system.
    D) A few thousand solar masses inside about 1 AU, the Earth's orbit.
    Answer: B Source: Section 11-1

22. A particular interstellar giant molecular cloud has a mass of 2,000,000 solar masses. What is the mass of hydrogen in this cloud?
    A) 1,960,000 solar masses.
    B) 1,500,000 solar masses.
    C) 1,000,000 solar masses.
    D) 40,000 solar masses.
    Answer: B Source: Section 11-1

23. A particular giant molecular cloud in interstellar space has a mass of 2,000,000 solar masses. What is the mass of helium in this cloud?
    A) 500,000 solar masses.
    B) 40,000 solar masses.
    C) 200,000 solar masses.
    D) 1,000,000 solar masses.
    Answer: A Source: Section 11-1

24. How do massive stars normally end their lives?
    A) They collapse and become black holes.
    B) We don't know, since their lifetimes are longer than the age of the universe.
    C) They explode.
    D) They gradually shrink to the size of the Earth.
    Answer: C Source: Section 11-2

25. The most likely places where stars and planetary systems are forming in the universe are
    A) in nebulae composed of gas and dust.
    B) in the rarified space between galaxies.
    C) in regions surrounding quasars.
    D) in the centers of galaxies.
    Answer: A Source: Section 11-2

26. New stars are formed from
    A) huge, cool dust and gas clouds.
    B) free space out of pure energy.
    C) activity in the centers of galaxies.
    D) hot supernova remnants.
    Answer: A Source: Section 11-2

27. How does the temperature of an interstellar cloud affect its ability to form stars?
    A) Higher temperatures help star formation.
    B) Star formation is independent of the temperature of the cloud.
    C) Higher temperatures inhibit star formation.
    D) Star formation is too complicated to be able to say how one quantity, such as temperature, affects it.
    Answer: C Source: Section 11-2

28. What determines whether a particular region of an interstellar cloud can collapse and form a star?
    A) The amount of gravity pulling inwards compared to gas pressure pushing outwards.
    B) Only the temperature, since higher temperatures act to prevent collapse.
    C) Only the amount of mass in the cloud, since this determines the strength of gravity.
    D) How the mass of the cloud compares to its diameter, since this determines how gravity compares to the distance needed to collapse.
    Answer: A Source: Section 11-2

29. What condition is considered sufficient for an interstellar cloud to collapse and form a star or stars (i.e., if this condition holds then the cloud has to collapse)?
    A) The cloud must be cooler than 100 K.
    B) Gravity must be strong enough to reach all parts of the cloud.
    C) Gravity must dominate gas pressure inside the cloud.
    D) The cloud must be alone in space (far from stars and other interstellar clouds).
    Answer: C Source: Section 11-2

30. Which of the following mechanisms is NOT considered to be one way in which stars are formed?
    A) Collisions between interstellar clouds.
    B) Heating of an interstellar cloud by young stars.
    C) Compression of an interstellar cloud by the shock waves from supernova explosion.
    D) Compression of an interstellar cloud by the pressure of light from nearby stars.
    Answer: B Source: Section 11-2

31. Which of the following mechanisms is thought to be ineffective and inefficient in the triggering of star birth in molecular clouds?
    A) Supernova explosions and the resultant shock waves.
    B) Gravitational contraction of a hot gas cloud.
    C) Pressure waves in the spiral arms of a galaxy.
    D) Collisions between two interstellar clouds.
    Answer: B Source: Section 11-2

32. Inside a giant molecular cloud, what is the typical temperature inside a dense core that is collapsing to form a star?
    A) Less than 1 K B) 10 K C) 1,000 K D) 100 K
    Answer: B Source: Section 11-2

33. When a giant molecular cloud collapses to form a group of stars, what is this stellar group called?
    A) A galaxy. B) An open cluster. C) A globular cluster. D) A constellation.
    Answer: B Source: Section 11-2

34. Which wavelength region is most useful for investigating dense cores inside giant molecular clouds?
    A) Ultraviolet B) X-ray C) Infrared D) Optical (visual)
    Answer: C Source: Section 11-2

35. Giant molecular clouds of $H_2$ and CO gas are found in which regions of our galaxy?
    A) Along the spiral arms.
    B) Above and below the plane of the spiral arms, over the galactic poles.
    C) At the center of the galaxy.
    D) They appear to be uniformly spread throughout the galaxy, both in and above and below the spiral arms.
    Answer: A Source: Section 11-2

36. How does an interstellar cloud collapse to become a star?
    A) The outer, less dense, part falls in faster, then its weight accelerates the collapse of the inner part.
    B) All parts of the cloud accelerate inwards more-or-less smoothly and evenly.
    C) The collapse is turbulent and chaotic, with no overall pattern.
    D) The innermost part collapses first, then the outer part accretes onto the inner part.
    Answer: D Source: Section 11-2

37. The characteristics of an open cluster of stars are
    A) a few hundred members, often very young and still embedded in the gas and dust from which they were formed.
    B) many thousand members, of different ages.
    C) a few dozen members, the remnant of a globular cluster of stars from which most of the members have escaped.
    D) hundreds of thousands of members, all very old, and no or very little interstellar gas and dust.
    Answer: A Source: Section 11-2

38. In which one of the following locations are clumps of gas most likely to be collapsing to form stars?
    A) In the outer part of our solar system.
    B) In a globular cluster.
    C) In hot, glowing H II regions.
    D) In giant molecular clouds.
    Answer: D Source: Section 11-2

39. What is a protostar?
    A) A star near the end of its life, before it explodes as a supernova.
    B) A sphere of gas after collapse from an interstellar cloud but before nuclear reactions have begun.
    C) A small interstellar cloud, before it collapses to become a star.
    D) A shell of gas left behind from the explosion of a star as a supernova.
    Answer: B Source: Section 11-2

40. At what stage in its life does a star pass through the protostar phase?
    A) After condensation but before nuclear reactions begin in its core.
    B) While it is converting hydrogen into helium in its core.
    C) After nuclear reactions end in its core, but before the red giant phase.
    D) When it is expanding in size as a red giant or supergiant.
    Answer: A Source: Section 11-2

41. Where in the universe would you look for a protostar?
    A) In dense dust and gas clouds.
    B) In globular clusters of stars.
    C) In the empty space between galaxies.
    D) Near black holes.
    Answer: A Source: Section 11-2

42. Protostars are
    A) very young objects, still contracting before becoming true stars.
    B) stars made almost entirely out of protons.
    C) objects with masses less than about 0.08 solar masses, which do not have enough mass to become true stars.
    D) old stars, contracting after using up all of their available hydrogen fuel.
    Answer: A   Source: Section 11-2

43. Star formation takes place in
    A) giant molecular clouds.
    B) blue reflection nebulae.
    C) H II regions.
    D) hot, turbulent gas thrown out in a supernova explosion.
    Answer: A   Source: Section 11-2

44. Protostars, when they first form from the interstellar medium, are usually
    A) easily detected because their light ionizes the surrounding interstellar gas, forming H II regions.
    B) very bright in ultraviolet light due to numerous flares (like solar flares but hotter and brighter).
    C) hidden from sight by dust clouds that emit infrared radiation.
    D) detected by emission lines in their visible spectra, emitted by gas being blown off their surfaces into space.
    Answer: C   Source: Section 11-2

45. Which range of electromagnetic radiation is useful for observing new-born protostars within their gas and dust nebulae?
    A) Radio.   B) Visible.   C) Infrared.   D) Highly penetrating x-ray.
    Answer: C   Source: Section 11-2

46. Which range of wavelengths of electromagnetic radiation is most effective in the study of newborn protostars in their dust clouds and nebulae?
    A) Infrared.   B) Gamma rays.   C) Ultraviolet.   D) Radio.
    Answer: A   Source: Section 11-2

47. The Orion Nebula is
    A) a supernova remnant (material thrown out by an exploding star).
    B) a red supergiant star.
    C) a large interstellar gas and dust cloud containing young stars.
    D) a spiral galaxy in the constellation Orion.
    Answer: C   Source: Section 11-2

48. Infrared stars within the Orion Nebula are examples of which stage of stellar evolution?
    A) Planetary nebula.
    B) Red giant.
    C) Supernova remnants.
    D) Protostar and young star.
    Answer: D Source: Section 11-2

49. The source of a protostar's heat is
    A) gravitational energy, released as the star contracts.
    B) nuclear reactions converting helium to carbon and oxygen in its core.
    C) gravitational energy, released as the protostar expands.
    D) nuclear reactions converting hydrogen into helium in its core.
    Answer: A Source: Section 11-2

50. The major source of energy in the pre-main-sequence life of the Sun was
    A) nuclear fusion.
    B) gravitational.
    C) nuclear fission.
    D) burning of carbon atoms.
    Answer: B Source: Section 11-2

51. Protostars are
    A) slowly contracting and heating up.
    B) slowly contracting and cooling.
    C) slowly heating up and expanding.
    D) slowly expanding at the surface while the core contracts.
    Answer: A Source: Section 11-2

52. What is believed to be the most important factor determining whether a collapsing region (dense core) in an interstellar cloud becomes a single star or a multiple star system?
    A) The mass of the collapsing region.
    B) The temperature.
    C) The amount of rotation (spin).
    D) The fraction of heavy elements in the cloud.
    Answer: C Source: Section 11-2

53. How much time does it take for a one solar mass protostar to finish accreting mass from the interstellar cloud?
    A) 10,000,000 years. B) 100,000 years. C) 10,000 years. D) 1 billion years.
    Answer: B Source: Section 11-2

54. How large was the Sun when it first formed as a protostar?
    A) About five times its present diameter.
    B) The same diameter that it has now.
    C) About fifty times its present diameter.
    D) About fifteen times its present diameter.
    Answer: A Source: Section 11-2

55. What is the most important process causing a protostar to stop accreting mass?
    A) Radiation and particles from the protostar push infalling matter away from the protostar.
    B) The dense core spins up as it collapses, and eventually the infalling matter is held away from the protostar by the centrifugal force.
    C) Other protostars formed in the vicinity are passing randomly through the infalling material and eventually disperse it.
    D) All of the infalling matter has been used up in the accretion.
    Answer: A Source: Section 11-2

56. What is a protostar called in the stage after it has finished accreting mass?
    A) A main sequence star.
    B) A pre-main sequence star.
    C) A white dwarf.
    D) A red giant (or supergiant).
    Answer: B Source: Section 11-3

57. How long does it take for a one solar mass star to pass through the pre-main-sequence phase?
    A) 10,000,000 years. B) 10 billion years. C) 10,000 years. D) 100,000 years.
    Answer: A Source: Section 11-3

58. In which region of the Hertzsprung-Russell diagram will a newly formed protostar first appear when it begins to shine at visible wavelengths?
    A) Bottom right corner; very low luminosity and cool.
    B) At the center of the main sequence, since all protostars begin their lives at this position and move away as time passes.
    C) The right side; relatively large luminosity but cool.
    D) Top left corner; at the top of the main sequence, down which it will progress with time.
    Answer: C Source: Section 11-3

59. A protostar of about 1 solar mass is gradually contracting and becoming hotter. This will cause its position in the Hertzsprung-Russell diagram to shift slowly
    A) upward and toward the right.
    B) downward and toward the right.
    C) upward and toward the left.
    D) downward and toward the left.
    Answer: D Source: Section 11-3

60. For any star on the main sequence, the same star when it was a protostar was
    A) hotter but not necessarily larger.
    B) larger and hotter.
    C) cooler but not necessarily larger.
    D) larger and cooler.
    Answer: D Source: Section 11-3

61. At what point in its evolution will a protostar stop shrinking and stabilize into a star?
    A) When nuclear processes generate enough energy and internal pressure to resist gravitational contraction.
    B) When nuclear reactions end in its core.
    C) When gravitational contraction has heated up the gas to the point where radiation pressure opposes gravity for the first time.
    D) When it has spun off enough of its matter and is spinning fast enough that centrifugal force opposes the gravitational contraction.
    Answer: A Source: Section 11-3

62. What event occurs at the end of the protostar stage of a star's life?
    A) Gas is spun off from its equator, from which planets may form.
    B) Nuclear reactions begin in its core, converting hydrogen into helium.
    C) It begins a long period of contraction, in which gravitational energy is converted into heat.
    D) It explodes, forming a supernova remnant.
    Answer: B Source: Section 11-3

63. If a protostar were able to contract (get smaller) without any change to its surface temperature, what would happen to its luminosity?
    A) It would remain the same, since the temperature does not change.
    B) It would increase, due to the compression of the gas.
    C) It is not possible to predict the change in luminosity, since other factors are involved.
    D) It would decrease, due to the smaller surface area of the protostar.
    Answer: D Source: Section 11-3

64. The lowest mass which a protostar can have and still become a star (i.e., start thermonuclear reactions in its core) is
   A) slightly less than 1/100 of a solar mass.
   B) about half a solar mass.
   C) slightly less than 1/10 of a solar mass.
   D) 8/10 of a solar mass.
   Answer: C Source: Section 11-3

65. The smallest mass which a main sequence star can have is about 0.08 solar mass. The reason for this is that
   A) thermonuclear reactions begin so suddenly in stars of less than 0.08 solar mass that the star is disrupted by an explosion.
   B) the temperature in a contracting protostar of less than 0.08 solar masses does not get high enough for nuclear reactions to start.
   C) protostars cannot form with masses less than 0.08 solar mass.
   D) protostars of less than 0.08 solar masses are not massive enough to contract.
   Answer: B Source: Section 11-3

66. What is the lowest mass that an object can have and still be a star?
   A) 0.02 solar masses.
   B) 0.80 solar masses.
   C) 0.002 solar masses (twice Jupiter's mass).
   D) 0.08 solar masses.
   Answer: D Source: Section 11-4

67. The lowest mass that a protostar can have and still become a star (i.e., start thermonuclear reactions in its core) is
   A) slightly less than 1/100 of a solar mass.
   B) slightly less than 1/10 of a solar mass.
   C) about half a solar mass.
   D) 8/10 of a solar mass.
   Answer: B Source: Section 11-4

68. What is the relationship between the mass of a protostar and the time needed for it to reach the main sequence, after it forms inside an interstellar cloud?
   A) More massive protostars reach the main sequence in a shorter time than less massive protostars.
   B) Less massive protostars reach the main sequence in a shorter time than more massive protostars.
   C) The time needed is independent of the mass of the protostar.
   D) The time needed is least for a protostar of approximately 4 solar masses, and longer for protostars of either greater or less mass.
   Answer: A Source: Section 11-4

69. How long did the Sun spend in the protostar phase of its life?
    A) 400 thousand years.
    B) 20 million years.
    C) 50 thousand years.
    D) 4.6 billion years.
    Answer: B  Source: Section 11-4

70. Suppose that an astronomy news item announces the discovery of a brown dwarf. What is it that has been discovered?
    A) A Pluto-like object in the Kuiper belt, beyond the edge of the planetary system.
    B) An object too large to be a planet but too small to be a star.
    C) A protostar still embedded in the cloud of gas and dust from which it formed.
    D) A pre-stellar object undergoing gravitational collapse.
    Answer: B  Source: Section 11-4

71. A brown dwarf is
    A) an object intermediate between a planet and a star, with not enough mass to begin nuclear reactions in its core.
    B) a general name for objects similar to the planet Jupiter.
    C) any star whose black-body spectrum peaks in the brown region of the visible spectrum
    D) a star of less than about 1½ solar masses at the very end of its life, after it has cooled to near-invisibility.
    Answer: A  Source: Section 11-4

72. Suppose that an astronomical observatory announces the discovery of an object with about 50 times the mass of Jupiter (this is not enough mass to be a star). What name would the observatory apply to this object?
    A) A red dwarf. B) A brown dwarf. C) A white dwarf. D) An infrared dwarf.
    Answer: B  Source: Section 11-4

73. An object which is too massive to be a planet but not massive enough to be a star is called
    A) a red dwarf. B) a white dwarf. C) a T Tauri star. D) a brown dwarf.
    Answer: D  Source: Section 11-4

74. What is believed to be the maximum mass that a star can have?
    A) About 50 solar masses.
    B) About 300 solar masses.
    C) About 100 solar masses.
    D) About 1000 solar masses.
    Answer: C  Source: Section 11-4

75. What is believed to prevent stars from having masses greater than about 100 solar masses?
    A) Stars of larger mass would collapse under their own gravity and become black holes.
    B) The temperature becomes so high that the excess mass is pushed back into space.
    C) No interstellar clouds have masses of more than 100 solar masses.
    D) The cores of larger-mass stars would run through their lives and explode before the stars finish contracting as protostars.
    Answer: B Source: Section 11-4

76. Which part of the Hertzsprung-Russell diagram is occupied by protostars?
    A) A band running from upper right to lower left.
    B) To the left of the main sequence.
    C) A band running from upper left to lower right.
    D) To the right of the main sequence.
    Answer: D Source: Section 11-4

77. Main sequence stars do not have masses larger than about 100 solar masses. The reason that stars of larger mass do not exist is that
    A) their temperature becomes so high that they are disrupted by the pressure of radiation inside them.
    B) such stars contract directly to become planet-like objects.
    C) the thermonuclear reactions in such stars proceed so rapidly that the star explodes.
    D) interstellar clouds of greater mass break up to become binary or multiple star systems, not single stars.
    Answer: A Source: Section 11-4

78. What point defines the end of the pre-main-sequence phase of a star's life and the start of the main-sequence phase?
    A) Nuclear reactions begin in its core.
    B) It stops accreting mass from the interstellar cloud.
    C) It begins to expand and become a red giant.
    D) Convection begins in its interior.
    Answer: A Source: Section 11-3

79. At what temperature do nuclear reactions begin in the core of a pre-main-sequence star?
    A) 100,000 K. B) 10 million K. C) 100 million K. D) 1 million K.
    Answer: B Source: Section 11-3

80. What is an H II region?
    A) A region of ionized hydrogen around one or more O and B stars.
    B) A region of molecular hydrogen inside a giant molecular cloud.
    C) A region of neutral, atomic hydrogen in interstellar space.
    D) A region of gas and dust formed by the explosion of a massive star.
    Answer: A Source: Section 11-5

81. Where are H II regions found?
    A) Around hot stars
    B) Around low-mass stars
    C) In or near old open clusters
    D) In globular star clusters
    Answer: A Source: Section 11-5

82. The bright stars at the center of an H II region (emission nebula) are
    A) young O and B stars.
    B) red supergiants.
    C) hot white dwarfs.
    D) T Tauri stars.
    Answer: A Source: Section 11-5

83. What radiation ionizes the hydrogen in an H II region?
    A) Ultraviolet radiation from O and B stars.
    B) X-ray from the coronas of solar-type stars.
    C) Infrared radiation from pre-main-sequence stars.
    D) Gamma rays from neutron stars.
    Answer: A Source: Section 11-5

84. What process makes an emission nebula glow?
    A) Electrons descending toward the ground state in hydrogen atoms.
    B) Free electrons emitting light as they pass close to positively charged ions.
    C) Electric currents in ionized neon gas.
    D) High-energy electrons spiraling along magnetic field lines.
    Answer: A Source: Section 11-5

85. An emission nebula shines predominantly by light
    A) in the Balmer Ha red line, from recombination of electrons with nuclei in ionized hydrogen.
    B) emitted over a wide range of wavelengths by dust grains, heated by radiation from embedded stars.
    C) scattered preferentially at blue wavelengths, originally emitted by stars within the nebula.
    D) emitted by molecules in the dense clouds of gas surrounding the stars in the nebula.
    Answer: A Source: Section 11-5

86. What is the dominant optical (visual) spectral emission line from an emission nebula?
A) Lyman La hydrogen line.
B) The "green line" of oxygen.
C) The Balmer Ha hydrogen line.
D) The sodium D lines.
Answer: C Source: Section 11-5

87. What is the characteristic color of an emission nebula?
A) Green B) Blue C) Red D) Yellow
Answer: C Source: Section 11-5

88. The predominant color of an emission nebula is
A) red, from the Balmer Ha line.
B) blue, from scattering of light from hot stars by dust particles.
C) yellow-green, from the 530.3 nm emission line of ionized iron, equivalent to that from the hot solar corona.
D) a continuum of all colors, the combined light from the stars in the nebula.
Answer: A Source: Section 11-5

89. The energy required to ionize the hydrogen gas in an emission nebula (H II region) comes from
A) hot O and B stars.
B) supernovae (exploding stars).
C) collisions between gas clouds in interstellar space.
D) T Tauri stars.
Answer: A Source: Section 11-5

90. Evidence of massive amounts of hydrogen gas surrounding some stars comes from
A) emission of characteristic red Balmer light from nebulosity around them.
B) theoretical calculations that correctly describe stellar formation by gravitational contraction of hydrogen gas.
C) the blue glow from scattered light within their reflection nebulae.
D) the reddening of the spectra of these stars because of absorption of blue light by hydrogen.
Answer: A Source: Section 11-5

91. Long-exposure color photographs of the night sky often show regions that glow red, such as the Eagle Nebula (Fig. 11-9). This distinctive red color is caused by
A) the emission of red and infrared light by warm dust grains.
B) the collective glow of many red giant stars in the region.
C) the ionization and recombination of hydrogen atoms.
D) scattering of incoming starlight by dust grains in the nebula.
Answer: C Source: Section 11-5

92. What causes the characteristic red color of an emission nebula?
   A) Electrons dropping from $n = 2$ to $n = 1$ in hydrogen atoms.
   B) Thermal (blackbody) radiation.
   C) Electrons dropping from $n = 3$ to $n = 2$ in hydrogen atoms.
   D) Scattering of starlight from dust grains in the nebula.
   Answer: C Source: Section 11-5

93. Infrared stars within the Orion nebula are examples of which stage of stellar evolution?
   A) Protostar and young star
   B) Planetary nebula
   C) Supernova remnants
   D) Red giant
   Answer: A Source: Section 11-6

94. What is a T Tauri star?
   A) A young G, K or M type star that is ejecting gas.
   B) A giant or supergiant star that varies regularly in brightness.
   C) A giant or supergiant star that varies randomly in brightness.
   D) A massing O or B type star that ionizes hydrogen in interstellar space.
   Answer: A Source: Section 11-6

95. What name is given to a young, cool star (spectral class G, K, or M) that is ejecting gas into the interstellar medium?
   A) A RR Lyrae star. B) A T Tauri star. C) A Cepheid variable. D) A flare star.
   Answer: B Source: Section 11-6

96. A T Tauri star is at what stage of its stellar evolution?
   A) Protostar, before main sequence phase.
   B) Just before red giant phase, when variability begins.
   C) At the end of its life, decaying away and cooling.
   D) A well-established main sequence star.
   Answer: A Source: Section 11-6

97. T Tauri stars are at what stage of stellar evolution?
   A) Horizontal branch phase.
   B) Main sequence, or "middle age".
   C) Post-main sequence, before helium shell flash.
   D) Early phases, after the formation of a protostar.
   Answer: D Source: Section 11-6

98. A T Tauri star is
A) a protostar that is ejecting mass near the end of its pre-main sequence lifetime.
B) a young, massive O or B star.
C) a young protostar embedded in a cocoon of dust clouds, visible only by infrared radiation.
D) a high-mass yellow giant star that pulsates regularly in size and brightness.
Answer: A Source: Section 11-6

99. Which of the following is NOT a characteristic of T Tauri stars?
A) Emission lines in the spectrum.
B) Irregular variations in brightness.
C) Nuclear reactions in the core.
D) Ejection of mass into space.
Answer: C Source: Section 11-6

100. Which of the following statements is characteristic of a T Tauri star?
A) Great age, near the end of its life as a star.
B) Nuclear reactions in the core.
C) Ejection of mass into space.
D) High mass, greater than about 3 solar masses.
Answer: C Source: Section 11-6

101. On a Hertzsprung-Russell diagram describing the stars in a young cluster, in which position would you expect to find the T Tauri stars?
A) Just above and slightly to the right of the main sequence.
B) Well above the main sequence and to the left of the diagram.
C) Well below the main sequence.
D) In the upper right of the diagram, well above the main sequence.
Answer: A Source: Section 11-6

102. If we plot the stars in a YOUNG star cluster on a Hertzsprung-Russell diagram, we would expect to see
A) the more massive stars on the main sequence and the less massive stars above the main sequence.
B) the more massive stars above the main sequence and the less massive stars on the main sequence.
C) all stars on the main sequence.
D) some stars on the main sequence and others above the main sequence, in random fashion depending on when each individual star condensed from the interstellar cloud.
Answer: A Source: Section 11-6

103. An astronomer plots the HR diagram of a star cluster and finds that it contains hot B-type stars on the main sequence and cooler G-and K-type stars noticeably above the main sequence. This cluster is
A) very young, because the G and K stars are still evolving toward the main sequence.
B) impossible, because one cannot have cool stars above the main sequence when hot stars are on the main sequence.
C) old, because the G and K stars are already evolving off (away from) the main sequence.
D) of indeterminate age, since one cannot estimate the age of the cluster from the information given.
Answer: A Source: Section 11-6

104. When plotted on an H-R diagram, a particular star cluster has its more massive stars on the main sequence and its less massive stars above the main sequence. How long is it since these stars condensed from the interstellar medium? (i.e., Can we find the age of this cluster?)
A) Yes, by finding the least massive protostar. The age of the cluster equals the time that this star takes to reach the main sequence.
B) Yes, by finding the least massive star which is on the main sequence. The age of the cluster equals the time that this star spent as a protostar.
C) No.
D) Yes, by finding the most massive star which is on the main sequence. The age of the cluster equals the time that this star spent as a protostar.
Answer: B Source: Section 11-6

105. What is the characteristic color of a reflection nebula?
A) Yellow. B) Blue. C) Green. D) Red.
Answer: B Source: Section 11-6

106. What causes the characteristic blue color of a reflection nebula?
A) Electrons dropping from n = 2 to n = 1 in hydrogen atoms.
B) Thermal (blackbody) radiation emitted by the hot gas.
C) Electrons dropping from n = 3 to n = 2 in hydrogen atoms.
D) Scattering of starlight from dust grains in the nebula.
Answer: D Source: Section 11-6

107. A reflection nebula is made visible by
    A) emission lines from hydrogen, which itself has been ionized by UV light from embedded stars.
    B) light from embedded stars reflected over a wide range of wavelengths toward Earth by crystals of water, methane and ammonia ices.
    C) thermal energy emitted as a continuous spectrum by the very hot gas, much like that emitted by a hot body on Earth.
    D) blue light preferentially scattered by dust grains.
    Answer: D Source: Section 11-6

108. The distinctive color of a reflection nebula is
    A) light of all colors, but predominantly in the red part of the spectrum, emitted by cool stars and reflected by crystals of water ice surrounding the stars.
    B) blue, caused by the scattering of light from dust grains.
    C) several specific colors, coming from fluorescence of atoms excited by ultraviolet radiation emitted by hot stars.
    D) red, coming from the emission of light from hydrogen gas.
    Answer: B Source: Section 11-6

109. The distinct blue color of the nebulosity around stars in young clusters such as the Pleiades is caused by
    A) atoms of gas emitting light by fluorescence, having been excited by ultraviolet radiation from hot stars.
    B) starlight reflected by blue-colored interstellar material.
    C) light emitted by interstellar gases but Doppler-shifted by motion toward the observer.
    D) starlight scattered and reflected by small dust grains in the interstellar material.
    Answer: D Source: Section 11-6

110. In photographs, the Pleiades open star cluster is surrounded by a bluish haze (see Fig. 11-1, Kaufmann & Comins, *Discovering the Universe*, 5th Ed.). What causes this blue light?
    A) Starlight scattered by the light-sensitive grains in the photographic plate when the picture was taken.
    B) Shock waves losing energy to interstellar gas in the star cluster, causing the atoms to emit light.
    C) Starlight absorbed and re-emitted by interstellar gas in the star cluster.
    D) Starlight scattered from interstellar dust in the star cluster.
    Answer: D Source: Section 11-6

111. What is the ultimate fate of an open star cluster?
   A) The shape of the cluster will remain more-or-less as it is at the present time as the stars in it age and die.
   B) Over time the stars will collide and merge, eventually creating a black hole.
   C) The stars will gradually sink toward the center, creating a globular cluster.
   D) The stars in it will escape one by one until the cluster no longer exists.
   Answer: D Source: Section 11-6

112. In the Hertzsprung-Russell diagram, how does the position of a typical star change while it is at the main sequence phase of its evolution?
   A) A star's position on the main sequence is determined only by its mass and not its age, and so, stars do not move along the main sequence during evolution.
   B) Massive stars (4 solar masses) move toward the upper left as their luminosity increases, while lower-mass stars move toward the lower right as their temperature decreases.
   C) Stars move from upper right to lower left while they are on the main sequence.
   D) Stars move from upper left to lower right while they are on the main sequence.
   Answer: A Source: Section 11-7

113. At what stage of its evolutionary life is the Sun?
   A) Main sequence, middle age.
   B) Pre-main sequence, variable star.
   C) Just before supernova stage (perhaps 5 years), late evolutionary stage.
   D) Post-main sequence, red giant (cool) phase.
   Answer: A Source: Section 11-7

114. What physical process is taking place inside stars that are on the main sequence in the Hertzsprung-Russell diagram?
   A) The gas is contracting gravitationally without nuclear reactions taking place.
   B) Hydrogen is being converted to helium in their cores.
   C) Hydrogen is being converted to helium in a shell around the helium-rich core.
   D) Helium is being converted to carbon in their cores.
   Answer: B Source: Section 11-7

115. Thermonuclear reactions convert hydrogen into helium in the core of a star during which phase of a star's life?
   A) The horizontal branch phase.
   B) The protostar phase.
   C) The main sequence phase.
   D) As the star moves up the red giant branch for the first time.
   Answer: C Source: Section 11-7

116. All stars on the main sequence
   A) generate energy by hydrogen fusion in their centers.
   B) are changing slowly in size, by gravitational contraction.
   C) have approximately the same age, to within a few million years.
   D) are at a late stage of evolution after the red giant stage.
   Answer: A Source: Section 11-7

117. What is happening in a star that is on the main sequence on the Hertzsprung-Russell diagram?
   A) The star is slowly shrinking as it slides down the main sequence from top left to bottom right across the H-R diagram.
   B) The star is generating internal energy by hydrogen fusion.
   C) Stars that have reached the main sequence have ceased nuclear "burning" and are simply cooling down by emitting radiation.
   D) The star is generating energy by helium fusion, having stopped hydrogen "burning".
   Answer: B Source: Section 11-7

118. What is the meaning of the phrase "zero-age main sequence"?
   A) The sequence of stars in the Hertzsprung-Russell diagram that have just collapsed from the interstellar medium.
   B) The sequence of stars in the Hertzsprung-Russell diagram that have just reached the main sequence.
   C) The sequence of stars in the Hertzsprung-Russell diagram that are just beginning to convert helium into carbon in their cores.
   D) The sequence of stars in the Hertzsprung-Russell diagram that contain no processed elements (elements heavier than hydrogen or helium).
   Answer: B Source: Section 11-7

119. Why is it that the majority of stars in the sky are in the main sequence phase of their lives?
   A) Because this is the only phase that is common to all stars.
   B) Because most stars die at the end of the main sequence phase.
   C) Because most stars in the sky were created at about the same time, so these are all in the same phase of their lives.
   D) Because this is the longest-lasting phase in each star's life.
   Answer: D Source: Section 11-7

120. How is the length of a star's lifetime related to the mass of the star?
   A) The lifetimes of stars are too long to measure, so it is not known how (or if) their lifetimes depend on mass.
   B) A star's lifetime does not depend on its mass.
   C) Lower-mass stars run through their lives faster and have shorter lifetimes.
   D) Higher-mass stars run through their lives faster and have shorter lifetimes.
   Answer: D Source: Section 11-7

121. In terms of the mass and lifetime of a star, which of the following statements is true?
   A) The more massive the star, the faster it will evolve through its life.
   B) The mass of a star has no bearing on the length of a star's life or the speed of its evolution.
   C) Stars of about one solar mass have the shortest lives; less massive stars evolve slowly and live a longer time, while more massive stars have long lives because of the large amount of fuel which they contain.
   D) The less massive the star, the shorter its life, because it has less hydrogen "fuel" to burn.
   Answer: A Source: Section 11-7

122. What is the most important quantity on which the lifetime of a star depends?
   A) The star's speed of rotation.
   B) The mass of the star.
   C) The abundance of heavy elements in the star.
   D) Its surface temperature.
   Answer: B Source: Section 11-7

123. How long will the Sun have spent as a main sequence star when it finally begins to evolve toward the red giant phase?
   A) $10^{10}$ years B) 1 billion years C) $10^{11}$ years D) 1 million years
   Answer: A Source: Section 11-7

124. The total time that the Sun will spend as a main sequence star is
   A) about 4.5 million years.
   B) at least 200 billion years ($2 \times 10^{11}$) years.
   C) about 1 million years.
   D) about 10 billion years ($10^{10}$ years).
   Answer: D Source: Section 11-7

125. Approximately what fraction of its main sequence lifetime has the Sun completed at the present time (see Fig. 11-17 and Table 11-2)?
A) About half B) About 3/4 C) Less than 10% D) About 1/4
Answer: A Source: Section 11-7

126. If you were to look at one kilogram of material taken from the surface of the Sun and one kilogram taken from the center, which of the following statements would be true of these two kilograms?
A) They both have the same amount of hydrogen, and are in fact mostly hydrogen.
B) Neither of them contain any hydrogen.
C) The kilogram from the surface contains more hydrogen than the one from the center.
D) The kilogram from the surface contains less hydrogen than the one from the center.
Answer: C Source: Section 11-7

127. Over which of the following stages of stellar evolution does the radius of a star remain approximately constant?
A) Asymptotic giant branch phase.
B) Red giant.
C) Main sequence phase.
D) Birth and initial formation.
Answer: C Source: Section 11-7

128. Why does the core of the Sun contain more helium and less hydrogen than does the surface of the Sun?
A) Helium condenses more easily so, when the Sun was forming the core became helium-rich; vast quantities of hydrogen were added only after the core became massive enough.
B) The hydrogen has been lifted out of the core by the Sun's magnetic field.
C) Helium is heavier than hydrogen, and has sunk toward the center in a process of chemical differentiation.
D) Thermonuclear reactions have converted much of the original hydrogen in the core into helium.
Answer: D Source: Section 11-7

129. Which of the following statements about the rate of stellar evolution is true?
A) The more massive the original star, the faster the evolution.
B) The chemical make-up of the original nebula is the major factor in deciding the rate of evolution, whatever the mass of the star.
C) Star mass has no bearing upon stellar evolution, since all stars evolve at the same rate, controlled by nuclear fusion.
D) The more massive the original star, the slower the evolution, since there is more material for thermonuclear burning.
Answer: A Source: Section 11-7

130. The evolution of a star depends predominantly upon
    A) its surface temperature.
    B) its initial mass.
    C) its chemical composition.
    D) its location in the galaxy.
    Answer: B Source: Section 11-7

131. What is the most important quantity upon which the lifetime of a star depends?
    A) The abundance of heavy elements in the star.
    B) The mass of the star.
    C) The star's speed of rotation.
    D) The temperature of the star's corona.
    Answer: B Source: Section 11-7

132. What particular feature of stellar behavior is associated with the fact that a star is on or close to the main sequence region of the Hertzsprung-Russell diagram?
    A) The star is generating internal energy by hydrogen fusion.
    B) The star is generating energy by helium fusion, having stopped hydrogen "burning".
    C) The star is slowly shrinking and thereby releasing gravitational potential energy, as it slides down the main sequence from top left to bottom right across the H-R diagram.
    D) Stars which have reached the main sequence have ceased nuclear "burning" and are simply cooling down by emitting radiation.
    Answer: A Source: Section 11-7

133. The main sequence lifetime of a star half the mass of the Sun
    A) is shorter than that of the Sun.
    B) is the same as that of the Sun, because mass does not affect the lifetime of a star.
    C) could be longer or shorter than that of the Sun; more information is required to answer this question.
    D) is longer than that of the Sun.
    Answer: D Source: Section 11-7

134. How long does a star of three times the Sun's mass live compared to the lifetime of the Sun? (See Table 11-1, Kaufmann & Comins, *Discovering the Universe*, 5th Ed.)
    A) It lives about 1/3 as long as the Sun
    B) It lives about 1/500 as long as the Sun.
    C) It lives about 3 times as long as the Sun.
    D) It lives about 1/20 as long as the Sun.
    Answer: D Source: Section 11-7

135. The present estimate of the age of the universe is about 15 billion years. How long will a half solar mass star spend on the main sequence, compared to the present age of the universe? (See Table 111, Kaufmann & Comins, *Discovering the Universe*, 5th Ed.)
A) More than 10 times the age of the universe.
B) About 1/2 of the age of the universe.
C) About 3 times the age of the universe.
D) Less than 1/10 of the age of the universe.
Answer: A Source: Section 11-7

136. The star zeta Pegasi (in the constellation Pegasus, the Flying Horse) has a spectral-luminosity class of B8 V, which gives it a surface temperature of about 11,000 K. According to Table 11-1 of Kaufmann and Comins, *Discovering the Universe*, 5th Ed., the expected main sequence lifetime of zeta Pegasi is
A) about 500,000 years.
B) much greater than 500 million years.
C) about 500 million years.
D) not defined in any way at all by this information.
Answer: C Source: Section 11-7

137. The spiral galaxy in which we live is roughly 15 billion years old. For the stars that formed when the Galaxy was very young (say, during the first billion years), which of the following statements is true? (See Table 11-1, Kaufmann & Comins, *Discovering the Universe*, 5th Ed.)
A) No stars are still on the main sequence-in this time all stars will have finished their lives.
B) All stars of less than 3/4 of the mass of the Sun are still on the main sequence.
C) All stars of more than 15 times the mass of the Sun are still on the main sequence.
D) All stars of 3 times the mass of the Sun are still on the main sequence.
Answer: B Source: Section 11-7

138. If you were able to return to the Earth 1 million years into the future, which of the following views of the sky would be most likely?
A) Nearby stars would have moved in position, and many blue stars would no longer be visible.
B) All the present stars, both blue and red, would be visible but nearby stars would have moved in position.
C) The sky would be very much as it is now, since 1 million years is a very short time in astronomical terms.
D) A few red stars would be missing since they would have evolved, but otherwise stars would be the same and in the same positions as today.
Answer: A Source: Section 11-7

139. What percentage of the mass of the Sun's core remains as hydrogen at the present time, after the thermonuclear furnace has transformed hydrogen into helium over its main sequence lifetime?
A) 35% B) 75% C) 25% D) Only a few %
Answer: A Source: Section 11-7

140. When a star leaves the main sequence and expands toward the red giant region, what is happening inside the star?
A) Hydrogen burning is taking place in a spherical shell just outside the core; the core itself is almost pure helium.
B) Helium is being converted into carbon and oxygen in the core.
C) Hydrogen burning is taking place in a spherical shell just outside the core; the core has not yet started thermonuclear reactions, and is still mostly hydrogen.
D) Helium burning is taking place in a spherical shell just outside the core; the core itself is almost pure carbon and oxygen.
Answer: A Source: Section 11-8

141. In terms of nuclear reactions, what is the next stage of a star's life after the end of hydrogen burning in the core?
A) Hydrogen burning in a thin shell around the core.
B) Helium burning in the core.
C) Carbon burning.
D) Death (it becomes either a supernova or a white dwarf).
Answer: A Source: Section 11-8

142. The next stage in a star's life after the main sequence phase is
A) the red giant phase.
B) the horizontal branch phase.
C) a protostar.
D) death (i.e., either a supernova or a white dwarf).
Answer: A Source: Section 11-8

143. What is a red giant?
A) A star that is burning hydrogen into helium in a shell around the core.
B) A protostar in the "upper right" part of the Hertzsprung-Russell diagram.
C) A large, red star that is burning hydrogen into helium in its core.
D) A large emission nebula.
Answer: A Source: Section 11-8

144. Which of the following are NOT very young stars or pre-stellar objects?
   A) T Tauri stars.
   B) Infrared emitting stars in gas and dust clouds.
   C) Red giants.
   D) Protostars.
   Answer: C Source: Section 11-8

145. Compared to the composition of the early Sun, the composition of the gas at the core of a star that has become a red giant for the first time is
   A) the same, with a high fraction of H compared to He, since these stars were produced with the same initial material.
   B) very different, since thermonuclear fusion has transformed all the H and He into heavier elements.
   C) very different, with lots of H but almost no He left after thermonuclear fusion.
   D) very different, since thermonuclear fusion has transformed all the H into He.
   Answer: D Source: Section 11-8

146. What makes a red giant star so large?
   A) Red giants are rapid rotators, and centrifugal force pushes the surface of the star outwards.
   B) The star has many times more mass than the Sun.
   C) The hydrogen-burning shell is heating the envelope and making it expand.
   D) The helium-rich core has expanded, pushing the outer layers of the star outwards.
   Answer: C Source: Section 11-8

147. What rates of mass loss are typical from red giant stars?
   A) Red giant stars do not suffer mass loss.
   B) $10^{-7}$ solar masses per year.
   C) 1 solar mass per year.
   D) $10^{-3}$ solar masses per year.
   Answer: B Source: Section 11-8

148. The majority of the elements heavier than hydrogen and helium in the Universe are believed to have originated
   A) in the central cores of stars.
   B) in the original Big Bang.
   C) in HII regions, under the action of Ha light.
   D) in giant molecular clouds.
   Answer: A Source: Section 11-8

149. Why does it require higher gas temperatures in the core of a star in order to produce nuclear fusion of helium compared to that required for hydrogen?
   - A) Because higher speeds are needed between two He atoms in order to overcome the shielding effect of the 2 electrons round the nucleus compared to the 1 electron per nucleus of H.
   - B) Because higher atomic speeds are required in order to strip off 2 electrons per helium atom rather than 1 electron per atom for hydrogen before fusion can take place.
   - C) Because the He nuclei need to be moving faster in order to avoid the more numerous and faster H nuclei, with which they can combine with no energy generation.
   - D) Because higher collision speeds are needed to overcome the extra electrostatic repulsion between doubly charged He nuclei.

   Answer: D Source: Section 11-8

150. What happens to the helium-rich core of a star after the core runs out of hydrogen?
   - A) It expands and cools down.
   - B) It heats up and expands.
   - C) It cools down and contracts.
   - D) It contracts and heats up.

   Answer: D Source: Section 11-8

151. How large will the Sun be as a red giant?
   - A) About 1.5 AU radius (out to Mars' orbit).
   - B) About 1/2 AU radius (beyond Mercury's orbit).
   - C) About 1 AU radius (out to the Earth's orbit).
   - D) About 1/10 AU radius (1/4 of Mercury's orbit).

   Answer: C Source: Section 11-8

152. When core helium burning begins in the Sun, the Sun will have a size (diameter) roughly equal to
   - A) twice its present size.
   - B) the size of the Earth.
   - C) the size of the Earth's orbit.
   - D) half its present size.

   Answer: C Source: Section 11-8

153. Nuclear fusion reactions which convert helium into carbon and oxygen in the central core of a star occur during which phase of a star's life?
   - A) During and immediately after the (first) red giant or supergiant stage.
   - B) In the red giant stage, before helium flash.
   - C) During the protostar stage.
   - D) After the main sequence phase, before the star becomes a red giant.

   Answer: A Source: Section 11-9

154. How hot does the core of a star have to be for helium burning to begin?
A) 1 million K B) 1 billion K C) 10 million K D) 100 million K
Answer: D Source: Section 11-9

155. The temperature at which thermonuclear reactions begin to convert helium into carbon (helium burning) is
A) 1 billion K. B) 100 million K. C) 15 million K. D) 1 million K.
Answer: B Source: Section 11-9

156. What is the dominant nuclear reaction during helium burning in a star?
A) 2He fusing into C.
B) 3He fusing into C.
C) He + 4H combining into C.
D) 4He fusing into C.
Answer: B Source: Section 11-9

157. During helium burning in a star's later life, the chemical element produced by the combination of helium nuclei is
A) beryllium, $^{8}Be$.
B) carbon, $^{12}C$.
C) the light isotope of helium, $^{3}He$.
D) heavy hydrogen, $^{2}H$
Answer: B Source: Section 11-9

158. During helium burning, some $^{4}He$ combines with $^{16}O$, much like the way it combines with $^{12}C$ to form $^{16}O$. What is produced by the $^{16}O + {}^{4}He$ reaction? (The periodic table shown in Chapter 4 of Kaufmann & Comins, *Discovering the Universe*, 5th Ed. may be useful.)
A) $^{18}O$ (heavy isotope of oxygen).
B) $^{20}Ne$ (regular isotope of neon).
C) $^{22}Na$ (light isotope of sodium).
D) $^{20}F$ (heavy isotope of fluorine).
Answer: B Source: Section 11-9

159. When a star leaves the main sequence and expands toward the red giant region, what is happening inside the star?
A) Hydrogen burning is taking place in a spherical shell just outside the core; the core itself is almost pure helium.
B) Helium burning is taking place in a spherical shell just outside the core; the core itself is almost pure carbon and oxygen.
C) Hydrogen burning is taking place in a spherical shell just outside the core; the core has not yet started thermonuclear reactions, and is still mostly hydrogen.
D) Helium is being converted into carbon and oxygen in the core.
Answer: A Source: Section 11-9

160. What are the products of helium burning in a star?
   A) Nitrogen and oxygen.
   B) Hydrogen and lithium.
   C) Magnesium and silicon.
   D) Carbon and oxygen.
   Answer: D Source: Section 11-9

161. How long does core helium burning last in a star, compared to core hydrogen burning?
   A) About 3 times as long as the time spent burning hydrogen.
   B) About 1/2 of the time spent burning hydrogen.
   C) About 1/5 of the time spent burning hydrogen.
   D) About 1/20 of the time spent burning hydrogen.
   Answer: C Source: Section 11-9

162. What is the Pauli exclusion principle?
   A) Photons of less than a certain wavelength cannot eject electrons from a metal.
   B) Two identical photons cannot be absorbed by the same atom.
   C) Two different atoms cannot have the same energy levels.
   D) Two identical particles cannot occupy the same place at the same time.
   Answer: D Source: Section 11-9

163. The Pauli exclusion principle states that two identical electrons (or other particles) cannot have
   A) the same position at the same time.
   B) the same speed at the same time.
   C) the same electrical charge at the same time.
   D) the same speed and direction at the same time.
   Answer: A Source: Section 11-9

164. According to the Pauli exclusion principle, no two identical particles (such as electrons) can have
   A) the same location in space and the same speed.
   B) the same speed.
   C) the same position in space and the same electrical charge.
   D) the same position in space.
   Answer: A Source: Section 11-9

165. When is electron degeneracy pressure important in a star?
   A) During core hydrogen burning.
   B) In a protostar evolving toward the main sequence.
   C) Just before the start of helium burning in the core.
   D) In a Cepheid variable that is burning helium in its core.
   Answer: C Source: Section 11-9

166. The Pauli exclusion principle
   A) limits the number of atoms which can become ionized in a star.
   B) prevents two stars of the same spectral class from occupying the same binary star system.
   C) prevents high-mass stars from forming in low-mass interstellar clouds.
   D) sets a limit to the crowding together of electrons into any given small volume of space.
   Answer: D Source: Section 11-9

167. What is the "safety valve" that prevents normal (nondegenerate) stars from self-destructing?
   A) If a part of a star where thermonuclear reactions are occurring is suddenly heated, it will expand and cool.
   B) If thermonuclear reactions proceed too quickly, the star will run out of fuel before anything drastic happens.
   C) If the temperature rises, the thermonuclear reaction rate will increase to match it.
   D) If the pressure rises, the volume occupied by the matter will decrease, reducing the nuclear reaction rate.
   Answer: A Source: Section 11-9

168. What is the "safety valve" which operates in normal (nondegenerate) stars?
   A) If the stellar gas is suddenly heated, it will expand and cool.
   B) If the pressure gets too high, electrons will combine with protons to relieve the pressure.
   C) If thermonuclear reactions proceed too quickly, the star will run out of fuel before anything drastic happens.
   D) If the star gets too big, it will collapse into a black hole.
   Answer: A Source: Section 11-9

169. A degenerate-electron gas, such as occurs in the core of a red giant star, lacks the "safety valve" of a normal gas. This is because
   A) a rise in temperature does not change the pressure, so the gas does not expand and cool.
   B) a rise in pressure reduces the number of electrons, causing the core to collapse.
   C) a rise in temperature lowers the pressure, causing the star to contract.
   D) a rise in pressure releases more electrons, thus increasing the pressure further.
   Answer: A Source: Section 11-9

170. Under what conditions does electron degeneracy occur?
A) When electrons become crowded too closely together.
B) When thermonuclear reactions release more electrons than protons.
C) When electrons and positrons annihilate, releasing energy.
D) When ultraviolet light from hot, young O and B stars ionizes the interstellar medium.
Answer: A Source: Section 11-9

171. Degeneracy occurs when
A) solar wind particles become trapped in the Earth's magnetic field.
B) magnetic fields inhibit the motion of charged particles in sunspots.
C) electrons inside a star resist being pushed closer together than a certain limit.
D) thermonuclear reactions halt the contraction of a protostar.
Answer: C Source: Section 11-9

172. If electrons are collectively compressed into a very small volume (e.g., within the core of a dying white dwarf star) where quantum mechanical considerations become important in preventing one electron from occupying space near to a second electron (Pauli exclusion principle), what is the result?
A) Half of the electrons are transformed into antimatter (positrons) which then annihilate electrons, producing a burst of energy and the explosion of the star.
B) The electrons generate a very large pressure to oppose further compression.
C) The electrons fall into orbit around one another in mutual pairs that reduce the restricted quantum space, allowing further shrinkage of the star.
D) Nuclear fusion occurs between electrons, producing energy and heating the star's core.
Answer: B Source: Section 11-9

173. Electron degeneracy, which is a result of the Pauli exclusion principle, is important
A) in a protostar evolving toward the main sequence.
B) in the core of a star just before the start of core helium burning.
C) in a Cepheid variable that is burning helium in its core.
D) in the core of a star during core hydrogen burning.
Answer: B Source: Section 11-9

174. The helium flash results from the
A) high temperature in the helium core of a blue (spectral class O or B) supergiant star.
B) electron degeneracy or quantum crowding in the core of a low-mass red giant star.
C) sudden onset of nuclear reactions at the end of the protostar.
D) sudden release of energy in strong magnetic fields near a sunspot.
Answer: B Source: Section 11-9

175. The helium flash is another name for
    A) a sudden onset of helium fusion reactions in the core of a low-mass red giant star.
    B) a sudden release of energy at the end of helium burning, due to core contraction.
    C) a sudden onset of helium fusion reactions in red giant and supergiant stars of any mass.
    D) the sudden appearance of helium during a supernova explosion (explosion of a star at the end of its life).

    Answer: A  Source: Section 11-9

176. Which stars undergo a helium flash in their cores?
    A) All stars of more than 2 solar masses.
    B) All stars that contain helium.
    C) All stars that have become red giants.
    D) All stars of less than 2 solar masses.

    Answer: D  Source: Section 11-9

177. What process is involved in the helium flash?
    A) An increase in temperature causes an increase in the pressure with no increase in the nuclear reaction rate.
    B) An increase in the pressure causes a decrease in the temperature and the nuclear reaction rate.
    C) An increase in temperature causes an increase in the nuclear reaction rate but has no effect on pressure.
    D) An increase in the nuclear reaction rate causes an increase in the temperature and the pressure.

    Answer: C  Source: Section 11-9

178. What happens to a star after the start of helium nuclear reactions in its core, compared to what it was like before these reactions began?
    A) The star is smaller and hotter.
    B) The star is smaller and cooler.
    C) The star is larger and hotter.
    D) The star is larger and cooler.

    Answer: A  Source: Section 11-9

179. After hydrogen burning ends in a star's core, its position on the Hertzsprung-Russell diagram moves in a direction toward the
    A) lower right.  B) upper right.  C) upper left.  D) lower left.

    Answer: B  Source: Section 11-10

180. On the Hertzsprung-Russell diagram, in which direction does the position occupied by a star move after helium burning begins in the star's core?
    A) Toward the upper right.
    B) Toward the lower right.
    C) Toward the lower left.
    D) Toward the upper left.

    Answer: C  Source: Section 11-10

181. The study of stars in clusters has helped astronomers to understand
A) the mechanism of mass loss in stars.
B) stellar evolution, the development of stars with time.
C) the reason for differences in surface temperatures of stars.
D) the action of nuclear fusion in stars.
Answer: B Source: Section 11-11

182. The main general features which make clusters of stars useful to astronomers are that
A) the stars are all at the same distance from Earth, have the same surface temperature, and joined the cluster at various times.
B) the stars all have the same apparent magnitude, the same surface temperatures, and the same sizes.
C) the stars are at the same distance from Earth, were formed at approximately the same time, and were made from same chemical mix.
D) the stars all have the same intrinsic brightness, but differ only in size and surface temperature.
Answer: C Source: Section 11-11

183. How many stars are there in a typical globular cluster?
A) About a thousand
B) Usually a few dozen
C) Up to 100 billion
D) About a million
Answer: D Source: Section 11-11

184. The characteristics of a globular cluster of stars are
A) many thousand members, of different ages.
B) a few dozen members, the remnant of a globular cluster of stars from which most of the members have escaped.
C) a few hundred members, often very young and still embedded in the gas and dust from which they were formed.
D) hundreds of thousands of members, all very old, and no or very little interstellar gas and dust.
Answer: D Source: Section 11-11

185. What would you expect to find in the population of stars in a globular cluster?
   A) Many red giants, white dwarfs and dim red stars, but no dust and gas.
   B) Stars along the entire main sequence from bright blue to dim red, with no bright red giant stars but significant amounts of dust and gas.
   C) Mainly white dwarf stars and the planetary nebular stages of dying stars, but no faint red stars, red giants or bright blue stars.
   D) Mostly blue giants and supergiants, with a few red giant stars, white dwarfs, and dim red stars.
   Answer: A Source: Section 11-11

186. Why does the Hertzsprung-Russell diagram of a globular cluster NOT contain any stars with high luminosity and high temperature on the main sequence?
   A) The stars that were in this position on the main sequence have undergone splitting into binary stars, and hence appear lower down on the diagram.
   B) Stars that will occupy this position on the main sequence have not yet evolved there from the protostar stage.
   C) This type of cluster contains only low-mass stars, and has never had high-luminosity stars on its main sequence.
   D) These high-mass stars evolved away from the main sequence long ago.
   Answer: D Source: Section 11-11

187. Which of the following statements is NOT true of a globular cluster?
   A) It has a round shape.
   B) It can contain up to a million stars.
   C) It contains many main sequence stars.
   D) It contains both high and low mass stars.
   Answer: D Source: Section 11-11

188. Which of the following statements is NOT true of a globular cluster?
   A) It does not contain main sequence stars.
   B) It has a round shape.
   C) It contains only low-mass stars.
   D) It can contain up to a million stars.
   Answer: A Source: Section 11-11

189. How do astronomers know that globular clusters are very old?
   A) Their stars have a high abundance of heavy elements.
   B) There are no main sequence stars in globular clusters.
   C) They do not contain any red giant stars.
   D) There are no massive main sequence stars in globular clusters.
   Answer: D Source: Section 11-11

190. An astronomer studying a globular cluster plots its stars on a Hertzsprung-Russell diagram and finds that certain stars in the cluster lie on the horizontal branch. What does this astronomer immediately know about these stars?
A) They are burning hydrogen into helium in a shell around their cores.
B) They are still contracting toward the main sequence.
C) They are burning hydrogen into helium in their cores.
D) They are burning helium into carbon and oxygen in their cores.
Answer: D Source: Section 11-11

191. At what stage of evolution are the horizontal-branch stars, which have similar luminosities between 50 and 100 times that of the Sun, but a range of temperatures?
A) Post-main sequence stage.
B) White dwarf stage.
C) Pre-main sequence stage, evolving toward the main sequence.
D) This region is part of the main sequence, so these are in the main sequence stage.
Answer: A Source: Section 11-11

192. Horizontal-branch stars, having a range of temperatures and with luminosities between 50 and 100 times that of the Sun, are in what stage of their lives?
A) Hydrogen shell burning, with a degenerate helium core.
B) Core hydrogen burning.
C) Gravitational contraction before the start of core hydrogen burning.
D) Core helium burning.
Answer: D Source: Section 11-11

193. Which of the following stars are metalrich?
A) Globular cluster stars.
B) Population I stars.
C) Population II stars.
D) Very old stars.
Answer: B Source: Section 11-11

194. How do the stars in a star cluster change with time?
A) The lowest-mass stars evolve the most quickly.
B) The stars with the greatest heavy-element content evolve the most quickly.
C) All stars in it evolve at the same rate.
D) The highest-mass stars evolve the most quickly.
Answer: D Source: Section 11-11

195. What is the best way to estimate the age of a star cluster?
   A) Count the number of T Tauri stars in the cluster.
   B) Measure the heavy-element abundance for the stars in the cluster.
   C) Plot the stars on a Hertzsprung-Russell diagram and see which stars are on the main sequence.
   D) Measure the motions of the stars in the cluster and compare to theoretical models.
   Answer: C Source: Section 11-11

196. Of the following astronomical objects or systems, which is likely to be the oldest?
   A) A globular cluster B) The Pleiades C) A T Tauri star D) The Sun
   Answer: A Source: Section 11-11

197. Of which of the following groups of stars is the Sun a member?
   A) Population I.
   B) Horizontal branch stars.
   C) Cepheid variables.
   D) Population II.
   Answer: A Source: Section 11-11

198. An astronomer plots the HR diagram of a star cluster and finds that it contains hot B-type stars on the main sequence and cooler G- and K-type stars noticeably above the main sequence. This cluster is
   A) very young, because the G and K stars are still evolving toward the main sequence.
   B) impossible, because one cannot have cool stars above the main sequence when hot stars are on the main sequence.
   C) of indeterminate age, since one cannot estimate the age of the cluster from the information given.
   D) old, because the G and K stars are already evolving off (away from) the main sequence.
   Answer: A Source: Section 11-11

199. In the HR diagram shown in Fig. 11-15, the brightest stars in the Pleiades cluster are not on the main sequence, but away from it toward the upper right. Why is this?
   A) These stars have already evolved through the red giant phase and have now returned to the blue giant phase on their way to the white dwarf phase.
   B) These blue supergiant stars have already begun to evolve toward the red supergiant phase.
   C) These stars have already become white dwarf stars, as shown by their position.
   D) These stars have not yet reached the main sequence, and are in the T Tauri phase.
   Answer: B Source: Section 11-11

200. What is the "turn-off" point for a star cluster?
   A) The point in the Hertzsprung-Russell diagram occupied by the highest-mass stars that have not yet reached the main sequence.
   B) The point in the Hertzsprung-Russell diagram occupied by the lowest-mass main sequence stars in the cluster.
   C) The point in the Hertzsprung-Russell diagram occupied by the highest-mass main sequence stars in the cluster.
   D) The point in the Hertzsprung-Russell diagram occupied by stars undergoing (or about to undergo) the helium flash.
   Answer: C Source: Section 11-11

201. For an astronomer, what is the significance of the turn-off point in the Hertzsprung-Russell diagram of a star cluster?
   A) It tells the astronomer which stars pulsate in brightness (variable stars).
   B) It tells the astronomer the metal content of the star cluster.
   C) It tells the astronomer the age of the star cluster.
   D) It tells the astronomer which stars might be about to explode as supernovae.
   Answer: C Source: Section 11-11

202. The age of a cluster can be determined by
   A) measuring its speed of motion relative to the Sun.
   B) observing its position in the sky with respect to the Sun.
   C) carrying out a number count of the stars in the cluster.
   D) determining the turnoff point on the main sequence of its HR diagram.
   Answer: D Source: Section 11-11

203. The age of a cluster of stars can be judged by
   A) the number of novae per year occurring within the cluster.
   B) the total number of stars within the cluster.
   C) the amount of radioactive elements detected on the surfaces of its stars.
   D) the turnoff point on the main sequence of its Hertzsprung-Russell diagram.
   Answer: D Source: Section 11-11

204. Suppose that, when the stars in a particular open star cluster are plotted in an H-R diagram, the turnoff point is at an absolute magnitude of +3. Approximately what is the age of this cluster? (See Fig. 11-20, Kaufmann & Comins, *Discovering the Universe*, 5th Ed.)
   A) 100 million years B) 500 million years C) 4 billion years D) 1 billion years
   Answer: C Source: Section 11-11

205. The stars at the turn-off point in the H-R diagram of the Hyades star cluster have an absolute magnitude of approximately M = +2, while those at the turnoff point in the cluster M41 have M = 0. From this information, we can say with certainty that
A) the Hyades cluster is older than M41.
B) the Hyades cluster is younger than M41.
C) the Hyades cluster is further away than M41.
D) the Hyades cluster has more stars in it than M41.
Answer: A Source: Section 11-11

206. Which of the following stars are metalpoor?
A) Population I stars.
B) Population II stars.
C) Very young stars.
D) Open cluster stars.
Answer: B Source: Section 11-11

207. Stars are formed from interstellar matter. Why then are stars in open clusters metalrich, while stars in globular clusters are metal-poor?
A) Because open clusters are young and stars have formed from material that has been enriched in metals by supernova explosions.
B) Because open clusters contain very hot stars that produce metals by nuclear reactions in their outer layers.
C) Because globular clusters are too young for their stars to have produced any significant amount of metals.
D) Because globular clusters have "burned" their heavy elements over their longer lifetime.
Answer: A Source: Section 11-11

208. Which of the following stars would you classify as a population II star?
A) A star with approximately the same abundance of heavy elements as we find in the Sun.
B) A star with very low abundance of heavy elements.
C) A star with much higher abundance of heavy elements than we find in the Sun.
D) Any member of an open star cluster.
Answer: B Source: Section 11-11

209. In connection with stars, what does the phrase "metal-poor" mean?
A) It might or might not have a low abundance of carbon in its spectrum, but is definitely weak in iron.
B) It has a low abundance of all elements heavier than hydrogen and helium in its spectrum.
C) It has a low abundance of all elements heavier than hydrogen in its spectrum.
D) It has a low abundance of all elements in its spectrum.
Answer: B Source: Section 11-11

210. When a star's evolutionary track in the Hertzsprung-Russell diagram carries it into the instability strip, what happens to the star?
A) It explodes.
B) It pulsates randomly in brightness.
C) It collapses and forms a black hole.
D) It pulsates regularly in brightness.
Answer: D Source: Section 11-12

211. What are the stars in the upper part of the instability strip called?
A) Cepheid variables B) RR Lyrae variables C) protostars D) T Tauri stars
Answer: A Source: Section 11-12

212. What are the stars in the lower part of the instability strip called?
A) T Tauri stars B) protostars C) RR Lyrae variables D) Cepheid variables
Answer: C Source: Section 11-12

213. RR Lyrae stars are
A) pulsating stars that vary irregularly, with periods of several hundred days.
B) pulsating stars that vary regularly, all with periods of less than one day.
C) eclipsing binary stars.
D) rotating neutron stars with very rapid and regular brightness fluctuations.
Answer: B Source: Section 11-12

214. What is a Cepheid variable star?
A) A low-mass, horizontal branch star that pulsates regularly in brightness.
B) One of several classes of stars that pulsate randomly in brightness.
C) A high-mass star that pulsates regularly in brightness.
D) A star that normally remains constant in brightness but occasionally flares up in brightness by several magnitudes.
Answer: C Source: Section 11-12

215. Cepheid stars are
   A) members of binary systems, in which one star periodically eclipses the other.
   B) stars at an early stage in stellar evolution, pre-main-sequence.
   C) white dwarf stars, late in their evolutionary life.
   D) giant stars that pulsate in brightness, size, and temperature.
   Answer: D Source: Section 11-12

216. A Cepheid variable
   A) is a high-mass giant or supergiant star that pulsates regularly in size and brightness.
   B) is a variable emission nebula near a T Tauri star.
   C) is a low-mass red giant that varies in size and brightness in an irregular way.
   D) is a type of eclipsing binary star.
   Answer: A Source: Section 11-12

217. What scientific method is used to observe the pulsation in size of a Cepheid variable star?
   A) The observed perturbations in the orbits of planets around these stars.
   B) This behavior has only been predicted theoretically, and has never been detected.
   C) Doppler shift of absorption lines in its spectrum.
   D) The observed increase and decrease in the size of the star's image.
   Answer: C Source: Section 11-12

218. What characteristic of Cepheid variables makes them extremely useful to astronomers?
   A) Their absolute magnitude is related directly to their surface temperature.
   B) Their absolute magnitude is related directly to their period of pulsation.
   C) Their absolute magnitude is related directly to their metal content (heavy element abundance).
   D) Their absolute magnitude is directly related to their diameter.
   Answer: B Source: Section 11-13

219. The period of variability of a Cepheid variable star, which is easily measured, is directly related to which stellar parameter, thereby providing a reliable method for the measurement of distance to stars?
   A) Velocity away from Earth.
   B) Surface magnetic field.
   C) Surface temperature.
   D) Luminosity.
   Answer: D Source: Section 11-13

220. What is the most important use of Cepheid variables for astronomers?
   A) The distance to a Cepheid variable can be found very easily.
   B) The diameter of a Cepheid variable can be found very easily.
   C) The metal content of a Cepheid variable can be found very easily.
   D) The characteristics of the pulsation of a Cepheid variable can be used to investigate conditions in the core of the star.

   Answer: A Source: Section 11-13

221. Which of the following major perturbations can occur to a close binary system and radically alter the evolution and behavior of the two individual stars?
   A) The heating of the localized areas of the atmosphere of one star by its companion.
   B) The transfer of matter from one star to its companion.
   C) The eclipsing of the light from one star by the other, when viewed from Earth.
   D) The gravitational disturbance of one star's motion by its companion, to force it to move in an orbit.

   Answer: B Source: Section 11-14

222. What particular and very important phenomenon frequently occurs in binary star systems where the stars are very close together?
   A) The less massive star spirals slowly into its more massive companion because of tidal interactions.
   B) The less massive star, in its elliptical orbit, will repeatedly pass through the thin, extended atmosphere of the second star, producing periodic rises and falls in light output from the star system.
   C) The radiation from the hotter star will slowly heat and evaporate away the cooler star.
   D) Mass lost from one star is deposited upon its companion.

   Answer: D Source: Section 11-14

223. The components of a binary star, particularly if they are close, can influence each other in various ways. Which of the following is NOT likely to be an effect of one star on its companion?
   A) The gravitational force of one star will make its companion move in an orbit, rather than remaining stationary.
   B) A very hot star can heat part of its cooler companion to produce a hot spot.
   C) Mass can be transferred from one star to its companion.
   D) Intense radiation from a hot star can produce nuclear reactions on the surface of a cooler companion and initiate a nova explosion.

   Answer: D Source: Section 11-14

224. The shape of the cross-section of the Roche lobes around a close binary star system, taken through the centers of both stars, is
A) two unequal ellipses that touch at the center of the lobes.
B) a sphere, centered on the center of mass of the star system.
C) a figure-eight.
D) an ellipse, with a star at each focus.
Answer: C Source: Section 11-14

225. The transfer of mass from the surface of one star onto the surface of another is most often observed in
A) overcontact binaries.
B) contact binaries.
C) detached binaries.
D) semidetached binaries.
Answer: D Source: Section 11-14

226. In some binary star systems, such as Algol, the less massive star is an old red giant star and the more massive star is on the main sequence. This is evidence that
A) the more massive star captured the other one into orbit some time after the two stars had formed.
B) the more massive star formed later, from a disk of gas surrounding the less massive star.
C) stars evolve differently in binary star systems, with less massive stars evolving faster than more massive stars.
D) mass transfer has occurred from one of the stars to the other.
Answer: D Source: Section 11-14

227. In some binary star systems, such as b (Beta) Lyrae, essentially no light is seen from the more massive star. This is because
A) the orbit of the more massive star keeps it hidden behind the larger but less massive star as seen from the Earth.
B) the more massive star is still hidden in the dust clouds from which it formed.
C) the more massive star is hidden by an accretion disk of material from the less massive star.
D) the more massive star is a black hole, from which light cannot escape.
Answer: C Source: Section 11-14

228. Which factor, more than any other, modifies the evolutionary tracks of stars in binary combinations compared to their single star counterparts?
   A) Tidal distortion of the shapes of the stars.
   B) Reduction of the quantum mechanical limitation on continued shrinking of one star by the gravitational field of the second star.
   C) Mass exchange between the stars.
   D) Radiation from one star heating the surface of the second star.
   Answer: C Source: Section 11-14

# Chapter 12: The Deaths of Stars

1. Which of the following objects is **NOT** an end-point of a star's evolutionary life?
   A) Black hole. B) Neutron star. C) Red giant. D) Supernova.
   Answer: C Source: Chapter 12

2. What are the main by-products of helium nuclear "burning" in red giant stars?
   A) Hydrogen nuclei, by nuclear fission.
   B) Energy, from the complete transformation of the mass of helium to energy.
   C) Carbon and oxygen nuclei.
   D) Iron nuclei.
   Answer: C Source: Introduction to Section 12-1

3. Helium nuclear reactions (helium burning) produces primarily
   A) oxygen and neon. B) carbon and oxygen. C) iron. D) carbon and silicon.
   Answer: B Source: Introduction to Section 12-1

4. Nuclear fusion reactions of helium (helium "burning") produces primarily
   A) nitrogen and neon nuclei.
   B) iron nuclei.
   C) carbon and oxygen nuclei.
   D) beryllium and lithium nuclei.
   Answer: C Source: Introduction to Section 12-1

5. The nuclear process in which helium "burning" occurs in the deep interiors of red giant stars produces
   A) carbon and oxygen nuclei.
   B) hydrogen nuclei by splitting of helium nuclei.
   C) pure energy from the nuclear mass.
   D) iron nuclei.
   Answer: A Source: Introduction to Section 12-1

6. The structure of the deep interior of a low-mass star, near the end of its life, is
   A) a carbon-oxygen core, a shell around it where helium nuclei are undergoing fusion, and a surrounding shell of hydrogen.
   B) an inactive hydrogen core and a helium shell undergoing nuclear fusion, surrounded by a carbon-oxygen shell.
   C) a turbulent mixture of hydrogen, helium, carbon and oxygen in which only helium continues to undergo nuclear fusion.
   D) a helium core surrounded by a thin hydrogen shell undergoing nuclear fusion, with very small concentrations of heavier nuclei.
   Answer: A Source: Section 12-1

7. During which phase of a low-mass star's life does helium shell burning occur?
   A) Horizontal branch.
   B) First red giant phase.
   C) Main sequence.
   D) Asymptotic giant branch.
   Answer: D Source: Section 12-1

8. Helium nuclear reactions take place in a shell around the core of a low-mass star during its
   A) asymptotic giant branch phase.
   B) horizontal branch phase.
   C) first red giant phase.
   D) main sequence phase.
   Answer: A Source: Section 12-1

9. A star on the asymptotic giant branch (AGB) is
   A) a red supergiant.
   B) a cool main sequence star.
   C) a blue supergiant.
   D) a star in its first red giant phase.
   Answer: A Source: Section 12-1

10. In a star's evolutionary life, the asymptomatic giant branch (AGB) is the
    A) helium core-burning phase.
    B) helium shell-burning phase.
    C) pre-main-sequence core hydrogen-burning phase.
    D) hydrogen shell-burning phase prior to helium ignition in the core.
    Answer: B Source: Section 12-1

11. A star ascending the red giant branch for the second time, in the asymptotic giant branch phase, will have
    A) no nuclear reactions occurring in the core, but hydrogen "burning" in a shell outside the core.
    B) hydrogen fusion reactions occurring in the core.
    C) no nuclear reactions in the core, but a helium "burning" shell outside the core, which itself is surrounded by a shell of hydrogen.
    D) no fusion reactions; the star has used up all of its nuclear fuel.
    Answer: C Source: Section 12-1

12. The characteristics of red supergiant stars are
    A) brightness of 10,000 Suns and a diameter of about Mars' orbit.
    B) brightness of the Sun and size of about Mercury's orbit.
    C) brightness of about 1 million Suns and a diameter of the whole solar system.
    D) brightness of about 10,000 Suns and a diameter of 1/10 of that of the Sun.
    Answer: A Source: Section 12-1

13. In what phase of its life does a low-mass star become a Mira (or long-period) variable?
    A) Horizontal branch.
    B) First red giant phase.
    C) Main sequence.
    D) Asymptotic giant branch.
    Answer: D Source: Section 12-1

14. In what phase of its life does a low-mass star become a Mira (or long-period) variable?
    A) Dwarf phase.
    B) Red giant phase.
    C) Blue supergiant phase.
    D) Red supergiant phase.
    Answer: D Source: Section 12-1

15. In what phase of its life does a low-mass star become a Mira (or long-period) variable?
    A) Helium core-burning phase.
    B) Helium shell-burning phase.
    C) Hydrogen core-burning phase.
    D) Hydrogen shell-burning phase prior to helium ignition in the core.
    Answer: B Source: Section 12-1

16. What is the cause of the brightness variations is a Mira (or long-period) variable?
    A) Large star-spots (similar to sunspots on the Sun), carried into and out of view by the star's rotation.
    B) A hotspot on the star's surface, created by an orbiting neutron star.
    C) Irregular energy production in the hydrogen- and helium-burning shells.
    D) Instabilities in the electron-degenerate carbon/oxygen core.
    Answer: C Source: Section 12-1

17. In the process of helium shell fusion in low-mass stars near the end of their lives, the star moves upwards and to the right on the asymptotic giant branch of the Hertzsprung-Russell diagram. In this process, the star is
    A) expanding, heating up and becoming more luminous.
    B) contracting, cooling and hence becoming less luminous.
    C) contracting, becoming hotter and much less luminous.
    D) expanding, cooling and becoming more luminous.
    Answer: D Source: Section 12-1

18. What will be the mass of the Sun at the end of its asymptotic giant branch (AGB) phase, due to mass loss to space by its stellar wind?
    A) Between 0.1 and 0.2 solar masses
    B) About 0.8 solar masses.
    C) About 0.5 solar masses.
    D) Still almost 1 solar mass, since mass loss is negligible for a low-mass star like the Sun.
    Answer: C Source: Section 12-1

19. What is the last nuclear burning stage in the life of a low-mass star like the Sun?
    A) Fusion of helium nuclei to form carbon and oxygen.
    B) Fusion of oxygen nuclei to form sulfur.
    C) Fusion of hydrogen nuclei to form helium.
    D) Fusion of silicon nuclei to form iron.
    Answer: A Source: Section 12-1

20. What happens to the surface of a low-mass star after the helium core and shell fusion stages are completed?
    A) It contracts back into the core and becomes hot enough to undergo further hydrogen fusion, leading to a very hot and active white dwarf star.
    B) It is propelled slowly away from the core to form a planetary nebula.
    C) It stabilizes at the size of a red giant star, radiation pressure from below balancing gravity from the core, and will slowly cool for the rest of its life.
    D) It is spun off into space to make a spiral structure known as a spiral galaxy.
    Answer: B Source: Section 12-1

21. Planetary nebulae are so-named because
    A) their spectra appear to be similar to the spectrum of the giant gas planets in our own solar system.
    B) they rotate slowly and condense into planetary objects around the remaining central star.
    C) the ejected material is rich in carbon and oxygen, necessary elements for the manufacture of planets in the nebulae surrounding stars.
    D) these extended objects, often green-colored, looked like planets to Herschel when he first observed them in the 1700s through his telescopes.
    Answer: D Source: Section 12-1

22. In astronomical terms, planetary nebulae are
    A) very long-lived objects, having been in existence since just after the Big Bang at the beginning of the Universe.
    B) relatively short-lived, existing around the central white dwarf star for millions of years before slowly spreading into space.
    C) very short-lived, with lifetimes of about 50,000 years.
    D) relatively long-lived, since they form when the original stars form and remain as slowly rotating shells for the whole of their lifetimes of several billion years.
    Answer: C Source: Section 12-1

23. Which of the following important components does a planetary nebula contribute to the interstellar medium?
   A) Rotational motion from the original star, which serves to concentrate interstellar matter into new stars and planetary systems.
   B) The nuclei of heavy elements, major components of planets such as our own.
   C) Molecules such as $NH_3$ and $CH_4$, which contribute to giant molecular clouds.
   D) UV light which photo-ionizes hydrogen. This hydrogen, upon recombination, produces the red Balmer, a light by which we see interstellar emission nebulae.
   Answer: B Source: Section 12-1

24. A planetary nebula is
   A) a disk-shaped nebula of dust and gas, photographed around a relatively young star, from which planets will eventually form.
   B) a contracting spherical cloud of gas surrounding a newly formed star in which planets are forming.
   C) the nebula caused by the supernova explosion of a massive star.
   D) an expanding gas shell surrounding a hot white dwarf star.
   Answer: D Source: Section 12-1

25. A planetary nebula is
   A) a gas cloud surrounding a planet after its formation.
   B) the formation stages of planets around stars other than the Sun dwarf star.
   C) the spherical cloud of gas produced by a supernova explosion.
   D) a shell of ejected gases, glowing because of ultraviolet light from a dying central star.
   Answer: D Source: Section 12-1

26. A planetary nebula is
   A) the nebula caused by the supernova explosion of a massive star.
   B) a gas shell, the atmosphere of a red giant star, slowly expanding away from a hot white dwarf, the core of the red giant.
   C) a contracting spherical cloud of gas surrounding a newly formed star in which planets are forming.
   D) a disk-shaped nebula of dust and gas rotating around a relatively young star, within which planets will eventually form.
   Answer: B Source: Section 12-1

27. A planetary nebula is
    A) a cloud of gas surrounding a very young star, in which planets are expected to form.
    B) a shell of gases ejected from the surface of a red giant star, glowing from fluorescence caused by UV light from the hot, central white dwarf star.
    C) the gas cloud surrounding a planet after its formation.
    D) the spherical, rapidly expanding cloud of gas produced by a supernova explosion.
    Answer: B Source: Section 12-1

28. The event which follows the asymptotic giant branch (AGB) phase in the life of a low-mass star is
    A) the ejection of a planetary nebula.
    B) core collapse and a supernova explosion.
    C) helium flash and the start of helium burning in the core.
    D) the onset of hydrogen burning in the core.
    Answer: A Source: Section 12-1

29. A planetary nebula is created
    A) over several hundred years, during mass transfer in a close binary star system.
    B) in hours or less, during the explosion of a massive star.
    C) over a few thousand years or more, in a slow expansion away from a low-mass star, driven by a series of thermal pulses from helium fusion.
    D) in seconds, during the helium flash in a low-mass star.
    Answer: C Source: Section 12-1

30. The major source of light in the expanding shell of gas in a planetary nebula is
    A) thermonuclear reactions in the hot gas, caused by the underlying explosion.
    B) reflection and scattering of light from the central white dwarf star by dust and gas in the shell.
    C) heating by friction of the gas as it travels rapidly outwards through the interstellar medium.
    D) fluorescence of the atoms, caused by UV light from the hot, white dwarf central star.
    Answer: D Source: Section 12-1

31. What physical process provides the energy for the ejection of a planetary nebula from a low-mass star?
    A) Helium shell flashes in the helium fusion shell.
    B) Collision with another star.
    C) Core collapse and the ensuing shock wave.
    D) Transfer of hydrogen-rich material onto the surface of a white dwarf, from its companion in a binary star system.
    Answer: A Source: Section 12-1

32. The diameter of a typical planetary nebula is
A) about 1 AU.
B) about 1000 light-years.
C) only about 3 to 5 stellar diameters.
D) about 1 light-year.
Answer: D Source: Section 12-1

33. The fraction of the mass of a red giant star that is ejected as a shell in a planetary nebula is
A) very small, close to 10%.
B) extremely small, less than 1 part in $10^4$, since it is only the star's atmosphere which has been ejected.
C) almost the entire star, more than 95%.
D) significant, up to 80%.
Answer: D Source: Section 12-1

34. The shell of a planetary nebula is measured by the Doppler shift of emission lines to be expanding outward at a speed of $10^4$m/s while its radius is measured to be 1 light year, or about $10^{16}$m. How long has the shell been expanding? (Hint: 1 year = $3.15 \times 10^7$ sec.)
A) 30,000 years B) $10^{12}$ years C) 30 years. D) 30 million years
Answer: A Source: Section 12-1

35. What percentage of all matter ejected into the interstellar medium in our galaxy each year by stars is contributed by planetary nebulae?
A) Between 10 and 20%. B) Almost 100%. C) About 50%. D) Less than 1%.
Answer: A Source: Section 12-1

36. Stars that have ejected a planetary nebula go on to become
A) supernovae. B) protostars. C) red giants. D) white dwarfs.
Answer: D Source: Section 12-2

37. The final remnant of the evolution of a red giant star that has ejected a planetary nebula is
A) a blue supergiant. B) a protostar. C) a white dwarf star. D) a supernova.
Answer: C Source: Section 12-2

38. The "star" that is seen at the center of a planetary nebula is
A) a planet in the process of formation.
B) a small, hot, and very dense, white dwarf star.
C) the accretion disk around a black hole.
D) composed almost entirely of neutrons, and is spinning rapidly.
Answer: B Source: Section 12-2

39. A white dwarf star, the surviving core of a low-mass star towards the end of its life, can be found on the Hertzsprung-Russell diagram
    A) at the bottom end of the main sequence, along which it has evolved throughout its life.
    B) above and to the right of the main sequence, since it evolved there after its hydrogen-burning phase.
    C) below and to the left of the main sequence.
    D) at the upper left end of the main sequence, since its surface temperature is extremely high.
    Answer: C Source: Section 12-2

40. The characteristics of interiors of white dwarf stars are
    A) mostly hydrogen nuclei, supported by normal gas pressure because of the very high gas temperature, in a volume about the size of the Earth.
    B) mainly carbon and oxygen nuclei, supported by electron degeneracy pressure in a volume about the size of the Earth.
    C) mainly carbon and oxygen nuclei, supported by electron degeneracy pressure in a volume about the size of the Sun.
    D) mainly helium nuclei, supported by electron degeneracy pressure in a volume with a radius about 11 times that of Earth, about the volume of Jupiter.
    Answer: B Source: Section 12-2

41. A white dwarf is
    A) a small, very hot, low-mass star.
    B) a hot, main sequence star.
    C) an object like Jupiter which was not quite massive enough to become a star.
    D) a type of small protostar.
    Answer: A Source: Section 12-2

42. At which phase of its evolutionary life is a white dwarf star?
    A) In its early phases, soon after formation.
    B) Just at main sequence, or hydrogen-burning phase.
    C) Very late for small mass stars, in the dying phase.
    D) Post-supernova phase, the central remnant of the explosion.
    Answer: C Source: Section 12-2

43. A white dwarf star is at what stage of its evolution?
    A) Protostar phase, just after formation, beginning to generate energy by nuclear fusion.
    B) Post-supernova stage, after the explosion of a star.
    C) Very late phase of evolution, no longer generating energy.
    D) Main sequence phase, "middle-aged," generating energy by fusion of hydrogen to helium.
    Answer: C  Source: Section 12-2

44. Our Sun will end its life by becoming
    A) a molecular cloud.  B) a pulsar.  C) a black hole.  D) a white dwarf.
    Answer: D  Source: Section 12-2

45. In which order will a single star of about 1 solar mass progress through the various stages of evolution?
    A) T Tauri, main sequence, planetary nebula, white dwarf.
    B) T Tauri, red giant, white dwarf, neutron star.
    C) planetary nebula, main sequence, red giant, white dwarf.
    D) planetary nebula, main sequence, neutron star, black hole.
    Answer: A  Source: Chapters 11-13

46. A white dwarf star is about the same size as
    A) the total solar system.  B) a major city.  C) the Earth.  D) the Sun.
    Answer: C  Source: Section 12-2

47. The energy generation process inside a white dwarf star is
    A) non-existent; a white dwarf star is simply cooling by radiating its original heat.
    B) the combining of protons and electrons to form neutrons within its core.
    C) hydrogen fusion.
    D) the helium flash; very efficient and rapid helium fusion.
    Answer: A  Source: Section 12-2

48. How does a white dwarf generate its energy?
    A) Nuclear fusion of hydrogen into helium is producing energy within its core.
    B) Nuclear fission of heavy elements in the central core is releasing energy.
    C) It no longer generates energy, but is slowly cooling as it radiates away its heat.
    D) Gravitational potential energy is released as the star slowly contracts.
    Answer: C  Source: Section 12-2

49. The one characteristic shared by all solitary white dwarf stars is that
    A) they are generating thermonuclear energy, but are maintaining a constant radius, and hence are not releasing gravitational energy.
    B) they have never generated either thermonuclear or gravitational energy, but are slowly cooling after their production in a supernova explosion.
    C) they have ceased to generate energy by thermonuclear processes or gravitational contraction, and are slowly cooling down.
    D) they have stopped generating thermonuclear energy but continue to shrink, thereby releasing gravitational energy as heat.
    Answer: C   Source: Section 12-2

50. What is it that keeps a white dwarf star from collapsing inward upon itself?
    A) Normal gas pressure.
    B) Convection currents or updrafts from the nuclear furnace.
    C) Electron degeneracy or "quantum crowding."
    D) The physical size of the neutrons of which this star is composed.
    Answer: C   Source: Section 12-2

51. White dwarf stars are supported from collapse by
    A) nuclear fusion reactions in a shell around the core.
    B) degenerate-electron pressure.
    C) nuclear fusion reactions in their cores.
    D) centrifugal force due to rapid rotation.
    Answer: B   Source: Section 12-2

52. A white dwarf star is supported from collapse under gravity by
    A) pressure of the gas, heated by nuclear fusion reactions in its core.
    B) pressure of the gas, heated by nuclear fusion reactions in a shell around its core.
    C) centrifugal force due to rapid rotation.
    D) degenerate-electron pressure in the compact interior.
    Answer: D   Source: Section 12-2

53. One distinctive physical characteristic of matter inside a white dwarf star is that it
    A) is composed only of electrons in a degenerate state.
    B) is composed only of neutrons.
    C) has extremely high density compared to ordinary stellar matter.
    D) is composed only of protons, with electrostatic repulsion preventing stellar collapse.
    Answer: C   Source: Section 12-2

54. The stars which eventually become white dwarfs are those which start life with masses less than
A) 25 solar masses. B) 8 solar masses. C) 3 solar masses. D) 1.4 solar masses.
Answer: B Source: Section 12-2

55. One peculiar feature of the evolution of a white dwarf star of a particular mass is that
A) its size or radius remains constant as it cools and dies.
B) it shrinks as it cools, eventually to become a cold black hole in space.
C) it heats up as it shrinks because of the release of gravitational energy, ending up as a very hot but very small star.
D) its size or radius slowly increases as it cools, until it ends up as a red giant star.
Answer: A Source: Section 12-2

56. As a white dwarf evolves, the direction of its motion on the Herzsprung-Russell diagram is from upper left to lower right, which means that
A) its size or radius remains constant as it cools and becomes less luminous.
B) it shrinks as it cools, eventually to become a cold, black hole in space.
C) its size or radius slowly increases as it cools, until it ends up as a red giant star.
D) it heats up as it shrinks because of the release of gravitational energy, ending up as a very hot but very small star.
Answer: A Source: Section 12-2

57. There is a mass limit for a star in the white dwarf phase, the Chandrasakhar limit, beyond which the electron degeneracy pressure can no longer support the star against its own gravity. This mass limit is
A) 30 solar masses. B) 14 solar masses. C) 1.4 solar masses. D) 0.2 solar mass.
Answer: C Source: Section 12-2

58. Which of the following types of stars or stellar remnants can have a mass **NO** larger than about 1.4 times the mass of the Sun, otherwise they will collapse under their own gravity?
A) Neutron stars. B) White dwarfs. C) Red giants. D) Black holes.
Answer: B Source: Section 12-2

59. The star Sirius (a Canis Majoris) is actually a binary star in which the brighter, more massive star (Sirius A) is an A-type main sequence star and the fainter, less-massive star (Sirius B) is a white dwarf. How is this system most likely to have formed?
    A) Sirius A formed as a single star, and later captured a passing white dwarf.
    B) Sirius A was always the more massive of the two and it became a red giant; then mass transfer to Sirius B accelerated the evolution of Sirius B and caused it to become a white dwarf.
    C) Sirius B formed first and evolved to become a white dwarf; capture of gas and dust from interstellar clouds then resulted in the formation of Sirius A.
    D) Sirius B was initially more massive than Sirius A and evolved faster; it then became less massive due to mass loss to space and mass transfer to Sirius A.
    Answer: D Source: Section 12-2

60. The nova phenomenon, an occasional and sometimes repeated intense brightening of a star by a factor of about $10^6$, is caused by
    A) the capture and rapid compression of matter by a black hole.
    B) hydrogen "burning" explosively on the surface of a white dwarf star after mass transfer from a companion star in a binary system.
    C) a beam of radiation from a nearby pulsar, illuminating the surface of a red giant star and inducing rapid and intense heating.
    D) the explosion of a single massive star at the end of its thermonuclear burning phases.
    Answer: B Source: Section 12-3

61. When a typical nova explodes, it brightens in a few hours by a factor of
    A) 10 to 100. B) $10^8$ to $10^{10}$. C) $10^4$ to $10^6$. D) 2 to 5.
    Answer: C Source: Section 12-3

62. The mechanism which gives rise to the phenomenon of the nova is
    A) material falling into a black hole and being condensed to the point where a thermonuclear explosion is produced.
    B) the impact and subsequent explosion of a large comet nucleus upon a star's surface.
    C) the complete disintegration of a massive star because of thermonuclear runaway in the star's interior.
    D) matter from a companion star falling onto a white dwarf in a close binary system, eventually causing a nuclear explosion on the dwarf's surface.
    Answer: D Source: Section 12-3

63. A nova is a sudden brightening of a star that occurs when
    A) the electron-degenerate iron core of a massive star collapses after its mass becomes greater than the Chandrasekhar mass limit.
    B) material is transferred onto the surface of a white dwarf from a companion star in a binary system, then subsequently blasted into space by a runaway thermonuclear explosion (leaving the white dwarf intact to repeat the process).
    C) material is transferred onto a neutron star from a companion star in a binary system, causing the neutron star to collapse into a black hole.
    D) material from a companion star is transferred onto the surface of a white dwarf star in a binary system, after which runaway carbon fusion reactions cause the entire white dwarf to be destroyed in an explosion.
    Answer: B Source: Section 12-3

64. Suppose you see a news announcement that astronomers have discovered a nova in a distant galaxy. What do you immediately know about the star that gave rise to this nova?
    A) It is a neutron star in a binary system in which the other is star a white dwarf.
    B) It is a neutron star in a binary system in which the other star fills its Roche lobe.
    C) It is a white dwarf in a binary system in which the other star fills its Roche lobe.
    D) It is a black hole in a binary system in which the other star fills its Roche lobe.
    Answer: C Source: Section 12-3

65. Which force induces the core to condense and collapse in massive stars at the conclusion of each episode of nuclear fusion, such as the carbon, oxygen and silicon fusion cycles?
    A) The nuclear attractive force between nuclei and between neutrons and protons.
    B) Gravity.
    C) Electron degeneracy pressure.
    D) Gas pressure produced by the very high gas temperatures.
    Answer: B Source: Section 12-4

66. Which nuclear fusion cycle follows the helium fusion phase as a massive star evolves?
    A) Iron "burning."
    B) Carbon "burning."
    C) Silicon "burning."
    D) Oxygen "burning."
    Answer: B Source: Section 12-4 and Table 12-1

67. A star of 25 solar masses spends roughly what percentage of its life as a main sequence star? (See Table 12-1, Kaufmann & Comins, *Discovering the Universe*, 5th Ed.)
    A) 93%. B) 58%. C) Very little, less than 1%. D) About 7%.
    Answer: A Source: Section 12-4

68. For a massive star (e.g., 25 solar masses), core hydrogen burning lasts for several million years. In contrast, core silicon burning lasts for only about
A) several thousand years. B) one year. C) one day. D) one minute.
Answer: C Source: Section 12-4

69. The main product of silicon fusion reactions in the core of a massive star is
A) carbon. B) helium. C) iron. D) magnesium.
Answer: C Source: Section 12-4

70. A high-mass star near the end of its life undergoes successive cycles of energy generation within its core in which gravitational collapse increases the temperature to the point where a new nuclear fusion cycle generates sufficient energy to stop the collapse. This process does not work beyond the silicon fusion cycle, which produces iron. Why is this?
A) Fusion of iron nuclei into heavier nuclei requires energy rather than producing excess energy and therefore will not produce the additional gas pressure to halt the collapse.
B) Electrostatic forces between the highly charged iron nuclei are sufficient to overcome the collapse and stabilize the stellar core.
C) Iron nuclei are so large that they occupy all remaining space and so the collapse cannot continue.
D) The pressure from high-energy photons and neutrinos at the very high core temperatures reached at this stage of development is finally sufficient to halt the collapse.
Answer: A Source: Section 12-5

71. Thermonuclear reactions release energy because
A) the product (ash) nucleus is more tightly bound than the original (fuel) nucleus.
B) the product (ash) nucleus is less tightly bound than the original (fuel) nucleus.
C) the product (ash) nucleus is moving faster than the original (fuel) nucleus, and this excess kinetic energy shows up as heat.
D) the product (ash) nucleus contains fewer protons than the original (fuel) nucleus, since these protons have been converted into energy.
Answer: A Source: Section 12-5

72. The sequence of thermonuclear "burning" and fusion processes inside massive stars can transform elements such as carbon, oxygen, etc. into heavier elements AND generate excess energy, until iron has been produced. Why is it not possible for fusion reactions to release energy from iron nuclei?
    A) The protons and neutrons in an iron nucleus are so tightly bound together that fusing other nuclei with iron absorbs energy, rather than releasing it.
    B) Iron has the largest nucleus of all elements, and fusing other nuclei with iron actually reduces the size of the nucleus.
    C) Iron is the heaviest naturally occurring element.
    D) The electrostatic charge of iron nuclei is so great that other nuclei cannot approach closely enough to react with them.
    Answer: A Source: Section 12-5

73. A sequence of thermonuclear "burning" and fusion processes inside massive stars can continue to transform the nuclei of elements such as carbon, oxygen, etc. into heavier nuclei AND also generate excess energy, up to a limit beyond which no further energy-producing reactions can occur. The element that is produced when this limit is reached is
    A) silicon. B) oxygen. C) uranium. D) iron.
    Answer: D Source: Section 12-5

74. In the core of a high-mass star just before a supernova explosion occurs, the density in its core is about that
    A) at the center of the Sun, about $1.5 \times 10^5$ kg/m$^3$.
    B) of nuclear matter in a normal nucleus, about $4 \times 10^{17}$ kg/m$^3$.
    C) of iron, $7.5 \times 10^3$ kg/m$^3$.
    D) of degenerate gases in white dwarf stars, about $10^3$ kg/m$^3$.
    Answer: B Source: Section 12-5

75. Which of the following will a high-mass star (say, 25 times the mass of the Sun) **NOT** do at or near the end of its life?
    A) Eject its outer layers and become a white dwarf.
    B) Eject its outer layers and become a neutron star.
    C) Emit copious amounts of neutrinos.
    D) Convert silicon into iron in its core.
    Answer: A Source: Section 12-5

76. After the material in the core of a massive star has been converted to iron by thermonuclear reactions, further energy can be released to heat the core **ONLY** by
    A) nuclear fission or splitting of nuclei.
    B) gravitational contraction.
    C) thermonuclear fusion of iron into heavier elements.
    D) the absorption of neutrinos.
    Answer: B Source: Section 12-5

77. From the start of collapse to the attainment of nuclear density, the process of core collapse at the end of the life of a massive star takes a time of about
    A) 1/4 second. B) 8 minutes. C) 1/2000 second. D) a few hours.
    Answer: A Source: Section 12-5

78. The core collapse phase at the end of the life of a massive star is triggered when
    A) nuclear fusion has produced a significant amount of iron in its core.
    B) the density reaches the threshold for electron degeneracy pressure to become important.
    C) the helium flash and thermal pulses have expelled the star's envelope.
    D) the core becomes as dense as an atomic nucleus.
    Answer: A Source: Section 12-5

79. Which of the following is NOT a consequence of core collapse at the end of the life of a massive star?
    A) Electrons combine with protons to form neutrons.
    B) The core density approaches the density of an atomic nucleus.
    C) Great numbers of neutrinos are produced.
    D) The silicon core is converted to iron by fusion reactions.
    Answer: D Source: Section 12-5

80. The very last nuclear process to occur at the centre of a massive star (at the end of its life) is
    A) silicon burning, resulting in the production of iron.
    B) the capture of electrons by protons to produce neutrons.
    C) the photo-disintegration of nuclei by gamma rays.
    D) the helium flash.
    Answer: B Source: Section 12-5

81. What is photo-disintegration?
    A) The heating and ejection of mass from the surface of a normal star by the radiation from an orbiting neutron star.
    B) The ejection of a neutron or proton from an atomic nucleus, accompanied by the emission of a gamma ray.
    C) The splitting apart of atomic nuclei by gamma rays.
    D) The destruction of a star by the pressure of the radiation inside it.
    Answer: C Source: Section 12-5

82. Which of the following processes is NOT involved in the supernova explosion of a massive star?
    A) Photo-disintegration of nuclei by gamma rays.
    B) Helium flash in the star's core.
    C) Passage of a shock wave through the star's envelope.
    D) Collapse of the star's core.
    Answer: B Source: Section 12-5

83. Which of the following is **NEVER** a consequence of a supernova explosion?
    A) The triggering of star formation by shock waves moving through interstellar space.
    B) The formation of a planetary nebula.
    C) The generation of a pulse of neutrino emission.
    D) The condensation of matter into a solid nuclear star composed entirely of neutrons.
    Answer: B Source: Chapter 12

84. What fraction of its mass does a 25 $M_{\odot}$ main sequence star eject into space during its lifetime?
    A) Almost all of it, greater than 80%.
    B) Only a small fraction, about 1/100.
    C) Between 1/4 and 1/2.
    D) Only its outer atmosphere, less than 1 part in $10^4$.
    Answer: A Source: Section 12-5

85. The luminosity of a typical supernova star, during the initial phases of the explosion, increases by a factor of
    A) $10^3$. B) $10^6$. C) $10^8$. D) $2^{-3}$, since the star is already very bright.
    Answer: C Source: Section 12-5

86. What is the source of most of the heavy elements on the Earth and in our own bodies?
    A) Cosmic ray interactions with hydrogen and helium nuclei in interstellar clouds.
    B) Nuclear reactions during the formation of the universe (the Big Bang).
    C) Thermonuclear fusion reactions in the cores of massive stars.
    D) Explosive nucleosynthesis during supernova explosions of massive stars.
    Answer: D Source: Section 12-5

87. The most recent supernova explosion known to have occurred in our own galaxy
    A) created the supernova remnant Cassiopeia A.
    B) gave rise to the Crab Nebula.
    C) was seen in 1987 (supernova 1987A).
    D) created the Gum Nebula.
    Answer: A Source: Section 12-6

88. What is remarkable about the supernova remnant Cassiopeia A?
    A) It contains a binary neutron star system, in which one neutron star is significantly more massive than the other one.
    B) It shows that a supernova occurred in our galaxy about 300 years ago, but there is no record of any supernova having been seen at that time.
    C) It is the nearest supernova remnant, located only about 300 light years from the Earth.
    D) It is bright at all wavelengths from radio through visible light to x-rays.
    Answer: B Source: Section 12-6

89. Measurements suggest that light first arrived at Earth from the Cassiopeia A supernova about 300 years ago and that this supernova is about 10,000 light years away from Earth. When did the explosion actually occur?
    A) It is not possible to say when it occurred from the information given.
    B) 9,700 years ago, or about 7700 BC.
    C) 300 years ago, or about 1700 AD.
    D) 10,300 years ago, or about 8300 BC.
    Answer: D Source: Section 12-6

90. What are cosmic rays?
    A) Atomic nuclei and other subatomic particles travelling through space at more than 90% of the speed of light.
    B) Beams of photons produced by rotating, magnetic neutron stars.
    C) The steady, low-energy flux of neutral atoms into the solar system due to the passage of the Sun through the interstellar medium.
    D) Neutron beams emitted along the rotational axes of accretion disks around neutron stars.
    Answer: A Source: Section 12-6

91. The main component of cosmic rays is
    A) gamma rays.
    B) electrons.
    C) protons.
    D) nuclei of heavy elements such as carbon and iron.
    Answer: C Source: Section 12-6

92. The main source of cosmic rays, the high-energy atomic nuclei that strike the Earth's atmosphere from interstellar space, is believed to be
    A) matter from accretion disks surrounding neutron stars and black holes.
    B) rotating, magnetized neutron stars.
    C) supernova explosions.
    D) acceleration of atoms by magnetic fields in interstellar clouds.
    Answer: C Source: Section 12-6

93. What is a cosmic ray shower?
    A) A shower of particles produced when a cosmic ray strikes atoms in the Earth's atmosphere.
    B) A burst of high-energy atomic nuclei arriving at the Earth from interstellar space.
    C) Another name for a meteor shower.
    D) A pulse of gamma rays arriving at the Earth from a rotating, magnetized neutron star.
    Answer: A Source: Section 12-6

94. Supernovae are detected
    A) only at X-ray wavelengths.
    B) only in elliptical galaxies, never in spirals.
    C) only in our Galaxy.
    D) in both our Galaxy and others.
    Answer: D Source: Section 12-7

95. The light from the most recent supernova, which was visible to the unaided eye, arrived at Earth in
    A) 1572 AD. B) 1604 AD. C) 1054 AD. D) 1987 AD.
    Answer: D Source: Section 12-7

96. Just before it exploded, the star that became supernova SN 1987A was
    A) an M2 I supergiant. B) a white dwarf. C) a pulsar. D) a B3 I supergiant.
    Answer: D Source: Section 12-7

97. The star that exploded to form supernova SN 1987A probably had a pre-explosion mass of about
    A) 1.4 solar masses.
    B) 20 solar masses.
    C) less than one solar mass.
    D) 40 to 50 solar masses.
    Answer: B Source: Section 12-7

98. Which of the following is NOT true for supernova SN 1987A?
    A) Observations of the star had been made before it blew up.
    B) It was a white dwarf exploding after mass transfer from a companion star in a binary star system.
    C) Bursts of neutrinos were detected at several sites from the supernova.
    D) It did not become as intrinsically bright as originally expected.
    Answer: B Source: Section 12-7

99. How did supernova SN 1987A differ from most other observed supernovae?
    A) It declined in brightness much faster than most supernovae.
    B) It occurred in an external galaxy, not our Milky Way Galaxy.
    C) The star was a blue supergiant when it blew up, rather than the usual red supergiant.
    D) It reached a maximum luminosity ten times that of a normal supernova.
    Answer: C Source: Section 12-7

100. The neutrino is
    A) another name for an anti-neutron, the anti-particle of the neutron, very difficult to detect.
    B) an elusive, subatomic particle, having no (or very small) mass, very difficult to detect.
    C) a heavy nuclear particle, easily detected.
    D) another name for an anti-electron or positron.
    Answer: B Source: Section 12-5

101. Which of the following properties does the neutrino NOT possess?
    A) Electrical charge equal to that of the electron.
    B) Zero or extremely small mass.
    C) High penetration through any matter.
    D) Travels at the speed of light.
    Answer: A Source: Section 12-5

102. One technique for the detection of neutrinos from outer space involves the reaction of neutrinos with protons in a water tank, with the resulting high-speed positrons, or positive electrons then generating Cerenkov radiation in the water. How is this Cerenkov radiation produced?
   A) By the passage of the charged positron through the water at a speed greater than the speed of light in a vacuum.
   B) By the annihilation of a negative electron in the water by the positron, to produce radiation.
   C) By the charged positron moving in the water faster than the speed of light in water.
   D) By the recombination of the positron with a proton to produce neutral hydrogen and Balmer series line emissions.
   Answer: C Source: Section 12-7

103. Cerenkov radiation, which was used to detect neutrinos from supernova SN 1987A, was produced by
   A) positive electrons (positrons) travelling faster than the speed of light in water.
   B) neutrons ejected from nuclei by the neutrinos.
   C) nuclear fusion produced in the detector by neutrinos.
   D) electrons and anti-electrons (positrons) colliding and annihilating.
   Answer: A Source: Section 12-7

104. What new method has recently provided astronomers with new information about the interiors of stars, for example, the collapse of the inner core of a star undergoing supernova explosion, or the interior of the Sun?
   A) Visible light spectroscopy.
   B) X-ray astronomy and photography.
   C) Neutrino astronomy.
   D) Radio astronomy.
   Answer: C Source: Section 12-7

105. A Type II supernova is
   A) the explosion of a red giant star as a result of a helium flash in the core.
   B) the collapse of a blue supergiant star to form a black hole.
   C) the explosion of a white dwarf in a binary star system after mass has been transferred onto it from its companion.
   D) the explosion of a massive star after silicon burning has produced a core of iron nuclei.
   Answer: D Source: Section 12-8

106. A Type I supernova is
   A) the collapse of a blue supergiant star to form a black hole.
   B) the explosion of a white dwarf in a binary star system after mass has been transferred onto it from its companion.
   C) the explosion of a massive star after silicon burning has produced a core of iron nuclei.
   D) the explosion of a red giant star as a result of a helium flash in its core.
   Answer: B Source: Section 12-8

107. What is the main observational difference between a Type I and a Type II supernova?
   A) The spectrum of a Type I supernova shows strong lines of both hydrogen and helium, whereas that of a Type II shows only hydrogen.
   B) Hydrogen lines are prominent in the spectrum of a Type I supernova, but absent in that of a type II.
   C) Hydrogen lines are prominent in the spectrum of a Type II supernova, but absent in that of a type I.
   D) The spectrum of a Type II supernova shows strong lines of both hydrogen and helium, whereas that of a Type I shows only hydrogen.
   Answer: C Source: Section 12-8

108. Type II supernovae show prominent lines of hydrogen in their spectra, whereas hydrogen lines are absent in spectra of Type I supernovae. Why is this? [HINT: Think about the type of star that gives rise to each of the two types of supernova.]
   A) White dwarfs have a thick surface layer of hydrogen, whereas neutron stars contain no hydrogen at all.
   B) Massive stars contain large amounts of hydrogen, whereas neutron stars contain no hydrogen at all.
   C) Massive stars have burned all of their hydrogen into heavier elements, whereas low-mass stars still have large hydrogen-rich envelopes.
   D) Massive stars contain large amounts of hydrogen, whereas white dwarfs are mostly carbon and oxygen.
   Answer: D Source: Section 12-8

109. Can a white dwarf explode?
   A) Yes, but only if another star collides with it; and stars are so far apart in space that this is unlikely to have ever happened in our galaxy.
   B) Yes, but only if it is in a binary star system.
   C) No; white dwarfs are held up by electron degeneracy pressure, and this configuration is stable against collapse or explosion.
   D) Yes, but only if nuclear reactions in the white dwarf core reach the stage of silicon burning, producing iron.
   Answer: B Source: Section 12-8

110. The estimated rate at which supernova explosions occur in a spiral galaxy such as our own is
A) about once every 5 years.
B) about once every 3,000 years.
C) about once every 300 years.
D) about one every 20 years.
Answer: D Source: Section 12-8

111. From observations of supernova explosions in distant galaxies, it is predicted that there should be about 5 supernovae per century in our galaxy, whereas we have seen only about 1 every 300 years from Earth. Why is this?
A) Most supernovae occur in the galactic plane where interstellar dust will have hidden them from our view from Earth.
B) The majority of supernovae produce no visible light, only radio and X-ray radiation, which we have only been able to observe for the past 3 decades.
C) Most supernovae occur within the Milky Way, which can be seen only from the southern hemisphere where there have been very few observers until recently.
D) The majority of stars are old, well beyond the supernova stages of evolution, in our galaxy.
Answer: A Source: Section 12-8

112. Measurements from distant galaxies indicate that supernovae should occur at a rate of 5 per century in a spiral galaxy such as the Milky Way but only 3 have been recorded in this galaxy in the past 1000 years. Why is this?
A) The majority of supernovae must have occurred in the plane of the Milky Way and hence were hidden from Earth by the dense gas and dust in the Milky Way plane.
B) Observers were not watching the sky carefully enough, particularly through the Dark Ages and over the past few centuries.
C) The Milky Way galaxy is somehow different, with much lower numbers of very massive stars in general, so many fewer stars have undergone supernova explosions.
D) Most supernovae produce X-rays and radio waves, and not visible light and were hence invisible to earlier observers.
Answer: A Source: Section 12-8

113. The explosion of a supernova appears to leave behind
A) a rapidly expanding shell of gas and a central neutron star.
B) a rapidly rotating shell of gas, dust and radiation, but no central object.
C) a rapidly expanding shell of gas and a compact white dwarf star at its center.
D) nothing; the explosion changes all the matter completely into energy which then radiates into space at the speed of light.
Answer: A Source: Sections 12-6 and 12-9

114. What important event was recorded by ancient Chinese astronomers in 1054 AD?
A) The appearance of a nearby nova explosion, in our Galaxy.
B) A bright binary star undergoing eclipse.
C) A brilliant world-wide auroral display.
D) The appearance of a supernova, visible in daylight.
Answer: D Source: Section 12-9

115. Which major astronomical event was apparently recorded faithfully by Chinese astronomers in the Sung Dynasty in 1054 AD?
A) The discovery of the planet Mercury.
B) The total eclipse of the Sun over China in that year.
C) A supernova explosion in our Galaxy, visible even in daylight.
D) The rare passage of the planet Venus across the face of the Sun, a solar transit.
Answer: C Source: Section 12-9

116. The Crab Nebula is
A) a planetary nebula surrounding a hot star.
B) a cool gaseous nebula in which stars are forming.
C) the active nucleus of a nearby spiral galaxy.
D) a supernova remnant.
Answer: D Source: Section 12-9

117. In terms of the evolutionary life of a star, at what stage is the Crab Nebula?
A) A black hole, very late stage of evolution.
B) Middle age, main sequence star, relatively near to the Sun.
C) Late, since it is the remnant of a star explosion or supernova.
D) The beginning, a nebula in which stars are forming.
Answer: C Source: Section 12-9

118. The Crab Nebula is a nearby example of what type of physical phenomenon?
A) The remnant of a supernova explosion.
B) A gas and dust cloud, the formation region for new stars.
C) A planetary nebula, a shell of gas leaving an old star.
D) A spiral galaxy, a collection of 100 billion stars.
Answer: A Source: Section 12-9

119. Where would you look for a pulsar, among the following locations in the universe?
   A) In the Orion Nebula.
   B) In the Crab Nebula.
   C) On someone's wrist, since it is a brand of wristwatch.
   D) At the center of our Galaxy.
   Answer: B Source: Section 12-9

120. The first pulsar was detected with
   A) Galileo's telescope.
   B) a British radio telescope.
   C) the infrared satellite IRAS.
   D) the 200-inch telescope at Mount Palomar.
   Answer: B Source: Section 12-9

121. The first pulsar was discovered by
   A) the Astronomer Royal in Newton's time, Sir Edmund Halley, in 1606.
   B) Galileo Galilei, in 1610.
   C) Albert Einstein, in 1905.
   D) an English graduate student, Jocelyn Bell in 1967.
   Answer: D Source: Section 12-9

122. The first pulsar was discovered in which year?
   A) 1967. B) 1930. C) 1054. D) 1978.
   Answer: A Source: Section 12-9

123. Which of the following astronomical objects is most closely associated with a pulsar?
   A) A neutron star. B) A black hole. C) A white dwarf star. D) A red giant star.
   Answer: A Source: Section 12-9

124. What is a pulsar?
   A) A pulsating white dwarf star, fluctuating rapidly in brightness.
   B) Very hot material orbiting a black hole.
   C) A Cepheid variable star with a period of a few days.
   D) A rapidly rotating neutron star, producing beams of radio energy and occasionally of X-rays and visible light.
   Answer: D Source: Section 12-9

125. A pulsar is
   A) a pulsating star, in which size, temperature, and light intensity vary regularly.
   B) an object at the center of each galaxy, supplying energy from its rapid rotation.
   C) a rapidly rotating neutron star, emitting beams of radio, and sometimes X-ray and visible, energy.
   D) a binary star in which matter from one star is falling onto the second star.
   Answer: C Source: Section 12-9

126. A pulsar is
   A) an accretion disk around a black hole, emitting light as matter is accumulated onto the disk.
   B) an very precise interstellar beacon manufactured by little green humanoid life forms.
   C) a type of variable star, pulsating rapidly in size and brightness.
   D) a rapidly spinning neutron star.
   Answer: D Source: Section 12-9

127. Pulsars, emitting very regular radio, and sometimes visible light pulses, are what type of object?
   A) Black holes, with material falling regularly into them.
   B) Rapidly rotating neutron stars.
   C) Rapidly rotating binary star systems in which the stars undergo regular eclipses as seen from Earth.
   D) Pulsating variable stars.
   Answer: B Source: Section 12-9

128. A pulsar is most probably formed
   A) at the center of a supernova explosion.
   B) in the high-temperature core of a star as it evolves through its main sequence phase.
   C) just after the formation of a protostar by gravitational condensation.
   D) within a huge gas cloud, by collisions between stars.
   Answer: A Source: Section 12-9

129. The period of pulsation for most pulsars are in the range
   A) between 1/1000 second and a few seconds.
   B) between $10^{-6}$ and $10^{-3}$ second.
   C) minutes to tens of hours.
   D) many hours to a few days.
   Answer: A Source: Section 12-9

130. The existence of stars composed almost entirely of neutrons was first predicted by
   A) Albert Einstein in 1908.
   B) Fritz Zwicky and Walter Baade in 1933.
   C) Stephen Hawking in 1985.
   D) Jocelyn Bell in 1967.
   Answer: B Source: Section 12-9

131. The diameter of a typical neutron star of 1 solar mass is predicted to be approximately
   A) that of an average city, a few km.
   B) 1 km.
   C) that of Earth, 12,800 km.
   D) that of the Sun.
   Answer: A Source: Section 12-9

132. Which of the following statements is NOT a property of neutron stars?
   A) They are composed almost entirely of neutrons.
   B) They emit relatively narrow beams of light and other radiation.
   C) They contain strong gravitational fields but weak magnetic fields.
   D) They rotate from one to thirty times each second.
   Answer: C Source: Sections 12-9 and 12-10

133. The very strong magnetic field on a neutron star is created by
   A) differential rotation of the neutron star, with the equator rotating faster than the poles, similar to sunspot formation.
   B) a burst of neutrinos produced by the supernova explosion, since this would be the equivalent of a very large electrical current flowing for a short time.
   C) the collapse of a star, which significantly intensifies the original weak magnetic field of the star.
   D) turbulence in the electrical plasmas during the collapse of a star; the original star would have had no magnetic field.
   Answer: C Source: Section 12-10

134. The pulsed nature of the radiation at all wavelengths, which is seen to come from a pulsar, is produced by
   A) the mutual eclipses of two very hot stars orbiting in a close binary system.
   B) the rapid pulsation in size and brightness of a small white dwarf star.
   C) extremely hot matter which is rapidly orbiting a black hole just prior to descending into it.
   D) the rapid rotation of a neutron star which is producing two oppositely directed beams of radiation.
   Answer: D Source: Section 12-10

135. The source of the beams of electromagnetic radiation (including light in some cases) emitted by pulsars is
   A) charged particles travelling along the magnetic axes of rotating neutron stars; the particles emit light as they are accelerated.
   B) jets of material flowing out along the rotation axis of the accretion disk around a black hole; collisions in the jets heat the material and produce light.
   C) the surface of a normal star which has a white dwarf companion; the white dwarf creates a hot spot on the normal star which emits a beam of light as the stars rotate around each other.
   D) electrons flowing out along the rotation axis of an accretion disk around a neutron star; the electrons are accelerated and hence emit light.
   Answer: A Source: Section 12-10

136. A neutron star will be detected from Earth as a pulsar by its regular radio pulses only if
   A) Earth lies within the path of the narrow beam of radiation being generated by the neutron star along its magnetic axis as the star rotates rapidly.
   B) Earth lies in the neutron star's "equator", the plane perpendicular to its spin axis.
   C) Earth lies in the plane of the neutron star's magnetic equator, halfway between its magnetic poles.
   D) Earth lies directly above the rotation axis of the rotating neutron star.
   Answer: A Source: Section 12-10

137. Visible pulses are seen to accompany radio pulses from a neutron star only
   A) from neutron stars which are high above the galactic plane, where visible light is not obscured by dust and gas.
   B) if the neutron star is part of a binary star system.
   C) if the neutron star is relatively young.
   D) if the neutron star is relatively old, since pulsars from recent supernova explosions produce only radio pulses.
   Answer: C Source: Section 12-10

138. The main reason for the observed slow-down of many pulsars is
   A) the slowing of rotation caused by slow expansion and redistribution of mass of the source, similar to that which occurs to a spinning skater.
   B) friction between the stellar surface and the surrounding nebular material.
   C) loss of rotational energy to provide energy for the emission of radiation.
   D) a slow build-up of the magnetic field, rotational energy being transferred to magnetic energy.
   Answer: C Source: Section 12-10

139. As time progresses, the pulse rate for most solitary pulsars is
   A) varying periodically as the neutron star undergoes periodic expansions and contractions.
   B) slowing down, since rotational energy is being used to generate the pulses.
   C) speeding up, as the neutron star slowly contracts under gravity.
   D) absolutely constant, pulsars providing ideal frequency standards or clocks.
   Answer: B Source: Section 12-10

140. Pulsating X-ray sources are believed to be
   A) white dwarf stars with intense magnetic fields; the X-rays are generated by flares like those on the Sun, but much stronger.
   B) black holes in binary systems, with the X-rays emitted from an accretion disk around the black hole.
   C) spinning neutron stars in binary systems, emitting X-rays because of mass transfer onto the neutron star from its normal companion.
   D) the same as regular (optical) pulsars, but observed in X-rays.
   Answer: C Source: Section 12-11

141. Pulsating X-ray sources with periods of a few seconds are caused by
   A) the eclipsing an X-ray-emitting star with a very hot surface by a cool companion in a close binary system.
   B) matter falling onto the surface of a very hot, rotating, white dwarf star from an ordinary companion star in a binary system, producing an X-ray-emitting hot spot which disappears periodically behind the white dwarf.
   C) matter falling violently onto the surface of a rotating neutron star from a close companion in a binary star system, causing an X-ray hot spot which disappears periodically behind the neutron star.
   D) the pulsation in radius, temperature and hence luminosity of a hot Cepheid variable star with a surface temperature hot enough to emit X rays.
   Answer: C Source: Section 12-11

142. One prominent feature recently identified within many energetic close binary star systems as a result of their mutual interaction and mass exchange is
   A) a cool dust cloud surrounding the whole star system, hiding it from visible view.
   B) planetary formation between the stars, emitting IR radiation from molecular constituents and dust.
   C) the beginnings of spiral arms, showing the possible origin of spiral arm galaxies.
   D) two oppositely directed high-speed jets of matter leaving the system.
   Answer: D Source: Section 12-12

143. SS433 is an object in our Galaxy that has two oppositely directed jets of material being ejected from it. These jets is believed to be
   A) material transferred onto the surface of a white dwarf star in a binary star system, then subsequently blasted into space by runaway thermonuclear fusion reactions.
   B) energetic, charged particles ejected along the magnetic axes of a rotating neutron star.
   C) material squirted out along the rotation axis of the accretion disk around a black hole in a binary star system.
   D) material squirted out along the rotation axis of an accretion disk around a neutron star in a binary star system.
   Answer: D Source: Section 12-12

144. X-ray bursters emit occasional and intense bursts of X-rays on top of to a steady low-level X-ray emission. These bursts of X-rays are believed to be caused by
   A) material transferred onto the surface of a neutron star in a binary system, then subsequently ignited in a thermonuclear explosion which leaves the neutron star intact to repeat the process.
   B) material transferred onto the surface of a neutron star, causing the neutron star to collapse suddenly into a black hole.
   C) material from a companion star pulled into an accretion disk around a black hole; periodic clumps of material falling from the disk into the black hole and being compressed produce the X-rays.
   D) material transferred onto the surface of a white dwarf in a binary star system, producing a thermonuclear explosion at the surface while leaving the white dwarf intact to repeat the process.
   Answer: A Source: Section 12-13

145. Which is the correct sequence for the following end-points of stellar evolution, in order of increasing mass?
   A) White dwarf, black hole, neutron star.
   B) White dwarf, neutron star, black hole.
   C) Black hole, neutron star, white dwarf.
   D) Neutron star, black hole, white dwarf.
   Answer: B Source: Chapter 12

146. The Oppenheimer-Volkov limit to the amount of mass in a neutron star before neutron degeneracy pressure is unable to withstand the force of gravity and the neutron star is crushed out of existence into a black hole is
   A) about 3 solar masses.
   B) 1.4 solar masses.
   C) 20 solar masses.
   D) about 100 solar masses.
   Answer: A Source: Section 12-13

147. A stellar core of about 2 solar masses collapses inwards in a supernova explosion. What will be the result of this collapse?
   A) A black hole, with infinitely small radius, since nothing can prevent such a mass from collapsing completely.
   B) A white dwarf star, supported from further collapse by electron degeneracy pressure.
   C) A neutron star, supported from further collapse by neutron degeneracy pressure.
   D) A star a little smaller than the size of the Sun, supported from further collapse by gas pressure from S the hot interior of the star.
   Answer: C   Source: Sections 12-9 and 12-13

# Chapter 13: Black Holes

1. Black holes are so named because
   A) all of their electromagnetic radiation is gravitationally redshifted to the infrared, leaving no light in the optical region.
   B) they emit a perfect blackbody spectrum.
   C) no light or any other electromagnetic radiation can escape from inside them.
   D) their only spectral lines are in the radio and infrared.
   Answer: C Source: Chapter 13: Introduction

2. A black hole is so named because
   A) it emits no visible light because it is cold, less than 10 K.
   B) its spectrum has the same shape as that of a laboratory blackbody, typically at a temperature of about 150 K.
   C) no light can escape from it because of its powerful gravitational field.
   D) the gravitational field is so high that the wavelength of the emitted light is shifted to radio wavelengths.
   Answer: C Source: Chapter 13: Introduction

3. What prevents a neutron star from collapsing and becoming a black hole?
   A) Gravity in the neutron star is balanced by an outward force due to neutron degeneracy.
   B) Neutron stars are held up by the centrifugal force due to their rapid rotation.
   C) Gravity in the neutron star is balanced by an outward force due to gas pressure, as in the Sun.
   D) Neutron stars are solid, and like any solid sphere are held up by the repulsive force between atoms in the solid.
   Answer: A Source: Chapter 13: Introduction

4. What is believed to be the maximum mass for a neutron star?
   A) 12 solar masses. B) 1.4 solar mass. C) 3 solar masses. D) 150 solar masses.
   Answer: C Source: Chapter 13: Introduction

5. What is the escape velocity from inside a black hole?
   A) Zero.
   B) Greater than the speed of light.
   C) Infinite.
   D) Twice that from a neutron star.
   Answer: B Source: Chapter 13: Introduction

6. One feature that distinguishes a black hole from all other objects in the universe is that
   A) its total mass exceeds 3 solar masses.
   B) the shape of its gravitational field is always different from that of an ordinary massive object, even at large distances from it.
   C) the escape velocity from inside this object is greater than the speed of light.
   D) it emits large quantities of x-rays.
   Answer: C Source: Chapter 13: Introduction

7. The escape velocity of matter from near the center of a black hole whose mass is 3 solar masses is
   A) quite small.
   B) about half the speed of light.
   C) much greater than the speed of light.
   D) always exactly equal to the speed of light.
   Answer: C Source: Chapter 13: Introduction

8. What is "special" about special relativity?
   A) It deals only with objects that are at rest relative to each other.
   B) It deals only with objects moving in a straight line at a constant speed.
   C) It deals with motion, including both constant and accelerated motion, but only in the absence of gravitational fields.
   D) It deals only with motion at speeds significantly less than the speed of light.
   Answer: B Source: Section 13-1

9. How must an object be moving for us to be able to use the theory of special relativity to describe the object?
   A) It must be moving close to the speed of light; if this is true then how speed and direction change is not important.
   B) It must be moving at a constant speed; whether the direction of motion changes is unimportant.
   C) It must be moving in a constant direction; how speed changes is unimportant.
   D) It must be moving at a constant speed in a straight line; how fast it is moving is not important.
   Answer: D Source: Section 13-1

10. Which of the following is a correct and complete statement of Einstein's first postulate of special relativity?
    A) Your description of physical reality is the same regardless of the velocity at which you move.
    B) Your description of physical reality is the same regardless of how you move.
    C) Your description of physical reality is the same regardless of the direction in which you move.
    D) Your description of physical reality is the same regardless of the speed at which you move.

    Answer: A Source: Section 13-1

11. Suppose you are in a jet airliner traveling at a constant speed of 400 km/h in a constant direction. All windows are blocked so you cannot see outside, and there are no vibrations from the engines. What experiment can you do to determine that you are in fact moving?
    A) Measure the speed of a sound wave traveling up the aisle (toward the nose of the aircraft) and another traveling down toward the tail, and calculate the difference between the two results.
    B) None; all experiments will give the same results as when you are at rest on the ground.
    C) Drop a small rock and measure the distance it moves backwards down the aisle while it falls.
    D) Suspend a ball by a thread from the ceiling and measure the angle the thread makes with the vertical.

    Answer: B Source: Section 13-1

12. Suppose you are in the space shuttle in orbit around the Earth at a speed of 7 km/s, and at some particular time your direction of travel is straight toward the Sun. The speed of light in a vacuum in 300,000 km/s. What speed will you measure for light from the Sun?
    A) 300,014 km/s, because your speed is added to that of the light and relativistic contraction has shortened the meter stick used in the measurement of the speed of the light.
    B) 300,000 km/s.
    C) 300,007 km/s, because your speed is added to that of the light.
    D) 300,007 km/s, because relativistic contraction has shortened the meter stick with which you measure the distance traveled by the light in order to measure its speed.

    Answer: B Source: Section 13-1

13. Two rocket ships are traveling past the Earth at 90% of the speed of light, in opposite directions (i.e., they are approaching each other). One turns on a searchlight, which is seen by scientists aboard the other. What speed do the scientists measure for this light? (c = speed of light in a vacuum)
A) 1.8 c (equal to $2 \times 0.9$ c) B) c C) 0.9 c D) 1.9 c (equal to c + 0.9 c)
Answer: B Source: Section 13-1

14. If you see an object moving past you at 90% of the speed of light, what do you see for the length of the object?
A) Its length looks unchanged from when it is at rest.
B) It looks shorter than at rest while it is coming toward you, and longer after it has passed you.
C) It looks shorter than if it were at rest.
D) It looks longer than if it were at rest.
Answer: C Source: Section 13-1

15. A person is standing beside the open doorway of an empty barn. A pole of the same length as the barn is lying on the ground outside the barn. A second person picks up the pole and runs toward the barn at 90% of the speed of light, with the pole pointing toward the barn (a very athletic person!). How long is the pole compared to the barn as seen by the person standing at the door?
A) 44% of the length of the barn.
B) 90% of the length of the barn.
C) 2.3 times the length of the barn.
D) 81% of the length of the barn.
Answer: A Source: Section 13-1

16. A pole is lying on the ground outside a barn. The length of the pole is the same as the length of the barn. A person picks up the pole and runs toward the barn at 90% of the speed of light, with the pole pointing toward the barn (a very athletic person!). How long is the barn compared to the pole as seen by the person running?
A) The barn appears to be 5.3 times longer than the pole.
B) The barn appears to be 2.3 times longer than the pole.
C) The barn appears to be 19% of the length of the pole.
D) The barn appears to be 44% of the length of the pole.
Answer: D Source: Section 13-1

17. At what speed would an object have to be traveling in order for its length to appear to a stationary observer to have decreased to 1/3 of its length when seen at rest?
A) 0.33c, or 1/3 of the speed of light.
B) 0.11c, or 11% of the speed of light.
C) 0.94c, or 94% of the speed of light.
D) 0.89c, or 89% of the speed of light.
Answer: C Source: Section 13-1

18. Suppose that a particular kind of subatomic particle lives on average for exactly 1.0 microsecond ($1.0 \times 10^{-6}$ s) before it decays, when the particles are created at rest in the laboratory. If these particles are created in a beam of particles traveling at 97% of the speed of light as seen by a scientist standing beside the beam, what lifetime does the scientist measure for these particles?
    A) 4.1 microseconds.
    B) 0.06 microsecond.
    C) 0.24 microsecond.
    D) 16.9 microseconds.
    Answer: A Source: Section 13-1

19. During a stop on Jupiter's satellite, Europa, to investigate a possible liquid ocean under the ice, you spot a spaceship passing Europa at 97% of the speed of light. All such spaceships are equipped with flashing external lights that flash once per second. How much time ticks by on your own clock while you watch 2 successive flashes, that is 1 of these time intervals?
    A) 0.24 second B) 1 second, of course. C) 33.3 seconds D) 4.1 seconds
    Answer: D Source: Section 13-1

20. During a stop on Jupiter's satellite, Europa, to investigate a possible liquid ocean under the ice, you spot a spaceship passing Europa at 97% of the speed of light. A person on the SPACESHIP sees 2.0 seconds tick past on YOUR clock. How much time does this person see elapse on the clock on board the spaceship?
    A) 8.2 seconds B) 0.48 seconds C) 0.12 seconds D) 33.8 seconds
    Answer: A Source: Section 13-1

21. Suppose you are in a rocket ship traveling toward the Earth at 95% of the speed of light. Compared to when your ship was at rest on Mars, what length do you measure for your spaceship?
    A) You can't tell; your life processes have slowed down too much for you to measure the length.
    B) Longer than when it was on Mars.
    C) Shorter than when it was on Mars.
    D) The same as when it was on Mars.
    Answer: D Source: Section 13-1

22. Fred and Joanne both measure the length of a particular spaceship to be 100 m when it is on the Earth. Joanne then gets in the spaceship and, after visiting the Moon, hurtles past the Earth at a speed close to the speed of light. Fred, still on the Earth, measures the length of the moving spaceship to be about 90m. At the same time, Joanne (using her own meter stick) measures the length of the spaceship to be
    A) 100 m, because she is "at rest" on the spaceship.
    B) We can't tell from the information given.
    C) about 110 m, because both she and the spaceship are moving.
    D) about 90 m, because of the motion of the spaceship.
    Answer: A Source: Section 13-1

23. Suppose you see a spaceship with a clock on it, hurtling past you at 80% of the speed of light. As it goes by, the second hand on the ship's clock ticks off five seconds. How much time elapsed on your clock while this was happening?
    A) less than five seconds.
    B) more than five seconds if the spaceship is approaching you, or less than five seconds if it is moving away from you.
    C) more than five seconds.
    D) five seconds; the same as on the ship's clock.
    Answer: C Source: Section 13-1

24. Suppose you are aboard a spaceship that is passing the Earth at 80% of the speed of light. You see a clock on the Earth tick off five seconds. How much time elapses on your own clock while this is happening?
    A) five seconds; the same as on the ship's clock.
    B) more than five seconds.
    C) less than five seconds.
    D) more than five seconds if you are approaching the Earth, or less than five seconds if you are moving away from the Earth.
    Answer: B Source: Section 13-1

25. A child on a playground swing is swinging back and forth (one complete oscillation) once every four seconds, as seen by her father standing next to the swing. At the same time a spaceship is hurtling by at a speed close to the speed of light. According to special relativity (and ignoring the Doppler effect for this question), a person on the spaceship finds that the time for one full swing is
    A) less than four seconds.
    B) equal to four seconds.
    C) less than four seconds when the spaceship is approaching the swing, and longer than four seconds when it is moving away.
    D) longer than four seconds.
    Answer: D Source: Section 13-1

26. If you stay on the Earth while a friend races off in a rocket at a speed close to the speed of light, then according to special relativity you will see a clock on the rocket appear to tick more slowly than the one on your wall. If your friend looks back at your clock, then according to the same theory the friend will see your clock appear to tick
    A) faster or slower than the clock on the rocket, depending on the direction of travel of the rocket compared to Earth.
    B) more slowly than the clock on the rocket.
    C) faster than the clock on the rocket.
    D) at the same speed as the clock on the rocket.
    Answer: B Source: Section 13-1

27. You are on Mars standing on the gangplank of your spaceship when you see an identical spaceship go past Mars at 90% of the speed of light. When you look closely at this spaceship, how do you find that it compares to your own spaceship?
    A) The moving spaceship is shorter than yours and time on it moves more quickly than on your ship.
    B) The moving spaceship is shorter than yours and time on it moves more slowly than on your ship.
    C) The moving spaceship is longer than yours and time on it moves more slowly than on your ship.
    D) The moving spaceship is longer than yours and time on it moves more quickly than on your ship.
    Answer: B Source: Section 13-1

28. In a TV tube, the picture is created by a beam of electrons that travel down the tube at a very high speed. What is the mass of these electrons, compared to electrons at rest?
    A) The moving electrons appear to have a greater mass.
    B) The moving electrons appear to have a smaller mass.
    C) The electrons appear to have a greater mass if you are in front of the tube (electrons approaching you) and a smaller mass if you are standing behind the tube (electrons moving away from you).
    D) The mass of an electron is measured to be the same regardless of how fast it is moving.
    Answer: A Source: Section 13-1

29. Which statement best describes the "fabric" of space and time as outlined by the classical physics of Newton?
    A) The shape of space and the rate of passage of time depends upon the relative velocities of observer and observed.
    B) Space becomes "curved" and time slows down near a source of gravity, as measured by a distant observer.
    C) Space is perfectly uniform, filling everywhere like a fixed network, while time passes at a uniform rate for all observers.
    D) Space is expanding uniformly, while time passes more slowly as the universe ages.
    Answer: C Source: Section 13-2

30. In what way is the general theory more general (deals with more situations) than the special theory of relativity?
    A) It includes the change in the rate of passage of time when objects are in motion.
    B) It includes gravitation and accelerated motion.
    C) It includes motion at the speed of light.
    D) It includes constant, unaccelerated motion.
    Answer: B Source: Section 13-2

31. Suppose a satellite were placed in orbit around (and very close to) a neutron star. Which theory would you need to use to describe how it moves?
    A) The general theory of relativity.
    B) The special theory of relativity.
    C) Newton's law of gravitation.
    D) Kepler's laws.
    Answer: A Source: Section 13-2

32. Light traveling away from the surface of a neutron star becomes strongly redshifted. What name is given to this effect?
    A) Zeeman effect.
    B) Gravitational redshift.
    C) Doppler shift.
    D) Cosmological redshift.
    Answer: B Source: Section 13-2

33. How does a gravitational field affect the passage of time?
A) Clocks in a gravitational field run slower than clocks outside the field.
B) Gravity has no effect on the passage of time.
C) Clocks in a gravitational field run faster than clocks outside the field.
D) Gravity makes time stop.
Answer: A Source: Section 13-2

34. Suppose you were far from a planet that had a very strong gravitational field, and you were watching a clock on the surface of the planet. During the time in which your own clock ticks out a time of 1 hour, how much time does the clock on the planet tick out?
A) No time at all.
B) Less than 1 hour (but more than zero).
C) More than 1 hour.
D) Exactly 1 hour, the same as yours.
Answer: B Source: Section 13-2

35. According to Einstein's General Theory of Relativity, a clock that ticks at a regular rate far from a source of gravity will appear to
A) tick faster, the closer it comes to the source of gravity.
B) tick slower, the closer it comes to the source of gravity.
C) tick at the same rate, wherever it is placed in a gravitational field.
D) tick at the same rate in a gravitational field if it is an atomic clock, but at a slower rate if it is a mechanical clock.
Answer: B Source: Section 13-2

36. According to Einstein's theory of general relativity, if you watch a clock from a distant location as it is moved closer to a source of gravity, you will see the clock
A) slow down.
B) maintain the same rate, since time is unaffected by gravity.
C) run faster.
D) only change its rate if it is moving rapidly, but maintain its standard rate if stationary in a gravity field.
Answer: A Source: Section 13-2

37. Suppose you were far from a planet that had a very strong gravitational field, and you were looking at a source of hydrogen (H-alpha) light on the surface of the planet. When you observe an H-alpha light source in your own spaceship, the wavelength is 656.3 nm. What wavelength do you see when you look at the light source on the planet?
A) Longer than 656.3 nm.
B) Shorter than 656.3 nm.
C) Infinite wavelength, since the source is in a gravitational field.
D) 656.3 nm, the same as from your light source.
Answer: A Source: Section 13-2

38. What happens to the wavelength of light as it travels outwards through the gravitational field of a planet or star, or other object?
    A) The wavelength decreases.
    B) The wavelength stays the same but the energy of each photon decreases.
    C) The wavelength increases.
    D) The wavelength stays the same but the intensity of the light decreases.
    Answer: C Source: Section 13-2

39. According to Newton's law of gravity, why does the Earth orbit the Sun?
    A) Space around the Sun is curved.
    B) The Sun exerts a gravitational force on the Earth across empty space.
    C) The Earth and the Sun are continually exchanging photons of light in a way that holds the Earth in orbit.
    D) Matter contains quarks, and the Earth and Sun attract each other with the "color force" between their quarks.
    Answer: B Source: Section 13-2

40. According to general relativity, why does the Earth orbit the Sun?
    A) The Earth and the Sun are continually exchanging photons of light in a way that holds the Earth in orbit.
    B) Matter contains quarks, and the Earth and Sun attract each other with the "color force" between their quarks.
    C) Space around the Sun is curved.
    D) The Sun exerts a gravitational force on the Earth across empty space.
    Answer: C Source: Section 13-2

41. What would happen to the gravitational force upon the Earth if the Sun were to be replaced by a 1 solar mass black hole?
    A) It would be much less, because the gravitational field of a black hole only exists very close to it.
    B) It would double in strength.
    C) It would become extremely high, sufficient to pull the Earth into it.
    D) It would remain as it is now.
    Answer: D Source: Section 13-2

42. If the Sun were replaced by a 1-solar-mass black hole, then the Earth would
    A) head off into interstellar space along a straight line tangent to its original orbit around the Sun.
    B) enter an elliptical orbit passing close to the black hole, with its furthest distance from the black hole equal to 1 AU.
    C) spiral quickly into the black hole.
    D) continue to orbit the black hole in precisely its present orbit.
    Answer: D Source: Section 13-2

43. Which of the following is not a test of general relativity, but rather is a test of special relativity?
    A) The perihelion position of Mercury proceeds more quickly than is predicted by Newtonian theory.
    B) Light travels in a curved path in a gravitational field.
    C) The length of a moving object decreases when observed by a stationary observer.
    D) The wavelength of light increases as it leaves a region of gravitational field.
    Answer: C Source: Section 13-2

44. A black hole can be thought of as
    A) densely packed matter inside a small but finite volume.
    B) strongly curved space.
    C) a star with a temperature of 0 K, emitting no light.
    D) the point at the center of every star, providing the star's energy by gravitational collapse.
    Answer: B Source: Section 13-2

45. Suppose that a neutron star of 2.8 solar masses is part of a binary star system in which the other star is a normal giant star. What would happen if half a solar mass of material were transferred onto the neutron star from its companion?
    A) The neutron star would explode as a supernova.
    B) The neutron star would collapse and become a black hole.
    C) The neutron degeneracy pressure inside the neutron star would increase to balance the increased gravitational force within the neutron star.
    D) The increased gravitational force would transform the neutrons into quarks, and the neutron star would re-establish equilibrium as a quark star of smaller diameter.
    Answer: B Source: Section 13-3

46. What is the likely final fate of a star whose mass is 15 solar masses after it has finished its nuclear burning phases?
    A) It will condense to the point where it is composed completely of neutrons, the degeneracy of which will prevent further shrinkage.
    B) It will collapse and become a black hole.
    C) The degeneracy of the electrons within the star will prevent collapse below the diameter of a white dwarf.
    D) It will immediately split into two and become a binary star system.
    Answer: B Source: Section 13-3

47. In a binary star system, an unseen component is found to have a mass of about 8 solar masses. It would be visible if this were a normal star, so it must be a collapsed object. Theoretical considerations tell us that it must be
    A) a brown dwarf. B) a neutron star. C) a black hole. D) a white dwarf.
    Answer: C Source: Section 13-3

48. In a binary star system, one component is found to have a mass of about 3 solar masses, and the other a mass of about 7 solar masses. The 3 solar mass star is visible from Earth but the 7 solar mass star is not. Theoretical considerations tell us that the 7 solar mass star must be
    A) a white dwarf. B) a black hole. C) a cool planetary object. D) a neutron star.
    Answer: B Source: Section 13-3

49. Which effects have been useful (and successful?) in the search for and identification of black holes in the universe?
    A) Their gravitational influence on nearby matter, particularly companion stars.
    B) Their magnetic fields and the influence on these fields upon nearby matter.
    C) The effect of their angular momentum or spin on nearby matter.
    D) The influence of their intense gravitational field on atoms that are emitting light from the event horizons of the black holes.
    Answer: A Source: Section 13-3

50. How has the diameter of the black hole candidate Cygnus X-1 been estimated?
    A) From its orbital period around its companion star.
    B) From its angular size in the sky and its known distance.
    C) From the length of time it blocks off the light from its companion star when it passes in front of (eclipses) its companion as seen from the Earth.
    D) From the time scale of flickering of the x-rays emitted by it.
    Answer: D Source: Section 13-3

51. The xray source Cygnus X-1 is a black hole candidate located in a binary star system. The x-ray source is believed to occupy a volume smaller than the Earth. This size is deduced from
    A) its apparent magnitude and distance.
    B) the shortness of its orbital period.
    C) its luminosity and spectral class.
    D) rapid flickering in its x-ray brightness.
    Answer: D Source: Section 13-3

52. One object that is believed to be a black hole in our Galaxy is
    A) the central star in the planetary nebula, the Ring Nebula in Lyra.
    B) the central star in the Crab Nebula.
    C) Cygnus X-1, a powerful x-ray source.
    D) the Vela pulsar.
    Answer: C Source: Section 13-3

53. How was the black hole candidate Cygnus X-1 first discovered?
    A) By the Hubble Space Telescope.
    B) By the Uhuru x-ray satellite.
    C) By the IRAS infrared satellite.
    D) By the VLA radio interferometric array in New Mexico.
    Answer: B Source: Section 13-3

54. Which of the following techniques have been successful in identifying good candidates for a black hole in our Galaxy?
    A) The gravitational lensing of light from a star to produce two identical images.
    B) The detection of x-rays from a binary star undergoing mass exchange, where masses of component stars have been determined.
    C) The detection of an extremely dark point in the sky, from which no light at all is seen.
    D) The detection of extremely redshifted starlight from a region in the nearby spiral arm of the Galaxy.
    Answer: B Source: Section 13-3

55. What method is used by astronomers to infer the existence in space of a dark object with a mass of about 5 solar masses such as a black hole?
    A) The measurement of the effect of its gravitational force upon a companion object in a binary system.
    B) The measurement of the gravitational red shift of spectral lines in the spectrum of the object.
    C) The photography or imaging of a region from which no light or radiation at all appears to come.
    D) The estimation of the luminosity of the object and the application of the mass-luminosity relationship.
    Answer: A Source: Section 13-3

56. I always thought nothing could escape from a black hole, yet astronomers are locating black hole candidates by the x-rays they emit. How can there be x-rays from a black hole?
    A) The black hole modifies space-time around it so much that particles and x-rays are created in the vacuum just outside the event horizon.
    B) The x-rays come from a highly compressed region in an accretion disk outside the event horizon of the black hole.
    C) X-rays are not light or matter and can therefore escape from inside the black hole.
    D) The x-rays are produced by vibrations of the black hole itself, and therefore do not come from inside the black hole.
    Answer: B Source: Section 13-3

57. X-rays that come from the vicinity of a black hole actually originate
    A) from well inside the event horizon.
    B) from just outside the event horizon, on the accretion disk.
    C) from its exact center, or singularity.
    D) from relatively far away from the black hole, where matter is quite cool.
    Answer: B Source: Section 13-3

58. If a black hole is truly black and the escape velocity associated with this black hole is greater than the speed of light such that no light can escape it, where do the x-rays come from in the black hole candidates so far identified?
    A) From stars behind the black hole, the light from which is focused and concentrated such that it becomes xray radiation by gravitational focusing.
    B) From the matter surrounding the black hole that is highly condensed and hence very hot because of the intense gravitational field.
    C) The black hole is only black to visible radiation but x-rays travel faster than the speed of light and so can escape.
    D) From the normal star accompanying the black hole, its ordinary light being blue-shifted into the x-ray spectral region by the intense gravity of the black hole.
    Answer: B Source: Section 13-3

59. The intense x-rays emitted by a suspected black hole are generated by what physical mechanism?
    A) Frictional and compressional heating as material moves into the hole.
    B) The deceleration of matter as it abruptly stops at the event horizon of the black hole.
    C) Light emitted by hot gas is extremely blueshifted by the motion of the gas into the black hole.
    D) Excitation and compressional heating as material moves into the hole.
    Answer: A Source: Section 13-3

60. How has the mass of the black hole candidate Cygnus X-1 been estimated?
    A) From the gravitational redshift, where the more massive the object the greater the redshift of its spectral lines.
    B) From the observed size and estimated density of the object.
    C) From the periodic wobble it produces in the spectral lines of its normal companion star.
    D) From the periodic wobble in its own spectrum.
    Answer: C Source: Section 13-3

61. Why is Cygnus X-1 thought to be a black hole?
    A) It emits x-rays that flicker on time scales of a hundredth of a second.
    B) It has pulled matter from its companion star into an accretion disk around itself.
    C) It is smaller than the Earth but its mass is too large to be a neutron star or white dwarf.
    D) No light has ever been observed to come from it.
    Answer: C Source: Section 13-3

62. What is believed to be the mass of the black hole candidate Cygnus X-1?
A) 30 solar masses. B) 1 solar mass. C) 7 solar masses. D) 120 solar masses.
Answer: C Source: Section 13-3

63. At present, the best candidate for a black hole in a binary star system (based on the accuracy of measuring its mass) is
A) LMC X-3. B) Cygnus X-1. C) HDE 226868. D) V404 Cygni.
Answer: D Source: Section 13-3

64. What is believed to be the mass of the black hole candidate at the center of the galaxy M87?
A) 3 million solar masses.
B) 3 billion solar masses.
C) 300 solar masses.
D) 300,000 solar masses.
Answer: B Source: Section 13-4

65. How was the mass of the candidate black hole at the center of the galaxy M87 estimated?
A) From the intensity of x-rays from it and the speed at which the x-rays flicker.
B) From the orbital speed of objects close to this center.
C) From the periodic shift in the wavelengths of spectral lines from a companion object around which the black hole is orbiting.
D) From the amount of mass that is disappearing into it per year.
Answer: B Source: Section 13-4

66. Where would you look for a supermassive black hole?
A) At the center of a supernova remnant.
B) Orbiting a normal star in our galaxy.
C) In the center of a galaxy.
D) At the center of the universe.
Answer: C Source: Section 13-4

67. What is a primordial black hole?
A) A black hole at the center of a galaxy.
B) Any black hole not in orbit around a normal star.
C) A black hole created during the formation of the universe.
D) A black hole created during the formation of the solar system.
Answer: C Source: Section 13-4

68. What name is given to any black hole that might have been created in the Big Bang (the creation of the universe)?
A) A primordial black hole
B) A relativistic black hole
C) A supermassive black hole
D) A Kerr black hole
Answer: A Source: Section 13-4

69. What separates a black hole from the rest of the universe?
    A) Its crystalline crust.
    B) The surface of the ergosphere.
    C) The event horizon.
    D) The singularity.
    Answer: C Source: Section 13-5

70. What is the event horizon of a black hole?
    A) The "surface" at which all events happen.
    B) The infinitesimally small volume at the center of the black hole that contains all of the black hole's mass.
    C) The "surface" from inside of which nothing can escape.
    D) The "surface" inside which any object entering will leave with a greater energy than that with which it entered.
    Answer: C Source: Section 13-5

71. At what location in the space around a black hole does the escape velocity become equal to the speed of light?
    A) At the point where escaping x-rays are produced.
    B) At the singularity.
    C) At the event horizon.
    D) At the point where clocks are observed to slow down by a factor of 2.
    Answer: C Source: Section 13-5

72. The escape velocity at the event horizon around a black hole is
    A) equal to the speed of light.
    B) much less than the speed of light.
    C) not quite but almost the speed of light.
    D) infinite.
    Answer: A Source: Section 13-5

73. If you were to pass through the event horizon of a black hole,
    A) you could escape again provided the black hole is spinning.
    B) you could, with a powerful rocket, move outward within the black hole (thereby avoiding the singularity until your fuel ran out),but you could never escape back out through the event horizon.
    C) you could avoid the singularity by going into orbit around it, but you could never move outward again from any particular orbit.
    D) there would be nothing you could do to prevent yourself from falling directly into the singularity at the center.
    Answer: D Source: Section 13-5

74. Where would you look for an event horizon?
    A) Near a black hole.
    B) At the edge of the visible universe.
    C) In the photosphere of a star (e.g., the Sun).
    D) In the magnetosphere of a neutron star.
    Answer: A Source: Section 13-5

75. Where is the event horizon of a black hole located?
    A) At the Schwarzschild radius away from its center.
    B) At the position of maximum x-ray emission.
    C) At the outer surface of the ergoregion.
    D) At the singularity.
    Answer: A Source: Section 13-5

76. What is it that is actually located at the event horizon of a black hole?
    A) An infinitely dense concentration of mass.
    B) A sphere of photons.
    C) A magnetic field of immense strength.
    D) Nothing specific.
    Answer: D Source: Section 13-5

77. In reference to black holes, a singularity is
    A) a place just outside the event horizon of a rotating black hole, where it is impossible to remain at rest.
    B) another name for an Einstein ring.
    C) a place where the escape velocity exactly equals the speed of light.
    D) a place where a non-zero mass occupies zero volume.
    Answer: D Source: Section 13-5

78. What is a singularity?
    A) A tunnel into another universe.
    B) Any point at the Schwarzschild radius of a black hole.
    C) A particleantiparticle pair.
    D) A point of infinite density.
    Answer: D Source: Section 13-5

79. What is the Schwarzschild radius of a black hole?
   A) The radius of the singularity.
   B) The distance from the singularity to the point at which the x-rays that we see from the Earth are emitted.
   C) The distance from the singularity to the point where nothing can escape from the black hole.
   D) The distance from the singularity to the point where any object entering will gain energy before leaving again.
   Answer: C Source: Section 13-5

80. The words "Schwarzschild radius" refer to
   A) the distance to which gas is ejected in a planetary nebula.
   B) half the diameter of the singularity in a black hole.
   C) half the diameter of a neutron star.
   D) the distance from the center of a black hole to the point at which the escape velocity becomes equal to the speed of light.
   Answer: D Source: Section 13-5

81. According to the equation in Toolbox 13-1 Kaufmann and Comins, *Discovering the Universe*, 5th Ed., what is the Schwarzschild radius of a 2 solar mass black hole?
   A) 6000 km B) 6 km C) 6 m D) 60 km
   Answer: B Source: Section 13-5

82. According to the equation in Toolbox 13-1, what happens to the Schwarzschild radius of a black hole if you double the amount of mass in the black hole?
   A) The Schwarzschild radius is quadrupled (4 times).
   B) The Schwarzschild radius is halved.
   C) The Schwarzschild radius is doubled.
   D) The Schwarzschild radius decreases by a factor of 4.
   Answer: C Source: Section 13-5

83. How does the diameter of a black hole (size of the event horizon) depend on the mass inside the black hole?
   A) The greater the mass the greater the diameter until the mass becomes relatively large, then the diameter decreases with increasing mass.
   B) The diameter does not depend on the mass.
   C) The greater the mass the greater the diameter.
   D) The greater the mass the smaller the diameter.
   Answer: C Source: Section 13-5

84. How many properties of the matter inside a black hole can be measured from outside the black hole?
A) 6 B) Only 1 C) 3 D) 4
Answer: C Source: Section 13-5

85. Which properties of the matter inside a black hole can be measured from outside the black hole?
A) The mass and the angular momentum.
B) Only the mass.
C) The mass, the angular momentum, the electric charge and the average atomic weight.
D) The mass, the angular momentum and the electric charge.
Answer: D Source: Section 13-5

86. The only physical properties that are necessary to describe a black hole and its interaction with the rest of the universe completely are
A) the size of the event horizon, strength of its magnetic field and size of its solid core.
B) its total mass, chemical or atomic structure of the matter within it and overall size.
C) its total mass, total electric charge and total angular momentum or spin.
D) its total mass, total angular momentum or spin and its temperature.
Answer: C Source: Section 13-5

87. Which of the following can you never know about a black hole?
A) Its angular momentum (spin).
B) The total amount of matter (the mass) inside it.
C) Its net electric charge.
D) The type of material inside it.
Answer: D Source: Section 13-5

88. Take two identical, non-rotating, 5-solar-mass black holes and place them side by side. Add one solar mass of pineapples to the left-hand one and one solar mass of uranium to the right-hand one (without changing the electrical charge or the rotation of either black hole). Afterwards, how will these two black holes differ?
A) The right-hand one will have a stronger gravitational field because of the denser material inside it.
B) The right-hand one will be radioactive, emitting alpha particles, electrons and gamma rays.
C) The left-hand one will smell better.
D) They won't differ at all.
Answer: D Source: Section 13-5

89. Place two identical 3-solar-mass black holes side by side. Add one solar mass of neutrons to the left-hand one, and one solar mass of protons to the right-hand one. Afterwards, how will these two black holes differ?
    A) The left-hand one will emit electrons and neutrinos as the neutrons inside it decay into protons.
    B) The left-hand one will be electrically neutral and the right-hand one will have an enormous electric charge.
    C) The left-hand one will have a stronger gravitational field than the right-hand one because a neutron is heavier than a proton.
    D) They won't differ at all.
    Answer: B Source: Section 13-5

90. Suppose that a large piece (e.g., five solar masses) of purple, magnetized iron is rotating five times per day. If this object were able to collapse gravitationally to form a black hole, which one of the following properties of the matter inside the black hole could an outside observer actually measure?
    A) Its rotation. B) Its color. C) Its magnetic field. D) Its composition.
    Answer: A Source: Section 13-5

91. The matter in an accretion disk is in orbit around a black hole, but friction within the disk causes the matter to spiral gradually into the black hole. What will change in terms of the observable properties of the black hole as this process continues?
    A) The mass, the angular momentum, and the rate of evaporation of the black hole will increase.
    B) Both the mass and the angular momentum of the black hole will increase.
    C) Nothing will change.
    D) Only the mass will increase.
    Answer: B Source: Section 13-5

92. What happens to the magnetic field of a star that collapses to become a black hole?
    A) The magnetic field becomes compressed and intensified by a factor equal to the ratio of the star's original diameter to the diameter of the event horizon.
    B) The magnetic field becomes weaker by the same ratio as the star's diameter decreases to the event horizon.
    C) The magnetic field becomes infinitely intensified.
    D) The magnetic field is radiated away; black holes never have magnetic fields.
    Answer: D Source: Section 13-5

93. Black holes cannot possess magnetic fields because
   A) any magnetic field would have been radiated away as gravitational waves during the collapse.
   B) the original object could not have collapsed to form a black hole if it had any magnetic field.
   C) magnetic fields are created by spinning charges, and black holes cannot spin.
   D) black holes cannot have electric charge, which is needed to create magnetic fields.
   Answer: A Source: Section 13-5

94. In general, how many fundamentally different types of black holes are there expected to be?
   A) Only one: all properties but mass are destroyed when a black hole is created.
   B) Two: those that rotate and those that don't.
   C) Three: atomic-mass black holes, stellar-mass black holes and supermassive black holes.
   D) Two: those that have electric charge and those that don't.
   Answer: B Source: Section 13-5

95. What is a Schwarzschild black hole?
   A) A hypothetical zero-mass black hole.
   B) Any uncharged black hole.
   C) Any non-rotating black hole.
   D) A supermassive black hole.
   Answer: C Source: Section 13-5

96. What name is given to a non-rotating black hole?
   A) A Hawking singularity.
   B) A Kerr black hole.
   C) A Schwarzschild black hole.
   D) A wormhole.
   Answer: C Source: Section 13-5

97. What name is given to a rotating black hole?
   A) A wormhole.
   B) A Kerr black hole.
   C) A Schwarzschild black hole.
   D) A Hawking singularity.
   Answer: B Source: Section 13-5

98. What is a Kerr black hole?
   A) Any uncharged black hole
   B) Any non-rotating black hole
   C) A hypothetical zero-mass black hole
   D) Any rotating black hole
   Answer: D Source: Section 13-5

99. How does a Kerr black hole differ from a Schwarzschild black hole?
    A) Kerr black holes have accretion disks, Schwarzschild black holes don't.
    B) Kerr black holes have net electric charge, Schwarzschild black holes don't.
    C) Kerr black holes rotate, Schwarzschild black holes don't.
    D) Kerr black holes have infinite mass, Schwarzschild black holes don't.
    Answer: C Source: Section 13-5

100. What is the ergoregion of a Kerr black hole?
    A) A region outside the event horizon where objects cannot remain at rest without falling into the black hole.
    B) The region between the event horizon and the singularity, from which nothing can escape.
    C) The inner part of the accretion disk, where x-rays are generated.
    D) A region outside the event horizon where virtual particles are created from the vacuum of space.
    Answer: A Source: Section 13-5

101. In reference to black holes, the word "ergosphere" refers to
    A) the region occupied by the accretion disk, where matter from a companion star collects around a black hole.
    B) the entire universe outside the black hole.
    C) a region just outside the event horizon of a rotating black hole, where it is impossible for anything to remain at rest.
    D) the entire region inside the event horizon.
    Answer: C Source: Section 13-5

102. One day, while straying dangerously close to a black hole, you notice that you must keep your spaceship moving. No matter how hard you try to remain at rest, you are inevitably drawn into the black hole unless you keep moving. What does this tell you about the black hole (other than that you should not be near it!)?
    A) It is rotating.
    B) It is electrically charged.
    C) It is supermassive.
    D) It is evaporating.
    Answer: A Source: Section 13-5

103. A space freighter accidentally drops a steel beam while passing a black hole, and the beam starts falling toward the black hole with the long direction of the beam pointing toward the black hole. What happens to the beam as it approaches the event horizon?
   A) It is compressed in both length and width.
   B) It is stretched in length and compressed in width.
   C) It is compressed in length and stretched in width.
   D) It begins to rotate faster and faster.
   Answer: B  Source: Section 13-6

104. Suppose it were possible to lower a yellow sodium lamp down toward the event horizon of a black hole. What would you see while watching from a safe distance?
   A) The light would remain yellow, but there would be less and less photons being emitted from it.
   B) The light would remain unchanged in either brightness or color.
   C) The light from the lamp would change to orange then red.
   D) The light from the lamp would change to green then blue.
   Answer: C  Source: Section 13-6

105. As you are investigating a black hole from a safe distance, a rivet pops out of the tailfin on your spaceship and falls toward the black hole. Will you ever see the rivet enter the event horizon?
   A) No; it will appear to stop and hover forever before entering the event horizon.
   B) Yes, but it will be so blueshifted that you would need x-ray eyes to see it.
   C) Yes; you will see it fall faster and faster until it disappears as it falls through the event horizon.
   D) No; it will be compressed to zero size and disappear from sight before it reaches the event horizon.
   Answer: A  Source: Section 13-6

106. A laborer repairing the clock tower on a space station orbiting a black hole accidentally drops the clock in such a way that it accelerates toward the black hole. What does this person see while watching the clock?
   A) The hands of the clock move slower and slower until they and the clock itself stop at the event horizon.
   B) The hands of the clock move faster and faster until the clock plunges through the event horizon.
   C) As the clock nears the event horizon the hands begin to move randomly as time becomes jumbled near the black hole.
   D) The hands of the clock keep normal time, since time is absolute and the same everywhere.
   Answer: A  Source: Section 13-6

107. What appears to happen to a clock as it approaches and reaches the event horizon around a black hole, when viewed by a remote observer?
   A) It ticks uniformly, since nothing changes the progress of time.
   B) It appears to slow down and stop.
   C) It speeds up because of the intensified gravitational field.
   D) Time appears to pass at a much faster rate, this rate becoming infinitely fast at the event horizon.
   Answer: B Source: Section 13-6

108. If you were watching a friend (or better still, an enemy!) who has fallen as far as the event horizon of a black hole, what would you measure as his heartbeat (apart from the effects caused by his adrenaline level)?
   A) It would appear to have speeded up to an incredible rate.
   B) It would appear to be zero, his heart would appear to have stopped.
   C) It would appear to be normal since gravity has no effect on time intervals.
   D) It would appear to have slowed down somewhat, but not much, because of the change of the speed of light in the gravity field.
   Answer: B Source: Section 13-6

109. In terms of black holes, what is a wormhole?
   A) A hole in a solid object, such as a planet, created by the passage of a small black hole through the object.
   B) A "tunnel" of undistorted space through an event horizon allowing objects to enter and leave a black hole without being torn apart.
   C) A direct connection from any black hole to another part of spacetime.
   D) A direct connection from a rotating black hole to another part of spacetime.
   Answer: D Source: Section 13-6

110. Which of the following statements correctly describes "cosmic censorship"?
   A) All properties of the matter inside a black hole are hidden by the event horizon, except for the total mass of the matter.
   B) Black holes cannot have magnetic fields.
   C) The only way into or out of a singularity is through an event horizon.
   D) The amount of mass in a black hole can never be measured.
   Answer: C Source: Section 13-6

111. What is a virtual particle?
    A) A particle whose existence is too short for us to know it ever existed.
    B) A particle that never does anything wrong.
    C) A particle which, if it comes in contact with ordinary matter, will annihilate to form pure energy.
    D) Any particle like a photon or a graviton that is made up of waves.
    Answer: A Source: Section 13-7

112. Sometimes particle-antiparticle pairs are created and then annihilate so quickly that we cannot know that they ever existed. What are these particles (or antiparticles) called?
    A) Field particles.
    B) Relativistic particles.
    C) Temporary particles.
    D) Virtual particles.
    Answer: D Source: Section 13-7

113. If nothing can ever leave a black hole, can the mass of a black hole ever decrease?
    A) Yes, if particle-antiparticle pairs are created outside the event horizon and one particle enters the event horizon while the other escapes.
    B) Yes, if the matter inside the black hole is radioactive (e.g., uranium), then alpha particles, electrons and gamma rays are constantly leaving the black hole.
    C) Yes, if antiparticles enter a black hole and annihilate with matter already inside the black hole.
    D) No.
    Answer: A Source: Section 13-7

114. When particle-antiparticle pairs are created just outside the event horizon of a black hole, one member can escape while the other enters the black hole. What is the name of this stream of particles leaving a black hole?
    A) Hawking radiation.
    B) Kerr radiation.
    C) Schwarzschild radiation.
    D) Planck radiation.
    Answer: A Source: Section 13-7

115. What is Hawking radiation?
    A) X-rays from an accretion disk around a black hole.
    B) A stream of particles and antiparticles from just outside the event horizon of a black hole.
    C) Electromagnetic radiation from electrons spiraling in the magnetosphere of a neutron star.
    D) Microwaves from the edge of the visible universe.
    Answer: B Source: Section 13-7

116. What mass would a primordial black hole have to have had originally in order to be just disappearing now, due to Hawking radiation?
   A) 10 thousand (10,000) kg.
   B) 10 kg.
   C) 10 billion (10,000,000,000) kg.
   D) 10 million (10,000,000) kg.
   Answer: C Source: Section 13-7

117. Which one of the following statements about the evaporation of black holes is correct?
   A) The rate at which a black hole evaporates is lower for a higher-mass black hole.
   B) The rate at which a black hole evaporates is higher for a higher-mass black hole.
   C) Black holes do not evaporate.
   D) The rate at which a black hole evaporates is independent of the mass of the black hole.
   Answer: A Source: Section 13-7

# Build Your Foundation IV: The Universe

1. The Milky Way galaxy
   A) is one of many billions of galaxies in the Universe.
   B) is unique in the Universe in showing definite spiral structure.
   C) contains the whole Universe; everything observable is within its volume.
   D) is one of only a few spiral galaxies; most other galaxies in the Universe are amorphous collections of stars shaped like ellipsoids.
   Answer: A Source: Section IV-1

2. The famous nineteenth-century observational astronomer, Lord Rosse, built the largest telescope of his time (and discovered the spiral nature of many so-called "nebulae") in which country?
   A) Ireland B) Germany C) England D) United States
   Answer: A Source: Section IV-1

3. The first suggestion that there were collections of stars beyond our Milky Way in the Universe was made by
   A) William Parsons, Earl of Rosse, in 1845.
   B) Edwin Hubble, in 1923.
   C) Sir Isaac Newton, in 1690.
   D) Immanuel Kant, in 1755.
   Answer: D Source: Section IV-1

4. The idea that some of the "nebulae" which are observed in the sky might be "island universes," immense collections of stars far beyond the Milky Way, was proposed by
   A) Heber Curtis in 1920.
   B) Albert Einstein in 1909.
   C) Immanuel Kant in 1755.
   D) Lord Rosse in 1845.
   Answer: C Source: Section IV-1

5. Variable stars, such as Cepheid variables and RR Lyrae stars, are used in what important measurement in astronomy?
   A) The measurement of the rotation speeds of galaxies.
   B) The keeping of accurate time.
   C) The measurement of the distances to stars.
   D) The measurement of the surface temperatures of stars.
   Answer: C Source: Section IV-2

6. What important role do Cepheid variables and RR Lyrae stars have in astronomy?
   A) The determination of speeds of stars in galactic arms from the Doppler shift of their spectra.
   B) The determination of stellar luminosities.
   C) Distance measurements to distant galaxies.
   D) The keeping of accurate time.
   Answer: C Source: Section IV-2

7. Cepheid-stars are useful to astronomers as indicators of
   A) the existence of black holes.
   B) distance, particularly to nearby galaxies.
   C) white dwarf star behavior.
   D) stars with very high speed motion.
   Answer: B Source: Section IV-2

8. Distances to a nearby galaxy can be determined most accurately by
   A) measuring the shifts of spectral lines from our Galaxy.
   B) measuring the total amount of energy received from the galaxy.
   C) measuring the chemical compositions of the brightest stars.
   D) using pulsating stars as beacons.
   Answer: D Source: Section IV-2

9. The intrinsic brightness (represented by luminosity) of a Cepheid variable star compared to that of the Sun is
   A) significantly less.
   B) several thousand times larger.
   C) about the same.
   D) about 10 times larger.
   Answer: B Source: Section IV-2 and Figure IV-4

10. The period-luminosity relationship for Cepheid variable stars, relating variability to absolute overall brightness, thereby providing identifiable beacons throughout our local space, was discovered by
    A) Henrietta Leavitt.
    B) Sir Isaac Newton.
    C) Harlow Shapley.
    D) Edwin Hubble.
    Answer: A Source: Section IV-2

11. Cepheid variable stars are invaluable in astronomy because of the close relationship between
    A) the peak wavelength of their spectra and their surface temperatures.
    B) the red shift of their spectrum and their distance from the Sun.
    C) their apparent magnitude and their pulsation period.
    D) their luminosity, or absolute magnitude, and their pulsation period.
    Answer: D Source: Section IV-2

12. The observational fact about a Cepheid variable star which leads to a measurement of its distance from us is that its period of variation is directly related to its
    A) position along the spiral arms of a galaxy.
    B) apparent magnitude.
    C) surface temperature.
    D) absolute magnitude.
    Answer: D Source: Section IV-2

13. The significant feature of a Cepheid variable is that there is a relationship between two intrinsic parameters, one of which can be easily measured while knowledge of the other parameter is required. These parameters are
    A) period of brightness variation and spectral color.
    B) variation of spectral color and distance to the star.
    C) amplitude of brightness variation and luminosity.
    D) period of brightness variation and luminosity.
    Answer: D Source: Section IV-2

14. A Cepheid variable star with a pulsation period of a few days is seen in the spiral arm of a galaxy. Its apparent brightness is measured as $10^4$ times fainter than an equivalent star 1000 ly away from the Sun in our galaxy. Assuming no light absorption between galaxies, what is the distance to the far Cepheid, and hence to the galaxy?
    A) 10,000 ly (10 times further away).
    B) $10^7$ ly ($10^4$ times further away).
    C) 100,000 ly, (100 times further away).
    D) 10 ly (100 times closer).
    Answer: C Source: Sections III-3 and IV-2

15. A classical Cepheid variable star is seen to vary regularly with a period of 25 days. How many times brighter than the Sun would this star appear to be if it were to replace the Sun in our solar system? (See Fig IV-4 of Kaufmann & Comins, *Discovering the Universe*, 5th Ed.)
    A) 10,000 B) $10^2$ C) 4 D) 1000
    Answer: A Source: Section IV-2

16. The famous Curtis-Shapley debate in 1920 concerned which fundamental astronomical question in astronomy?
   A) Whether the spiral "nebulae" were part of the Milky Way Galaxy or more distant, separate entities.
   B) Whether the universe was expanding outward in all directions.
   C) Whether the Sun was at the center of the Milky Way Galaxy.
   D) Whether all stars were like the Sun, or fundamentally different.
   Answer: A Source: Section IV-2

17. The event that settled the Shapley-Curtis debate about "spiral nebulae" was
   A) Edwin Hubble measuring the distance to the Andromeda galaxy.
   B) Edwin Hubble showing that the universe was expanding.
   C) Albert Einstein showing that gravity can bend the path of light.
   D) Arno Penzias and Robert Wilson detecting the cosmic microwave background radiation.
   Answer: A Source: Sections IV-2 and IV-3

18. Determination of the distance to the Andromeda "nebula," (which finally resolved the Curtis-Shapley debate on the nature of "spiral nebulae"), was carried out by Hubble by observing
   A) Cepheid variable stars.
   B) the Main Sequence of stars in the "nebula."
   C) pulsars.
   D) the Doppler shift of stars in the nebula.
   Answer: A Source: Sections IV-2 and IV-3

19. The observation by Hubble which demonstrated for the first time that the Andromeda "nebula" was at a very large distance from the Sun, and outside our galaxy, was
   A) that globular clusters appeared to be distributed in a halo around the "nebula," a sure sign of a separate galaxy.
   B) that stars with characteristics similar to those of our Sun appeared to be absent in this "nebula."
   C) that the "nebula" appeared to be rotating night by night around a center, which was not the center of our galaxy.
   D) that Cepheid variable stars appeared to be very faint in the "nebula."
   Answer: D Source: Section IV-2

20. The method used by Hubble to determine the distance to the Andromeda galaxy (M31), thereby establishing the concept of separate and individual galaxies throughout the universe, was the
    A) observation of the brightnesses of novas.
    B) measurement of stellar parallax, or apparent motion of stars because of Earth's orbital motion.
    C) measurement of the redshift of the whole galaxy.
    D) observation of Cepheid variable stars.
    Answer: D Source: Section IV-2

21. The Andromeda Galaxy (M31) is best described as
    A) a gaseous nebula, extending for 6° across our sky.
    B) a spiral collection of stars, dust, and gas, 2 million light-years away.
    C) a vortex surrounding a black hole.
    D) an extension of the Milky Way.
    Answer: B Source: Section IV-2

22. What is it that makes the study of the structure of our own galaxy more difficult than that of much more distant spiral galaxies?
    A) Most of our galaxy is hidden behind dense gas and dust clouds in the galactic plane.
    B) The galactic center is visible only from the southern hemisphere where, until recently, no major telescopes were available for the study of galactic structure.
    C) Our galaxy is too close, such that photographs or images cannot be taken of the whole galaxy at any one time.
    D) Our star is within the galaxy and its motion confuses the interpretation of the motion of other parts of the galaxy.
    Answer: A Source: Section IV-3

23. Galaxies throughout the Universe appear to be distributed
    A) mostly in a single spherical shell surrounding a void in space, presumed to have been caused by a vast
    B) uniformly throughout space.
    C) in groups and surfaces surrounding vast voids, much like the surfaces of giant bubbles.
    D) around a single point in space, the presumed location of the original Big Bang, which created the Universe.
    Answer: C Source: Section IV-3

# Chapter 14: The Milky Way Galaxy

1. Who was the first to look at the Milky Way with a telescope?
   A) Galileo Galilei.
   B) Sir Isaac Newton.
   C) Johannes Kepler.
   D) Sir William Herschel.
   Answer: A   Source: Introduction to Chapter 14

2. The plane of the Milky Way is
   A) perpendicular to the spin axis if Earth.
   B) almost in the plane of the ecliptic.
   C) almost perpendicular to the celestial equator.
   D) almost perpendiclur to the ecliptic plane.
   Answer: D   Source: Introduction to Chapter 14

3. In the 1780s, Sir William Herschel tried to measure the Sun's position in our Galaxy by
   A) counting the density of stars in different directions along the Milky Way.
   B) measuring distances to star clusters and H II regions in the disk of the Galaxy.
   C) measuring the locations of globular clusters around the Galaxy.
   D) comparing our Galaxy to photographs of the Andromeda galaxy.
   Answer: A   Source: Introduction to Section 14-1

4. The factor that misled Herschel into concluding that the stars of the Milky Way were distributed with the Sun at the center of the galaxy was
   A) gravitational bending of light by the mass of the galaxy, distorting the relative positions of the stars.
   B) that most of the "stars" that he measured were in fact distant galaxies that are distributed uniformly around the Sun.
   C) interstellar dust, which obscured the more distant stars and thereby localized his observations.
   D) hot hydrogen gas in the galaxy, its emission hiding the more distant stars.
   Answer: C   Source: Section 14-1

5. In the eighteenth century, Sir William Herschel used star counts in different regions of the sky along the Milky Way to estimate the position of the center of the Milky Way. He incorrectly concluded that the Sun was close to that center. The reason for this erroneous conclusion was
   A) that the redshift of the more distant stars made them invisible to Herschel.
   B) that Herschel counted all "stars" in each star field, and included many galaxies were outside our Galaxy, thus confusing the distribution.
   C) the large quantity of absorbing dust between stars, which obscured the more distant regions of the Galaxy.
   D) that emissions from hot hydrogen gas clouds served to hide the more distant stars, localizing his search.
   Answer: C Source: Section 14-1

6. Interstellar dust obscures our view of distant regions of space at optical wavelengths. In this regard, which of the following statements is true?
   A) The obscuration is severe only in the plane of the Galaxy.
   B) The obscuration is very clumpy and random over the whole sky, the individual absorbing dust clouds showing no preference for one particular direction or plane.
   C) The obscuration is roughly uniform over the whole sky.
   D) The obscuration is the least in the plane of the Galaxy, and is strongest when we look out into the galactic halo, at right angles in this plane.
   Answer: A Source: Section 14-1

7. Which component of our galaxy accounts for interstellar extinction, the dimming of light from distant objects?
   A) Molecules such as $H_2$ and CO, which are strong absorbers, in molecular clouds.
   B) The so-called hidden or missing matter, since its absorbing properties render it invisible in the galaxy.
   C) Dust.
   D) Cool hydrogen gas.
   Answer: C Source: Section 14-1

8. The one component of the material of the Milky Way Galaxy that prevents us from seeing and photographing the galactic center at optical wavelengths is
   A) very cold hydrogen gas.
   B) interstellar dust.
   C) the glare of light from nearby stars.
   D) hot hydrogen gas.
   Answer: B Source: Section 14-1

9. Interstellar matter obscures our view of the disk of our Galaxy
   A) more-or-less equally at all wavelengths, from radio waves to light waves.
   B) most at radio wavelengths, where hydrogen absorbs radio waves efficiently, and least at optical wavelengths.
   C) more at optical wavelengths, less in the infrared, and not at all at radio wavelengths.
   D) very little at any wavelength.
   Answer: C Source: Section 14-1

10. What useful purpose did RR Lyrae stars serve for Harlow Shapley in locating the galactic center?
   A) They are important spiral arm tracers, and thus defined the shape of the galaxy.
   B) They are concentrated in the galactic center and so defined its direction.
   C) They emit copious amounts of infrared radiation, and are thus visible through interstellar dust which obscured visible light.
   D) Their brightness variations allowed accurate distances to be measured.
   Answer: D Source: Section 14-1

11. The method used by Harlow Shapley in 1917 to estimate the Sun's location in our Galaxy was the measurement of
   A) the locations of globular clusters around the Galaxy.
   B) the density of stars in different directions along the Milky Way.
   C) distances to open star clusters and H II regions in the disk of the Galaxy.
   D) the structure of the Andromeda galaxy and a comparison of this to the structure of our galaxy.
   Answer: A Source: Section 14-1

12. Harlow Shapley first located the center of our Galaxy in 1917 by
   A) measuring the positions of supernova explosions throughout the galaxy.
   B) observing the distribution of globular clusters in the galactic halo.
   C) observing the distribution of hydrogen gas, measured by 21-cm radio emission.
   D) redshift measurements on stars in the galactic plane and disk.
   Answer: B Source: Section 14-1

13. When distances were carefully measured from Earth to globular clusters above and below the Milky Way plane, (where our view of them is not obscured by interstellar dust and gas) their distribution was found to be
    A) in a relatively flat disk almost perpendicular to the plane of the galaxy, with relatively higher density of clusters towards its center.
    B) uniformly distributed throughout space, with no concentration in any area of the Milky Way.
    C) spherically symmetric about a point in the constellation Sagittarius and concentrated in that direction.
    D) concentrated in the plane of the Milky Way and clustered around the Sun's position, indicating that the Sun is close to the galaxy's center.
    Answer: C Source: Section 14-1

14. In which constellation in our sky is the center of our Milky Way Galaxy located?
    A) Lyra. B) Hercules. C) Ursa Major. D) Sagittarius.
    Answer: D Source: Section 14-1

15. The dimensions of the disk of our Milky Way Galaxy are
    A) diameter 10,000 light-years; thickness, 28,000 light-years.
    B) diameter 2000 light-years; thickness, 100,000 light-years.
    C) diameter 28,000 light-years; thickness 2000 light-years.
    D) diameter 100,000 light-years; thickness, 2000 light-years.
    Answer: D Source: Section 14-2

16. The ratio of thickness to diameter of the Milky Way galaxy is
    A) 1/1000. B) 1/50. C) 1/500. D) 1/5.
    Answer: B Source: Section 14-2

17. The diameter of our Galaxy is about
    A) 31 kpc. B) 100 kpc. C) 2 kpc. D) 3.1 kpc.
    Answer: A Source: Section 14-2

18. Where in space would you look for a globular cluster?
    A) In the Milky Way galactic halo, orbiting the galactic center in a long elliptical orbit around the galactic center.
    B) Only in elliptical galaxies, since they are composed of old stars and do not exist in young systems like spiral galaxies.
    C) In the Milky Way disk, moving in a circular orbit around the galactic center.
    D) In the asteroid belt.
    Answer: A Source: Section 14-2

19. What would you expect would be the overall color of a globular cluster of stars?
    A) Red, because of the emission of light by the hydrogen gas in HII regions surrounding the stars in the cluster.
    B) Blue, because of the contribution from young and very hot stars in the cluster.
    C) Blue, because of the scattering of starlight from the dust surrounding the stars in the cluster.
    D) Red, because of the older population of stars in the cluster.
    Answer: D Source: Sections 14-2 and 11-11

20. Where are the majority of older, metal-poor stars found in the Milky Way galaxy?
    A) In globular clusters in the galactic halo. C) In the disk and spiral arms.
    B) Throughout the whole galaxy. D) At the galactic center.
    Answer: A Source: Sections 14-2 and 11-11

21. In our Galaxy, young metal-rich stars are found
    A) in the disk and spiral arms.
    B) everywhere in the galaxy.
    C) in the globular clusters, in the galactic halo.
    D) only at the galactic center.
    Answer: A Source: Sections 14-2 and 11-11

22. Which of the following is NOT useful for mapping the locations and shapes of the spiral arms of our Galaxy?
    A) The distribution of globular clusters.
    B) The distribution of O and B stars.
    C) The distribution of emission nebulae (H II regions).
    D) The distributions of giant molecular clouds.
    Answer: A Source: Section 14-2

23. What is the distribution of giant molecular clouds in our galaxy and other similar galaxies?
    A) They occur primarily in the spiral arms.
    B) They are distributed uniformly throughout the disk.
    C) They are distributed throughout the halo, with greater density towards the center.
    D) They are concentrated close to the galactic center.
    Answer: A Source: Section 14-2

24. The Milky Way is an example of which type of galaxy?
    A) Elliptical. B) Spiral. C) Irregular. D) Lenticular, S0 type.
    Answer: B Source: Section 14-2

25. The Milky Way in which the Sun resides is an example of which type of galaxy?
    A) It is not a galaxy at all, but a large cluster of stars.
    B) A normal spiral galaxy.
    C) An irregular galaxy.
    D) An elliptical galaxy.
    Answer: B Source: Section 14-2

26. How is cool, neutral hydrogen gas, H I, detected in the spiral arms of galaxies?
    A) By its Balmer line emissions.
    B) By the absorption of infra-red radiation from extragalactic sources.
    C) By its 21 cm line radio emissions.
    D) By its ultraviolet, Lyman a, hydrogen line emissions.
    Answer: C Source: Section 14-2

27. Radio waves of 21-cm wavelength originate from which component of the interstellar medium?
    A) Cool, carbon monoxide, CO.
    B) Cold, molecular hydrogen, $H_2$.
    C) Hot, ionized atomic hydrogen.
    D) Cool, neutral atomic hydrogen.
    Answer: D Source: Section 14-2

28. The spiral-arm structure of the Milky Way Galaxy has been measured and evaluated most effectively by observations of
    A) globular clusters in the halo of the Galaxy.
    B) 21-cm radiation from interstellar hydrogen and the distribution of young stars.
    C) Lyman UV radiation from hot hydrogen gas.
    D) Balmer emission lines of visible radiation from hydrogen.
    Answer: B Source: Section 14-2

29. Which type of radiation has been most effective in evaluating the spiral arm structure of our galaxy?
    A) 21-cm radio emission from electron "spin-flip" transitions in cool hydrogen gas.
    B) Synchrotron radiation from electrons spiralling in magnetic fields within the spiral arms.
    C) Lyman a ultraviolet emission from hot hydrogen gas.
    D) Neutrinos from exploding stars in the spiral arms since they can penetrate dust and gas easily.
    Answer: A Source: Section 14-2

30. What happens when the electron in a hydrogen atom flips its direction of spin, from parallel to antiparallel to that of the proton?
    A) The atom emits a photon of 121.5 nm wavelength (La), in the UV region of the spectrum.
    B) The atom emits a photon of 21 cm wavelength, in the radio region of the spectrum.
    C) Nothing; this is a forbidden transition and never occurs.
    D) The atom emits a photon of 656.3 nm wavelength (Ha), in the red region of the spectrum.
    Answer: B Source: Section 14-2

31. What atomic transition occurs in the atoms of hydrogen gas in the spiral arms of our galaxy to produce the 21-cm radio emission?
    A) The transition from the n = 2 to n = 1 levels in atomic hydrogen.
    B) The inversion of the electron spin relative to the proton spin, from parallel to anti-parallel.
    C) The change in rotation of the molecule $H_2$ about an axis perpendicular to the molecular axis.
    D) The change in the vibrational state of the H atoms in the $H_2$ molecule.
    Answer: B Source: Section 14-2

32. What quantum transition occurs inside a hydrogen atom to produce a 21-cm radio photon?
    A) An electron in the ground atomic state reverses its direction of spin with respect to that of the proton.
    B) An electron falls from the level n = 100 to the level n = 99 in the atom.
    C) The electron combines with the proton in the nucleus to become a neutron, producing energy.
    D) An electron reverses the direction of its motion in orbit around the proton.
    Answer: A Source: Section 14-2

33. The major advantages of the 21-cm radio emission from hydrogen gas for investigating the spiral structure of our galaxy are
    A) that it is relatively easily absorbed by hydrogen gas in the Milky Way, so that measurements are not confused by emission of this radiation from other galaxies beyond the Milky Way. It originates only from cold hydrogen gas, and can be used to map this important component.
    B) that radio waves easily penetrate the Milky Way dust and gas and it is a very narrow line emission, thus its Doppler shift can be used to measure gas motions.
    C) that Doppler shift of this narrow-wavelength line emission is caused by the temperature of the hot hydrogen gas and therefore can be used to measure the distribution and temperature of this important component of the Milky Way.
    D) that this emission can easily penetrate the Milky Way gas and dust and comes only from hot gas, hence can be used to map the distribution of hot hydrogen gas.
    Answer: B Source: Section 14-2

34. When we measure the narrow line emissions of hydrogen at 21-cm radio wavelengths along a particular line of sight through the disk of our Galaxy, we can tell the distances to different hydrogen clouds because
    A) clouds that are further away have smaller angular sizes.
    B) absorption of extragalactic radiation at this wavelength will be greater, the further away the absorber is from the Sun.
    C) the emission is weaker from clouds which are further away.
    D) clouds at different distances have different Doppler shifts because of the rotation of the Galaxy.
    Answer: D Source: Section 14-2

35. Which of the following components of the galaxy best outline the spiral arms of the Galaxy?
    A) Young O and B stars, dust and gas.
    B) Globular clusters.
    C) Predominantly solar-type stars.
    D) White dwarf stars.
    Answer: A Source: Section 14-2

36. The stellar components of the galaxy which act as tracers for the mapping of spiral arm structure in the Milky Way are
    A) old, red giant stars and white dwarfs.
    B) globular clusters.
    C) bright, population I stars and emission nebulae surrounding them.
    D) supernova explosions, since they are very luminous and can be seen through considerable dust and gas.
    Answer: C Source: Section 14-2

37. Observation of the different components of the Milky Way galaxy indicates that the **spiral arms** contain very different populations of stars and other material to those in **globular clusters**. In what way are they different?
    A) Spiral arms contain older, more developed and hence brighter and bluer stars, while globular clusters are composed largely of young, red stars in the early stages of formation and development.
    B) Spiral arms contain young stars, dust and gas within which star formation continues, whereas globular clusters contain older star populations, with no dust and gas and no on-going star formation.
    C) Globular clusters contain dust and gas and are the only locations where star formation continues in the galaxy at the present time. The older stars in the spiral arms have no surrounding dust or gas.
    D) Both spiral arms and globular clusters contain about the same populations of stars both young and old but, in contrast to the spiral arms, there is no dust and gas, no star formation and there are no nova explosions in globular clusters.
    Answer: B Source: Section 14-2

38. Where is the solar system located in our galaxy?
    A) In the galactic disk.
    B) In the galactic halo.
    C) It is not in a galaxy, but in the intergalactic space between galaxies.
    D) In the galactic nucleus.
    Answer: A Source: Section 14-2

39. The Sun's position in the galaxy is
    A) we cannot tell where we are located because our view is too severely restricted by interstellar dust.
    B) in the disk of the Galaxy, inside a spiral arm or segment of a spiral arm.
    C) in the disk of the Galaxy, between and well away any spiral arm.
    D) in the spherical halo, somewhat above and outside of the spiral arms.
    Answer: B Source: Section 14-2

40. Where is the Sun located in our galaxy? (i.e., What is our address in the Universe!) (See Fig. 14-9 of Kaufmann & Comins, *Discovering the Universe*, 5th Ed.)
    A) In the Perseus Arm, between the Orion and Cygnus Arms.
    B) In or close to the Orion Arm, which is between the Sagittarius and Perseus Arms.
    C) In the Sagittarius Arm, which is between the Centaurus and Orion Arms.
    D) In the Centaurus Arm, between the galactic center and the Orion Arm.
    Answer: B Source: Section 14-2

41. A map of our Galaxy deduced from radio observations of the 21-cm line emission from cool hydrogen gas reveals
    A) at least four spiral arms.
    B) one spiral arm, which wraps around the Galaxy several times.
    C) a smooth distribution of stars, characteristic of an elliptical Galaxy.
    D) two spiral arms, one on each side of the Galaxy.
    Answer: A Source: Section 14-2

42. The Milky Way galaxy appears to have a spiral structure with
    A) two major arms, wound twice around the nucleus.
    B) three loosely wound arms.
    C) four separate major arms.
    D) one "arm," wound round the nucleus four times.
    Answer: C Source: Section 14-2

43. If a spacecraft were to travel outwards from the Sun in a direction opposite to the galactic center, how many and which spiral arms would it have to cross before reaching intergalactic space? (See Fig. 14-9 of Kaufmann & Comins, *Discovering the Universe*, 5th Ed.)
    A) None, since this direction takes the spacecraft out of the plane of the galaxy.
    B) Two, the Perseus and Cygnus arms.
    C) Two, the Sagittarius and Centaurus arms.
    D) Four, the Sagittarius, Centaurus, Cygnus arms, and then the Centaurus arm again.
    Answer: B Source: Section 14-2

44. Recent observations seem to indicate that, rather than being a spiral galaxy, the Milky Way may be
    A) a barred spiral, with a definite, straight bar across its center.
    B) an irregular galaxy, with chaotic distribution of matter within it.
    C) an elliptical galaxy with little structure.
    D) two elliptical galaxies colliding with each other, in view of the very active star formation within the galactic plane, brought about by the vastly increased density during the collision.
    Answer: A Source: Section 14-2

45. The speed of the Sun in its orbit around the Galaxy is deduced by reference to and observations of
    A) Cepheid variables, which provide a distance standard.
    B) the galactic center, about which the Sun is orbiting.
    C) globular clusters and halo stars.
    D) the orbital motions of stars near the Sun.
    Answer: C Source: Section 14-3

46. The most important reason why globular clusters are useful for finding the speed of the Sun in its orbit around the Galaxy is that
    A) globular clusters on average rotate at the same speed as the Sun around the center of the galaxy.
    B) globular clusters are bright and easily seen at large distances.
    C) globular clusters on average do not rotate around the center of the galaxy.
    D) globular clusters are distributed uniformly around the galaxy.
    Answer: C Source: Section 14-3

47. Which of the following statements correctly describes the rotation of our galaxy?
    A) The disk rotates like a solid object (objects at all distances take the same time to complete an orbit), and the halo objects have random orbits with no net rotation of the halo about the center of the galaxy.
    B) Objects in the disk have random orbits with no net rotation of the disk about the center of the galaxy, and the halo rotates differentially (objects further from the center take longer to complete an orbit than objects closer to the center).
    C) The disk rotates differentially (objects further from the center take longer to complete an orbit than objects closer to the center), and the halo objects have random orbits with no net rotation of the halo about the center of the galaxy.
    D) The disk rotates differentially (objects further from the center take longer to complete an orbit than objects closer to the center), and the halo rotates differentially (objects further from the center take longer to complete an orbit than objects closer to the center).
    Answer: C Source: Section 14-3

48. The time taken for the Sun to orbit the galactic center once in its motion in the Galaxy is
    A) 28, 000 years.
    B) 230 million years.
    C) 2.3 million years.
    D) about 1/2 million years.
    Answer: B Source: Section 14-3

49. In its orbit around the center of our Galaxy, the Sun moves a distance equal to the diameter of the Earth in a time of about (see Section 14-3 of Kaufmann & Comins, *Discovering the Universe*, 5th Ed. and the Appendix)
A) one second. B) one hour. C) one minute. D) one day.
Answer: C Source: Section 14-3

50. Which two parameters of star motion in the Milky Way are represented by its rotation curve?
A) Orbital period of the stars as a function of their distance from the galactic center.
B) Star position above or below the galactic plane as a function of distance from the galactic center.
C) Orbital speed as a function of star distance from the galactic center.
D) Orbital speed of the stars as a function of their individual masses.
Answer: C Source: Section 14-3

51. If the Sun were to be travelling around the galactic center along with companion stars as depicted in Fig. 14-10 of Kaufmann and Comins, *Discovering the Universe*, 5th Ed., from which of these neighboring stars and direction(s) would you measure a Doppler shift of their light? (Hint: Think about relative velocities and orbital velocities at different orbital distances from the galactic center.)
A) Only those in the direction of a line joining the Sun to the galactic center.
B) Only those in directions at 45° angles from the Sun's direction of motion.
C) Only those at the same orbital distance as the Sun.
D) None of them, since they are all moving along with the Sun and have no relative velocity with respect to it.
Answer: B Source: Section 14-3

52. Which parameter is plotted as a function of distance from the galactic center in a rotation curve of a galaxy?
A) The speed of stars orbiting the galactic center.
B) The thickness of the galactic disk.
C) The mass of matter inside the distance from the galactic center.
D) The mass of cool hydrogen gas.
Answer: A Source: Section 14-3

53. How do we obtain an estimate of the amount of mass that is inside the Sun's orbital path in our galaxy?
    A) By counting stars, assuming an average stellar mass and calculating the total mass.
    B) By observing the movement of the galaxy towards neighboring galaxies because of mutual gravitational attraction.
    C) By applying Kepler's Law to the motion of the Sun and other nearby stars.
    D) By observing the bending of light from distant galaxies as this light passes near the Milky Way center.
    Answer: C Source: Section 14-3

54. The present estimate for the total mass of our galaxy in units of the solar mass is about
    A) $2.3 \times 10^8$. B) $10^{66}$. C) $1.1 \times 10^{11}$. D) $10^{12}$.
    Answer: D Source: Section 14-4

55. Much of the mass of our Galaxy appears to be in the form of "dark" matter of unknown composition. At present, this matter can be only detected because
    A) its gravitational pull affects orbital motions in the Galaxy.
    B) it blocks out the light from distant stars in the plane of our Galaxy.
    C) it bends light from distant quasars.
    D) it emits synchrotron radiation at radio wavelengths.
    Answer: A Source: Section 14-4

56. The presence of a very large amount of unseen ("dark") matter in the halo of our Galaxy is deduced from
    A) the unexpected absence of luminous matter (stars, etc.) beyond a certain distance from the galactic center.
    B) the rotation curve of our Galaxy, in which orbital speeds of stars in the outer regions of the Galaxy are significantly higher than is predicted by Kepler's Law in which the value for the observed mass in the galaxy is used.
    C) the rotation curve of our Galaxy, in which orbital speeds of stars appear to obey Kepler's Law.
    D) the high amount of interstellar absorption in certain directions.
    Answer: B Source: Section 14-4

57. What fraction of the mass of our galaxy is in a form that we have been able to see?
    A) About 10%.
    B) About 50%.
    C) 100%; who ever heard of matter which can't be seen?
    D) About 90%.
    Answer: A Source: Section 14-4

58. What fraction of the mass of our galaxy appears to be in the form of dark matter, which we cannot see but can detect through its gravitational influence?
    A) About 10%.
    B) 0%; who ever heard of matter which can't be seen?
    C) About 90%.
    D) About 50%.
    Answer: C Source: Section 14-4

59. What is microlensing?
    A) The beaming of radiation by accretion disks.
    B) The use of small telescopes to enhance contrast by eliminating scattered light.
    C) The focusing of starlight by the gravitational fields of "small" objects like planets or brown dwarfs.
    D) The focusing of starlight by planetary atmospheres.
    Answer: C Source: Section 14-4

60. What is microlensing?
    A) A gradual reduction in brightness of a star as an object passes in front of it.
    B) A minute shift in the apparent position of a star as an object passes in front of it.
    C) The temporary disappearance of a star as an object passes in front of it.
    D) A slow brightening of a star as an object passes in front of it.
    Answer: D Source: Section 14-4

61. What physical process has allowed astronomers to measure the overall number and distribution of brown dwarfs in our Galaxy?
    A) Molecules in their atmospheres emit radio waves that can be detected from the Earth.
    B) Their gravitational fields can bend the light from background stars.
    C) They occasionally eclipse (block the light from) more distant stars.
    D) They emit copious amounts of infrared radiation, which can penetrate the interstellar medium.
    Answer: B Source: Section 14-4

62. What has microlensing told us about the role of brown dwarfs in the galaxy?
    A) They are one of the most important components of the galaxy's dark matter.
    B) It has told us almost nothing as yet; microlensing is too difficult to observe.
    C) They cannot be a major component of the dark matter in the galaxy.
    D) They are a major component of the matter which is falling into the supermassive black hole in the galactic centre.
    Answer: C Source: Section 14-4

63. One curious fact about the Milky Way galaxy, discovered in the past few years, is that
   A) a large black hole is slowly clearing out the mass near the galactic center, leaving a cold, dark void.
   B) enormous amounts of energy are pouring out of a compact but very massive source at its center.
   C) the majority of its mass is within the spiral arms extending out into space.
   D) a significant fraction of its mass is in the form of globular clusters, distributed in a spherical halo centered on the galactic center.
   Answer: B Source: Section 14-5

64. If the Sun were to be at or close to the galactic center, the intensity of starlight in the night-time sky upon Earth would be
   A) about twice as bright as at present, since neighboring stars would be mostly bright, young blue stars in about the same numbers as the present, older, and less bright red giant neighbors to the Sun.
   B) about the same as it is now, since neighboring stars would still be relatively far away.
   C) extremely intense from the dense field of stars, equivalent to about 200 full Moons.
   D) very much fainter than at present because neighboring stars would be obscured by dense dust and gas clouds.
   Answer: C Source: Section 14-5

65. Which kind of stars are the major source of energy for the heating of the dust clouds and the HII emission nebulae within the planes of the Milky way and other galaxies?
   A) Hot, young O and B stars, via their UV radiation.
   B) The numerous old, red giant K and M stars, via their IR heat radiation.
   C) Very hot white dwarf stars, the remnants of planetary nebulas in the gas clouds.
   D) The very many nova and supernova explosions of stars within the gas and dust clouds.
   Answer: A Source: Section 14-5

66. The center of our Milky Way Galaxy can be observed most easily at which of the following electromagnetic wavelengths?
   A) Ultraviolet radiation.
   B) Infrared and radio waves.
   C) Visible light.
   D) High-energy gamma rays.
   Answer: B Source: Section 14-5

67. What is the significance of the object Sagittarius A* ("Sagittarius A-star") in our galaxy?
    A) It is a bright, high-speed cloud of gas close to the galactic nucleus, which allows the mass of the nucleus to be calculated.
    B) It appears to be the actual nucleus of the galaxy.
    C) It is a globular cluster passing close to the galactic nucleus, and the RR Lyrae stars in it allow the distance to the galactic center to be calculated.
    D) It appears to be a jet of material ejected from an accretion disk around a supermassive black hole in the galactic nucleus.
    Answer: B Source: Section 14-5

68. Positrons (i.e., positively charged electrons) are being produced in the galactic center. How have these positrons been detected?
    A) They reach the Earth as cosmic rays from the direction of the galactic center.
    B) They annihilate with ordinary electrons, producing gamma rays with a characteristic energy.
    C) They spiral in magnetic fields, producing a strong source of synchrotron radiation.
    D) They create arches of magnetic field which are visible at radio wavelengths.
    Answer: B Source: Section 14-5

69. What type of object has been proposed to explain the tremendous activity detected at the center of our Galaxy?
    A) A supermassive black hole.
    B) A rapidly rotating neutron star.
    C) A giant molecular cloud.
    D) A supernova explosion.
    Answer: A Source: Section 14-5

70. In which of the following sites in our Universe has a supermassive black hole been proposed to account for recent observations?
    A) At the center of the universe, where the Big Bang occurred at the beginning of the Universe.
    B) At the center of our Galaxy.
    C) At the center of the Crab Nebula, an old supernova remnant.
    D) At the center of the Ring Nebula, a planetary nebula in Lyra.
    Answer: B Source: Section 14-5

71. What evidence now exists for a supermassive black hole at the center of our galaxy?
    A) A very dark void in an otherwise bright region of space near the galactic center, indicating the presence of a black hole.
    B) Very bright X-ray emissions from the galactic center.
    C) Very rapid motion of matter close to the nucleus of the galaxy, requiring a very massive body to hold it in orbit.
    D) Observations of intense inflow of matter towards the center of the galaxy, as seen by light, Doppler-shifted towards the red, emitted by this matter.
    Answer: C Source: Section 14-5

72. What is the evidence that indicates to some astronomers that a super-massive black hole exists at the center of our galaxy?
    A) No electromagnetic radiation at all comes from the precise position of the galactic center and it just looks like a dark void in space.
    B) The Sun's motion in space shows that, if Kepler's Law holds for its orbit around the galactic center, there MUST be a very massive object at this center.
    C) Measurement of gas clouds orbiting the galactic center at very high speeds, which would rapidly move out of the galaxy unless some very massive object holds them in orbit.
    D) Doppler shift of light from stars in the near neighborhood of the galactic center which indicates that the stars are falling inwards at very high speeds.
    Answer: C Source: Section 14-5

73. The possible presence of a supermassive black hole at the center of our Galaxy has been deduced from
    A) the number of globular clusters which concentrate towards the galactic center.
    B) gravitational radiation being emitted by stars as they are swallowed by the black hole.
    C) the very high orbital speed of ionized gas clouds close to the galactic center.
    D) powerful magnetic fields in the huge filaments arching away from (or toward) the center.
    Answer: C Source: Section 14-5

74. What appear to be the characteristics of the object at the center of our galaxy?
    A) Five billion solar masses in a volume smaller then Jupiter's orbit.
    B) Several trillion solar masses in a volume two light-years in diameter.
    C) Two million solar masses in a volume the size of our solar system.
    D) Twenty solar masses in a volume the size of the Sun.
    Answer: C Source: Section 14-5

75. If the galactic center is now thought to contain a supermassive black hole, why is the Sun not falling into it under the black hole's extreme gravity?
    A) Because the mutual gravitational forces of local stars in the Orion spiral arm are sufficient to overcome the strong inward force and keep the Sun moving in its orbit.
    B) Because the inward force exerted upon the Sun from the black hole is offset by the force exerted outward by the hidden "dark" matter beyond the Sun's orbit.
    C) Because it has sufficient velocity that it can orbit the galactic center in a circle.
    D) Because its mass is so small that even this extreme mass concentration at the galactic center will not exert a significant force upon it.

    Answer: C Source: Section 14-5

# Chapter 15: Galaxies

1. A particular galaxy has a nuclear region of more-or-less uniform brightness from which long lanes of stars curve outwards. What type of galaxy is this?
   A) Quasar B) Irregular C) Spiral D) Elliptical
   Answer: C Source: Section 15-1

2. What is the basic shape of a spiral galaxy?
   A) A round, flat disk with long lanes of stars that curve outwards from a round, nuclear region of uniform brightness.
   B) Approximately spherical with long lanes of dark dust clouds curving through it in a spiral pattern.
   C) A round, flat disk with long lanes of stars that curve outwards right from the center of the galaxy.
   D) A round, thin disk of uniform brightness with its edges bent up and down into a spiral shape.
   Answer: A Source: Section 15-1

3. Who developed the classification system that divides galaxies into spiral, elliptical and irregular, and classifies spirals by the size of their nuclear region and the tightness of winding of their arms?
   A) Edwin Hubble B) Sir John Herschel C) Olaus Roemer D) Ejnar Hertzsprung
   Answer: A Source: Section 15-1

4. In the Hubble classification scheme for spiral galaxies, the tightness of the winding of the spiral arms appears to be directly related to
   A) the size of the central bulge of the galaxy.
   B) the age of the galaxy, as determined from the age of its individual stars.
   C) the number of globular clusters in the halo of the galaxy.
   D) the overall intrinsic size of the galaxy, or the diameter across the spiral arms.
   Answer: A Source: Section 15-1

5. What is the Hubble classification for a spiral galaxy with a large nuclear region and tightly wound arms?
   A) Sc B) Sb C) SBc D) Sa
   Answer: D Source: Section 15-1

6. What is the Hubble classification for a spiral galaxy with a moderate-sized nuclear region and moderately wound arms?
   A) Sb B) SBc C) Sc D) Sa
   Answer: A Source: Section 15-1

7. According to the Hubble classification scheme, an Sc galaxy has
   A) a round or spherical appearance with a smooth light distribution.
   B) an irregular shape with no obvious disk or spiral arms.
   C) a small central bulge and loosely wound spiral arms.
   D) a large central bulge and tightly wound spiral arms.
   Answer: C Source: Section 15-1

8. What is the Hubble classification for a spiral galaxy with a small nuclear region and loosely wound arms?
   A) Sc B) Sb C) SBa D) Sa
   Answer: A Source: Section 15-1

9. According to the Hubble classification scheme, an Sa galaxy has
   A) a round or spherical appearance with a smooth light distribution
   B) a large central bulge and tightly wound spiral arms.
   C) an irregular shape with no obvious disk or spiral arms.
   D) a small central bulge and loosely wound spiral arms.
   Answer: B Source: Section 15-1

10. The Andromeda Galaxy (M31) is best described as
    A) a vortex surrounding a black hole.
    B) a spiral collection of stars, dust, and gas, 200,000 light-years across.
    C) a gaseous nebula, extending for 6° across our sky.
    D) an extension of the Milky Way.
    Answer: B Source: Section 15-1

11. An astronomer studying a distant cluster of galaxies finds that several of the galaxies are spiral-shaped, with a large nuclear region and tightly wound arms. How should the astronomer classify these galaxies?
    A) Sa B) Sb C) Sc D) SBb
    Answer: A Source: Section 15-1

12. An astronomer studying a distant cluster of galaxies finds that several of the galaxies are spiral-shaped, with a nuclear region of moderate size and moderately wound arms. How should the astronomer classify these galaxies?
A) Sa B) Sc C) SBa D) Sb
Answer: D Source: Section 15-1

13. An astronomer studying a distant cluster of galaxies finds that several of the galaxies are spiral-shaped, with a small nuclear region and loosely wound arms. How should the astronomer classify these galaxies?
A) SBb B) Sc C) Sa D) Sb
Answer: B Source: Section 15-1

14. The typical diameter of a spiral galaxy is about
A) 100 light-years. B) $10^7$ light-years. C) 1 light-year. D) $10^5$ light-years.
Answer: D Source: Section 15-1

15. What does a spiral galaxy look like when seen edge-on?
A) A thick, flat line with a bulge in the center.
B) Round but without spiral arms because they are hidden.
C) A thick, flat line.
D) A thick line curved into a spiral shape.
Answer: A Source: Section 15-1

16. A particular galaxy appears round, with a nuclear region of uniform brightness and an outer region that is broken up into curved but fuzzy and poorly defined lanes of stars. How would this galaxy be classified?
A) Flocculent spiral. B) Irregular. C) Granddesign spiral. D) Elliptical.
Answer: A Source: Section 15-1

17. A particular galaxy appears round, with a nuclear region of uniform brightness and an outer region that is broken up into long, curved, welldefined lanes of stars. How would this galaxy be classified?
A) Irregular. B) Eliptical. C) Gand-design spiral. D) Focculent spiral.
Answer: C Source: Section 15-1

18. What mechanism is believed to produce flocculent spiral galaxies?
A) Shock waves from explosive star formation in the nuclear bulge.
B) Self-propagating star formation, where star formation occurs in bursts.
C) Satellite galaxies plunging through the disk of the nuclear spiral galaxy.
D) Density waves in the interstellar medium.
Answer: B Source: Section 15-1

19. What mechanism is believed to produce grand-design spiral galaxies?
    A) Density waves in the interstellar medium.
    B) Self-propagating star formation, where star formation occurs in bursts.
    C) Shock waves from explosive star formation in the nuclear bulge.
    D) Satellite galaxies plunging through the disk of the nuclear spiral galaxy.
    Answer: A Source: Section 15-1

20. How many stars per cubic parsec are there in the spiral arms of a spiral galaxy, compared to the regions between the spiral arms?
    A) There are more than 100 times as many stars in the spiral arms as in the regions between the arms.
    B) There are a lot of stars in the spiral arms and none at all in the regions between the arms.
    C) There are about 5% more in the spiral arms than in the regions between the arms.
    D) There are about twice as many stars in the spiral arms than in the regions between the arms.
    Answer: C Source: Section 15-2

21. Why do the spiral arms show up so clearly in spiral galaxies?
    A) Stars are spread almost uniformly over the galaxy (outside the nuclear bulge), but the brightest stars occur only in the spiral arms, making the arms stand out.
    B) Stars occur only in the spiral arms (and the nuclear bulge), with essentially none between the arms, making the arms stand out brightly.
    C) The number of stars in the arms is several times larger than in the regions between, so they stand out brightly.
    D) Stars are spread uniformly over the galaxy but the dust forms a spiral pattern, absorbing starlight; the spiral arms are the dust-free regions between the dust lanes.
    Answer: A Source: Section 15-2

22. Why is it strange to find spiral arms in spiral galaxies (see Fig. 15-4, Kaufmann & Comins, *Discovering the Universe*, 5th Ed.)?
    A) Spiral arms require new stars, and previous new stars should have already used up all of the interstellar medium; no new stars should be forming now.
    B) There is no known mechanism to generate spiral arms.
    C) Their motion and differential rotation should have wound up their arms and made them blend and disappear.
    D) Galaxies are not rotating fast enough to form the observed spiral arms.
    Answer: C Source: Section 15-2

23. What is a barred spiral galaxy?
    A) A galaxy with a bar extending across an entire diameter and the arms starting at various positions along the bar.
    B) A galaxy with a bar through the nuclear bulge, and the spiral arms starting from the ends of the bar.
    C) A galaxy in which the arms form straight bars instead of curved lines.
    D) A spiral galaxy with a straight bar instead of a nuclear bulge.
    Answer: B Source: Section 15-3

24. What is the name given to a galaxy with a large nuclear bulge, and tightly wound arms starting from a bar through the nuclear bulge?
    A) Sb B) Sa C) SBa D) SBc
    Answer: C Source: Section 15-3

25. What is the name given to a galaxy with a small nuclear bulge, and loosely wound arms starting from a bar through the nuclear bulge?
    A) Sb B) SBc C) Sc D) SBa
    Answer: B Source: Section 15-3

26. What is an SBc galaxy?
    A) A galaxy with a large nuclear bulge, and tightly wound arms starting from a bar through the nuclear bulge.
    B) A galaxy with a small nuclear bulge and loosely wound arms coming from the nuclear bulge.
    C) A galaxy with a small nuclear bulge, and loosely wound arms starting from a bar through the nuclear bulge.
    D) A galaxy with a moderate nuclear bulge, moderately wound arms, and a bright core.
    Answer: C Source: Section 15-3

27. How many barred spirals are there, compared to ordinary spirals?
    A) Half as many are barred as are ordinary.
    B) The numbers are essentially equal.
    C) Twice as many are barred as are ordinary.
    D) About one in ten spirals is barred.
    Answer: A Source: Section 15-3

28. How does the number of barred spirals in the universe compare to the number of ordinary spirals?
    A) Ordinary spirals outnumber barred spirals.
    B) The question is meaningless; barred spirals are simply ordinary spirals seen edge-on.
    C) Barred spirals outnumber ordinary spirals.
    D) There are about equal numbers of barred spirals and ordinary spirals.
    Answer: C Source: Section 15-3

29. How do spiral galaxies rotate?
    A) They don't rotate; if they did the spiral pattern would soon disappear.
    B) The arms lead the rotation (point forwards).
    C) We don't know; they rotate too slowly for us to have seen any motion in the time since galaxies were discovered.
    D) The arms trail the rotation (point back).
    Answer: D Source: Section 15-3

30. What is an elliptical galaxy?
    A) A galaxy with an elliptical outline and a smooth distribution of brightness (no spiral arms).
    B) A spiral galaxy seen from an angle, giving it an elliptical profile.
    C) Any galaxy with an elliptical halo when observed at radio wavelengths.
    D) A spiral galaxy with an elliptically shaped nuclear bulge and the spiral arms starting from the ends of the ellipse.
    Answer: A Source: Section 15-4

31. What name is given to a galaxy with a smooth distribution of brightness and a round shape?
    A) SBa B) Sa C) E0 D) E7
    Answer: C Source: Section 15-4

32. What name is given to a galaxy with a smooth distribution of brightness and a very elongated shape?
    A) SBc B) E0 C) Sc D) E7
    Answer: D Source: Section 15-4

33. What is an E3 galaxy?
    A) A galaxy with a smooth light distribution and a moderately elongated elliptical shape, without a disk or central bulge.
    B) A galaxy with a smooth light distribution and a very elongated elliptical shape, without a disk or central bulge.
    C) A galaxy with a smooth light distribution and a moderately elliptical shape, having a pronounced disk and central bulge.
    D) A galaxy with an irregular light distribution and a very elongated shape.
    Answer: A Source: Section 15-4

34. According to the Hubble classification scheme, an E4 galaxy has
    A) a disk and central bulge, with a smooth light distribution and no spiral arms.
    B) a round or spherical shape with a smooth light distribution and no disk or central bulge.
    C) an elliptical shape (flattened circle) with a smooth light distribution.
    D) an irregular shape.
    Answer: C Source: Section 15-4

35. According to the Hubble classification scheme, an E3 galaxy
    A) has a shorter central bar in its disk than an E5 galaxy.
    B) has more tightly wound spiral arms than an E5 galaxy.
    C) is rounder-looking than an E5 galaxy.
    D) is flatter-looking than an E5 galaxy.
    Answer: C Source: Section 15-4

36. According to the Hubble classification scheme, an E6 galaxy
    A) has more tightly wound spiral arms than an E2 galaxy.
    B) has a shorter central bar in its disk than an E2 galaxy.
    C) is flatter-looking than an E2 galaxy.
    D) is rounder-looking than an E2 galaxy.
    Answer: C Source: Section 15-4

37. Which one of the following statements does NOT correctly describe a typical elliptical galaxy?
    A) They have a central bulge and a disk, but no spiral arms.
    B) They have a smooth light distribution with various degrees of flattening from a circular shape.
    C) They contain primarily low-mass stars.
    D) They cover the entire range of masses from the smallest to the biggest galaxies in the universe.
    Answer: A Source: Section 15-4

38. Which of the following statements is NOT characteristic of elliptical galaxies?
    A) They are almost devoid of interstellar gas and dust.
    B) They have a disk and central bulge, but no spiral arms.
    C) They stopped forming stars billions of years ago.
    D) Different elliptical galaxies appear to be flattened by different amounts.
    Answer: B Source: Section 15-4

39. There is little or no interstellar dust or gas in which of the following galaxy types?
    A) Spirals. B) Barred spirals. C) Ellipticals. D) Irregulars.
    Answer: C Source: Section 15-4

40. In which of the following types of galaxy is star formation no longer occurring?
    A) Elliptical galaxies.
    B) Irregular galaxies.
    C) Spiral galaxies.
    D) Barred spiral galaxies.
    Answer: A Source: Section 15-4

41. Which are the largest galaxies in the universe?
    A) Lenticular galaxies.
    B) Large spiral galaxies like the Milky Way Galaxy.
    C) Irregular galaxies.
    D) Giant elliptical galaxies.
    Answer: D Source: Section 15-4

42. At visible wavelengths, which galaxies are the brightest in the universe?
    A) Giant elliptical galaxies.
    B) Starburst galaxies.
    C) Lenticular galaxies.
    D) Large spiral galaxies like the Milky Way Galaxy.
    Answer: A Source: Section 15-4

43. The biggest and intrinsically brightest galaxies in the universe are members of which group?
    A) Irregular galaxies. B) Barred spirals. C) Large spirals. D) Ellipticals.
    Answer: D Source: Section 15-4

44. The largest range of sizes of galaxies is found in which class of galaxies?
    A) Irregular galaxies.
    B) Elliptical galaxies.
    C) Spiral galaxies.
    D) Starburst galaxies.
    Answer: B Source: Section 15-4

45. Which class of galaxies has the greatest range of sizes from largest in the universe to smallest in the universe?
A) Lenticular galaxies. B) Irregular galaxies. C) Spirals. D) Ellipticals.
Answer: D Source: Section 15-4

46. Which of the following types of galaxies contains primarily population II, low-mass, long-lived stars?
A) Starburst galaxies.
B) Barred spiral galaxies.
C) Flocculent spiral galaxies.
D) Elliptical galaxies.
Answer: D Source: Section 15-4

47. What kind of stars does an elliptical galaxy typically contain?
A) Stars of all ages from young, metal-rich stars to old, metal-poor stars.
B) Stars of all ages, but all metal-poor.
C) Primarily young, metal-rich stars.
D) Primarily old, metal-poor stars.
Answer: D Source: Section 15-4

48. An astronomer studying a galaxy finds that its spectrum shows only old, low-mass, population II stars, and photographs of the galaxy show little or no interstellar gas or dust. What kind of galaxy is this astronomer studying?
A) A barred spiral galaxy.
B) An irregular galaxy.
C) A spiral galaxy.
D) An elliptical galaxy.
Answer: D Source: Section 15-4

49. What is a lenticular (or S0) galaxy?
A) A galaxy with a central bulge and a disk like a spiral galaxy, but with no spiral arms.
B) A galaxy with a smooth brightness profile, and lacking the central bulge and disk of a spiral galaxy.
C) A galaxy with a lot of gas and dust and no particular structure.
D) A spiral galaxy with fuzzy and poorly formed spiral arms.
Answer: A Source: Section 15-4

50. The Hubble classification scheme for an S0 galaxy is
A) a large central bulge with tightly wound spiral arms.
B) a disk and central bulge, with a smooth light distribution and no spiral arms.
C) a small central bulge with loosely wound spiral arms.
D) a round or spherical shape, with a smooth light distribution and no disk or central bulge.
Answer: B Source: Section 15-4

51. An astronomer studying a cluster of galaxies finds a galaxy that is round and has a disk and central bulge like a spiral galaxy, but has no spiral arms. How should the astronomer classify this galaxy?
A) E7 B) Sa C) S0 D) E0
Answer: C Source: Section 15-4

52. Who developed the "tuning-fork" diagram connecting the different shapes of elliptical and spiral galaxies?
A) Edwin Hubble.
B) Stephen Hawking.
C) Ejnar Hertzsprung.
D) Martin Schwarzschild.
Answer: A Source: Section 15-5

53. How are the Magellanic Clouds, the two nearby satellite galaxies of our own Galaxy, classified? (See Figs. 15-11 and 15-12, Kaufmann & Comins, *Discovering the Universe*, 5th Ed.)
A) grandd-esign spiral galaxies
B) irregular galaxies
C) elliptical galaxies
D) flocculent spiral galaxies
Answer: B Source: Section 15-5

54. The Magellanic clouds seen from the southern hemisphere are examples of what type of objects?
A) Planetary nebulae.
B) Irregular galaxies.
C) Supernova remnants.
D) Spiral galaxies.
Answer: B Source: Section 15-5

55. The overall distribution of galaxies through space is now found to be
A) galaxies clustered together in several high-density centers, with very little matter linking them together.
B) galaxies distributed uniformly throughout space, out to the furthest distances.
C) galaxies concentrated around one position in space, presumably the original site of the Big Bang.
D) galaxies concentrated on the surface of huge open spaces or voids, like soap bubbles.
Answer: D Source: Section 15-5

56. How are galaxies spread through the universe?
    A) They are grouped into clusters that in turn are grouped into clusters of clusters (superclusters).
    B) Galaxies are spread more-or-less evenly throughout the universe.
    C) They are grouped into clusters that are spread more-or-less evenly throughout the universe.
    D) Galaxies are densest near the Milky Way Galaxy and become less and less numerous the further we look out into the universe.
    Answer: A Source: Section 15-6

57. What is a supercluster of galaxies?
    A) It is a cluster of galaxies that is packed much more densely than normal clusters, giving it a higher mass.
    B) It is a phrase describing all the galaxies in the universe as a single system.
    C) It is a cluster of galaxies that is spread out over a larger-than-normal volume of space.
    D) It is a cluster of galaxy clusters.
    Answer: D Source: Section 15-6

58. How are superclusters spread through the universe?
    A) They are distributed in long lines that cover the universe like a gigantic string collection.
    B) They are distributed more-or-less evenly (i.e., at random) through the universe.
    C) They are grouped into clusters of superclusters that in turn are grouped into clusters of clusters, etc.
    D) They are distributed over the surfaces of large voids like a collection of gigantic soap bubbles.
    Answer: D Source: Section 15-6

59. In discussing galaxies and the universe, astronomers often talk about "voids". What are voids?
    A) Holes in some galaxies in dense clusters of galaxies where a collision by another, smaller, galaxy has removed the stars in that region.
    B) The large expanses of space between galaxies, where there is essentially no gas or dust.
    C) Volumes of space hundreds of millions of light-years across which contain almost no galaxies.
    D) Regions in clusters of galaxies that seem to be devoid of matter.
    Answer: C Source: Section 15-6

60. In discussing distant reaches of the universe, astronomers sometimes talk about the "Great Wall". What it this?
    A) The edge of the visible universe, beyond which the matter in the universe is opaque to electromagnetic radiation.
    B) A sheet of galaxies more than 500 million light-years long.
    C) The distance at which galaxies become too faint to see, so we cannot study anything beyond it.
    D) The sheet of dust in the disk of our Galaxy, which prevents us from seeing distant galaxies when we look along the plane of this sheet.
    Answer: B Source: Section 15-6

61. Our Galaxy is
    A) an isolated galaxy, not a member of any cluster.
    B) one member of a small cluster of galaxies.
    C) one member of a large, regular cluster of thousands of galaxies.
    D) one member of a large, irregular cluster of thousands of galaxies.
    Answer: B Source: Section 15-7

62. What is the Local Group?
    A) A group of galaxies clustered around the Andromeda Galaxy M31, apparently gravitationally bound to it but separate from the Milky Way.
    B) A cluster of about 30 galaxies of which the Milky Way is a member.
    C) A group of about 100 stars within 20 light-years of the Sun, which appear to have been formed at about the same time from similar material.
    D) The stars which occupy the same spiral arm as the Sun.
    Answer: B Source: Section 15-7

63. The Local Group is
    A) a cluster of galaxies in which the Milky Way is located.
    B) the family of planets around the Sun.
    C) the name of the spiral arm of our Galaxy in which the Sun is located.
    D) a star cluster to which the Sun belongs.
    Answer: A Source: Section 15-7

64. In the Local Group of galaxies, how many known galaxies are within $10^6$ light-years of our own Milky Way Galaxy? (See Figure 15-16, Kaufmann & Comins, *Discovering the Universe*, 5th Ed.)
    A) 5 B) 14 C) 11 D) 22
    Answer: C Source: Section 15-7

65. What is a rich cluster of galaxies?
    A) A cluster with more spiral galaxies than ellipticals.
    B) A cluster with a high metal content.
    C) A cluster containing thousands of galaxies.
    D) A cluster (like our Local Group) that contains at least two large galaxies.
    Answer: C Source: Section 15-7

66. What is a poor cluster of galaxies?
    A) A cluster containing a few dozen galaxies.
    B) A cluster with very little gas or dust.
    C) A cluster with very few spiral galaxies compared to other types.
    D) A cluster with a low metal content.
    Answer: A Source: Section 15-7

67. What is the famous Virgo cluster of galaxies?
    A) The nearest cluster beyond the Local Group, with about 3 dozen galaxies in it.
    B) A rich, regular cluster of thousands of galaxies.
    C) A rich, irregular cluster of over 1,000 galaxies.
    D) A cluster of unknown type centered on a quasar near the edge of the visible universe.
    Answer: C Source: Section 15-7

68. Why is the Coma cluster of galaxies famous?
    A) It contains two quasars.
    B) It is the nearest rich, irregular cluster.
    C) It is the cluster in which our Milky Way Galaxy is situated.
    D) It is the nearest rich, regular cluster.
    Answer: D Source: Section 15-7

69. The estimated distance from the Earth to the Coma cluster of galaxies is about (see Fig. 15-18, Kaufmann & Comins, *Discovering the Universe*, 5th Ed.)
    A) 300 million light-years.
    B) 2 million light-years.
    C) 10 light-years.
    D) 200 billion light-years.
    Answer: A Source: Section 15-7

70. The light which arrives at Earth from the Coma cluster of galaxies has traveled for approximately how long?
    A) 92 million years.
    B) 30 million years.
    C) 300 million years.
    D) 300 billion years.
    Answer: C Source: Section 15-7

71. Suppose we were to detect radio signals from an intelligent civilization in the Coma cluster of galaxies. If we sent a message to this civilization, how long would it take for us to get a reply?
   A) 150 million years.
   B) 75 million years.
   C) 600 million years.
   D) 300 million years.
   Answer: C Source: Section 15-7

72. The total number of galaxies in a rich, regular cluster of galaxies such as the Coma cluster is believed to be
   A) several dozen galaxies.
   B) over a million galaxies.
   C) about 1,000 galaxies.
   D) about 10,000 galaxies.
   Answer: D Source: Section 15-7

73. What is the galaxy content of rich, regular clusters like the Coma cluster?
   A) Mostly ellipticals and S0 galaxies, and relatively few spirals and irregulars.
   B) Entirely elliptical galaxies.
   C) More-or-less even distribution of spirals, ellipticals, irregulars and S0 galaxies.
   D) Mostly spirals and irregulars with relatively few ellipticals and S0 galaxies.
   Answer: A Source: Section 15-7

74. What is the galaxy content of rich, irregular clusters of galaxies, such as the Hercules cluster?
   A) More than half ellipticals, less than 1/4 spirals.
   B) More than half spirals, less than 1/4 ellipticals.
   C) About the same number of spirals as ellipticals.
   D) Entirely spirals.
   Answer: B Source: Section 15-7

75. A rich, regular cluster of galaxies differs from a rich, irregular cluster in that it
   A) has its galaxies distributed in a regular, highly flattened system (like a disk).
   B) contains fewer galaxies than an irregular cluster.
   C) has less spirals and more ellipticals and S0 galaxies than an irregular cluster.
   D) lacks the giant elliptical galaxies often found in irregular clusters.
   Answer: C Source: Section 15-7

76. What would happen if the Andromeda galaxy (a spiral about the same size as ours) collided with our own Milky Way Galaxy?
    A) The two galaxies would pass through each other, with the stars sailing past each other unharmed but the interstellar gas and dust clouds would collide.
    B) All of the gas and dust clouds and a great many of the stars would collide with each other, stopping both galaxies and creating a galactic merger.
    C) The two galaxies would pass through each other almost unchanged, with essentially no interactions at all.
    D) The two galaxies would shatter or even explode, essentially destroying their stars and any life forms that there may have been (including us).
    Answer: A Source: Section 15-8

77. Which of the following statements is most likely to be true, when discussing galactic motions and interactions?
    A) Galaxies are so widely separated that they never interact or collide.
    B) Galaxies are so closely packed in the universe that they are always interacting with one another.
    C) The universe is composed of one giant galaxy of which all observed stars are members, thus, the question of interaction between galaxies is irrelevant.
    D) Galaxies occasionally collide with one another, particularly within clusters of galaxies.
    Answer: D Source: Section 15-8

78. What is a starburst galaxy?
    A) A galaxy that is still in the process of formation from the intergalactic medium and undergoing its first star formation.
    B) A galaxy with streams of stars arching out from one place as if from an explosion.
    C) A galaxy with an unusually large number of supernovae (exploding stars).
    D) A galaxy with an unusually large number of newborn stars.
    Answer: D Source: Section 15-8

79. What is believed to be the origin of starburst galaxies?
    A) A recent series of supernovae have compressed the interstellar medium and started a wave of star formation.
    B) The galaxies are newly formed and are undergoing their initial, rapid star formation.
    C) A recent collision with another galaxy has triggered a wave of star formation.
    D) The galaxies are slower-rotators than other galaxies, so the slower-speed collisions between interstellar clouds produce more star formation.
    Answer: C Source: Section 15-8

80. A starburst galaxy appears to be a galaxy in which
   A) essentially all of the star formation took place in a single, billion-year-long burst early in the life of the galaxy.
   B) most of the energy from the galaxy is being produced by supernovae.
   C) a collision with another galaxy has produced a burst of star formation.
   D) a central, supermassive black hole is throwing jets of gas and stars out of the nucleus into intergalactic space.
   Answer: C Source: Section 15-8

81. One of the consequences of the collision of two galaxies appears to be
   A) a very large explosion, similar to but much larger than a supernova.
   B) a burst of vigorous star birth.
   C) the disappearance of one of them into the central black hole of the other.
   D) almost nothing, since stars are widely separated in each galaxy and the probability of star-star collisions is very small.
   Answer: B Source: Section 15-8

82. Which one of the following statements best describes galactic motions and interactions?
   A) Galaxies occasionally collide with one another, particularly within clusters of galaxies.
   B) The universe is composed of one giant galaxy of which all observed stars are members, thus the question of interaction between galaxies is irrelevant.
   C) Galaxies are so widely separated that they never interact or collide.
   D) Galaxies are so closely packed in the universe that they are always interacting with one another.
   Answer: A Source: Section 15-8

83. What is believed to be the origin of giant elliptical galaxies?
   A) They grew by devouring smaller galaxies.
   B) They have grown continuously since their formation, by accretion of intergalactic gas.
   C) They formed that way and have remained unchanged ever since.
   D) Collisions of galaxies in the cluster produces a smooth distribution of stars through the cluster, which sink to the center of the cluster and form giant galaxies there.
   Answer: A Source: Section 15-8

84. Which single major problem perhaps puzzles astronomers the most as they attempt to interpret the properties and behavior of clusters of galaxies?
   A) The presence of star formation in many galaxies, long after it is expected to have died out.
   B) The missing-mass problem, with at least 10 times more than the observed mass needed for galactic cluster stability.
   C) The structure and motion of spiral arms in galaxies in many clusters.
   D) The rate of supernova occurrence in galaxies in some clusters.
   Answer: B Source: Section 15-9

85. Which one of the following statements is true about clusters of galaxies?
   A) The observed mass of the galaxies of the cluster is many thousands of times too small to hold the cluster together.
   B) The observed mass of the galaxies of the cluster is almost exactly equal to that needed to hold the cluster together.
   C) The observed mass of the galaxies of the cluster is about ten times too small to hold the cluster together.
   D) The observed mass of the galaxies of the cluster is about ten times larger than the amount needed to hold the cluster together.
   Answer: C Source: Section 15-9

86. Which of the following has NOT provided a means by which astronomers can infer the presence of dark matter in the universe?
   A) Measurement of the orbital speeds of stars near possible supermassive black holes at the centers of galaxies.
   B) Observation of the bending of light from remote galaxies as it passes through intervening clusters of galaxies.
   C) Measurement of the orbital speeds of stars in the outer regions of galaxies.
   D) Measurement of the line-of-sight speeds of individual galaxies in clusters of galaxies.
   Answer: A Source: Section 15-9

87. The rotation curve of a galaxy is a graph showing the galaxy's speed of rotation at different distances from the center. The observed rotation curve in the OUTER PARTS of a typical large spiral galaxy
   A) decreases smoothly with increasing distance from the center.
   B) is quite flat (roughly the same speed at all distances).
   C) increases drastically with increasing distance from the center.
   D) decreases suddenly to zero at the outer edge of the visible galaxy.
   Answer: B Source: Section 15-9

88. If most of the mass of a galaxy is located near the center of the galaxy, then in the OUTER PART of this galaxy we would expect the orbital speeds of stars to decrease with increasing distance from the center. This is an example of
A) Wien's law. B) Hubble's law. C) Kepler's third law. D) Newton's third law.
Answer: C Source: Section 15-9

89. Most of the light from a galaxy comes from the inner parts. IF this means that most of the galaxy's mass is also in the inner region, then how would we expect the galaxy's speed of rotation to behave in its outer region?
A) The rotation speed should decrease sharply to zero at the outer edge of the visible galaxy.
B) The rotation speed should decrease smoothly with increasing distance from the center.
C) The rotation speed should increase with increasing distance from the center.
D) The rotation speed should not change appreciably with increasing distance from the center (i.e., a "flat" rotation curve).
Answer: B Source: Section 15-9

90. Dark, unknown forms of matter appear to make up about what fraction of the mass of a typical rich cluster of galaxies?
A) Half B) 90% C) 10% D) Much less than 1%
Answer: B Source: Section 15-9

91. The mass of the presently observable matter in a typical rich cluster of galaxies appears to be about what fraction of the mass required by the cluster's observed gravity (as shown by the speeds of galaxies in the cluster)?
A) 90% B) 10% C) 100%, or all of the galaxy D) About 1%
Answer: B Source: Section 15-9

92. One of the big puzzles about the properties and behavior of large clusters of galaxies is that
A) they appear not to take part in the general expansion of the Universe, in contrast to single separate galaxies, probably because they are gravitationally bound to one another.
B) they appear to be spread uniformly throughout space in all directions, which is difficult to explain with the Big Bang Theory.
C) there appears to be insufficient mass in the luminous matter (stars etc) to hold the cluster together gravitationally.
D) each one appears to consist of the same type of galaxy, some made up totally of spiral galaxies while others contain only ellipticals.
Answer: C Source: Section 15-9

93. What is the dominant radiation that we see from the intergalactic matter in rich clusters of galaxies?
    A) 21-cm radio radiation from cool, neutral hydrogen gas.
    B) Infrared radiation from dust.
    C) Ultraviolet light from electrons spiraling in magnetic fields.
    D) X-rays from very hot gas.
    Answer: D  Source: Section 15-9

94. What is the typical temperature of the intergalactic gas in rich clusters of galaxies?
    A) 10,000 to 20,000 K.
    B) 1 to 10 billion K.
    C) 10 to 100 K.
    D) 10 to 100 million K.
    Answer: D  Source: Section 15-9

95. Many rich, regular clusters of galaxies contain substantial amounts of very hot, intergalactic gas. Where is this gas believed to have come from?
    A) Bursts of star formation in merging galaxies.
    B) Jets of gas ejected by supermassive black holes at the centers of the galaxies.
    C) Collisions between galaxies in the cluster.
    D) Supernovae (exploding stars).
    Answer: C  Source: Section 15-9

96. The source of the hot, intergalactic gas in many rich, regular clusters of galaxies appears to be
    A) bursts of star formation in merging galaxies.
    B) supernovae (exploding stars).
    C) jets of gas ejected by supermassive black holes at the centers of the galaxies.
    D) collisions between galaxies in the cluster.
    Answer: D  Source: Section 15-9

97. How has the hot, x-ray emitting intergalactic gas in galaxy clusters affected the "missing-mass" problem for these clusters?
    A) It supplies about 1/10 of the "missing mass," leaving about 9/10 still missing.
    B) It is very tenuous gas and supplies almost no mass at all; therefore it has almost no effect on the "missing mass" problem.
    C) It supplies the mass needed to solve the problem, which is now no longer a problem!
    D) It supplies about ten times too much mass; now, the problem is how clusters so massive are not much more compact.
    Answer: A  Source: Section 15-9

98. One of the recently discovered components of clusters of galaxies that may have some bearing on the "missing-mass" problem is
   A) x-ray measurements of substantial amounts of very hot intergalactic gas in clusters.
   B) very large, cool molecular clouds within and surrounding the galaxies, detected by radio astronomy.
   C) large numbers of cool, dark "brown dwarf" stars, previously unknown, within the galaxies.
   D) large numbers of small but massive black holes within every galaxy.
   Answer: A Source: Section 15-9

99. Who first discovered that the majority of galaxies are moving away from the Earth?
   A) Edwin Hubble. B) Albert Einstein. C) Karl Jansky. D) V. M. Slipher.
   Answer: D Source: Section 15-10

100. Who first showed that the recessional speeds of galaxies increase with increasing distance from the Earth?
   A) Edwin Hubble. B) V. M. Slipher. C) Karl Jansky. D) Albert Einstein.
   Answer: A Source: Section 15-10

101. What is the Hubble flow?
   A) Distant galaxies are all moving away from us, with speed increasing with increasing distance.
   B) Distant galaxies are all moving away from us, with speed decreasing with increasing distance.
   C) Distant galaxies are all moving away from us, and at approximately the same speed.
   D) Distant galaxies are all moving towards us on one side and away from us on the other, as part of a universal flow of galaxies through the universe.
   Answer: A Source: Section 15-10

102. The primary evidence for the expanding universe concept is
   A) observation of supernova explosions.
   B) the slow increase in the Earth-Moon separation with time, about 4 cm per year.
   C) the discovery of black holes in binary stars.
   D) the redshift of light from distant galaxies, which increases with distance of the galaxy from Earth.
   Answer: D Source: Section 15-10

103. Which two quantities are shown to be related to one another in Hubble's Law?
   A) Distance and brightness.
   B) Distance and recession velocity.
   C) Brightness and recession velocity.
   D) Brightness and the width of the 21-cm radio emission line of hydrogen.
   Answer: B Source: Section 15-10

104. The Hubble relation links which two characteristics of distant objects in the universe?
   A) Stellar mass and luminosity.
   B) State of organization and age of clusters of stars.
   C) Distance and velocity of recession.
   D) Luminosity and surface temperature.
   Answer: C Source: Section 15-10

105. The Hubble Law, representing observations of distant objects in the universe, relates which two parameters?
   A) The distance to a distant object and its recession velocity.
   B) The mass of a distant object and its recession velocity.
   C) The mass of an object and its luminosity.
   D) The luminosity and position of stars in the galaxy.
   Answer: A Source: Section 15-10

106. The Hubble distance-velocity relation states that
   A) mutual gravitational attraction of all objects in the universe means that all objects appear to be moving toward the Sun, the closest ones traveling fastest.
   B) the further an object is from the Sun, the faster it appears to be moving away from the Sun.
   C) all distant objects are moving toward the Sun, the most distant objects fastest.
   D) all objects appear to have the same velocity away from the Sun, irrespective of distance from the Sun.
   Answer: B Source: Section 15-10

107. The Hubble distance-velocity relationship states that
   A) the more distant a galaxy is from the Sun, the faster it is traveling towards the Sun.
   B) the more distant a galaxy is from the Sun, the faster it is traveling away from the Sun.
   C) all galaxies are being pulled towards a central gravitational attractor in the universe, with the ones closest to the center traveling fastest.
   D) all galaxies are being repelled from the central "engine" of the universe, with the ones closest to the center traveling fastest.
   Answer: B Source: Section 15-10

108. Because of the expansion of space, we see all distant galaxies moving away from us, with more distant galaxies moving faster. An observer in one of these distant galaxies would see
   A) all galaxies on one side of the observer moving towards the observer and all galaxies on the other side moving away from the observer, with more distant galaxies moving faster.
   B) all galaxies moving away from the observer, with more distant galaxies moving faster.
   C) all galaxies moving away from the observer, with closer galaxies moving faster.
   D) all galaxies moving towards the observer, with more distant galaxies moving faster.
   Answer: B Source: Section 15-10

109. For which objects in the universe has the Hubble relation been shown to hold experimentally?
   A) Stars in the distant spiral arms of our Galaxy.
   B) Stars in the near neighborhood of the Sun, in our Galaxy.
   C) Galaxies in the Local Group, in the near vicinity of the Milky Way.
   D) Distant galaxies.
   Answer: D Source: Section 15-10

110. Which of the following speeds are described by Hubble's law?
   A) The speeds of individual galaxies in the Local Group and other nearby galaxy clusters.
   B) The speeds of superclusters of galaxies.
   C) The speeds of stars in our own Galaxy.
   D) The speeds of individual clusters of galaxies in our own local supercluster.
   Answer: B Source: Section 15-10

111. The expansion of the universe takes place
   A) primarily in the huge voids between clusters of galaxies: "small" objects like galaxies or the Earth do not expand.
   B) only over distances about the size of a galaxy or larger; consequently, our Galaxy expands but the solar system does not.
   C) only between objects separated by a vacuum; as a result, our bodies do not expand but the Earth-Moon system does.
   D) between all objects, even between the atoms in our bodies, although the expansion of a person is too small to be measured reliably.
   Answer: A Source: Section 15-10

112. In our universe, we can consider four different regimes of space in which distances between objects might be changing as a result of the general expansion of the universe. These are
    1. distances between different parts of the Earth.
    2. distance between planets in our solar system.
    3. distances between stars in our Galaxy.
    4. distances between clusters of galaxies.

    In which of these regimes are the distances changing because of the universal expansion?
    A) 4, 3 and 2 B) 4 only C) 4 and 3 D) 4, 3, 2 and 1
    Answer: B Source: Section 15-10

113. Hubble's law for distant galaxies, with $H_0$ = Hubble's constant, tells us that
    A) recessional speed x distance = $H_0$.
    B) $H^{-}0$ x recessional speed = distance.
    C) recessional speed = $H_0$ = constant.
    D) recessional speed = $H_0$ x distance.
    Answer: D Source: Section 15-10

114. Which of the following is the correct form of Hubble's law? ($H_0$ = Hubble's constant, r = distance, v = recessional velocity)
    A) $v\,r = H_0$. B) $H_0\,v = r$. C) $v = H_0$ = constant. D) $v = H_0\,r$.
    Answer: D Source: Section 15-10

115. The mathematical form of the Hubble law for the expanding universe concept relates the velocity of recession, v, to the distance of the observed object r (with $H_0$ the Hubble Constant) as follows:
    A) $v = H_0\,r$ B) $v = H_0/r$ C) $v = H_0/r^2$ D) $r = H_0\,v$
    Answer: A Source: Section 15-10

116. On the basis of the distance to the Coma cluster of galaxies, and the Hubble relation (using an intermediate value of the Hubble constant $H_0$ equal to 80 km/s per million parsecs; see Figure 15-18, Kaufmann & Comins, *Discovering the Universe*, 5th Ed.), what would be the approximate wavelength shift of the Balmer $H_a$ spectral line at 656.3 nanometers emitted by a galaxy in the cluster because of the general expansion of the universe? (See also Astronomer's Toolbox 15-1 of *Discovering the Universe*.)
    A) $1.61 \times 10^6$ nm B) 16.1 nm C) 1.61 nm D) 161.0 nm
    Answer: B Source: Section 15-10

117. The Hercules cluster of galaxies shown in Fig. 15-19 of Kaufmann and Comins, *Discovering the Universe*, 5th Ed., is at a distance of 500 million light years from our galaxy. Using a Hubble constant of 23 km/s/Mly, at what wavelength will the Balmer Ha spectral line, with rest wavelength = 656.3 nm, be seen in the spectrum of a galaxy in this cluster?
A) 656.7 nm B) 25.2 nm C) 681.6 nm D) 631.1 nm
Answer: C Source: Section 15-10

118. What is the present range of measured values of Hubble's constant, from various techniques?
A) Between 28 and 32 km/s per megaparsec.
B) Somewhere between 50 and 90 km/s per megaparsec.
C) Somewhere between 200 and 1,000 km/s per megaparsec.
D) Somewhere between 80 and 300 km/s per megaparsec.
Answer: B Source: Section 15-10

119. One astronomer (astronomer A) claims that the Hubble constant is 84 km/s/Mpc, while another (astronomer B) claims that it is 63 km/s/Mpc. If, based on the Hubble constant, astronomer A claims that a particular galaxy is 4 billion light-years away, then astronomer B would claim that it is
A) 3 billion light-years away.
B) 2 billion light-years away.
C) 6 billion light-years away.
D) 5 1/3 billion light-years away.
Answer: D Source: Section 15-10

120. What method is used to determine the distances of very remote galaxies?
A) Measurement of the angular size of the galaxy and an assumption about the actual physical size of the galaxy.
B) Comparison of their apparent and absolute magnitudes.
C) Measurement of the apparent brightness and period of Cepheid variable stars within the galaxies.
D) Use of their spectral red shifts and the Hubble law.
Answer: D Source: Section 15-10

121. Suppose a galaxy is 400 million parsecs from the Earth. What is the recessional velocity of this galaxy? Assume Hubble's constant to be 80 km/s/Mpc.
A) 5 million km/s. B) 200,000 km/s. C) 5 km/s. D) 32,000 km/s.
Answer: D Source: Section 15-10

122. An astronomer studying a distant galaxy finds that its recessional velocity is 14,000 km/s. What is the distance to the galaxy? Take Hubble's constant to be 70 km/s/Mpc.
A) 200 Mpc. B) 2,000 Mpc. C) 98 Mpc. D) 980,000 Mpc.
Answer: A Source: Section 15-10

123. Suppose an astronomer discovers a distant quasar whose recessional velocity is 1/3 the speed of light. If Hubble's constant is 50 km/s/Mpc, how far away is the quasar?
A) 200 Mpc. B) 50 Mpc. C) 2000 Mpc. D) 5,000,000 Mpc.
Answer: C Source: Section 15-10

124. If the elliptical galaxy in Hydra, shown at the bottom of Fig. 15-30, were to be at a distance of 2.5 billion light-years, what would be the value of Hubble's constant $H_0$? (Be careful with units.)
A) About 95 km/s/Mpc
B) About 60 km/s/Mpc
C) About 25 km/s/Mpc
D) About 80 km/s/Mpc
Answer: D Source: Section 15-10

125. In the verification of the Hubble law for the expansion of the universe and the determination of the constant $H_0$, the greatest difficulty has been
A) allowance for the fact that high gravitational fields at distant galaxies can redshift light (gravitational redshift).
B) the accurate determination of distances to very distant galaxies.
C) the identification of distant objects as galaxies, rather than very bright, nearby stars.
D) the measurement of recession velocities of distant galaxies by Doppler shift.
Answer: B Source: Section 15-10

126. To an astronomer, what is a "standard candle"?
A) Any type of object whose absolute magnitude is known.
B) A standard light source that is placed in a telescope, to which the brightness of stars and other objects can be compared.
C) An accurately defined brightness scale for stars and galaxies, such as the magnitude scale.
D) Any galaxy whose redshift has been measured accurately.
Answer: A Source: Section 15-10

127. "Standard candles," which are important for finding distances to remote galaxies, are
    A) stars and other objects of known intrinsic brightness.
    B) standard laboratory light sources with which the brightness of a galaxy can be compared.
    C) heat sources used for calibrating infra-red observations of galaxies.
    D) standard bars of known length with which the size of a galaxy can be measured.
    Answer: A Source: Section 15-10

128. To what distance in parsecs can Cepheid variables now be seen with the Hubble Space Telescope?
    A) About 60 pc. B) About 600 Mpc. C) About 60 Mpc. D) About 200 Mpc.
    Answer: C Source: Section 15-10

129. What is the brightest "standard candle" found so far, and therefore the one that is visible to the greatest distance?
    A) The brightest globular cluster of each galaxy.
    B) The brightest H II regions in each galaxy.
    C) Cepheid variables.
    D) Supernovae.
    Answer: D Source: Section 15-10

130. Which of the following objects are NOT used as "standard candles" for distance measurement to distant galaxies?
    A) Cepheid variable stars.
    B) Globular clusters.
    C) Supernova explosions.
    D) Hot white dwarf stars.
    Answer: D Source: Section 15-10

131. What is the Tully-Fisher relation?
    A) The more distant the galaxy, the fainter it appears.
    B) The brighter the Cepheid, the longer the pulsation period.
    C) The more distant the galaxy, the greater the recessional velocity.
    D) The brighter the galaxy, the wider the 21-cm radio emission line.
    Answer: D Source: Section 15-10

132. In the 1970s it was discovered that, among spiral galaxies, the wider the 21-cm radio emission line, the brighter is the galaxy. What is the name of this relation?
    A) The Hawking effect.
    B) The Hubble law.
    C) The Tully-Fisher relation.
    D) The mass-luminosity law.
    Answer: C Source: Section 15-10

133. The Tully-Fisher relation provides a method of determining distances to galaxies by estimating the galaxy luminosity from a measurement of which parameter relating to the 21-cm atomic hydrogen radio emission line?
A) Its intensity.
B) The split between its two components.
C) Its position and hence Doppler shift from the rest wavelength.
D) Its width.
Answer: D Source: Section 15-10

# Chapter 16: Quasars and Active Galaxies

1. Astronomy with a radio telescope was initiated by
   A) the National Science Foundation and the American Astronomical Society.
   B) Marconi in Europe.
   C) an amateur astronomer Grote Reber, after Jansky has detected radio energy from the galaxy.
   D) the British Broadcasting Corporation in England.
   Answer: C Source: Section 3-14 and Intro. to Section 16-1

2. One of the three astronomical objects which were first detected at radio wavelengths by Grote Reber in the late 1930s was
   A) Jupiter. B) the galactic center. C) the Moon. D) the Sun.
   Answer: B Source: Introduction to Section 16-1

3. The major surprise about Cygnus A, one of the first three sources of strong radio emission detected by Grote Reber in the late 1930s, when examined with large optical telescopes, was that
   A) it corresponded to no optical source at all.
   B) it was at the center of our galaxy.
   C) it was a supernova remnant.
   D) it was a very distant object which was relatively faint at optical wavelengths, yet extremely luminous at radio wavelengths.
   Answer: D Source: Introduction to Section 16-1

4. The discovery of the peculiar galaxy Cygnus A was a surprise to astronomers because
   A) it was first discovered at X-ray wavelengths and only later detected optically.
   B) it was so bright at optical wavelengths that no one expected it to be a galaxy.
   C) it had a spectral redshift as high as those of the most distant quasars.
   D) it was very faint at visible wavelengths but extremely bright at radio wavelengths.
   Answer: D Source: Introduction to Section 16-1

5. Quasars all appear to be
   A) moving towards us at high speeds, as high as 90% of the speed of light.
   B) moving across our line of sight at very high speeds, as seen from time-lapse photographs.
   C) moving away from us at very high speeds, at up to 90% of the speed of light.
   D) extremely massive objects in our galaxy, their intense surface gravity having red-shifted their spectra.
   Answer: C Source: Section 16-1

6. The specific characteristics that identify most quasars are
   A) that they look like elliptical galaxies, but with high spectral redshifts.
   B) starlike appearance, and very high spectral blueshift, indicating that they are approaching the Sun very fast.
   C) spiral galaxy appearance, and very high spectral blueshift, indicating that they are coming towards the Sun at high speed.
   D) starlike appearance, very high redshifts and hence very large distances, indicating very energetic sources.
   Answer: D Source: Section 16-1

7. The typical optical spectrum of a quasar shows
   A) a weak continuum of radiation with very redshifted emission lines upon it.
   B) a continuum with a sequence of redshifted absorption lines upon it.
   C) a continuum with a sequence of blueshifted absorption lines upon it.
   D) very blueshifted emission lines, with no continuum.
   Answer: A Source: Section 16-1

8. Astronomers initially had difficulty identifying the emission lines in quasar spectra at optical wavelengths because
   A) no one expected to see ultraviolet spectral lines in the visible region.
   B) the lines are created by elements which do not exist on Earth.
   C) they were emission lines from ionized atoms which had not been seen before.
   D) the emission lines were smeared out by the extremely high speed of the quasars, making them hard to measure.
   Answer: A Source: Section 16-1

9. The emission lines in quasar spectra were difficult to identify initially because
   A) they appeared to be created by elements which did not exist on Earth.
   B) no one expected violet and ultraviolet spectral lines to be shifted so far toward the red.
   C) they were very faint and could not be measured accurately.
   D) emission lines of such intensity were not expected from astronomical sources.
   Answer: B Source: Section 16-1

10. Which of the following astronomical objects can be described as follows: "Starlike in appearance, showing very high red-shift, energy output of at least 100 galaxies from a small region about 1 light year across"?
    A) A quasar.
    B) A red supergiant star.
    C) A supernova explosion in a neighboring galaxy.
    D) The center of our galaxy.
    Answer: A Source: Section 16-1

11. The observed characteristics of a quasar are
    A) starlike image, extremely blue-shifted spectrum, often an intense radio emitter.
    B) starlike image, spectrum highly red-shifted, often an intense radio emitter.
    C) starlike image, with a highly variable Doppler spectrum which shifts from red to blue, and often a very bright radio source.
    D) diffuse circular image, no red shift of the spectrum, often a very bright radio source.
    Answer: B Source: Section 16-1

12. Which observations of quasars convinced astronomers that they were very distant objects?
    A) Extreme red shift of visible Balmer and UV Lyman hydrogen emission lines, indicating high recessional velocities and hence, by the Hubble Law, very large distances.
    B) Their extremely red spectrum, reddened by extreme interstellar absorption of the blue part of the spectrum, meaning that the source must be a very long way away.
    C) They appeared as point-like star images under the highest magnification, meaning that they must be very far away.
    D) Their extreme faintness at visible wavelengths, the inverse square law for visible light showing that they must be very distant.
    Answer: A Source: Section 16-1

13. The unusual feature which was noted in the optical spectra of the faint star-like objects that coincided in position with the intense sources of radio energy known as quasars was
    A) the extreme red-shift of emission lines which indicated high recessional velocities and hence great distances, requiring extremely high energy output in order to be detected.
    B) the periodic variation of the Doppler shift from red to blue, indicating a light source oscillating back and forth over a few weeks.
    C) sets of spectral lines which indicated simultaneous motion of sources towards and away from the Sun, possibly a rapidly expanding shell of material around the radio source.
    D) the extreme blue-shift, meaning that these stars in our galaxy are coming towards the Earth at very high velocities.
    Answer: A Source: Section 16-1

14. An intense radio source is found to coincide with a starlike object whose spectrum contains a pattern of intense emission lines in the visible range which matches that of the Lyman UV hydrogen spectral lines, but is very red-shifted. What is this object?
    A) The exploding shell of a supernova.
    B) A quasar.
    C) A pulsar.
    D) A black hole.
    Answer: B Source: Section 16-1

15. A starlike object seen on deep sky photographs coincides with an intense radio source and has a spectrum in which the characteristic Lyman pattern of hydrogen spectral lines have been shifted from the ultraviolet to the visible spectral range. What is this object?
A) A quasar. B) A supernova explosion. C) A pulsar. D) A black hole.
Answer: A Source: Section 16-1

16. Quasars appear to be
A) relatively close, very bright objects moving away from Earth.
B) very distant and intrinsically bright objects moving in random directions at high speeds.
C) very distant, intrinsically faint objects, moving towards Earth very rapidly.
D) very distant, intrinsically bright objects, moving away from Earth at very high speeds.
Answer: D Source: Section 16-1

17. All quasars appear to be
A) relatively close, very bright objects.
B) moving away from Earth at very high speeds.
C) moving towards the Earth at very high speeds.
D) very distant, intrinsically faint objects.
Answer: B Source: Section 16-1

18. The distance to the bright quasar, 3C 273, is estimated to be
A) 3 million light-years.
B) 2 billion light-years.
C) just beyond our Milky Way.
D) 20,000 light-years.
Answer: B Source: Section 16-1

19. The highest recession velocities which have recently been detected for quasars is
A) almost half the speed of light.
B) 70 - 80% of the speed of light.
C) about one-quarter of the speed of light.
D) more than 90% of the speed of light.
Answer: D Source: Section 16-1

20. The extreme redshifts of quasar spectra are caused by
A) absorption of all but the red parts of the quasar spectrum by intergalactic matter.
B) very high recession speeds of the sources away from our Galaxy.
C) Zeeman effects from the very intense magnetic fields in the vicinity of the source.
D) high gravitational fields at the surfaces of these quasars (gravitational redshift).
Answer: B Source: Section 16-1

21. The energy output of a bright quasar is equivalent to
    A) 100 bright galaxies.
    B) $10^6$ spiral galaxies.
    C) 1000 Suns.
    D) that of the Milky Way Galaxy.
    Answer: A Source: Section 16-2

22. The fact that quasars can be detected from distances where even the biggest and most luminous galaxies cannot be seen means that
    A) their spectra have not been as red-shifted by their motion as those of galaxies, and hence they can still be seen.
    B) they must be in directions where gravitational focusing by the masses of nearer galaxies makes them visible from Earth.
    C) they must be in directions where intergalactic absorption by dark matter is minimum, allowing us to see them.
    D) they must be far more luminous than the brightest galaxies.
    Answer: D Source: Section 16-2

23. A quasar is now thought to be
    A) a distant but very luminous and active star in our galaxy.
    B) a very luminous object a very large distance away from the Sun.
    C) a long-lived supernova explosion.
    D) a nearby star, ejected with great violence out of a galaxy.
    Answer: B Source: Section 16-2

24. Evidence obtained over the last few years indicates that quasars are most probably
    A) the remnant cores of exploding stars or supernovae.
    B) evidence of very intense star-building activity in certain distant dust and gas clouds.
    C) the focused image of a distant galaxy by the gravitational lens effect of a closer galaxy.
    D) the central nuclei of very distant, very active galaxies.
    Answer: D Source: Section 16-2

25. The typical fluctuation time for the brightnesses of quasars can be as short as
    A) a few seconds. B) about a month. C) about a day. D) about a year.
    Answer: C Source: Section 16-2

26. Variations in the output of quasars and BL Lacertae objects within a few hours is an indication
    A) that these objects are actually binary systems in which we see mutual eclipses.
    B) of the rapid rotation of the sources.
    C) of the relatively small size of the emitting regions.
    D) of objects moving in front of them, from our point of view.
    Answer: C Source: Section 16-2

27. How can astronomers determine the size of an emission region in a very distant and unresolvable source?
    A) By measuring brightness variability, since an object cannot vary more rapidly than the time taken for light to cross the source.
    B) By measuring the object's mass and by using a reasonable value for the average density for matter, calculate its volume and hence its diameter.
    C) By measuring the red shift of its spectrum, since this will be dependent upon the source size.
    D) By using radio interferometry, since this technique can resolve far greater detail than optical imaging.

    Answer: A Source: Section 16-2

28. What observations convince us that the energy source of a quasar is physically very small?
    A) The instant disappearance of the quasar when occulted by the Moon's edge as the Moon moves in front of a quasar, indicating a very small source size.
    B) The extremely small image size of a quasar, even from Hubble Space Telescope images and radio interferometry measurements.
    C) The sharpness of the emission lines in their optical spectra, since a large source size would smear out the line shapes.
    D) Rapid fluctuations in brightness, since variations over 1 day mean that the source must be less than 1 light-day across.

    Answer: D Source: Section 16-2

29. What observational fact convinces astronomers that the source of energy in a typical quasar is physically very small?
    A) Their starlike appearance in our sky.
    B) The rapid variation in the energy output of the source.
    C) The narrowness of the emission lines in their spectra.
    D) The extreme distance of all quasars.

    Answer: B Source: Section 16-2

30. Which observations of the radiation from quasars indicate that they are physically very small objects compared to galaxies?
    A) The very high redshift of their light.
    B) Their starlike appearance on photographs, showing no structure.
    C) Their emission mostly being in the infrared and radio range.
    D) The rapid fluctuations in output, often in less than one day.

    Answer: D Source: Section 16-2

31. Why is it that we will not see fluctuations in light output in times shorter than about one day when we observe an extragalactic source whose diameter is about one light-day?
    A) Because it is inconceivable that a source of this size could vary on such short time scales.
    B) Because the light from different parts of the source will be Doppler-shifted by different amounts, allowing us to see only an average shift.
    C) Because absorption of light by intergalactic matter will smooth out rapid fluctuations within the beam.
    D) Because arrival times will be different from different parts of the source, and this will smooth out short-term fluctuations.
    Answer: D Source: Section 16-2

32. The surprising observational fact about quasars is that they appear
    A) to be associated with ancient supernova explosions.
    B) to be moving rapidly towards us, while emitting large amounts of energy.
    C) to be the largest known structures in the universe, although they produce only modest amounts of energy.
    D) to produce the energy output of greater than 100 galaxies in a volume similar to that of our planetary system.
    Answer: D Source: Section 16-2

33. Which of the following amazing observational facts about quasars and their behavior is perhaps the most extraordinary?
    A) They are incredibly powerful radio emitters, such that simple receivers can detect them from vast distances across the Universe.
    B) They exist at distances from the Sun of up to 18 billion light-years.
    C) Their recession velocities can be up to 9/10 of the speed of light.
    D) Their energy output, equivalent to more than 100 ordinary galaxies, from a volume as small as our planetary system.
    Answer: D Source: Section 16-2

34. Seyfert galaxies are
    A) active galaxies, with bright, starlike nuclei.
    B) very small elliptical galaxies.
    C) the largest galaxies in the universe.
    D) irregular galaxies with no shape or structure.
    Answer: A Source: Section 16-3

35. Seyfert galaxies are a distinct class of galaxies because
    A) they are in the constellation of Seyfert in the southern hemisphere sky.
    B) they have hot and very bright and variable, starlike central cores.
    C) they are completely devoid of structure, appearing to be amorphous spheres of gas and dust.
    D) they are very close to the Milky Way and appear to be gravitationally bound to it.
    Answer: B Source: Section 16-3

36. Seyfert galaxies are
    A) supergiant elliptical galaxies which are periodically disturbed by supernova explosions within them.
    B) elliptical galaxies with extremely bright nuclei.
    C) active galaxies which shine mainly by radiation from two relatively widely spaced radio lobes.
    D) spiral galaxies with extremely active cores.
    Answer: D Source: Section 16-3

37. A spiral galaxy with a bright, starlike nucleus showing strong emission lines is called
    A) a Seyfert galaxy.
    B) a gravitational lens.
    C) a quasar.
    D) a BL Lacertae object.
    Answer: A Source: Section 16-3

38. BL Lacertae objects are
    A) elliptical galaxies with bright, starlike nuclei.
    B) active galaxies, most of whose energy is emitted by two widely spaced radio lobes.
    C) spiral galaxies with bright, starlike nuclei.
    D) giant irregular galaxies with neither spiral arms nor the smooth shape of elliptical galaxies.
    Answer: A Source: Section 16-3

39. A BL Lacertae object is
    A) a rapidly spinning neutron star.
    B) an active galactic nucleus.
    C) an eclipsing binary star with a black hole as one component.
    D) an emission nebula containing a young T Tauri star.
    Answer: B Source: Section 16-3

40. The spectrum of a BL Lacertae object shows
   A) no features at all other than the continuum.
   B) strong, highly red-shifted emission lines.
   C) absorption lines of highly-ionized atoms.
   D) doubled emission lines, split by the Doppler shift of oppositely-directed jets of material.
   Answer: A Source: Section 16-3

41. A starlike object showing a completely featureless continuum spectrum would be
   A) a quasar.
   B) a BL Lacertae object.
   C) a gravitational lens.
   D) a Seyfert galaxy.
   Answer: B Source: Section 16-3

42. Observations indicate that BL Lacertae objects are
   A) distant spiral galaxies with high rates of supernova explosions within them.
   B) radio galaxies whose jets and radio lobes point almost directly at the Earth.
   C) black holes in binary star systems, where matter pulled from the companion stars forms a hot accretion disk around the black holes.
   D) quasars colliding with smaller galaxies and initiating vigorous star formation.
   Answer: B Source: Section 16-6

43. Observationally, the biggest difference between quasars and other active galaxies such as Seyferts and BL Lacertae objects appears to be that
   A) the brightness of Seyferts and BL Lacertae objects does not vary with time.
   B) Seyferts and BL Lacertae objects are less powerful than quasars.
   C) quasars appear to be located inside elliptical galaxies, whereas Seyferts and BL Lacertae objects are all spirals.
   D) Seyferts and BL Lacertae objects do not have the bright, starlike nuclei of quasars.
   Answer: B Source: Section 16-3

44. Which of the following objects is NOT classified as an active galaxy?
   A) A Seyfert galaxy. B) A barred spiral. C) A quasar. D) A BL Lacertae object.
   Answer: B Source: Section 16-3

45. An electron moving in a magnetic field in space is forced to move a spiral pattern. As it does so it emits
   A) nothing, since such electrons are in equilibrium.
   B) X rays.
   C) visible light, mostly blue in color.
   D) synchrotron radiation, mostly radio waves.
   Answer: D Source: Section 16-4

46. A moving electron in a magnetic field in space follows a spiral pattern, emitting what type of radiation as it does so?
    A) Cerenkov radiation.
    B) Synchrotron radiation.
    C) No radiation at all, since they are moving smoothly without acceleration.
    D) Lyman radiation.
    Answer: B Source: Section 16-4

47. Synchrotron radiation is produced whenever
    A) electrons jump from level to level in an atom.
    B) electrons are accelerated as they circle in spirals in a magnetic field.
    C) electrons move in a transparent medium at a speed faster than the speed of light in that material.
    D) atoms in a molecule vibrate back and forth.
    Answer: B Source: Section 16-4

48. Which of the following will produce synchrotron radiation?
    A) The radioactive decay of an atomic nucleus.
    B) The accelerated motion of high-speed electrons as they spiral in a magnetic field.
    C) The heating of matter by compression as it spirals into a black hole.
    D) The slowing down of charged particles as they enter a dense medium such as the atmosphere of a star or the Earth.
    Answer: B Source: Section 16-4

49. What mechanism appears to produce the double radio sources seen in intergalactic space?
    A) Two radio galaxies orbiting each other, much like two binary stars.
    B) Two black holes on either side of a small galactic nucleus.
    C) Two pulsars on opposite sides of a quasar.
    D) Two oppositely directed jets of matter, ejected from a small source.
    Answer: D Source: Section 16-4

50. The mechanism that appears to generate two extensive regions of radio emission near active galaxies is
    A) the double image of a single source behind the galaxy, produced by gravitational lensing by the galaxy.
    B) two very hot gas clouds, emitting 21-cm radio waves.
    C) two oppositely directed jets of energetic particles.
    D) two small black holes orbiting the center of the galaxy.
    Answer: C Source: Section 16-4

51. The radio emission from the jets in a double radio source is
    A) synchrotron radiation from relativistic electrons spiralling in magnetic fields.
    B) recombination radiation from electrons recombining with protons as the gas cools in the jets.
    C) 21-cm emission from neutral hydrogen atoms.
    D) thermal emission from very hot matter ejected from the accretion disk around the central black hole.
    Answer: A Source: Section 16-4

52. Many bright radio sources at very large distances from the Sun appear to emit energy from two relatively widely spaced sources. What mechanism is thought to produce these double radio emission regions?
    A) Two oppositely directed plasma jets from a central source produce radio energy as the plasma slows down.
    B) Two oppositely directed beams of light and radio energy illuminating the intergalactic dust and gas.
    C) They are in fact a single source and its reflection in a dense intergalactic gas and dust "mirror" or plasma sheet.
    D) These sources are two galaxies orbiting each other, like binary stars.
    Answer: A Source: Section 16-4

53. What is now considered to be the mechanism for the production of the double lobes of radio emission that appear on either side of many galaxies?
    A) The supernova explosion of a massive star and the ejection outwards in two opposite directions of very hot gas which emits thermal energy.
    B) Material streaming in towards a black hole, perpendicular to and on opposite sides of the accretion disk.
    C) Two oppositely directed jets of relativistic particles spiralling around magnetic fields.
    D) Radio emission from the two halves of galactic halos containing many globular clusters that generate radio-noise.
    Answer: C Source: Section 16-4

54. What recent evidence seems to indicate that several nearby galaxies may contain supermassive black holes at their centers?
   A) Extreme redshift of light from stars near the centers of these galaxies, caused by gravitational redshift from a very massive object.
   B) Spectroscopic observations of stars near the centers of these galaxies, showing extremely fast orbital velocities. A large mass is needed to keep these stars in orbit.
   C) The rotation curve of these galaxies, which shows no decrease in orbital velocity of stars as the radius of orbit increases out to the observable limit of the galaxy, indicating an unusual source of gravity.
   D) Observation that these galaxies are rushing rapidly toward each other (and toward the Milky Way!) because of the gravitational attraction between these very large masses.

   Answer: B Source: Section 16-5

55. What observational evidence seems to indicate the presence of a supermassive black hole at the center of our neighboring galaxy, M31, in Andromeda?
   A) The slow but measurable motion of our galaxy toward M31 under the intense gravitational force of its black hole.
   B) Spectroscopic measurements of extremely high and symmetric Doppler shifts from material orbiting very rapidly round a very massive object near the galactic center.
   C) A small but very dark region on detailed photographs of M31, showing an area from which light cannot escape.
   D) Very intense X-ray emission from a very small central core, indicating a very hot source.

   Answer: B Source: Section 16-5

56. What appears to be the central energy-generating system or "engine" that is producing prodigious amounts of energy in the centers of galaxies, active galaxies and quasars?
   A) A supermassive black hole, where matter is compressed upon falling into the hole and heated to extremely high temperatures.
   B) A very rapidly rotating core of matter, where friction between it and the surrounding matter causes tremendous heat and energy output.
   C) There is no central "engine" in these sources. Their high gravity has focused radiation from many sources beyond them by gravitational lensing and they thus appear to be very bright.
   D) a steady series of supernova explosions, the late evolutionary stages of massive stars.

   Answer: A Source: Section 16-5

57. The "central engine" of an active galaxy appears to be
    A) supernova explosions in an extremely dense star cluster at the center of the galaxy.
    B) the violent merger of two galaxies, in which the collision throws out jets of matter along the rotation axis of the larger galaxy.
    C) a supermassive black hole at the center of an accretion disk, with material being projected out perpendicular to the disk.
    D) stars falling into a supermassive black hole, their remnants being thrown out in all directions.
    Answer: C Source: Sections 16-5 and 16-6

58. What is an Einstein ring?
    A) An example of a gravitational lens.
    B) A ring of material in some spiral galaxies, created by a collision with a smaller galaxy.
    C) An arc of relativistic particles ejected from a galactic nucleus.
    D) An accretion disk around a black hole.
    Answer: A Source: Section 16-6

59. What produces the arcs of light which appear to be centered upon several clusters of galaxies, such as those shown in Fig. 16-18 of Kaufmann and Comins, *Discovering the Universe*, 5th Ed.?
    A) These arcs represent hot gas which is being heated by friction as it orbits a supermassive black hole in the center of the cluster of galaxies.
    B) Gravitational bending of light from a single distant source by both the hidden and visible matter of the intervening cluster of galaxies.
    C) The arcs are reflections of light from the galaxies by sheets of intergalactic dust and gas, similar to the sundogs and halos around the Sun caused by ice crystal clouds.
    D) These arcs represent the ends of plasma jets, where the material in the jets has been stopped as it ploughs into dark matter within the cluster of galaxies.
    Answer: B Source: Section 16-6

60. In which direction(s) are the jets of matter ejected from a supermassive black hole in order to produce the many effects surrounding quasars and active galaxies?
    A) Tangential to the accretion disk in two opposite directions, 180° apart, the matter being spun off by the rapid disk rotation.
    B) In a single direction above the accretion disk, the precession of which sweeps the jet along the surface of a cone.
    C) In two opposite directions: across the accretion disk and rotating with it.
    D) In two opposite directions, perpendicular to the accretion disk of the black hole.
    Answer: D Source: Section 16-6

61. I thought black holes gobbled up matter! If the central engine of a double-lobed radio source is a black hole swallowing matter from an accretion disk, where do the jets of matter come from that we see traveling OUTWARD from the galaxy?
    A) They are accelerated in the ergoregion of the rotating black hole and ejected outwards in the black hole's equatorial plane.
    B) The jets arise in the weak galactic magnetic field, not in the region near the black hole.
    C) They are squirted out by high pressure in the accretion disk before the matter reaches the black hole.
    D) They are accelerated outwards by intense magnetic fields around the black hole, much like the beams of particles ejected from the magnetic poles of a pulsar.
    Answer: C Source: Section 16-6

62. If double radio sources, quasars and BL Lacertae objects are considered to be the same basic object, why do they appear to us to be very different?
    A) Because they are of different ages.
    B) Because they are at different distances from us and so we can see more detail and different properties in those which are closer to us.
    C) Because the relativistic particles ejected in the double jets are different in each case; electrons in double radio sources, protons in quasars and quarks in BL Lacertae objects.
    D) Because we view them at different angles to the line of the double jets which are emitted from their cores. Face-on the source will look like a BL Lac object, edge-on a double radio source and between these positions, a quasar.
    Answer: D Source: Section 16-6

63. In the "unified model" of active galaxies, the main difference between quasars, BL Lacertae objects and radio galaxies appears to be that
    A) the mass of the central black hole is different in each case: largest in quasars, less in BL Lacertae, and least in radio galaxies.
    B) we see the accretion disk around the central black hole from a different angle in each case: face-on for BL Lacertae, edge-on for radio galaxies, and in between for quasars.
    C) the galaxy type with which they are associated is different in each case: spiral for BL Lacertae, elliptical for quasars, and irregular for radio galaxies.
    D) the rate at which matter is falling into the central black hole is different in each case: largest in quasars, less in BL Lacertae objects, and least in radio galaxies.
    Answer: B Source: Section 16-6

64. Quasars, double radio sources and BL Lacertae objects may well be the same kind of object, the different appearance of them being simply caused by
   A) their size: quasars are star-sized, double radio sources larger and BL Lac objects galaxy-sized.
   B) the orientation of our line of sight to the axis of their ejected jets of matter.
   C) their position in the Universe: quasars are in our galaxy, double radio sources are associated with other galaxies and BL Lac objects are within the vast voids of space between galaxies.
   D) their age: BL Lac objects evolving through the double radio source phase end as a quasar.

   Answer: B Source: Section 16-6

65. What is the observed distribution of gamma-ray bursters in the sky?
   A) Concentrated primarily along the plane of the Milky Way, indicating an origin within our galaxy.
   B) Clumpy, but not coinciding with any known galaxy clusters, indicating an origin in a new kind of astronomical object.
   C) Clumpy; approximately coinciding with large clusters of galaxies such as the Coma cluster.
   D) Uniform over the entire sky, indicating an origin at "cosmological" distances.

   Answer: D Source: Section 16-7

66. What is the typical duration of a gamma-ray burst?
   A) Less than a couple of minutes, indicating a source smaller than Mercury's orbit.
   B) Several hours, indicating a source somewhat larger than our solar system.
   C) Up to about an hour, indicating a source less than half the size of our solar system.
   D) Several days, indicating a size much smaller than the distance from the Earth to the nearest star beyond the Sun.

   Answer: A Source: Section 16-7

67. What property of gamma-ray bursters indicates that they are located at large ("cosmological") distances from the Earth?
   A) The faintness of the bursts.
   B) Highly redshifted emission lines from the burster's nucleus.
   C) Absorption lines in their spectra, due to intergalactic clouds.
   D) The small apparent sizes of the objects producing the bursts.

   Answer: C Source: Section 16-7

68. How much energy does a gamma-ray burster emit in 100 seconds?
    A) As much as the Sun since the demise of the dinosaurs.
    B) As much as the Sun in 100 years.
    C) As much as the Sun over its entire lifetime.
    D) As much as the Sun in a month.
    Answer: C Source: Section 16-7

# Chapter 17: Cosmology

1. What do cosmologists study?
   A) The origin, structure, and evolution of the universe.
   B) The formation, structure, and evolution of galaxies.
   C) The formation, structure, and evolution of stars.
   D) The origin, structure, and evolution of the solar system.
   Answer: A Source: Section 17-1

2. Which scientist discovered that the equations he had derived predicted an expanding universe, then modified his equations to eliminate this expansion?
   A) Edwin Hubble B) Isaac Newton C) Albert Einstein D) Stephen Hawking
   Answer: C Source: Section 17-1

3. In relation to the universe, what does "isotropy" mean?
   A) The expansion is the same in all directions.
   B) The speed of expansion is the same at all distances.
   C) The speed of expansion at any given distance is the same at all times.
   D) The universe is the same everywhere, neither expanding nor contracting.
   Answer: A Source: Section 17-1

4. Where are we?
   A) Near the edge of an expanding universe, as shown by the microwave radiation coming to us from the edge.
   B) At the exact center of an expanding universe, as shown by the universal expansion away from us in all directions.
   C) Somewhere in an expanding universe, but not in any special part of it.
   D) Near, although probably not right at, the center of the universe, as shown by the fact that the edge is so far away.
   Answer: C Source: Section 17-1

5. Why is the universe expanding?
   A) Because the energy from all the stars is heating the universe, making it expand like a gas that is heated.
   B) Because an infinitely dense clump of matter exploded, sending the galaxies (or superclusters of galaxies) hurtling out through space.
   C) Because space time itself is expanding, carrying the galaxies (or superclusters of galaxies) with it.
   D) It's not expanding; it is we who are getting smaller, making the universe seem bigger and bigger.

   Answer: C Source: Section 17-1

6. Which one of the following statements is a correct description of the expansion of the universe?
   A) Space is a vacuum, but the vacuum has real properties; as galaxies (or superclusters of galaxies) hurtle outwards, the expansion is gradually slowing down by the resistance of space to the passage of the galaxies.
   B) Space time is static, but exerts an outwards pressure on the galaxies in it; this pressure is accelerating the galaxies (or superclusters of galaxies) outwards through space time and away from each other.
   C) Space time is something real, with galaxies inside it; as space time expands, the galaxies (or superclusters of galaxies) are carried along by the expansion.
   D) Space is a vacuum, which is really nothing at all; the galaxies (or superclusters of galaxies) are hurtling outwards through this nothingness.

   Answer: C Source: Section 17-1

7. What is the "cosmological redshift"?
   A) The stretching of the wavelengths of photons by the Doppler shift, because they are emitted by galaxies that are moving away from us.
   B) The stretching of the wavelengths of photons as they pass through absorbing matter in galaxies between us and the emitting galaxy.
   C) The stretching of the wavelengths of photons as they travel through expanding space.
   D) Photons lose energy by interacting with virtual particles in the vacuum, so their wavelength gradually increases as they travel toward us through space.

   Answer: C Source: Section 17-1

8. The further away a galaxy is, the more its light is redshifted, as seen by us on the Earth. This relationship between redshift and distance is caused by
   A) the Doppler shift of light leaving a moving object. More distant galaxies are moving faster through space, so their light is more strongly Doppler-shifted (redshifted).
   B) the expansion of space itself, which stretches the wavelength of the photon. The longer the time that the photon has traveled, the more space has expanded and therefore the more the photon has been redshifted.
   C) the gravitational redshift. Photons leaving a more distant galaxy have traveled further through the galaxy's gravitational field, so they have lost more energy and are more redshifted.
   D) energy losses. The universe does not really expand; photons simply lose energy (wavelength lengthens) as they travel. Photons from more distant galaxies have traveled further and so are more redshifted.

   Answer: B Source: Section 17-1

9. The cosmological red shift of light from distant galaxies is explained by which of the following effects?
   A) The light spreads out over larger areas as distance increases according to $1/(\text{distance})^2$, which causes the wavelength to increase in proportion to distance.
   B) The light from more distant galaxies has traveled through the gravitational fields of more galaxies in getting to us, and is therefore more gravitationally redshifted.
   C) A photon's wavelength is a distance, and is therefore lengthened by the general expansion of the universe, making the light appear reddened.
   D) The light that we see was Doppler-shifted to longer wavelengths by the motion of the objects (e.g., galaxies) away from us.

   Answer: C Source: Section 17-1

10. The cosmological redshift of the light from very distant galaxies is caused by
    A) absorption of blue light by interstellar dust between us and the galaxy, so that only the red wavelengths reach us.
    B) the rotation of the universe around its center (faster at greater distances from us).
    C) the Doppler shift, in which the photon's wavelength is stretched by the galaxy's motion through space, away from us, while the photon is being emitted.
    D) the expansion of space, stretching the photon's wavelength while the photon is traveling toward us.

    Answer: D Source: Section 17-1

11. According to Hubble's law, how old is the universe? ($H_0$ = Hubble's constant)
    A) Age = $1/H_0$
    B) Age = $r/H_0$ (where r = distance in Mpc)
    C) Age = $v/H_0$ (where v = recession velocity in km/s)
    D) Age = $H_0$
    Answer: A Source: Section 17-1

12. In cosmology, the constant that is intimately related to the present "age" of the universe is
    A) $1/H_0$, the inverse of the Hubble constant of expansion.
    B) G, the universal gravitational constant.
    C) the Planck time, $10^{-43}$ s, in which space and time came into existence.
    D) the constant in Wien's law of radiation.
    Answer: C Source: Section 17-1

13. For any object moving uniformly, velocity = distance/time. If so, in the Hubble relationship for the expansion of the universe, $v = H_0 r$, what is the significance of the constant $1/H_0$?
    A) It is merely a constant of proportionality, to allow for the different units in v and r.
    B) It represents the time since the expansion began, or the age of the universe.
    C) It is the inverse of the velocity that the object would have at a standard distance of 10 parsecs.
    D) It represents the average spacing between objects in the universe at the present time.
    Answer: B Source: Section 17-1

14. What major problem would arise if the value of Hubble's constant turned out to be 100 km/s/Mpc?
    A) Galaxies would be traveling too fast for the universe to be gravitationally bound.
    B) Galaxies would be traveling faster than observations allow.
    C) The age of the Universe would be less than the ages of some of the stars in it.
    D) Some galaxies would be further away than the edge of the Universe.
    Answer: C Source: Section 17-1

15. If the elliptical galaxy in Hydra, shown at the bottom of Fig. 15-28 in Kaufmann and Comins, *Discovering the Universe*, 5th Ed, were to be at a distance of 3 billion light-years, what would be an upper limit to the age of the universe? (Be careful with units.)
    A) 15 million years B) 15 billion years C) 0.47 billion years D) 4.5 billion years
    Answer: B Source: Section 17-1

16. If Hubble's constant is 75 km/s/Mpc then the age of the universe is 13 billion years. Suppose it were discovered that Hubble's constant is actually larger than 75 km/s/Mpc. What effect would this have on the calculated age of the universe?
   A) It would have no effect on the calculated age.
   B) It could increase or decrease the calculated age, depending on the recession velocity of the galaxy being investigated.
   C) It would increase the calculated age.
   D) It would decrease the calculated age.
   Answer: D Source: Section 17-1

17. One astronomer (astronomer A) claims that the Hubble constant is 84 km/s/Mpc, while another (astronomer B) claims that it is 63 km/s/Mpc. The age of the universe calculated by astronomer A would be (see Astronomer's Toolbox 17-1, Kaufmann & Comins, *Discovering the Universe*, 5th Ed.)
   A) 1.33 times that calculated by astronomer B.
   B) 3/4 of that calculated by astronomer B.
   C) 1.25 times that calculated by astronomer B.
   D) 2/3 of that calculated by astronomer B.
   Answer: B Source: Section 17-1

18. Calculate the age of the universe if Hubble's constant, $H_0$, is 90 km/s/Mpc (see Astronomer's Toolbox 17-1, Kaufmann & Comins, *Discovering the Universe*, 5th Ed.).
   A) 15 billion years B) 11 billion years C) 9 billion years D) 17 billion years
   Answer: B Source: Section 17-1

19. Calculate the age of the universe if Hubble's constant, $H_0$, is 60 km/s/Mpc (see Astronomer's Toolbox 17-1 Kaufmann & Comins, *Discovering the Universe*, 5th Ed.).
   A) 60 billion years
   B) 16.6 billion years
   C) 11.1 billion years
   D) 15.0 billion years
   Answer: B Source: Section 17-1

20. Who first proposed that the observed motion of galaxies away from us originated with a colossal explosion at the beginning of the universe?
   A) Robert Dicke. B) Fred Hoyle. C) Edwin Hubble. D) George Gamow.
   Answer: D Source: Section 17-2

21. What is the "cosmic particle horizon"?
    A) It is the maximum distance to which our own radio and television signals can have traveled through the universe.
    B) It is the distance at which absorbing matter in the universe prevents us from seeing any further away.
    C) It is the distance from which light will have traveled over the finite age of the universe.
    D) It is the distance at which (because we see back in time as we look out into space) galaxies are just being formed.
    Answer: C Source: Section 17-2

22. Why does the observable universe have an "edge"?
    A) Because we cannot see any further out into space than the distance that light has traveled over the lifetime of the universe.
    B) Because absorbing matter prevents us from seeing beyond a certain distance.
    C) Because there are so many galaxies in the universe that every line of sight eventually hits a galaxy, stopping us from seeing any further.
    D) Because the density of neutrinos at the "edge" becomes so large that photons cannot penetrate this barrier and this prevents us from seeing beyond this point.
    Answer: A Source: Section 17-2

23. Our view of the universe is a limited one because of what fundamental fact?
    A) The matter expanding into space from the Big Bang only extends so far since matter can only travel at or below the speed of light.
    B) Intergalactic space contains absorbing material that blocks our view of more distant objects.
    C) Light from objects further away than a certain distance, defined by the travel time of light in the lifetime of the universe, has not yet reached us.
    D) The cosmological red shift has moved the light from very distant objects out of our detectable range, making these objects invisible to us.
    Answer: C Source: Section 17-2

24. Who developed the steady-state theory of the universe?
    A) Alpher and Gamow
    B) Dicke and Peebles
    C) Hoyle, Bondi, and Gold
    D) Penzias and Wilson
    Answer: C Source: Section 17-2

25. How does the universe behave, according to the steady-state theory?
    A) As the universe expands, new matter is created from which new galaxies form, thus maintaining a "steady state."
    B) The universe is static, not expanding or contracting, but new matter is being created so that as old galaxies die, new galaxies form to take their place.
    C) The universe is static, neither expanding nor contracting, thus maintaining a "steady state" in which no change takes place.
    D) New matter is being continuously created, which adds to the absorption of light in the universe and makes distant galaxies seem further and further away.
    Answer: A Source: Section 17-2

26. Which single observation is perhaps the strongest argument against the steady-state model of the universe, and for the Big Bang model?
    A) The universe is bathed in a sea of microwaves coming from the edge of the visible universe.
    B) The number of supporters of the steady-state model is less than the number of supporters of the Big Bang model.
    C) The universe is expanding.
    D) We have not observed matter being created from nothing in the space around us.
    Answer: A Source: Section 17-2

27. What is the cosmic microwave background radiation?
    A) An almost-uniform background of radiation from distant, unresolved, overlapping galaxies.
    B) A uniform background of radiation from electrons spiraling in weak intergalactic magnetic fields.
    C) Radiation left over from the Big Bang, after the universe expanded and cooled.
    D) Radiation from a very tenuous, ionized gas that fills the universe equally in all directions.
    Answer: C Source: Section 17-2

28. The cosmic background radiation is
    A) the beam of atomic nuclei known as cosmic rays that continuously rain down upon the Earth from all directions in space.
    B) the radio noise generated by Earth-bound transmitters, spreading out into space since about 1920.
    C) low intensity radio noise, with a 3 K blackbody temperature, almost uniform in intensity in all directions.
    D) the flux of visible radiation in empty space, contributed by all visible stars in the universe.
    Answer: C Source: Section 17-2

29. The cosmic background radiation is
    A) the electromagnetic remnants of the explosion in which the universe was born.
    B) the radio noise from hot gas in rich clusters of galaxies.
    C) the faint glow along the elliptic, caused by sunlight scattering from dust particles.
    D) the result of the radioactive decay of heavier, unstable elements produced in supernova explosions.
    Answer: A Source: Section 17-2

30. Good evidence for an original Big Bang that "created" our universe comes from
    A) a background "glow" of microwaves, with blackbody temperature of about 3 K.
    B) the amount of gas and dust in the solar neighborhood.
    C) the rapid motions of some nearby stars, such as Barnard's Star.
    D) the measurement of the rotation of our galaxy.
    Answer: A Source: Section 17-2

31. The cosmic microwave background was discovered by
    A) the IRAS satellite, which produced an all-sky infrared survey.
    B) the Voyager 2 spacecraft during one of its "coasting" periods between planetary encounters.
    C) scientists testing a new antenna and receiver for satellite communications.
    D) rocket-borne telescopes that also discovered x-ray sources in space.
    Answer: C Source: Section 17-2

32. Who discovered the cosmic microwave background radiation, the radiation left over from the Big Bang?
    A) Robert Dicke and P.J.E. Peebles.
    B) Ralph Alpher and George Gamow.
    C) Anthony Hewish and Jocelyn Bell.
    D) Arno Penzias and Robert Wilson.
    Answer: D Source: Section 17-2

33. How was the cosmic microwave background radiation discovered?
    A) Using a horn antenna on the Earth's surface.
    B) Using the Ulysses spacecraft observing from above the Sun's north pole.
    C) Using the COsmic Background Explorer (COBE) spacecraft.
    D) Using a microwave detector on the Hubble Space Telescope.
    Answer: A Source: Section 17-2

34. What is the temperature of the blackbody radiation that we receive from the Big Bang (the cosmic microwave background radiation)?
    A) 30 K B) 3 K C) 30 billion K D) 3000 K
    Answer: B Source: Section 17-2

35. The cosmic background radiation, left over after the Big Bang of the universe and pervading all observable space, has an effective blackbody temperature of approximately
A) 10 K. B) 0 K. C) 273 K. D) 3 K.
Answer: D Source: Section 17-2

36. I thought the Big Bang was hot! If the cosmic microwave background radiation is the radiation left over from the Big Bang, why is it only 3 K?
A) The Big Bang wasn't hot, but a cold explosion; its temperature was the same as we observe now from the cosmic background radiation.
B) The universe has expanded and cooled to the temperature we now observe for the background radiation.
C) The cosmic microwave background radiation we observe is not cool; it is very hot, just like the Big Bang.
D) It isn't from the Big Bang at all; its from cold, intergalactic hydrogen clouds that cover the sky.
Answer: B Source: Section 17-2

37. What was the COBE satellite designed to measure?
A) 21-cm radio radiation from intergalactic hydrogen.
B) X rays from quasars and other objects at cosmological distances.
C) The cosmic microwave background radiation.
D) Redshifts of objects at cosmological distances, to obtain an accurate measurement of Hubble's constant.
Answer: C Source: Section 17-2

38. When the intensity of the cosmic microwave background radiation is plotted against wavelength, what is the shape of the resulting curve?
A) A composite of many overlapping blackbody curves from gas clouds of different temperatures, peaking in the microwaves.
B) A blackbody curve modified by many deep, overlapping absorption lines and several emission lines.
C) Emission lines, strongest and most densely concentrated in the microwave region.
D) Essentially a perfect blackbody spectrum peaking in the microwave region.
Answer: D Source: Section 17-2

39. The cosmic microwave background radiation is not uniform over the sky; it is slightly hotter toward the constellation Leo and slightly cooler in the opposite direction, toward Aquarius. Why?
    A) The Earth is slightly off-center in the universe, so one side of the universe is a bit closer and the other is a bit further away.
    B) That's the way the universe began: hotter in one direction and cooler in the other.
    C) The background radiation really is uniform; the observed difference is due to the Earth's motion through the universe.
    D) The difference is probably a statistical fluctuation, and therefore not real.
    Answer: C   Source: Section 17-2

40. Why does the cosmic microwave background appear to be slightly warmer in one direction in the sky and slightly cooler in the opposite direction?
    A) Because the universe is younger in one direction and therefore warmer.
    B) Because the radiation in one direction is Doppler-shifted to shorter wavelengths by the Earth's motion in space and to longer wavelengths in the other direction.
    C) Because large amounts of matter in that direction in space have focused the radiation slightly by gravitational lensing, making this direction appear hotter.
    D) Because this is the direction in which the Big Bang occurred, hence we are seeing the remnant of the explosion in this direction.
    Answer: B   Source: Section 17-2

41. The effect that makes the cosmic microwave background appear slightly warmer in one direction and cooler in the opposite direction is
    A) a basic asymmetry in the background radiation, related to its origin.
    B) microwave emission from cool, primordial (pre-galactic) clouds of gas and dust in that direction.
    C) the Doppler shift, caused by motion of our galaxy through space toward the constellation Leo.
    D) the presence of large clusters of galaxies in one direction.
    Answer: C   Source: Section 17-2

42. Opposite sides of the universe have the same temperature, yet according to the standard Big Bang theory these points are too far apart for light to have traveled from one to the other in the age of the universe; i.e., they cannot have exchanged heat to even out their temperature. Why, then, do they have the same temperature?
    A) Light (and heat) could travel much faster in the early universe, allowing them to exchange heat while the universe was young.
    B) Pure coincidence.
    C) The expansion of the universe has always been the same everywhere; therefore all parts of the universe have the same temperature regardless of whether they have ever exchanged heat or not.
    D) They were originally close together and evened out their temperature, then a rapid inflation of the universe carried them far apart.
    Answer: D Source: Section 17-2

43. The cosmic microwave background is found to be extremely uniform throughout space, with only very small fluctuations in intensity. The event that produced this remarkable smoothness in the early universe was
    A) the start of the production of matter in the universe, which smoothed out the irregularities in space.
    B) the fact that the Big Bang occurred everywhere in space at the same time.
    C) a sudden but brief period of rapid expansion of the universe during the general expansion of the early universe.
    D) Heisenberg's uncertainty principle, which prevented the concentration of radiant energy in localized volumes of space.
    Answer: C Source: Section 17-2

44. The isotropy of the cosmic microwave background radiation (same temperature in all directions) indicates that
    A) the universe did not begin to expand significantly until after the era of recombination.
    B) the universe had an early period of inflation in which regions initially in contact were carried out of contact with each other.
    C) the universe has always been dominated by matter.
    D) regions that appear to us to be on opposite sides of the visible universe are in fact in close contact with each other.
    Answer: B Source: Section 17-2

45. In cosmology, what is the "inflationary epoch"?
A) A period when the cost of living rose faster than astronomers' salaries.
B) The period of universal expansion from the Big Bang to the present.
C) A short period of extremely rapid expansion when the universe was very young.
D) The first 300,000 years of the life of the universe, when matter and radiation interacted vigorously.
Answer: C Source: Section 17-3

46. How many fundamental forces are known in science?
A) Three. B) Four. C) Five. D) Six.
Answer: B Source: Section 17-4

47. How many fundamental forces are there in nature?
A) Three: strong, electromagnetic, and gravitational.
B) Four: strong, weak, electromagnetic, and gravitational.
C) Five: strong, weak, magnetic, electric, and gravitational.
D) Six: color, strong, weak, magnetic, electric, and gravitational.
Answer: B Source: Section 17-4

48. What is the range of the electromagnetic force (the maximum distance over which it acts)?
A) Infinity.
B) $10^{-15}$ m (1 femtometer, or roughly the size of a proton).
C) A few thousand meters, or roughly the size of the Earth.
D) $10^{-9}$ m (1 nanometer, or roughly the size of a hydrogen atom).
Answer: A Source: Section 17-4

49. Which of the four fundamental forces holds the electrons in the atom?
A) The gravitational force.
B) The strong nuclear force.
C) The electromagnetic force.
D) The weak nuclear force.
Answer: C Source: Section 17-4

50. What is the range of the gravitational force (the maximum distance over which it acts)?
A) $10^{26}$ m, or roughly the distance to the farthest quasars.
B) Infinity.
C) $10^{21}$ m, or roughly the size of the Milky Way Galaxy.
D) $10^{13}$ m, or roughly the size of the solar system.
Answer: B Source: Section 17-4

51. The forces of gravity and electromagnetism are long-range forces, extending in principle from their source (mass and electric charge respectively) to infinity. Why is it that, in our universe, only gravity extends to infinity, whereas electromagnetic forces are much more limited in extent?
    A) Electromagnetic forces from positive charges are canceled by negative charges, whereas there are no negative "masses" to cancel the gravitational force.
    B) All atoms are electrically neutral, so in reality the electromagnetic force never reaches beyond the size of an atomic nucleus.
    C) Electromagnetic forces from charged particles will move other charged particles around to produce a uniform charge distribution and therefore zero electromagnetic forces, whereas gravity concentrates mass and enhances the overall gravity force.
    D) Gravity and electromagnetism are one and the same force, with electromagnetic effects extending over limited spatial ranges and transforming into gravitational forces at large distances from matter.
    Answer: A Source: Section 17-4

52. The one physical force that extends farthest in our universe, and is not canceled out by other effects, is
    A) the gravitational force.
    B) the strong nuclear force.
    C) the electromagnetic force.
    D) the weak nuclear force.
    Answer: A Source: Section 17-4

53. Gravity holds galaxies together; what does the weak nuclear force hold together?
    A) Nothing.
    B) Nuclei.
    C) Leptons (particles including electrons and neutrinos).
    D) The quarks inside protons and neutrons.
    Answer: A Source: Section 17-4

54. When is the weak nuclear force encountered?
    A) When a neutron is transformed into a proton with the ejection of an electron and a neutrino.
    B) When a positively charged nucleus repels another positively charged nucleus in the core of a star like the Sun.
    C) When an atom absorbs a photon and one of the electrons in the atom is sent into a higher energy level.
    D) When two quarks interact inside a proton or neutron.
    Answer: A Source: Section 17-4

55. The weak force
    A) acted only during the Big Bang, and has no known role in the universe at the present time.
    B) attracts the electrons to the nucleus, holding the atom together.
    C) holds the quarks together inside a proton or neutron.
    D) acts during certain kinds of radioactive decay.
    Answer: D Source: Section 17-4

56. The physical force that controls the structure of the nucleus and binds together protons and neutrons is the
    A) electromagnetic force.
    B) weak nuclear force.
    C) gravitational force.
    D) strong nuclear force.
    Answer: D Source: Section 17-4

57. What is the range of the strong nuclear force (the maximum distance over which it acts)?
    A) $10^{-15}$ m (1 femtometer, or roughly the size of a proton).
    B) $10^{-9}$ m (1 nanometer, or roughly the size of a hydrogen atom).
    C) Infinity.
    D) A few thousand meters, or roughly the size of the Earth).
    Answer: A Source: Section 17-4

58. What is the range of the strong nuclear force compared to the size of the nucleus, $10^{-14}$ m?
    A) 10 times smaller than the size of an atomic nucleus.
    B) Infinite; it has no limit.
    C) The same, since it is the strong force that holds the nucleus together.
    D) 10 times larger than the size of an atomic nucleus.
    Answer: A Source: Section 17-4

59. The four physical forces at work in the universe are gravitation, electromagnetic, strong nuclear and weak nuclear forces. Which two of these are very short-ranged, extending over distances or only about $10^{-15}$ m?
    A) Strong and weak nuclear forces.
    B) Strong nuclear and electromagnetic forces.
    C) Gravitation and electromagnetic forces.
    D) Electromagnetic and weak nuclear forces.
    Answer: A Source: Section 17-4

60. What are quarks?
    A) Antielectrons (the antimatter form of electrons).
    B) The component particles making up electrons.
    C) Particles of zero electric charge and zero mass that are emitted by nuclear reactions in the Sun's core.
    D) The component particles making up protons and neutrons.
    Answer: D Source: Section 17-4

61. What are the particles that make up protons and neutrons?
    A) Quarks. B) Neutrinos. C) Muons. D) Gravitons.
    Answer: A Source: Section 17-4

62. In modern particle physics, the proton and the neutron are now thought to be composed of more fundamental particles called
    A) quarks. B) photons. C) neutrinos. D) gluons.
    Answer: A Source: Section 17-4

63. How many different types of quarks make up ordinary matter, such as neutrons and protons?
    A) Six B) Four C) Two D) Three
    Answer: C Source: Section 17-4

64. How many quarks are there in a proton or a neutron?
    A) Six B) Four C) Three D) One
    Answer: C Source: Section 17-4

65. What is the difference between a proton and a neutron, in terms of their constituent quarks?
    A) A proton is made of two "up" quarks and a "down" quark, and a neutron is made of two "down" quarks and an "up" quark.
    B) A proton is made of two "down" quarks and an "up" quark, and a neutron is made of two "up" quarks and a "down" quark.
    C) A proton is made of three "up" quarks and a neutron is made of three "down" quarks.
    D) A proton is made of three "down" quarks and a neutron is made of three "up" quarks.
    Answer: A Source: Section 17-4

66. How do we know that the fundamental forces become unified as the energy of particle interactions increases?
    A) Because we have actually seen the start of it in high-energy particle accelerators, where the electromagnetic and weak forces become unified.
    B) Only because the theory that says that they do is a very elegant one, and it would be nice if it were right.
    C) Because experiments have actually been done with high-energy particle accelerators where all four forces have been unified.
    D) Because, although it is not possible to see unification in the laboratory, the theory of unification agrees with many of the observed properties of the universe.
    Answer: A Source: Section 17-4

67. In what way does the study of the collisions of very high speed nuclear particles with other matter at particle accelerator laboratories help in the understanding of the early universe?
    A) Collisions in nuclear accelerator laboratories produce large numbers of neutrinos and hence mimic the conditions that were thought to exist in the early universe, where neutrinos occupied most of the space.
    B) Motion of particles in circular orbits around the original black hole produced by the Big Bang were similar to the particle motions in an accelerator.
    C) Nuclear accelerators generate enormous quantities of microwaves similar to those that existed in the early universe, thus allowing the study of the interaction of microwaves with matter.
    D) The temperature of the early universe was extremely high, such that mutual collisions of particles occurred at energies equivalent to those in nuclear particle accelerators.
    Answer: D Source: Section 17-4

68. The electromagnetic and weak nuclear forces are predicted to have been indistinguishable at some stage in the early universe. What conditions were required at this stage?
    A) Very low density of matter, such that the weak force became as strong as the electromagnetic force in this region of space.
    B) Extremely high temperatures, producing very energetic collisions between components of matter.
    C) Very high density of matter such as the interior of the nucleus of an atom or a neutron star, where particles are so close that weak forces become equivalent to electromagnetic forces.
    D) Extremely low temperatures, where collisions of particles were of low energy and extremely infrequent, such that electromagnetic and weak forces were equivalent.
    Answer: B Source: Section 17-4

69. At what times during the Big Bang were all four fundamental forces unified?
    A) During the first 300,000 years from the start of the Big Bang, when the universe was dominated by radiation.
    B) Until $10^{-24}$ s after the start of the Big Bang, when the inflationary epoch ended.
    C) Until $10^{-6}$ s after the start of the Big Bang, when era of quark confinement ended.
    D) During the Planck time, up to $10^{-43}$ s after the start of the Big Bang.
    Answer: D Source: Section 17-4

70. The Planck time refers to
    A) the time at which the expanding universe became transparent to radiation.
    B) the time of extremely rapid inflation that started $10^{-35}$ second after the universe began.
    C) the present age of the universe as given by Hubble's law.
    D) the first $10^{-43}$ second of time, when all four fundamental forces were united.
    Answer: D Source: Section 17-4

71. The first $10^{-43}$ second of the age of the universe, during which all four fundamental forces were united, is called
    A) the Planck time.
    B) the event horizon.
    C) the Hubble time.
    D) the inflationary epoch.
    Answer: A Source: Section 17-4

72. Within which time frame from the initial Big Bang do we believe all four fundamental forces of nature were united into a single force?
    A) $t = 0$ to $10^{-35}$ s, when the strong nuclear force "froze out" of the universe.
    B) $t = 0$ to $10^{-43}$ s, the Planck time, when gravity "froze out" of the universe.
    C) $t = 0$ to 1 second, when photons interchanged freely with electron-positron pairs.
    D) $t = 0$ to $10^{-6}$ years, when radiation dominated the universe.
    Answer: B Source: Section 17-4

73. The predicted temperature of the early universe at which the four forces of nature would have been unified is
    A) $10^{6}$ K. B) $10^{15}$ K C) $10^{27}$ K. D) $10^{32}$ K.
    Answer: D Source: Section 17-4

74. At the start of the Big Bang, all of the fundamental forces were unified, and behaved like a single force. As the universe expanded and cooled during the Big Bang, which was the first force to "freeze out" as a separate force?
    A) The strong nuclear force.
    B) Gravity.
    C) The weak nuclear force.
    D) The electromagnetic force.
    Answer: B Source: Section 17-4

75. When the Big Bang began, all of the fundamental forces were unified, and behaved like a single force. As the universe expanded and cooled during the Big Bang, when did the strong nuclear force "freeze out" and become a separate force?
    A) When pair production ceased, when the universe was about 1 second old.
    B) At the start of the inflationary epoch, when the universe was about $10^{-35}$ s old.
    C) At the Planck time, when the universe was about $10^{-43}$ s old.
    D) At the start of quark confinement, when the universe was about $10^{-6}$ s old.
    Answer: B   Source: Section 17-4

76. When did the inflationary epoch begin?
    A) When the strong force "froze out" as a separate force.
    B) When pair production ceased.
    C) When the electromagnetic force "froze out" as a separate force.
    D) When gravity "froze out" as a separate force.
    Answer: A   Source: Section 17-4

77. In high energy physics, when two gamma ray photons meet, they can
    A) produce a huge number of low-energy photons.
    B) disappear, creating two negative electrons.
    C) disappear, creating a particle-anti-particle pair.
    D) disappear completely, leaving nothing behind.
    Answer: C   Source: Section 17-5

78. When two gamma ray photons undergo pair production, what are the final products?
    A) An electron and a positron (the antielectron).
    B) An electron and an antineutrino.
    C) A proton, an electron, and an antineutrino.
    D) Two x-ray photons, each with half the energy of the gamma rays.
    Answer: A   Source: Section 17-5

79. What experimental evidence do we have for the direct transformation of energy into matter?
    A) Conversion of gamma ray photons into electrons and antielectrons (i.e., positrons).
    B) The blueshift of light because of the motion of its source.
    C) The increase in the force upon a body when it is accelerated upward (e.g., in an elevator).
    D) The relativistic increase in mass of an object when it is moving very rapidly.
    Answer: A   Source: Section 17-5

80. Where was most of the helium in the universe created?
A) By nuclear reactions in the cores of stars, and was then thrown out into space by supernovae.
B) By nuclear reactions during the Big Bang.
C) By the collision of cosmic rays with hydrogen nuclei in interstellar gas clouds.
D) By high-energy processes during the collapse of pre-galactic clouds during the formation of galaxies.
Answer: B Source: Section 17-5

81. Over what time interval was helium created in the Big Bang?
A) During the first $10^{-43}$ second.
B) During the first 300,000 years.
C) During the first $10^{-6}$ second.
D) During the first three minutes.
Answer: D Source: Section 17-5

82. Which elements were created during the Big Bang?
A) Hydrogen, helium, and lithium.
B) Hydrogen and helium.
C) Hydrogen, helium, lithium, and beryllium.
D) Hydrogen only.
Answer: A Source: Section 17-5

83. In the present universe, the "mass density" of radiation, using Einstein's relation $E = mc^2$ for the photons, is
A) much greater than the mass density of matter, leading to a radiation-dominated universe.
B) much less than the mass density of matter, leading to a matter-dominated universe.
C) equal to the mass density of matter, since these parameters have remained balanced throughout the evolution of the universe as a consequence of the equipartition of energy.
D) essentially zero, since there is very little radiation left in the universe at the present time.
Answer: B Source: Section 17-5

84. What significant event occurred 300,000 years after the start of the Big Bang?
A) The formation of helium stopped.
B) Electrons and nuclei combined to form neutral atoms.
C) Quarks became confined.
D) All of the galaxies we see today formed.
Answer: B Source: Section 17-6

85. In the present theory of the Big Bang, what significant event occurred at about 300,000 years after the universe started expanding?
    A) The universe became transparent to photons of radiation.
    B) The universe became transparent to neutrinos.
    C) The primordial helium in the universe was produced.
    D) The temperature of the cosmic background radiation had cooled to its present level of about 3 K.
    Answer: A Source: Section 17-6

86. Which of the following statements correctly describes the universe for the entire first 300,000 years of its life?
    A) It was filled with free quarks (not confined inside neutrons or protons).
    B) It was a filled with a sea of nuclear particles undergoing violent reactions.
    C) All of the fundamental forces of nature were unified into one force.
    D) It was opaque.
    Answer: D Source: Section 17-6

87. At an age of 300,000 years, the temperature of the universe had fallen to 3000 K, and electrons could then combine with protons to produce neutral hydrogen gas. What major transition took place in the universe at this time?
    A) It became transparent to light for the first time.
    B) Nuclear fusion no longer occurred below this temperature
    C) The present laws of physics began to apply for the first time.
    D) The universe suddenly lost its electrical charge and become neutral.
    Answer: A Source: Section 17-6

88. As we look at more distant regions of space, we see those regions as they existed at earlier times, but our furthest views are blocked by a "wall" beyond which the universe is opaque. What event occurred at the time marked by this wall?
    A) Quarks combined to form neutrons and protons.
    B) Electrons and protons combined to form neutral hydrogen atoms.
    C) Gravity froze out as a separate force.
    D) Protons combined with neutrons to form helium nuclei.
    Answer: B Source: Section 17-6

89. Which one of the following statements does NOT correctly describe the universe at the era of recombination?
    A) The temperature of the universe was about 3 K.
    B) The universe became transparent to radiation.
    C) The universe was about 300,000 years old.
    D) Electrons and protons combined to form neutral hydrogen atoms.
    Answer: A Source: Section 17-6

90. What was the temperature of the universe at the time the universe became transparent to radiation?
    A) 300,000 K B) 3 K C) 30 billion K D) 3000 K
    Answer: D Source: Section 17-6

91. When did the universe cool to a temperature of 3 K?
    A) 300,000 years after the Big Bang, when the universe became transparent to radiation.
    B) 1 second after the start of the Big Bang, when pair production ceased.
    C) 3 minutes after the start of the Big Bang, when primordial nuclear reactions ceased.
    D) Very recently.
    Answer: D Source: Section 17-6

92. At what point in time did the universe cool to a temperature of about 3 K?
    A) At the era of recombination.
    B) At the end of the Planck time.
    C) At the end of the inflationary epoch.
    D) Very recently.
    Answer: D Source: Section 17-6

93. Because of the travel time of light, we see more distant parts of the universe as they were when the universe was younger; but we cannot see back "into" times when the universe was opaque and light could not travel freely. Using photons, then, what is the furthest back in time that we can see as we look out into the universe?
    A) To a time 3000 years after the start of the Big Bang.
    B) To a time 3 minutes after the start of the Big Bang.
    C) To a time 300,000 years after the start of the Big Bang.
    D) To a time 1/1,000,000 second after the start of the Big Bang.
    Answer: C Source: Section 17-6

94. During the formation of the universe, where did the density enhancements come from which subsequently collapsed to form superclusters of galaxies?
   A) From turbulence during the time when the universe was opaque, which left denser regions that were free to collapse when the universe became transparent.
   B) From quantum fluctuations in the density of matter during the early Big Bang, which were later expanded by inflation.
   C) From spatial variations in the rate heating from pair annihilation after the first second of the Big Bang.
   D) From a random gravitational fragmentation of the matter in the universe after the era of recombination.
   Answer: B Source: Section 17-6

95. How do the youngest, most distant galaxies compare to the older galaxies we see closer to us today?
   A) The youngest galaxies are bluer and brighter than the older ones.
   B) The youngest galaxies have the same color as the older ones, but are significantly brighter.
   C) There is no observable difference between the youngest galaxies and the older ones.
   D) The youngest galaxies are redder and fainter than the older ones.
   Answer: A Source: Section 17-7

96. Why are the youngest galaxies that we see in distant parts of the universe bluer and brighter than the older galaxies we see closer to us?
   A) Young galaxies have more dust, and dust scatters blue light better than red.
   B) Young galaxies have a burst of star formation that produces many hot, bright, blue stars.
   C) The red light has been partially absorbed by gas and dust between us and the galaxies.
   D) The light from the most distant galaxies has been bent by gravitational lenses and focused on the Earth.
   Answer: B Source: Section 17-7

97. For how long did vigorous star formation last in elliptical galaxies?
   A) For about the first eight to ten billion years of the galaxy's life.
   B) For about the first million years of the galaxy's life.
   C) It has continued throughout the galaxy's life, right to the present day.
   D) For about the first billion years of the galaxy's life.
   Answer: D Source: Section 17-7

98. For how long did continuous star formation last in spiral galaxies?
    A) For about the first billion years of the galaxy's life.
    B) It has continued throughout the galaxy's life, right to the present day.
    C) For about the first eight to ten billion years of the galaxy's life.
    D) For about the first million years of the galaxy's life.
    Answer: B Source: Section 17-7

99. In a primordial, pre-galactic gas cloud, what is believed to have been the most important condition that caused the cloud to become a spiral galaxy?
    A) The initial rate of star formation was low.
    B) The cloud started off flattened and disk-shaped before it collapsed.
    C) The initial rate of star formation was high.
    D) The cloud started off with a lot of dust and heavy elements.
    Answer: A Source: Section 17-7

100. Recent results from very bright supernovae in very distant galaxies seem to indicate that the expansion of the universe
    A) is continuing at a constant rate and has done so since just after the Big Bang.
    B) has now stopped and the universe will shortly begin to contract again toward a Big Crunch.
    C) is decelerating (slowing down).
    D) is accelerating (speeding up).
    Answer: D Source: Section 17-9

101. Why would we expect the rate of expansion of the universe to be slowing down?
    A) The greater the distance between two objects (such as galaxies or superclusters), the harder it is to push them further apart.
    B) Galaxies feel a kind of friction as they move through space, and this slows them down.
    C) The gravitational pull of all objects in the universe on each would lead to slow-down.
    D) All expansions after explosions just naturally slow down with time.
    Answer: C Source: Section 17-9

102. What condition is necessary for the universe to be unbounded (infinite in extent)?
    A) They density of the universe must be greater than some critical value.
    B) The density of the universe must be equal to some critical value.
    C) The density of the universe must be equal to or less than some critical value.
    D) The universe must have no mass in it.
    Answer: C Source: Section 17-9

103. What condition is necessary for the universe to be bounded (limited in extent)?
A) The density of the universe must be equal to some critical value.
B) They density of the universe must be equal to or less than some critical value.
C) The universe must have no mass in it.
D) The density of the universe must be greater than some critical value.
Answer: D Source: Section 17-9

104. What will happen if the universe is bounded?
A) The universe will expand past its maximum size, then fragment into mini-universes.
B) The universe will eventually fall back in on itself, heading toward a "Big Crunch."
C) The universe will reach a maximum size and remain there, like a balloon being blown up.
D) The universe will expand forever.
Answer: B Source: Section 17-9

105. Which parameter of the present universe, more than any other, is considered to be critical in determining the ultimate fate of the universe?
A) The average density of neutrinos.
B) The average density of matter.
C) The average density of photons of radiation.
D) The average density of black holes throughout the universe.
Answer: B Source: Section 17-9

106. The future of the overall universe, in terms of its ultimate evolution and whether it will expand forever or eventually contract again, is determined by which of its parameters?
A) The intensity of cosmic microwave background radiation.
B) The temperature of the gas within it.
C) The average density of matter within it.
D) The present volume of the universe.
Answer: C Source: Section 17-9

107. In cosmology, the phrase "critical density" refers to
A) the density needed to produce precisely flat space.
B) the smallest density that will produce inflation.
C) the density above which the universe is opaque to radiation.
D) the density below which stars will never form.
Answer: A Source: Section 17-9

108. In cosmology, to what does the phrase "critical density" refer?
   A) The density for the universe above which the universe is bounded and below which it is unbounded.
   B) The density for the universe above which matter is ionized and the universe is opaque.
   C) The density for the universe below which the universe is bounded and above which it is unbounded.
   D) The density for the universe above which the universe cannot be expanding.
   Answer: A Source: Section 17-9

109. How does the observed average density of presently measured matter in the universe compare to the critical density required to just close the universe?
   A) The observed density is about twice the critical density.
   B) The observed density is about 1/500 of the critical density.
   C) The observed density is about 1/50 of the critical density.
   D) The observed density equals the critical density within observational uncertainty.
   Answer: C Source: Section 17-9

110. Which of the following statements correctly describes our current state of knowledge about the future expansion of the universe?
   A) The density of luminous matter is only 1/50 of the critical density, but including the dark matter makes the density more than critical so the universe will eventually collapse.
   B) The density of all matter, both luminous and dark, is only 1/50 of the critical density, so the universe will expand forever.
   C) The density of luminous matter is only 1/50 of the critical density, but we do not know how much dark matter there is, so the future expansion of the universe cannot yet be predicted.
   D) The density of luminous matter is 50 times more than the critical density, so the universe will eventually collapse whether dark matter exists or not.
   Answer: C Source: Section 17-9

111. The future of our universe, continuous expansion or eventual contraction, can be determined by observing the rate at which cosmological expansion is changing because of gravitational attraction between masses in the universe. How is this effect of deceleration measured, in terms of the Hubble relationship between speed of recession and distance to distant galaxies?
   A) The relationship will be a straight line at closer distances, but will become slightly curved and less steep at the largest distances.
   B) The relationship will be a straight line at closer distances, but will become slightly curved and steeper at the largest distances.
   C) The line representing the relationship will slope upward at closer distances, but will level out and then slope downward again at the farthest distances.
   D) The relationship will be will be curved at relatively close distances but straighten out at farthest distances.
   Answer: B Source: Section 17-10

112. What method is being used to discover whether we live in an unbounded universe, in which expansion will continue forever, or a bounded universe, in which expansion will eventually turn into contraction and lead us to the Big Crunch?
   A) Measurement of the curvature of the Hubble relationship, $v = H_0 r$, at large distances.
   B) Careful and very accurate monitoring of the Moon-Earth distance to detect slowdown of the expansion of the universe by laser ranging.
   C) Measurement of the bending of light by distant galaxies as the light follows the curvature of space.
   D) Measurement of the deviation from uniformity of the cosmic background radiation.
   Answer: A Source: Section 17-10

113. It is not known yet whether we live in an unbounded universe that will expand forever or in a bounded universe that will eventually collapse into a Big Crunch. How are astronomers trying to settle this question?
   A) By accurate measurement of the speed of a galaxy of known distance at two different times, to measure the deceleration directly.
   B) By measuring the curvature of space by tracing photon paths to the Earth from distant galaxies.
   C) By determining how much galaxy speeds depart from a straight-line Hubble relationship, $v = H_0 r$, at large distances.
   D) By careful observations of the sizes of the bumps in the cosmic microwave background radiation.
   Answer: C Source: Section 17-10

114. To what does a deceleration parameter of $q_0 = 0$ correspond?
A) A completely empty universe (no matter in it).
B) It does not correspond to anything; a deceleration parameter of 0 is meaningless.
C) A universe with exactly the critical density needed to close the universe (or equivalently have it just open).
D) A universe of infinite density.
Answer: A Source: Section 17-10

115. To what does a deceleration parameter of $q_0 = 1/2$ correspond?
A) A completely empty universe (no matter in it).
B) A universe which will just barely continue to expand forever.
C) A universe of infinite density.
D) It does not correspond to anything; a deceleration parameter greater than zero is meaningless.
Answer: B Source: Section 17-10

116. What will happen to the universe if the deceleration parameter $q_0$ is between 0 and 1/2?
A) The universe will expand to a maximum size and then it will begin to collapse into a Big Crunch (bounded universe).
B) The universe will expand forever (unbounded universe).
C) The universe will just barely continue to expand forever, reaching zero expansion speed after infinite time (marginally bounded universe).
D) Who knows? Deceleration parameters less than 1/2 are impossible.
Answer: B Source: Section 17-10

117. What will happen to the universe if the deceleration parameter $q_0$ is greater than 1/2?
A) Who knows? Deceleration parameters greater than 1/2 are impossible.
B) The universe will expand forever (unbounded universe).
C) The universe will expand to a maximum size and then it will begin to collapse into a Big Crunch (bounded universe).
D) The universe will just expand forever, reaching zero expansion speed after infinite time (marginally bounded universe).
Answer: C Source: Section 17-10

118. What value of the deceleration parameter, $q_0$, is required for the universe to be just bounded (density = critical density)?
A) 1/2 B) 0 C) 1/4 D) 1
Answer: A Source: Section 17-10

119. What kind of curvature (geometry of space) does the universe have if the universe is unbounded (density less than critical)?
A) Parabolic. B) Hyperbolic. C) Flat. D) Spherical.
Answer: B Source: Section 17-11

120. What kind of curvature (geometry of space) does the universe have if the universe is just bounded (or just unbounded; i.e., exactly critical density)?
A) Flat. B) Parabolic. C) Spherical. D) Hyperbolic.
Answer: A Source: Section 17-11

121. What kind of curvature (geometry of space) does the universe have if the universe is bounded (density greater than critical)?
A) Flat. B) Spherical. C) Parabolic. D) Hyperbolic.
Answer: B Source: Section 17-11

122. If space is flat, what is the future of the universe?
A) It will expand forever, not stopping even when infinite time has elapsed.
B) It will expand to a maximum size and then collapse into a Big Crunch.
C) It will barely expand forever, reaching zero expansion speed after infinite time.
D) The future of the universe is not related to the geometry of space.
Answer: C Source: Section 11-17

123. If the geometry of space is hyperbolic, what is the future of the universe?
A) It will expand to a maximum size and then collapse into a Big Crunch.
B) The future of the universe is not related to the geometry of space.
C) It will barely expand forever, reaching zero expansion speed at infinite time.
D) It will expand forever.
Answer: D Source: Section 17-11

124. If the geometry of space is spherical, what is the future of the universe?
A) The future of the universe is not related to the geometry of space.
B) It will barely expand forever, reaching zero expansion speed at infinite time.
C) It will expand to a maximum size and then collapse into a Big Crunch.
D) It will expand forever, not stopping even when infinite time has elapsed.
Answer: C Source: Section 17-11

125. If space has a hyperbolic geometry (unbounded universe), what will happen to two initially parallel laser beams as they traverse billions of light years of space?
A) They will gradually diverge (move apart) to a maximum separation, then gradually approach and cross.
B) They will remain parallel.
C) They will gradually diverge (move apart).
D) They will gradually approach each other and eventually cross.
Answer: C Source: Section 17-11

126. The flatness problem in cosmology asks the question:
A) Why are the four forces (strong, weak, electromagnetic, and gravitational) not unified in the present-day universe?
B) Why is temperature of the cosmic back-ground radiation so smooth (isotropic) around the sky?
C) Why was the density of the universe so close to the critical density just after the Big Bang?
D) Why is the night sky dark?
Answer: C Source: Section 17-12

127. Which of the following statements correctly describes the "flatness problem" in cosmology?
A) The universe appears to have a hyperbolic geometry to within observational error, yet the universe is expanding, and expanding universes have to be flat.
B) The universe appears to be flat to within observational error, yet the universe is expanding, and it is impossible for an expanding universe to be flat.
C) Observations of the distant universe indicate that the universe is at least moderately flat, yet matter creates lumps in the geometry of space-time. Therefore it is hard to account for the observed flatness.
D) The density of the universe must have been equal to the critical density to 50 decimal places in order for us to see the universe we do today. This astounding flatness is hard to account for.
Answer: D Source: Section 17-12

128. If the universal expansion is found to be decelerating, how does this affect the age of the universe derived from Hubble's law?
A) Deceleration has no effect on the derived age.
B) The derived age is higher than if the expansion is not decelerating.
C) The derived age is lower than if the expansion is not decelerating.
D) It can either increase or decrease the derived age, depending on the density of the universe and the resulting curvature of space.
Answer: C Source: Section 17-12

129. The Hubble age for the universe would be 18 billion years if the value of the Hubble constant were 51 km/s/Mpc. In that case, if the deceleration parameter, $q_0$, is 1/2 (giving flat space and a marginally open universe), then the actual age of the universe would be
A) 12 billion years.
B) 24 billion years.
C) 9 billion years.
D) 18 billion years (equal to the Hubble age).
Answer: A Source: Section 17-12

# Chapter 18: The Search for Extraterrestrial Life

1. Which of the following ideas has been borne out by astronomical observations over the past few centuries?
   A) The probability of life existing elsewhere in the Universe, even in our solar system, is infinitesimally small.
   B) Our Sun is an unremarkable star in a commonplace galaxy, and many such stars exist in the Universe around which life could evolve on planets.
   C) All the observational evidence so far suggests that conditions for the evolution of life exist only on our Earth, in its position near to our Sun.
   D) We appear to occupy a unique position in the Universe, unrepeated anywhere, since our Sun is unique in properties and position in a remarkable galaxy.
   Answer: B Source: Section 18-1

2. One of the great lessons being learned from modern astronomy is that
   A) we live at the center of a very massive black hole and all the observed cosmological effects such as redshift and cosmic background radiation and even the evolution of life are a consequence of this unique position which we occupy.
   B) the chemistry, geology, and physics upon Earth are unique to our planet and the behavior of matter anywhere else appears to be significantly different from that upon Earth.
   C) we occupy a unique position in the universe and nowhere else do we find conditions equivalent to those in our solar system.
   D) our position and circumstances in the universe are quite ordinary and certainly not unique.
   Answer: D Source: Section 18-1

3. Which of the following events of the past century do you think will have announced our presence upon Earth most effectively to extraterrestrial watchers?
   A) The appearance of artificial satellites orbiting Earth, after 1957.
   B) The slow build-up of radio transmissions after the invention of radio, with modulated signals carrying sound and visual television images.
   C) Nuclear weapons explosions, producing extremely intense but brief flashes of light and electromagnetic radiation.
   D) Slow changes in vegetation patterns and the appearance of man-made structures such as road systems and cities upon Earth.
   Answer: C Source: Section 18-1

4. Why is the strategy of searches for extraterrestrial life usually based upon a carbon chemistry?
   A) Carbon dioxide is the main ingredient of planetary atmospheres, both terrestrial and Jovian.
   B) Carbon is abundant and is versatile in forming complex, long-chain molecules.
   C) No other atom can combine easily with the abundant hydrogen and helium to form long molecules in interstellar gas.
   D) Most meteorites that reach Earth are composed of carbon.
   Answer: B Source: Section 18-1

5. Why is it highly likely that life, should it exist elsewhere in the Universe than upon the Earth's surface, will be based upon carbon chemistry?
   A) Because carbon releases more energy than most other atoms when it combines with oxygen, thereby providing the energy for life processes in living organisms.
   B) Because carbon is expected to be far more abundant than silicon or other like elements which can combine to produce complex molecules.
   C) Because carbon combines more readily than other atoms with nitrogen, the major component of atmospheres such as that of Earth, to produce complex molecules.
   D) Because carbon can bond with many more atomic species in a wider variety of complex forms than other equivalent elements such as silicon.
   Answer: D Source: Section 18-1

6. In which type of meteorite has evidence been found for organic molecules, some of which, in contrast to most life molecules on Earth, have both right and left hand spiral structure?
   A) Iron meteorite.
   B) Carbonaceous chondrite.
   C) Stony-iron meteorite.
   D) Stony meteorite.
   Answer: B Source: Section 18-1

7. Which type of meteorite was found to contain large organic molecules that make up the building blocks of life and provide strong evidence for their extraterrestrial production?
   A) Stony-iron meteorites.
   B) Carboneous chondrites.
   C) Iron meteorites.
   D) Stony meteorites.
   Answer: B Source: Section 18-1

8. Several lines of evidence now suggest that large and complex organic molecules can exist or could evolve in outer space, from which the building blocks of life could be formed. Which of the following is NOT one of these observational findings?
   A) Laboratory experiments in which electrical sparks passing through a combination of simple gases such as $H_2O$, $H_2$, $N_2$ and $CO_2$ produced large organic molecules.
   B) The discovery of organic molecules inside certain meteorites that arrived upon Earth.
   C) Radio astronomical observations of large organic molecules in giant molecular clouds.
   D) The discovery of large organic molecules in the tails of comets, many of which have hit the Earth in past times.

   Answer: D Source: Section 18-1

9. In which of the following environments have long-chain, carbon-based molecules not been found?
   A) On the surface of Earth.
   B) Inside meteorites.
   C) On the surfaces of nearby planets, such as Venus and Mars.
   D) In interstellar gas clouds.

   Answer: C Source: Section 18-1

10. What source of energy was used to trigger the manufacture of complex organic compounds in laboratory simulations of conditions in primordial planetary atmospheres?
    A) Electric discharges, to simulate lightning.
    B) UV and visible radiation, to simulate the intensity of sunlight at earlier times in planetary life.
    C) Heat from chemical reactions in early reactive planetary atmospheres.
    D) Thermal heating, to simulate volcanic heating.

    Answer: A Source: Section 18-1

11. The Urey-Miller "experiment" consisted of
    A) monitoring tens of millions of radio frequencies at once in an effort to detect extraterrestrial radio communications.
    B) attaching a metal plaque to the Voyager spacecraft to tell extraterrestrial beings about us, should they ever recover the spacecraft.
    C) sending a coded message via radio toward nearby stars which are similar to the Sun and which may have planets.
    D) passing an electrical arc through a mixture of hydrogen, ammonia, methane and water, and then look for resulting organic compounds.

    Answer: D Source: Section 18-1

12. The basic chemical molecules which are thought to have been present in abundance in the early planetary system, and which were used in the classical Urey-Miller laboratory experiments in which complex compounds essential to life were formed by electric discharges, were
A) $H_2$ (hydrogen), He (helium), Ar (argon), Ne (neon).
B) $CO_2$ (carbon dioxide), $H_2O$ (water), $H_2$ (hydrogen).
C) $H_2$ (hydrogen), $O_2$ (oxygen), $CO_2$ (carbon dioxide).
D) $CH_4$ (methane), $NH_3$ (ammonia), $H_2O$ (water), $H_2$ (hydrogen).
Answer: D Source: Section 18-1

13. The classical laboratory experiments performed by Urey and Miller in order to explore the necessary conditions for the production of organic molecules (the building blocks of living things) in the solar system involved the passing of electrical discharges through what mixure of gases?
A) Nitrogen, oxygen, water vapor and carbon dioxide.
B) Ammonia, methane, water vapor and hydrogen.
C) Carbon dioxide, water vapor and dust.
D) Hydrogen and helium.
Answer: B Source: Section 18-1

14. Modern laboratory experiments, which repeated those of Urey and Miller in exploring the possibility of producing organic molecules (the building blocks of life) from mixtures of gases expected to exist in the early planetary system, passed electrical discharges through which mixture of gases?
A) Hydrogen and helium.
B) Ammonia, methane, water vapor and hydrogen.
C) Carbon dioxide, water vapor and dust.
D) Nitrogen, oxygen, water vapor and carbon dioxide.
Answer: D Source: Section 18-1

15. Which of the following observations regarding the likelihood of life existing elsewhere in the Universe has NOT yet been made?
A) Discovery of long-chain, carbon-based molecules in interstellar clouds by radio astronomers.
B) Discovery of assemblies of organic molecules into cell-like, self-replicating structures in the soils of Mars and the atmosphere of Venus.
C) The manufacture of organic compounds in laboratory simulations of primordial planetary atmospheres.
D) Discovery of long-chain, amino acid protein molecules in meteorites.
Answer: B Source: Section 18-1

16. If we succeed in being able to communicate with other civilizations in space, which method of communication will prove to be the fastest?
   A) It does not matter what is used since the speed of light can never be exceeded. by anything in our Universe, thereby setting the communication speed limit .
   B) Laser light, since this single wavelength light can be directed into an extremely narrow and intense beam and can therefore be made to travel much faster than ordinary light.
   C) Nuclear powered rockets, since we can then use unlimited power to accelerate these systems to almost infinite speeds.
   D) Neutrinos, because a modulated beam of these particles can penetrate almost anything very easily and can be made to travel almost infinitely fast.
   Answer: A Source: Sections 13-1 and 18-1

17. If we do eventually make contact with other civilizations across space, which of the following conditions will inevitably hold?
   A) The chemistry and biology of the life-forms is likely to be very different from ours, for example, silicon-based, rather than carbon-based.
   B) The civilization is likely to be very much more primitive than ours, in terms of technology.
   C) We will not be able to understand the messages.
   D) Conversations will take a very long time.
   Answer: D Source: Chapter 18

18. Why was the discovery of pulsars initially misinterpreted as evidence of intelligent life elsewhere in the Universe?
   A) It was not thought possible for a "natural" radio source to produce the rapid and extremely regular radio pulses detected from space.
   B) Radio telescopes occasionally detected the same sequence of 5 musical notes that were then whistled or hummed regularly by all the people associated with these telescopes. (D, E, C, lower C and G on a piano)
   C) Pulses arriving from several nearby star systems showed Doppler shifts of frequency apparently caused by orbital motion of their source around the central stars, as if coming from planets.
   D) The rate of pulses detected from space appeared to contain primitive coding similar to a crude Morse Code.
   Answer: A Source: Section 12-9

19. What is the current status of our search for inhabitable, Earth-like planets circling other stars?
    A) Planets have been detected orbiting other stars, but none of these appear to be suitable for life.
    B) No extra-solar planets of any kind have yet been confirmed.
    C) Two stars have been found with an Earth-mass planet orbiting at a distance suitable for liquid water and life, but we cannot yet see if they have oxygen-rich atmospheres.
    D) Several planets have been found with a mass similar to that of the Earth, but they are either too close to their star or too far away from it to have liquid water or life.
    Answer: A Source: Sections 18-1 and II-6

20. To what do the letters SETI refer?
    A) Search for Extra-Terrestrial Invaders.
    B) Sourcebook of Extrasensory Transient Incidents.
    C) Search for Evidence of Terrestrial-planet Inhabitants.
    D) Search for Extra-Terrestrial Intelligence.
    Answer: D Source: Section 18-1

21. The Drake equation attempts to predict
    A) the probability of primitive life existing elsewhere in our galaxy.
    B) the number of technically advanced civilizations in our galaxy.
    C) the number of inhabitable planets around stars in our galaxy.
    D) the number of intelligent civilizations that exist in the whole Universe.
    Answer: B Source: Section 18-2

22. In what way does the Drake equation combine the various factors (e.g., fraction of stars with planets, fraction of planets which can support life, etc.) in an attempt to determine the number of technically advanced civilizations existing elsewhere in our Galaxy?
    A) The product of all the probabilities is calculated.
    B) The sum of each of the factors is determined.
    C) The sum of three factors is divided by the sum of the other four factors.
    D) The sum of the last six factors is subtracted from the initial factor, the rate of solar-type star formation.
    Answer: A Source: Section 18-2

23. In the Drake equation for estimating the possible number of technically advanced civilizations in our Galaxy, the factor for the rate at which solar-type stars form in a galaxy excludes massive stars with masses greater than about 1.5 times that of the Sun. Why?
    A) These stars would never develop nuclear processes that could produce heavy elements (e.g., iron) that are needed for planetary formation.
    B) Such stars are prone to repeated and violent supernova explosions throughout their lives, which would destroy any developing life forms.
    C) Such stars never develop a nuclear furnace in their interiors and hence can never heat any planet sufficiently to sustain life.
    D) Such stars have much shorter lifetimes than it took for intelligent life to develop upon Earth, and hence should probably not be considered.
    Answer: D Source: Section 18-2

24. Around which types of stars are we most likely to find planets supporting our kind of life?
    A) Low-mass main sequence stars.
    B) Red giant stars.
    C) High-mass main sequence stars.
    D) Very low-mass stars.
    Answer: A Source: Section 18-2

25. It is unlikely that planets near to stars much more massive than our Sun would develop life because
    A) these stars would have evolved to red giant and even supernova stages before life could evolve.
    B) planets would have to be too close to these cool stars in order to be sufficiently warm for life to evolve and they would become tidally linked, resulting in no night and day.
    C) there would be no region around the star where UV, visible and IR light intensities would be suitable for the evolution of life.
    D) no Moon would form around a planet near to such a star and a Moon is considered to have been essential for the evolution of life because of tidal variations on Earth.
    Answer: A Source: Section 18-2

26. It is unlikely that intelligent life would develop on a planet circling a star of significantly less mass than the Sun because
    A) the planet would likely become tidally locked to the star, making one side of the planet too hot and the other side too cold.
    B) the lifetime of such a star on the main sequence is too short; life forms on such a planet would not have time to evolve intelligence.
    C) there would be no region around the star where UV, visible and IR light intensities would be suitable for the evolution of life.
    D) no Moon would form around a planet near to such a star and a Moon is considered to have been essential for the evolution of life because of tidal variations on Earth.
    Answer: A Source: Section 18-2

27. How often do Sun-like stars (of a type considered likely to be circled by an inhabitable planet) form in our galaxy, on average?
    A) We cannot yet answer this question.
    B) Less than one such star forms per thousand years.
    C) Hundreds of such stars form per year.
    D) About one such star forms per year.
    Answer: D Source: Section 18-2

28. One of the important numbers in determining how many extraterrestrial civilizations there may be at the present time in our galaxy is, out of those stars which have planets, what fraction, $f_s$, of those stars have planets suitable for life? (This is similar to the number $n_e$ in the Drake equation.) Based on the extrasolar planets already found by astronomers (see Section II-6 of Kaufmann & Comins, *Discovering the Universe*, 5th Ed.), what is a good guess for $f_s$? [NOTE: If all stars with planets have at least one planet suitable for life, then $f_s = 1$; if no such star anywhere in the galaxy has a planet suitable for life, then $f_s = 0$.)
    A) $f_s$ less than about 1/10, since most stars so far found to have planets have high-mass planets (similar to Jupiter or larger) orbiting close to the star, where they would disrupt the orbit of an inhabitable planet.
    B) $f_s = 0$, since no star has yet been found to have an inhabitable planet.
    C) Close to $f_s = 1/2$, since about half of the stars so far found to have planets have planets similar in mass to the Earth orbiting the star at about the Earth's distance from the Sun.
    D) Close to $f_s = 1$, since most stars so far found to have planets have high-mass planets (similar to Jupiter or larger) orbiting far from the star, indicating that our solar system is probably typical of planetary systems in general.
    Answer: A Source: Sections 18-2 and II-6

29. One of the first people to look for radio signals from extraterrestrial civilizations was
    A) Jocelyn Bell. B) Arno Penzias. C) Martin Schwarzschild. D) Frank Drake.
    Answer: D Source: Section 18-3

30. The earliest searches for radio signals from extraterrestrial civilizations were made about
    A) 1995. B) 1960. C) 1973. D) 1948.
    Answer: B Source: Section 18-3

31. The first successful detection of signals from extraterrestrial civilizations was accomplished in which year?
    A) 1960 B) 1985 C) 1999 D) Never; no such signals have been detected as yet.
    Answer: D Source: Section 18-3

32. Most searches of space for evidence of intelligent life concentrated upon radio wavelengths because
    A) radio signals can carry the greatest amount of information per unit time, so information transfer will be most efficient at these wavelengths.
    B) our atmosphere is most transparent at these wavelengths and such signals will be more easily detected from Earth.
    C) it is likely that extraterrestrial beings will have developed radio transmitters before more complex lasers or IR light transmitters.
    D) radio energy is least affected by dust and gas in the interstellar medium.
    Answer: D Source: Section 18-3

33. The so-called "water-hole," a region of the radio spectrum chosen for searches for signals from intelligent life because galactic and Earth-based noise and atmospheric absorption are at a minimum, is so named because
    A) water vapor, $H_2O$, has an intense laser-like emission line at this wavelength which extraterrestrials might use to communicate with us.
    B) water vapor absorption in our atmosphere reaches a sharp minimum at this wavelength.
    C) water vapor emissions from planets at this wavelength will be a good indicator of life on other planets, since water is essential for life as we know it.
    D) two astronomically important wavelengths, the 21-cm line of H and a line from the hydroxyl radical OH are in this region, the letters H and OH signifying water.
    Answer: D Source: Section 18-3

34. What strategies and electromagnetic frequencies are thought to be the most logical for long range communication across the universe with other intelligent beings?
    A) Explosion of nuclear devices at specific intervals and in specific patterns across the Earth when we are closest to a nearby star.
    B) X-ray surveys of space at appropriate times (e.g., when Earth is closest to nearby stars in its orbit), in view of the penetrability of space at these wavelengths.
    C) Continuous radio and microwave listening and transmitting at frequencies at which the natural radio sky noise background is low.
    D) Night-by-night photography of nearby stars at hydrogen Balmer wavelengths.
    Answer: C Source: Section 18-3

35. What conditions restrict the suitable wavelength range for communication with extraterrestrial intelligence to the "water hole" in the high-frequency radio range of the electromagnetic spectrum? (See Fig. 18-3 of Kaufmann & Comins, *Discovering the Universe*, 5th Ed.)
    A) Cosmic background radio noise at low frequencies and radio noise from the Sun and Jupiter at high frequencies.
    B) Local radio noise from TV, radio and aircraft across the band, except a narrow range in which transmission is prohibited by international agreement in order to permit such extraterrestrial communication.
    C) Atmospheric absorption by the ionosphere at low frequencies and by $CO_2$ at high frequencies in the long infrared range.
    D) High galactic radio background noise at low frequencies and high atmospheric absorption by $H_2O$ and $O_2$ at high frequencies, with a suitable region in between.

    Answer: D Source: Section 18-3

36. The so-called "water hole," a region of the radio spectrum chosen for searches for signals from intelligent life because galactic and Earth-based noise and atmospheric absorption are at a minimum, is so named because
    A) water vapor absorption in our atmosphere reaches a sharp minimum at this wavelength.
    B) water vapor emissions from planets at this wavelength will be a good indicator of life on other planets, since water is essential for life as we know it.
    C) two astronomically important wavelengths, the 21 cm line of H and a line form the hydroxyl radical OH are in this region, the letters H and OH signifying water.
    D) water vapor, $H_2O$, has an intense laser-like emission line at this wavelength that extraterrestrials might use to communicate with us.

    Answer: C Source: Section 18-3

37. Why would intelligent alien beings wanting to communicate with us probably choose the 21-cm atomic hydrogen radio wavelength?
   A) Because this radiation at this particular wavelength is very weak from natural sources in space and messages would be easily distinguished from other sources.
   B) Because they would expect that many of our telescopes would already be tuned to this precise wavelength for scientific work.
   C) Because this wavelength shows a very strong Doppler effect when its source is moving, and they would know that we would be able to detect the orbital motion of their home planet round their star by this method.
   D) Because they would have detected this particular wavelength from our transmitters on Earth since it is used extensively for satellite communications, and they would know that we could detect them easily.

   Answer: B Source: Section 18-3

38. What is the current state of the search for extraterrestrial radio communications?
   A) Ten million or more frequencies in the "water hole" are being monitored, but there has been no major effort as yet to send out continuous signals from the Earth at these frequencies.
   B) Several extraterrestrial civilizations have been found, but they are not intelligent enough for us to bother with and the search is continuing.
   C) Occasional single-frequency searches have been made for extraterrestrial signals, but funds are still being sought for the first major continuous monitoring effort.
   D) Continuous transmissions are being sent out from the Earth at several frequencies in the "water hole," and tens of millions of other frequencies are being monitored.

   Answer: A Source: Section 18-3